VHDL for Engineers

Kenneth L. Short

Stony Brook University

Upper Saddle River, NJ 07458

Library of Congress Cataloging-in-Publication Data

Short, Kenneth L.
VHDL for engineers / Kenneth L. Short.
p. cm.
ISBN-13: 978-0-13-142478-4
ISBN-10: 0-13-142478-5
1. VHDL (Computer hardware description language) 2. Programmable logic devices. 3. Digital electronics. I. Title.
TK7885.7.S525 2009
621.39'2--dc22

2008003321

Vice President and Editorial Director, ECS: *Marcia Horton*
Associate Editor: *Alice Dworkin*
Editorial Assistant: *William Opaluch*
Director of Team-Based Project Management: *Vince O'Brien*
Senior Managing Editor: *Scott Disanno*
Production Liaison: *Jane Bonnell*
Production Editor: *Kate Boilard, Pine Tree Composition, Inc.*
Manufacturing Manager: *Alan Fischer*
Manufacturing Buyer: *Lisa McDowell*
Marketing Manager: *Tim Galligan*
Marketing Assistant: *Mack Patterson*
Art Director, Cover: *Jayne Conte*
Cover Designer: *Margaret Kenselaar*
Art Editor: *Greg Dulles*
Media Project Manager: *John M. Cassar*
Full-Service Project Management/Composition:
Pine Tree Composition, Inc./Laserwords, Pte. Ltd.–Chennai
Printer/Binder: *Edwards Brothers Malloy*

About the cover: "Path to Understanding," Gros Morne National Park of Canada, Newfoundland, July 2004. Photograph courtesy of Patricia L. Long. Used by permission.

Pearson Education Ltd., London
Pearson Education Singapore, Pte. Ltd.
Pearson Education Canada, Inc.
Pearson Education–Japan
Pearson Education Australia PTY, Limited
Pearson Education North Asia, Ltd., Hong Kong
Pearson Educación de Mexico, S.A. de C.V.
Pearson Education Malaysia, Pte. Ltd.
Pearson Education, Upper Saddle River, New Jersey

10 9 8 7 6 5 4 3 2

ISBN-13: 978-0-13-142478-4
ISBN-10: 0-13-142478-5

To my father, Robert F. Short, who always inspires me.

Contents

List of Figures

List of Programs

Preface

VHDL/PLD DESIGN METHODOLOGY

This book is about designing and implementing digital systems. The hardware description language VHDL is used to both describe and verify a system's design. Programmable logic devices (PLDs) are used for hardware implementation. To proceed from a system's description in VHDL to its realization in a PLD requires the use of a number of computer-aided engineering (CAE) software tools. These tools automate much of the traditional design and debugging of a digital system. The methodology that uses these techniques and tools is referred to in this book as the VHDL/PLD design methodology.

VHDL

VHDL is an industry standard hardware description language that is widely used for specifying, modeling, designing, and simulating digital systems. You can use VHDL to describe a system structurally or behaviorally at any of several different levels of abstraction. VHDL's features allow you to quickly design large and complex systems.

VHDL was originally developed as a language for describing digital systems for the purpose of documentation and simulation. A system's description could be functionally simulated to verify its correctness. An advantage of using VHDL is that a testbench, a description of input sequences and expected responses needed to drive the simulation of a system, can also be written in VHDL. Thus, it is not necessary to learn a different language for verifying a design.

As VHDL became widely accepted, synthesizer tools were developed that could synthesize digital logic directly from a VHDL design description. This resulted in significant automation of the design process and a corresponding substantial reduction in design time.

PLDs

A PLD is a digital integrated circuit (IC) that is programmed by the user to define its functionality. Programming a PLD defines how on-chip programmable interconnects connect the chip's logic elements. Modifications of a PLD's functionality can be made by erasing and reprogramming these interconnections.

A large variety of PLD architectures are available from different IC vendors. These architectures include simple PLDs (SPLDs), complex PLDs (CPLDs), and Field Programmable Gate Arrays (FPGAs). The logic capacities of PLDs allow fast and economic implementation of systems requiring from 100 gates to more than 8 million gates.

CAE Tools

The practical application of the design methodology described in this book is predicated on the existence of powerful and relatively inexpensive CAE tools. The types of tools used in the design flow for this methodology include a compiler, simulator, synthesizer, and place and route (fitter).

The development of synthesis tools was followed by the development of place-and-route tools that could fit synthesized logic to the logic elements available in a particular PLD's architecture. A place-and-route tool automatically produces a timing model that can be simulated to determine if a particular design will meet a system's timing requirements. A place-and-route tool also produces a configuration file containing the information used to program the target PLD.

VHDL/PLD Design Methodology Advantages

There are many advantages to the VHDL/PLD design methodology over traditional digital design techniques. These advantages include:

- The designer can experiment with a design without having to build it.
- Synthesis of a design into hardware is automatic.
- The same language is used to describe and to verify a system.
- VHDL is a nonproprietary standard, so designs are portable to other vendors' software tools and/or PLDs.

Systems being designed in industry are becoming more complex, and less time is available to bring a design to market. Use of the VHDL/PLD design methodology allows a digital system to be designed, verified, and implemented in the least possible time.

FOCUS OF THIS BOOK

This book focuses on writing VHDL design descriptions that are synthesizable into logic implemented in the form of a PLD. It also focuses on writing VHDL testbenches to verify a design's functionality and timing before it is mapped to a target PLD.

Fundamental VHDL concepts and the relation of VHDL constructs to synthesized hardware are stressed. It is not the intention of this book to show the use of every possible VHDL construct in every possible way. Constructs that are not useful for writing synthesizable design descriptions or testbenches are not covered.

This book also focuses on the design flow or steps in the VHDL/PLD design methodology and the tools used in each step. The basic architectures and distinguishing features of the various classes of PLDs are covered. Essentially the same process used to synthesize and test PLDs is used to synthesize and test application-specific integrated circuits (ASICs). Differences occur primarily at the end of the design flow. So the material presented here is also a useful starting point for VHDL-based ASIC design.

The design methodology and examples presented in this book are independent of any particular set of VHDL software tools or target PLD devices. Design examples in the early chapters are intentionally kept simple to illustrate one or more specific concepts. Some later designs reuse earlier designs as components to produce more complex systems.

OBJECTIVES

The objectives of this book are to help the reader:

- Understand the VHDL design process from design description through functional simulation, synthesis, place and route, timing simulation, and PLD programming.
- Understand the structure and operation of VHDL programs.
- Comprehend the association between physical logic structures in digital systems and their representation in the VHDL language.
- Understand the basic architecture and operation of PLDs.
- Gain experience in designing and verifying digital systems using synthesis and simulation.
- Write VHDL code that is efficiently synthesized and realized as a PLD.
- Write VHDL testbenches for verifying the functionality and timing of designs.

STRATEGY

VHDL is a complex language; it includes constructs not typically found in traditional programming languages, such as signals, concurrency, and time.

However, not all of VHDL's language features are useful for synthesizing digital systems, and not all of its useful synthesis features need to be understood before simple systems can be designed. Language constructs and their syntax and semantics are introduced in this book as they are needed. We start with simple designs, then increase design complexity as the language is mastered. The order of introduction of language features is motivated by the need to specify particular operational aspects of the digital systems being designed.

A much wider range of VHDL language features is useful for writing testbenches. Here, too, the approach is to first introduce only those features needed to write simple testbenches. Later, more powerful features for creating more complex testbenches are introduced.

Language concepts are introduced in an order that allows the reader to begin producing synthesizable designs as soon as possible. This means that all aspects of a single topic are not necessarily covered in a single chapter. For example, object types (bit, std_logic, integer, physical, floating, and so on) are not all introduced in the same chapter. Instead, they are introduced throughout a number of different chapters, as they are needed to design or verify systems. Simple combinational designs can be completed after reading the first few chapters.

A number of program examples are provided. These programs are written to be compliant with various IEEE standards, such as IEEE standards 1076 and 1076.6, that are applicable to VHDL programs and VHDL synthesis. Important aspects of these standards are covered in this book. Use is also made of many subprograms from IEEE standard packages. Their use makes programs more portable for processing by CAE tools from various tool and device vendors.

ASSUMED BACKGROUND OF READER

This book is written for students and practitioners alike. It can be used in a classroom environment or for self-study. The reader is expected to have a basic knowledge of digital logic and of a block-structured high-level programming language, such as C or Pascal. No prior knowledge of VHDL or PLDs is assumed.

BOOK ORGANIZATION

This book consists of 16 chapters.

The first three chapters provide an overview of the VHDL/PLD design methodology, the basic structure and coding styles of design descriptions and testbenches, and the importance of signals in VHDL.

- Chapter 1, "Digital Design Using VHDL and PLDs," describes the design flow of the VHDL/PLD design methodology and goes through each phase of the design of a simple combinational system.
- Chapter 2, "Entities, Architectures, and Coding Styles," examines VHDL's entity and architecture library units in detail and briefly introduces VHDL's other library units. The concept of a design entity as the basic unit of a hardware design in VHDL is explained. VHDL's primary coding styles, dataflow,

behavioral, and structural, are introduced and example descriptions of the same design are written in each style.

- In Chapter 3, "Signals and Data Types," signals, a concept unique to hardware description languages, are discussed in detail. They are like wires in a hardware system. In a VHDL program, signals connect concurrent statements. Scalar and array (bus) signals are described. Signals, like other objects in VHDL, have a type. The type std_logic is introduced; it is commonly used for signals in design descriptions that are synthesized.

Chapters 4–7 focus primarily on the design and verification of combinational systems.

- Chapter 4, "Dataflow Style Combinational Design," introduces several forms of concurrent signal assignment statements and their use in dataflow style descriptions of combinational systems. A number of common combinational logic functions are described in dataflow style. "Don't care" input and output conditions and their use in design descriptions are discussed. The description of three-state outputs in VHDL is also covered.
- Chapter 5, "Behavioral Style Combinational Design," focuses on describing combinational systems using behavioral style VHDL descriptions. Behavioral style uses a process construct. Within a process construct is a sequence of sequential statements. Sequential statements are like the statements in conventional high-level programming languages. Sequential assignment, if, case, and loop statements are defined and used to describe combinational systems. Variables, which are declared and used in processes, are introduced in this chapter. When a process is synthesized, logic is generated that has a functional effect equivalent to the overall effect of the sequential execution of the statements in the process.
- Chapter 6, "Event-Driven Simulation," describes the operation of an event-driven simulator. The VHDL Language Reference Manual (LRM) describes the semantics of VHDL in terms of how the language's constructs should be executed by an event-driven simulator. Understanding how an event-driven simulator executes VHDL's statements leads to a clearer understanding of the language's semantics.
- Chapter 7, "Testbenches for Combinational Designs," introduces single process and table lookup testbenches for functionally testing combinational designs. Assertion and report statements, which are used in testbenches to automatically check a design description's simulated responses to input stimulus against the expected responses, are presented.

Chapters 8–11 focus primarily on the design of sequential systems.

- Chapter 8, "Latches and Flip-flops," introduces various latches and flip-flops and their description in VHDL. Timing requirements and synchronous input data are considered.

- In Chapter 9, "Multibit Latches, Registers, Counters, and Memory," descriptions of multibit latches, shift registers, counters, and memory are covered. A microprocessor-compatible pulse width modulator (PWM) design is presented.
- In Chapter 10, "Finite State Machines," descriptions of both Moore- and Mealy-type finite state machines (FSMs) are presented. The development of state diagrams to abstractly represent a FSM's algorithm is discussed. Use of enumerated states in FSM descriptions as well as state encoding and state assignment are covered.
- Chapter 11, "ASM Charts and RTL Design," discusses an alternative way to graphically represent a FSM using an algorithmic state machine (ASM) chart. An ASM chart is often used in register transfer level design, where a system is partitioned into a data path and control. The data path consists of design entities that perform the system's tasks, and the control consists of a design entity that is the controlling FSM.

Chapters 12–14 introduce subprograms and packages. These language features promote design modularity and reuse. Testbenches for sequential systems that make use of subprograms and packages are discussed.

- In Chapter 12, "Subprograms," use of subprograms, functions, and procedures allows code for frequently used operations or algorithms to appear at one place in a description but be invoked (used) from many different places in the description. This reduces the size and improves the modularity and readability of descriptions.
- In Chapter 13, "Packages," subprograms are placed in packages so that they are available for use in different designs. Packages allow subprograms to be easily reused in later designs. A number of standard packages that provide prewritten subprograms, which can be used to speed up the development of a design, are discussed.
- In Chapter 14, "Testbenches for Sequential Systems," because a sequential system's outputs are a function of both its input values and the order in which they are applied, development of testbenches for sequential systems is more complex than for combinational systems. Several approaches for generating stimulus and verifying outputs for sequential systems are presented.

Chapter 15 focuses on the modular design of systems and the use of hierarchy to manage complexity. Chapter 16 provides a few larger design examples that bring together many of the concepts presented in the book.

- In Chapter 15, "Modular Design and Hierarchy," partitioning and design hierarchy make the description of complex designs manageable. Structural VHDL descriptions allow a complex design to be described and verified as a hierarchy of subsystems. The same hierarchical concepts are also applicable to testbench design.

- In Chapter 16, "More Design Examples," three larger designs are discussed as is the parameterization of design entities to increase their reuse.

USE OF THIS BOOK IN A CE OR EE CURRICULUM

This book is suitable for use in a one- or two-semester course. At Stony Brook University this book is used in a one-semester course for computer and electrical engineering majors (ESE 382, Digital Design Using VHDL and PLDs). As prerequisites students taking the course at Stony Brook must have previously completed a traditional one-semester digital logic design course and a programming course using a high-level language.

For the laboratory portion of the course students write design descriptions and testbenches for designs and verify their designs using functional and timing simulations. Each design is then programmed into a PLD and bench tested in a prototype.

USE FOR SELF-INSTRUCTION

This book is also appropriate for self-study. To obtain maximum benefit, example designs should be simulated and synthesized. Those end-of-chapter problems that involve writing a VHDL design description and/or testbench should be completed and the description functionally simulated. This requires the reader to have access to a VHDL simulator. If a synthesizer is available the designs should also be synthesized. To accomplish timing simulations for synthesized designs, a place-and-route tool is required.

ACTIVE-HDL STUDENT EDITION

The designs in this text were created and simulated using Aldec's Active-HDL design and simulation environment. Synplicity's Synplify synthesizer was used to synthesize the designs. Various PLD vendors' place-and-route tools were used to obtain timing models for timing simulations and configuration files to program target PLDs.

Aldec has developed a student version of its simulator. Active-HDL Student Edition retains much of the same functionality, menus, and icons as the Professional or Expert Editions of this product but is limited in features and simulation capacity/performance. Version 7.2 of Active-HDL Student Edition is included with this text. Readers without access to the Professional or Expert versions of this simulator can use the Student Edition to create and simulate their designs. This tool provides an excellent environment for learning VHDL and verifying your designs.

ACKNOWLEDGEMENTS

I would first like to express my thanks to Tom Robbins who, some years ago, as publisher of electrical and computer engineering at Prentice Hall, signed me to do this book. I greatly appreciated his patience as the completion of this endeavor took much longer than anyone expected.

Thanks to those at Prentice Hall involved in the book's production: Jane Bonnell, Scott Disanno, Alice Dworkin, and William Opaluch; and to Kate Boilard at Pine Tree Composition.

Comments by reviewers of a draft of the text were very helpful to me in making decisions on revisions and are much appreciated. I would like to thank Nazmul Ula, Loyola Marymount University, Electrical Engineering and Computer Science, Richard C. Meitzler, Johns Hopkins University, Engineering Professionals Program, and an anonymous reviewer for their time and effort.

Thanks to Aldec Incorporated for allowing Version 7.2 of Active-HDL Student Edition to be included in this book,

A special thanks to Scott Tierno at the Department of Electrical and Computer Engineering at Stony Brook. Scott and I have worked together for over 25 years on the development of the embedded system design and VHDL courses and laboratories that I teach at Stony Brook. Scott read early drafts of this book and provided valuable suggestions and critical feedback. He was insistent that the book be titled, *VHDL for Engineers*.

Draft copies of this text have been used at Stony Brook University in numerous offerings of the undergraduate course ESE 382 Digital Design Using VHDL and PLDs. As a result comments by students have been considered in revisions of the draft text and their contributions in this regard are appreciated.

Thanks to my friend Patricia Long for the wonderful photograph on the cover.

Chapter 1

Digital Design Using VHDL and PLDs

VHDL is a programming language for designing and modeling digital hardware systems. Using VHDL with electronic design automation (EDA) software tools and off-the-shelf user programmable logic devices (PLDs), we can quickly design, verify, and implement a digital system.

We refer to this approach as the VHDL/PLD design methodology. It is applicable to designs ranging in complexity from a single gate to a microcontroller system.

Because of the ease in creating and modifying a design using this methodology, products can be rapidly brought to market. Subsequent design changes to upgrade a system can often be made by simply modifying the VHDL program and reprogramming the PLD.

This chapter provides an overview of the VHDL/PLD design methodology and the EDA tools used.

1.1 VHDL/PLD DESIGN METHODOLOGY

A *design methodology* is a coherent set of methods, principles, or rules for designing a system. Embodied in a design methodology are the phases (steps) in the design process and the EDA tool(s) used in each phase. The complete sequence of phases is referred to as the *design flow*.

Systems of interest in this book are electronic digital systems. While the term "circuit" carries the connotation of a network that is simpler than a system, we often use these terms interchangeably.

The VHDL/PLD design methodology uses:

- VHDL to describe both the system being designed and the testbench used to verify the design
- A software simulator tool to simulate the design to verify its functionality and timing
- A software synthesis tool to create the logic described by the VHDL description
- A software place-and-route tool to map the synthesized logic to the target PLD and to generate a timing model and a configuration file
- A PLD to physically implement the design
- Information in the configuration file to program the PLD

Hardware Description Languages (HDLs)

A programming language *construct* is a building block of the language. It is formed from basic items such as keywords and other language building blocks and expresses a particular operation or effect.

A *hardware description language (HDL)* is a high-level textual programming language that includes specialized constructs to describe or model the behavior or structure of a hardware system. An HDL allows a system's behavior or its structure to be described at an abstract and technology independent level. An HDL program appears similar to a program written in a conventional high-level programming language. However, there are important differences, such as the inclusion of constructs to support the description of concurrency, hardware structure, and timing.

VHDL and Verilog HDLs

There are two industry standard HDLs, *VHDL* and *Verilog*. While their syntax and semantics are quite different, they are used for similar purposes. The syntax of VHDL is closer to that of the programming language ADA, while the syntax of Verilog is closer to that of the programming language C.

VHDL contains more constructs for high-level modeling, model parameterization, design reuse, and management of large designs than does Verilog. Many EDA tools are designed to work with either language or both languages together. This text focuses on VHDL exclusively.

VHDL Model

A VHDL *model* is a textual description of a system that, when simulated, behaves like the intended or actual system. The terms "description" and "model" are used interchangeably. Different models may be created to represent the same system at different levels of abstraction (detail). A model at a particular level of abstraction represents all the information that is relevant at that level and leaves out all irrelevant details.

Events and Concurrent Events

In digital systems, logic signals transfer information from one component to another, and from one system to another. A change in a signal's value is called an *event* on that signal. One critical feature of an HDL is its ability to describe events that occur simultaneously. Simultaneous events are intrinsic to digital system operation and are called *concurrent events.*

VHDL as an IEEE Standard

VHDL is a nonproprietary HDL defined by an *Institute of Electrical and Electronic Engineers (IEEE)* standard. Because it is nonproprietary, designs written in VHDL can be processed using EDA tools from different software tool vendors. Furthermore, VHDL designs that conform to another standard, the IEEE synthesis standard, can be synthesized and mapped to PLDs from various PLD vendors.

EDA Software Tools

Contemporary digital design is very dependent on EDA software tools. These include simulators, synthesizers, and place-and-route tools. When using the VHDL/PLD design methodology, the combined purpose of all these tools is to convert a VHDL design description to a stream of bits used to program a PLD.

EDA tools automate the more labor-intensive and error-prone design tasks. For example, optimizations that require so much time when using traditional design methodologies, such as logic minimization (to minimize the number of gates in a design) and state minimization (to minimize the number of flip-flops), are done automatically.

Because EDA tools substantially shorten the design process, they make it practical for us to explore different approaches to solving a particular design problem to find the highest performance or lowest cost solution. Learning the more advanced optimization features of EDA tools and the architectural features of PLDs helps us to create better designs.

Programmable Logic Devices (PLDs)

A *programmable logic device (PLD)* is an off-the-shelf (premanufactured) integrated circuit (IC) that we can program to perform a logic function. Once programmed, a PLD provides an immediate physical implementation of our design. PLDs with high logic capacities allow complex designs to be implemented in a single IC.

Most PLDs are reprogrammable. The configuration information used to program a PLD can be erased and the PLD reprogrammed. Many reprogrammable PLDs are in-circuit reprogrammable. This allows the system containing the PLD to be upgraded in the field without removing the PLD from its printed circuit board.

A VHDL timing model of a design mapped to a specific PLD is generated by a place-and-route tool. This timing model allows us to conduct timing simulations to verify whether the PLD, when it is later programmed, will meet the system's timing specifications.

A system implemented using PLDs may be comprised of:

- A single PLD that realizes all of the system's digital logic
- Multiple interconnected PLDs

- One or more PLDs interconnected with fixed function analog and digital ICs
- A PLD integrated as part of a configurable system-on-chip (CSoC) or system-on-reprogrammable-chip (SoRC) IC

VHDL/PLD Design Flow

The VHDL/PLD design flow is illustrated in Figure 1.1.1. For purposes of explanation, it is separated into groups of related phases. All phases in a column are closely related. Shaded phases involve verifying some previous phase(s) of the design. A description of each phase is provided in the sections that follow.

While Figure 1.1.1 represents the design flow as an idealized linear flow, in practice, changes in design requirements, coding errors, timing constraint violations, and other issues often require some phases to be repeated. This results in an iterative implementation of the design flow.

In a university environment, the VHDL/PLD methodology makes it possible for students to design and implement digital systems of substantial complexity in a relatively short time.

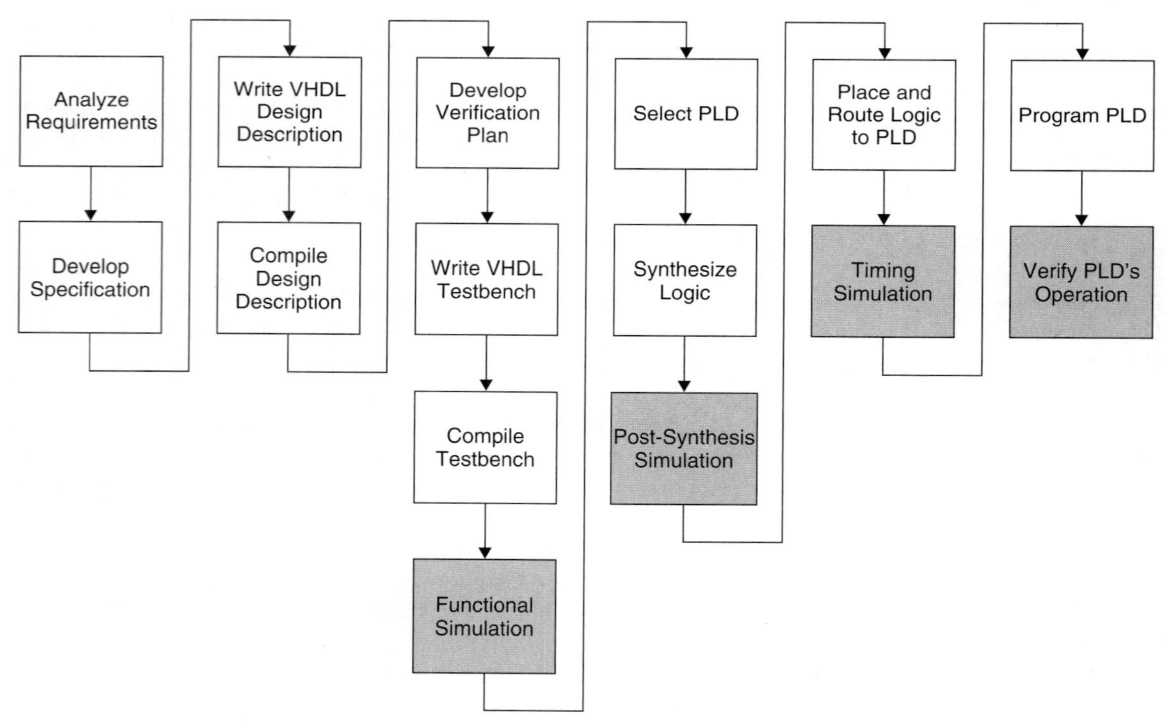

FIGURE 1.1.1
Design flow for the VHDL/PLD design methodology.

In industry, the continual increase in the complexity of the systems being designed and the need to design and implement them in less time makes use of a design methodology like that in Figure 1.1.1 essential.

1.2 REQUIREMENTS ANALYSIS AND SPECIFICATION

The first two phases in Figure 1.1.1:

- Analyze requirements
- Develop specification

are required for any design effort irrespective of the methodology used.

Analyze Requirements

During the *requirements analysis* phase, the problem that the system is to solve is analyzed to establish a clear problem definition. Any constraints imposed by the environment in which the system must operate are also determined. A written requirements document is produced that completely describes the problem and its constraints.

Develop Specification

Using the requirements analysis document as a basis, a *specification* is written that defines the system's interface to its environment and the functions the system must accomplish to solve the problem. System functions are specified in terms of desired system behavior, not how this behavior is to be implemented. A system specification also includes any performance requirements and constraints, such as speed of operation and maximum power consumption.

Traditionally, a specification is written in a natural language. However, it is possible to write a specification directly in VHDL. This latter approach has the advantage of eliminating ambiguities that are typical in a natural language specification. A specification written in VHDL is an executable *system model* that represents the algorithms to be performed by the system, independently of any particular hardware implementation. Such a specification can be simulated to verify its correctness.

However, in practice, specifications are not commonly written in VHDL form. Although it is an executable model, a specification written in VHDL may not be synthesizable. If it is not, a synthesizable VHDL description must be written from the nonsynthesizable specification. This topic is discussed further in Section 1.14.

Half-Adder Requirements and Specification

A very brief requirements definition and specification for a half adder serve as very simple examples of these two phases.

A natural language requirements definition for a half adder might be:

"We need to be able to add two one-bit binary numbers."

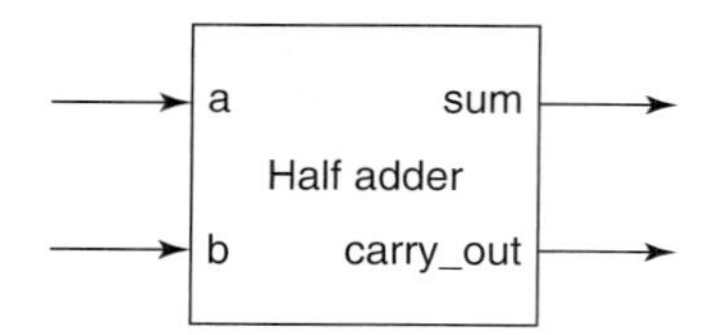

FIGURE 1.2.1
Block diagram of a half adder.

From this requirements definition we can write a natural language specification: "The circuit must add two binary inputs, producing two outputs. One output is the sum from the addition and the other is the carry."

A timing constraint that requires the addition to be completed in a specified time might be included in the specification. A block diagram, Figure 1.2.1, that shows the half-adder's inputs and outputs might also be included.

1.3 VHDL DESIGN DESCRIPTION

The next two phases in the design methodology are:

- Write VHDL design description
- Compile design description

Design Description

Using the specification as a basis, a design description is written. The term *design description,* as used in this book, refers to a VHDL program that can be synthesized into hardware.

A design description can be behavioral or structural. A behavioral description specifies the computation of output values as a function of input values. In contrast, a structural description is a hierarchical description of interconnected components. For a structural description to be complete, each component must have an associated behavioral description.

From the half-adder's specification and the definition of binary addition, we know that the Boolean equations for the outputs are:

$$sum = \overline{a} \cdot b + a \cdot \overline{b}$$
$$carry\_out = a \cdot b$$

If we did not know the appropriate equations, we could create a truth table, Table 1.3.1, and determine them.

For a function that is specified by a truth table, we can simply write the canonical sum-of-products (sum of minterms) equation for each output. Unlike in traditional design, we would not bother to simplify these equations, because the synthesizer will automatically do this for us.

Table 1.3.1
Half-adder truth table and minterms.

a	b	sum	carry_out	sum minterms	carry_out minterms
0	0	0	0		
0	1	1	0	$\bar{a}b$	
1	0	1	0	$a\bar{b}$	
1	1	0	1		ab

Design File

Once we know the functions a system must accomplish to meet its specification, we can write its VHDL design description. This description is placed in a *design file* (source file). The text of the design file is a sequence of *lexical elements*, each of which is composed of characters. The lexical elements represent delimiters, keywords, identifiers, literals, and comments.

VHDL Editor

For a design file to be readable by a VHDL compiler it must be an unformatted (plain text) file. Any *plain text editor* or word processor that can output a file in plain text form, can be used to create a design file. However, special *VHDL editors* designed for creating and editing VHDL source code are preferable.

VHDL editors color code and indent the language's syntactical constructs when the file is displayed on a computer screen. This makes many syntax errors immediately obvious while we are creating a design file. It also makes the file's text easier to read and comprehend.

For instructional purposes, programs in this book are not printed in plain text format. Instead, they are formatted to make their keywords and constructs more distinctive.

Keywords

Keywords (reserved words) are words that have special meaning in VHDL. They can only be used for the purposes defined by the language. In this book, keywords in program listings are **boldfaced.**

Statements

A design description is composed of *statements* (instructions). A statement is a construct that specifies one or more actions that are to take place. The end of each statement is delimited (marked) by a semicolon.

Comments

Comments are used to clarify a program. They are invaluable to others who must read and understand our programs and to ourselves when we return to a program after a considerable amount of time. A comment starts with two hyphens and continues to the end of the line. A comment can appear on any line. There is no block comment feature in VHDL. Comments in programs in this book are italicized.

LISTING 1.3.1
Design description of a half adder.

```
library  ieee;                              -- Context clause
use ieee.std_logic_1164.all;

entity half_adder is                        -- Entity declaration
   port (a, b : in std_logic;
         sum, carry_out: out std_logic);
end half_adder;

architecture dataflow of half_adder is    -- Architecture body
   begin
   sum <= (not a and  b) or (a and not b);
   carry_out <= a and b;
end dataflow;
```

A *formatted design description* of a half adder is given in Listing 1.3.1. While a brief description of this program and the other programs in this and the next chapter are given to provide an overview, the reader is not yet expected to understand the details of the syntax and semantics of the VHDL language constructs.

Context Clause

The two statements at the beginning of this program form a *context clause,* consisting of a library clause and a use clause. A *library clause* is required when objects that are not predefined in the VHDL language, but are defined in a library, are used. This half-adder program uses a data type called std_logic for its input and output signals. This data type is defined in the package `STD_LOGIC_1164` in the library `ieee`. The first statement in the program, starting with the keyword **`library`**, names the library.

In addition to other things, a library can contain packages. A *package* is a collection of commonly used data type declarations and/or subprograms. The statement starting with the keyword **`use`** is a *use clause* that specifies which package in the `ieee` library is to be made available to this program.

Design Entity

The simplest VHDL design descriptions consist of a single design entity. A *design entity* can represent all or a portion of a design. A design entity has well-defined inputs and outputs and performs a well-defined function. A design entity consists of two parts, an entity declaration and an architecture body. These two parts can be placed in the same file, as in Listing 1.3.1, or they can be placed in separate files.

Entity Declaration

An *entity declaration* gives a design entity its name and describes its interface (input and output signals). The entity declaration in Listing 1.3.1 provides information similar to that provided by the block diagram of Figure 1.2.1. An entity declaration starts with the keyword **`entity`**, followed by the entity's name. In this example the name is `half_adder`.

Ports

A design entity's input and output signals are called *ports* and are listed in the entity declaration following the keyword **port**. The half-adder's inputs are a and b. Its outputs are sum and carry_out.

Port Data Types

The type of data each port transfers is also specified. A signal's *data type* determines the values the signal may have and the operations that can be performed on those values. The data carried by the half-adder's input and output ports is type std_logic.

Type std_logic provides nine different values to represent a logic signal. This allows a more detailed representation of a logic signal's state than does type bit. Both types include the values '0' and '1'. Type std_logic is discussed in detail in Chapter 3. Until then we will only be using the '0' and '1' values from std_logic's nine possible values.

Architecture Body

An *architecture body* starts with the keyword **architecture** followed by the name given to the architecture. In this example, the architecture's name is dataflow. Following the keyword **of** is the name of the entity declaration with which the architecture is associated.

A design entity's architecture describes either the entity's behavior or its structure. In Listing 1.3.1, the architecture is written in a style called *dataflow*. Dataflow style describes a system's behavior in terms of how data flows through the system. Each output in Listing 1.3.1 is defined by a concurrent signal assignment statement:

```
sum <= (not a and b) or (a and not b);
carry_out <= a and b;
```

The **and**, **or**, and **not** keywords in these statements are VHDL logical operators, not actual gates. In a physical implementation, each logical operation might, in fact, be synthesized using AND, OR, and NOT gates or by other types of logic gates (such as NAND gates) or by other logic elements, such as multiplexers.

Concurrent Signal Assignment Statements

Signal assignment statements are unique to HDLs. Each of the previous *concurrent signal assignment statements* assigns a new value to the signal on the left-hand side of the *signal assignment symbol* (<=) whenever there is an event on one of the signals on the right-hand side of the assignment symbol. The signal assignment statement is said to be *sensitive* to the signals on the right-hand side of the statement.

The symbol <= is read as "gets." So, the second assignment statement in Listing 1.3.1 is read as "carry_out gets a **and** b." Delimiters that consist of two adjacent special characters, such as the signal assignment symbol (<=), are called *compound delimiters*.

Execution Order of Concurrent Statements

A concurrent signal assignment statement executes only in response to an event on one or more of the signals to which it is sensitive. The order of execution of concurrent signal assignment statements is solely determined by the order of events on the signals to which the concurrent statements are sensitive.

In Listing 1.3.1, if there is an event on signal `a`, both assignment statements are executed simultaneously to determine new values for `sum` and `carry_out`. These statements are called concurrent because they allow the possibility of simultaneous execution, which they must if they are going to model the operation of logic circuits. In our half adder, gates compute the two outputs simultaneously, not sequentially.

The concept of concurrent statements in VHDL is critically important. It creates a distinction between VHDL, or any HDL, and the sequential execution of statements in a conventional programming language. This distinction must always be kept in mind when interpreting or writing a VHDL description.

Sequential Statements

VHDL also has *sequential statements,* like those in conventional programming languages. VHDL's sequential statements can only appear inside a process statement or a subprogram. As a result, when reading a program it is easy to determine from its location whether a statement is concurrent or sequential. A process statement is a concurrent statement that contains a sequence of sequential statements. An example of a process statement is seen in Section 1.5 of this chapter and process statements are discussed in detail in Chapter 5. The half-adder program of Listing 1.3.1 contains no sequential statements.

Design Description Compilation

A VHDL design description is compiled like a program written in a conventional programming language, such as C++ or Ada. *VHDL compilers* are sometimes called *analyzers*. The term "analyze" means to check a design file for syntax and static semantic errors before it is simulated. This term is sometimes used instead of the term "compile," because compiling typically means to generate executable object code. In contrast, code generation for VHDL was initially associated only with simulation. Analyzed VHDL code is stored in an intermediate form in a library for subsequent simulation. However, the distinction between the terms compile and analyze is often not made. In this book the two terms are used interchangeably.

Syntax Errors

Syntax errors are violations of the language grammar. For example, each entity declaration must be terminated with an end statement.

Semantic Errors

Every programming language attaches specific meanings to its language constructs. *Semantic errors* are violations of the meaning of the language.

Static Semantic Errors

Static semantic errors are errors in meaning that can be detected by a compiler before a simulation is performed. For example, if we have a signal assignment statement that assigns the value of a signal of type bit to a signal of type std_logic, we have a type-conflict static semantic error.

Dynamic Semantic Errors

In contrast, *dynamic semantic errors* can only be detected during simulation. For example, if we have a signal assignment statement where the left-hand and right-hand sides are the same type, but during execution the computed value of the right-hand side is outside the allowed range of values for that type, we have a dynamic error.

We must find and remove all syntax and static semantic errors from a design description during compilation. This is accomplished by correcting the offending statements using an editor. Afterward, dynamic semantic errors are detected and corrected during simulation.

1.4 VERIFICATION USING SIMULATION

Before considering the next group of design flow phases, lets examine the importance of simulation. When using VHDL, a simulator is the primary development, verification, and debugging tool.

Verification

Verification is the process used to demonstrate the correctness of a design. We verify that our design description meets its functional and timing specifications before the target PLD is programmed. Generally speaking, verification using simulation can show the existence of errors in a design, but cannot prove that a design is correct. The process of verification often consumes a significant percentage of the total design effort.

Simulation for Verification and Debugging

A VHDL *simulator* can execute a design description's source code. The simulator allows input values to be applied to the design description. In response, the simulator computes the corresponding output values by executing statements in the design description.

When errors are encountered during a simulation, we can use the simulator's debugging features to find and correct the source of the errors in the design description. Once modified, a design description must be compiled and simulated again to verify that the errors have indeed been corrected.

Kinds of Simulation

Simulation is performed during three different phases of the design flow (Figure 1.1.1). Each simulation phase has a particular purpose and uses a different VHDL model of the same design entity:

- *Functional simulation:* we simulate our VHDL design description. This simulation is performed to verify that the system we described meets the functional requirements of its specification.

- *Post-synthesis (gate-level) simulation:* we simulate the synthesized VHDL netlist. This netlist is a gate-level VHDL model of the synthesized logic. It is automatically generated by the synthesizer. This simulation is performed to verify that the synthesized logic is functionally correct.
- *Timing (post-route) simulation:* we simulate the VHDL timing model of the synthesized logic mapped to the target PLD. The VHDL timing model is automatically generated by the place-and-route tool. This simulation is performed to verify that the synthesized logic, as mapped to the target PLD, meets all timing constraints.

Later in this chapter, a separate section is devoted to each of these kinds of simulation. Prior to those sections, we consider how stimulus is provided to the model of the design entity during a simulation.

Simulation Stimulus

A simulator requires a sequence of input values (*stimulus*) to apply to the VHDL model being verified. If an appropriate sequence of input values is created, it can be used for all three kinds of simulation. There are three approaches to generating stimulus: interactive (manual), command line, and testbench.

Interactive Simulation

An *interactive simulator* allows us to manually apply a stimulus. The simulator executes the design description using the input values applied and computes the corresponding output values.

Waveform Viewer

Waveforms representing the simulation model's input and output values as a function of time are displayed using the simulator's waveform viewer. The *waveform viewer* allows us to see transitions of multiple signals over time. We can zoom in or out to view selected time periods and measure the time between signal transitions. We can display bus signals as bits, hexadecimal numbers, or symbolic values. We can verify that the simulated design meets its specification by visually examining the waveforms.

For simple designs, interactive simulation is adequate. For designs with more than a few inputs, applying a comprehensive stimulus interactively becomes tedious and time-consuming. Furthermore, if the design description is modified, the entire stimulus has to again be manually applied to the modified design.

Command Line Simulation

Most simulators include a command line stimulus language. Commands in the stimulus language allow us to specify input waveforms. These commands can be entered interactively or placed in a file and executed. Using commands placed in a file makes it easy to reapply a stimulus after a design is modified. However, this approach is inefficient if the stimulus to be applied is complex. In addition, a simulator's stimulus language is simulator specific and, therefore, not portable to other simulators.

The most powerful and portable approach to generating a stimulus is to use a testbench written in VHDL. This approach is described in the next section.

1.5 TESTBENCHES

We use the next four phases to verify the functional correctness of our design description:

- Develop verification plan
- Write VHDL testbench
- Compile testbench
- Functional simulation

Verification Plan

Verification requires planning based on the system's requirements and specification documents. The results of this planning are documented in a verification plan. For complex systems this plan is a substantial document requiring significant effort.

Even a very simple system requires some thought as to how it should be verified. Often, the greatest difficulty in conducting a successful verification is determining an appropriate stimulus to apply and the output responses to be expected. The stimulus values applied must cover all of the circumstances in which the design will be used.

Testbench

A *testbench* is a VHDL program that applies a predetermined stimulus to a design entity during simulation. A testbench can be written to also verify the output values produced by the design entity. If there is a discrepancy between the expected and the actual output values, the testbench generates an error message. A testbench is basically a stimulus–response system.

UUT

The design entity being verified is usually called the *unit under test (UUT).* Strictly speaking, *testing* refers to the process of determining whether a design was manufactured correctly. For example, each system manufactured would be individually tested. In contrast, *verification* is performed to determine whether a design meets its specifications, before it is manufactured in quantity. However, use of the term UUT, from manufacturing testing, has been carried over to verification. Other terms used are device under test (DUT), design under test (DUT), device under verification (DUV), and module under test (MUT).

A testbench is also a design entity, it consists of a testbench entity declaration and its associated architecture body. A simple testbench architecture body contains:

- An instance of the UUT
- Signals mapped to the UUT's ports
- A process statement that applies stimulus to the UUT, via the signals

A testbench completely encloses its UUT and simulates the environment in which the design will operate.

The relationship between the half-adder UUT and its testbench is illustrated in Figure 1.5.1.

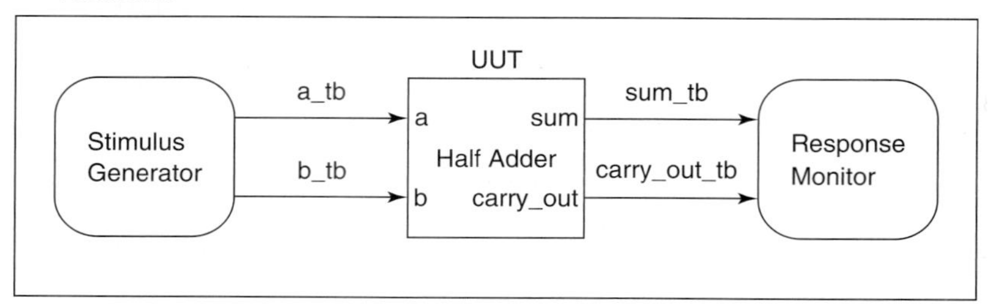

FIGURE 1.5.1
Conceptual relationship of the half-adder UUT to its testbench.

A testbench is entirely self-contained; it has no inputs or outputs. Statements in the testbench's architecture body apply stimulus values to signals that connect to the UUT's inputs. Other statements monitor the output values from the UUT.

UUT Models for Different Kinds of Simulations

For a functional simulation, our VHDL design description is the UUT. For a post-synthesis simulation, the VHDL netlist model (the synthesized gate-level logic) is the UUT. For a timing simulation, the VHDL timing model (produced by the place-and-route tool) is the UUT.

Half-adder Testbench

A testbench for the half adder of Listing 1.3.1 is given in Listing 1.5.1. This program contains a design entity named `testbench` with an architecture body named `behavior`. Its entity declaration indicates that design entity `testbench` has no inputs or outputs.

LISTING 1.5.1
Testbench for a half adder.

```
library ieee;
use ieee.std_logic_1164.all;

entity testbench is
end testbench;

architecture behavior of testbench is

   -- Declare signals to assign values to and to observe
   signal a_tb, b_tb, sum_tb, carry_out_tb :  std_logic;

   begin
   -- Create an instance of the circuit to be tested
   uut: entity half_adder port map(a => a_tb, b => b_tb,
      sum => sum_tb, carry_out => carry_out_tb);
```

```
  -- Define a process to apply input stimulus and test outputs
  tb : process
     constant period: time := 20 ns;

     begin -- Apply every possible input combination

     a_tb <= '0'; --apply input combination 00 and check outputs
     b_tb <= '0';
     wait for period;
     assert ((sum_tb = '0') and (carry_out_tb = '0'))
     report "test failed for input combination 00" severity error;

     a_tb <= '0'; --apply input combination 01 and check outputs
     b_tb <= '1';
     wait for period;
     assert ((sum_tb = '1') and (carry_out_tb = '0'))
     report "test failed for input combination 01" severity error;

     a_tb <= '1'; --apply input combination 10 and check outputs
     b_tb <= '0';
     wait for period;
     assert ((sum_tb = '1') and (carry_out_tb = '0'))
     report "test failed for input combination 10" severity error;

     a_tb <= '1'; --apply input combination 11 and check outputs
     b_tb <= '1';
     wait for period;
     assert ((sum_tb = '0') and (carry_out_tb = '1'))
     report "test failed for input combination 11" severity error;

     wait;  -- indefinitely suspend process
  end process;
end;
```

In the declarative part of the architecture body (between keyword **architecture** and the first **begin** keyword), signals used to provide connections between the half adder and the testbench are declared. In this example, these signals are named using the name of the UUT port they connect to followed by the suffix "`_tb`."

Following the **begin** keyword, UUT `half_adder` is instantiated and its connections to the signals of the testbench are defined by a port map.

Sequential Statements in a Process

Starting with the keyword **process**, a process named `tb` is defined. This process consists of signal assignment statements interspersed with wait statements and assertion statements. The statements in a process are sequential statements. They are executed in sequence, like statements in a conventional programming language. The process statement and the sequential statements used in a process are discussed in detail in Chapter 5. The brief discussion of these constructs at this point is just to provide some insight into the nature of a testbench.

Process `tb` simply applies, over time, four combinations of values to the signals connected to the UUT's two inputs. These are the four possible combinations of the two values '0' and '1' from the set of nine std_logic values. After each input combination is applied, the process waits 20 ns before verifying the outputs and applying the next input combination. This wait period would normally be determined based on timing constraints stated in the system specification.

Using Assertion Statements to Verify Outputs

Process `tb` uses assertion statements, which start with the keyword **`assert`**, to verify that the UUT's outputs have their expected values. If there is a difference between a simulated output value and the expected output value, an assertion statement causes the simulator to display an error message.

For example, after the input combination `a_tb` = '0' and `b_tb` = '1' is applied, an assertion statement verifies that `sum_tb` = '1' and `carry_out_tb` = '0'. If, during the simulation, this is not actually the case, the assertion statement causes the simulator to display the message "`test failed for input combination 01`".

A testbench should be written for every VHDL design. The testbench can even be written prior to writing the design description and serve as part of the specification.

As seen in Figure 1.1.1, once the design description and testbench have been written, the remaining phases in the design flow, excluding PLD selection, are all, essentially, automated. Compared to traditional digital design methodologies, this significantly simplifies and speeds up the design process.

While it is appropriate that the testbench be written in parallel with, or even before, writing the design description, this book focuses first on writing design descriptions. After an adequate subset of VHDL language constructs have been covered, the writing of testbenches is discussed in several chapters.

1.6 FUNCTIONAL (BEHAVIORAL) SIMULATION

The same simulator is used to perform all three kinds of simulation. But, as previously stated, the VHDL model used as the UUT is different for each. When the original VHDL design description is used, the simulation is a functional simulation.

Functional Simulation

A *functional simulation* simulates the design description to verify its logical correctness. None of the propagation delays that affect signals in a target PLD can be observed in a functional simulation. Functional simulation is a necessity for complex designs, because early detection of logical errors saves significant time and expense. Functional simulation of simple designs is an extremely valuable aid in learning VHDL.

The design flow for the VHDL/PLD design methodology is presented in greater detail in this section and Sections 1.9 and 1.10. Each of these sections has a figure that represents a portion of the design flow from Figure 1.1.1 shown in greater detail.

Each figure includes one of the three types of simulation. Taken together, these three figures form a detailed flowchart of the entire design flow.

The first portion of the detailed design flow is shown in Figure 1.6.1. Phases are shown as rectangles. The software tool used in each phase is shown as a shaded and rounded rectangle to the right of the associated phase.

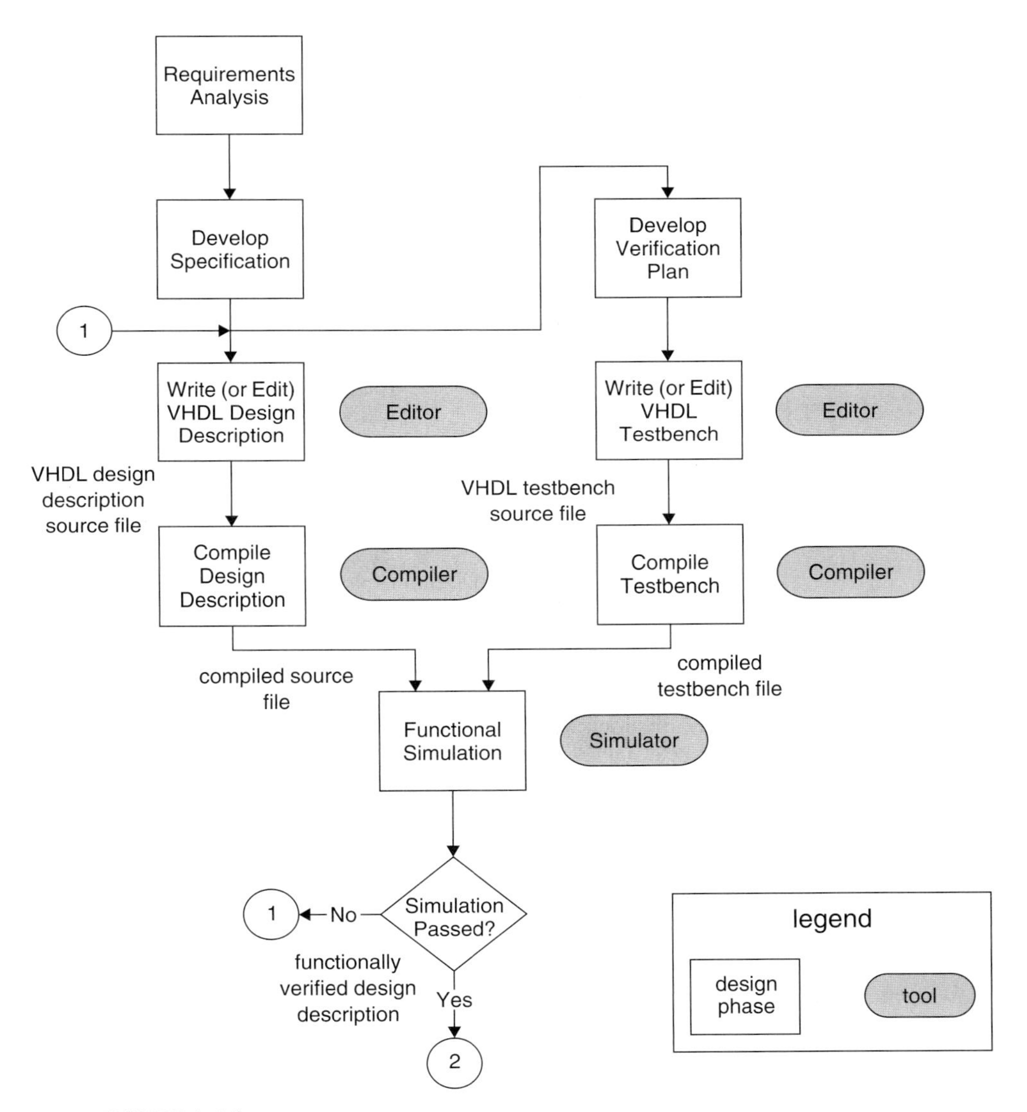

FIGURE 1.6.1
Detailed view of functional simulation portion of VHDL/PLD design flow.

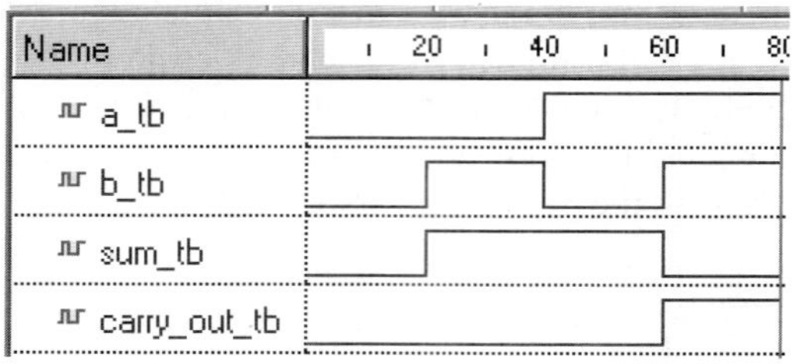

FIGURE 1.6.2
Waveforms from the functional simulation of the half-adder description.

In Figure 1.6.1, the testbench is shown as being written in parallel with the design description. Design teams often write the design description and testbench in parallel.

Detecting Dynamic Semantic Errors

Functional simulation is the time for logically debugging our design description and detecting dynamic semantic errors. Since the design description is the UUT, changes in signal values during the functional simulation map directly to statements in our source code.

Simulator Debugging Capabilities

Typical debugging capabilities provided by a simulator allow us to:

- Single step through source code statements
- Set breakpoints on source code statements
- Set breakpoints on signal or variable changes in value
- Monitor (watch) signal and variable values

Simulator Output Waveforms

Waveforms from the functional simulation of the half adder are shown in Figure 1.6.2. The signals displayed are those used to connect the UUT to the testbench. Because this is a functional simulation, the waveforms do not show any delay from an input change to a corresponding output change. That is, as soon as a UUT input changes, its outputs respond instantaneously.

1.7 PROGRAMMABLE LOGIC DEVICES (PLDS)

To proceed further in the design flow we must select a target PLD. This section provides a brief introduction to PLDs.

Fixed-Function ICs

Prior to the advent of PLDs, digital systems were implemented by interconnecting off-the-shelf fixed-function ICs. Small-scale integration (SSI) ICs provided discrete gates and flip-flops. Medium-scale integration (MSI) and large-scale integration

(LSI) ICs provided more complex combinational and sequential functional subsystems. The specific functions implemented by a *fixed-function IC* are determined by the IC manufacturer and cannot be changed by the user.

PLDs

The advent of *programmable logic devices (PLDs)* provided a more flexible way to implement a digital system. PLDs come in a wide variety of architectures and logic capacities. The term PLD is used generically in this book and, therefore, includes all ICs used for implementing digital hardware that are programmed by the user.

SPLDs, CPLDs, and FPGAs

PLDs are broadly classified based on their architectures as *simple PLDs (SPLDs)*, *complex PLDs (CPLDs)*, and *field programmable gate arrays (FPGAs)*. Each of these classes of devices has manufacturer-specific variants.

Logic Capacity

A PLD's *logic capacity* is usually specified in terms of the number of two-input NAND gate functional equivalents it provides.

A SPLD has a basic AND/OR array architecture and a maximum logic capacity of approximately 500 gates. A CPLD consists of multiple SPLD blocks on a single IC. Also included on a CPLD is a programmable interconnect matrix that allows the integrated SPLD blocks to be interconnected. The maximum logic capacity for a CPLD is approximately 60,000 gates.

FPGA architectures differ significantly from those of SPLDs and CPLDs. FPGA gate capacities start as low as 10,000 gates. Some FPGAs' logic capacities exceed 8,000,000 gates.

Logic capacity ranges for the different classes of PLDs tend to overlap near their boundaries. For CPLDs and FPGAs, the upper bounds of these ranges continue to increase over time as IC process technology advances.

AND/OR Arrays

To illustrate the concept of a PLD we consider a combinational SPLD. Conceptually, this SPLD consists of an array of AND gates followed by an array of OR gates. The logical structure of such an architecture is shown in Figure 1.7.1(a).

Buffered Inputs

Inputs to the SPLD are internally buffered, producing the normal and complement forms of each input signal. These signals appear on the vertical lines to the left of the AND gates in Figure 1.7.1(a).

AND Array

The vertical lines from the input buffers intersect horizontal lines that represent inputs to the AND gates. Each AND symbol has a single horizontal input line called a product line. The vertical lines from the input buffers, the product lines, and AND gates comprise the *AND array*.

Product Line

A *product line* is a single line that concisely represents all of the inputs to an AND gate. For example, each AND gate in Figure 1.7.1 is a four-input AND. Therefore, each product line actually represents four inputs.

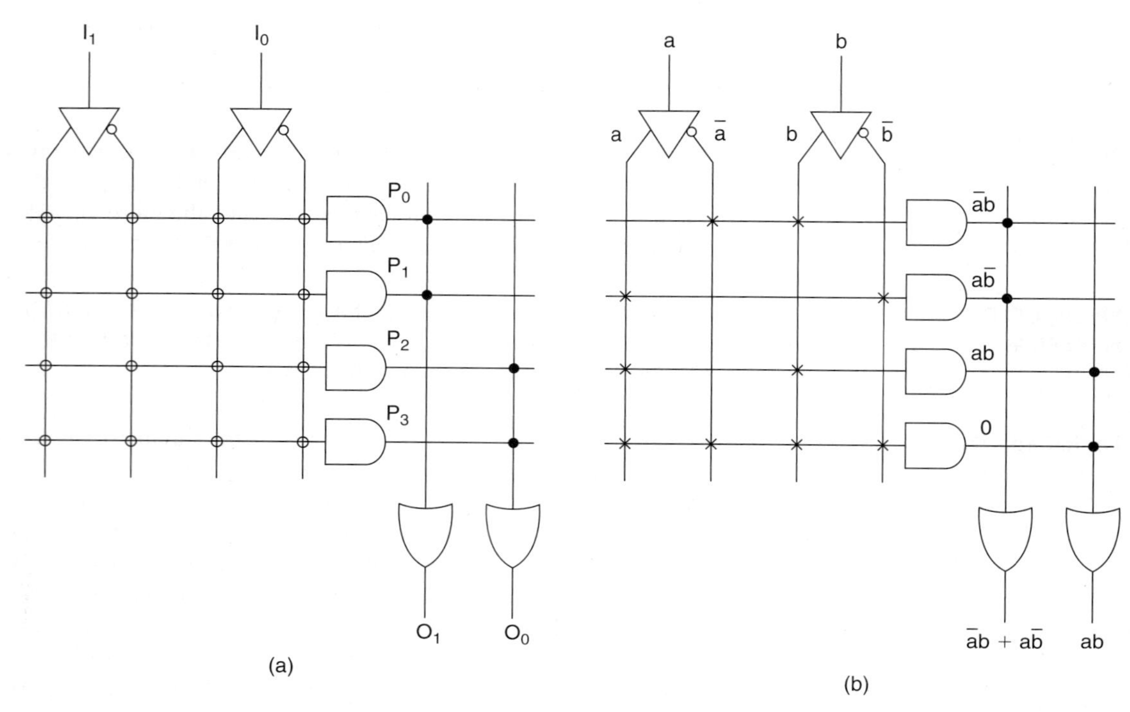

FIGURE 1.7.1
A SPLD logic diagram: (a) unprogrammed; (b) programmed to implement a half adder.

Programmable Interconnects

At the intersection of each vertical and horizontal line in the AND array in Figure 1.7.1(a) is a *programmable interconnect.* Each programmable interconnect can be programmed to either provide a connection between its associated vertical and horizontal lines or to not provide such a connection. Programmable interconnects are shown as small open circles on the SPLD logic diagram.

It is the programming of these interconnects that determines the function implemented by this SPLD. Interconnects programmed to make a connection at an intersection are represented by Xs in Figure 1.7.1(b). Each X represents the connection of a buffered external input (or its complement) to a single input of an AND gate. Intersections without an X represent interconnects programmed so that there is no connection. The logic value at an unconnected AND gate input defaults to a 1.

OR Array

The horizontal lines representing AND gate outputs intersect vertical lines that represent inputs to OR gates. Each OR symbol has a single input line called a sum line. The AND gate output lines, the sum lines, and OR gates comprise the *OR array.* The SPLD's OR gates produce the logical sum of the outputs of selected AND gates.

Sum Line

A *sum line* is a single line that concisely represents all of the inputs to an OR gate. The particular SPLD of Figure 1.7.1 has a fixed OR array that cannot be programmed. The fixed connections in this OR array are indicated by solid circles at the intersections of its horizontal and vertical lines.

The SPLD in Figure 1.7.1(b) represents the SPLD in Figure 1.7.1(a) programmed to implement a half adder. The physical programming of a SPLD is accomplished using information in the configuration file produced by the place-and-route tool. The AND/OR architecture of this PLD is ideal for implementing the required sum-of-products Boolean functions.

Programmable Array Logic (PAL)

A SPLD with a programmable AND array and a fixed OR array, as in Figure 1.7.1, is more specifically referred to as a *programmable array logic (PAL)*.

1.8 SPLDS AND THE 22V10

Designs in the early chapters of this book are of low logic complexity and can be implemented using a SPLD. This section provides an overview of an actual SPLD that could be used as the target PLD for these designs. Since the majority of today's SPLDs are based on the PAL architecture, this section focuses on the 22V10 PAL-type SPLD.

Contemporary SPLDs

There are two major differences in today's PAL-type SPLDs and the conceptual PAL shown in Figure 1.7.1(a).

Erasable PLDs

First, today's SPLDs use electrically erasable and reprogrammable interconnects. Such an interconnect uses a floating gate transistor structure similar to that used in an EEPROM.

Erasing returns each programmable interconnect to its unprogrammed state. *Erasable PLDs* have the advantage that if a logic error is found in a design after the PLD is programmed or if features need to be added, the previously loaded configuration information can be erased. After the design error is corrected or features are added in the design description, the modified design is verified. The PLD is then reprogrammed using the new configuration information. Use of reprogrammable PLDs allows us to rapidly and economically iterate design changes.

Electrically erasable interconnects have fast erase times, typically less than 100 ms. The minimum number of program/erase operations that can be performed on such a SPLD is typically greater than 10,000. Data retention typically exceeds 20 years. A SPLD might typically be programmed from one to 20 times during development. During production, a SPLD is normally programmed only once.

Output Logic Macrocells

The second major difference in today's PAL-type SPLDs and the conceptual PLD shown in Figure 1.7.1(a) is that a programmable *output logic macrocell (OLMC)* is connected to the output of each OR gate. The output of each OLMC connects to a

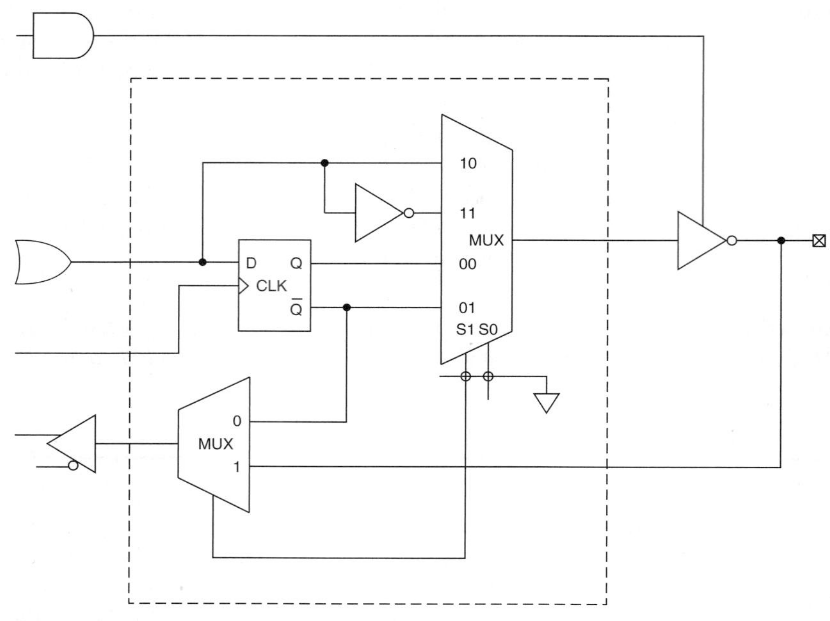

FIGURE 1.8.1
Output logic macrocell.

pin of the PAL. Use of OLMCs allows each pin to be independently programmed to implement any one of three different I/O structures. An OLMC can be configured to make the pin that it connects to an output, input, or bidirectional. An OLMC also contains a flip-flop, so that the output of the OR gate can be stored. This allows a PAL to be used to implement either combinational or sequential systems.

An OLMC is represented in Figure 1.8.1 as the logic inside the dashed block. The data input to the OLMC is the output from the OR gate that sums the product terms from the AND array. The AND gate at the top of the diagram is an output enable product term that controls the three-state inverting output buffer. The product line for this AND gate can be programmed so that its output is fixed at 0, disabling the three-state buffer, to configure the I/O pin as an input. In contrast, if this gate's output is fixed at 1, enabling the three-state buffer, the I/O pin is configured as an output. Alternatively, since this AND gate's product line is part of the AND array, its output can be a function of the PAL's inputs, allowing the I/O pin to be bidirectional.

Combinational Macrocell Output

In the OLMC of Figure 1.8.1, the configuration of the output is controlled by the macrocell's four-input multiplexer. This multiplexer allows the D flip-flop in the macro cell to be bypassed. It also allows the output's polarity to be specified. The multiplexer's select inputs are from two programmable connections, S_1 and S_0. These connections are called *configuration bits.* Each OLMC has its own configuration bits.

For a *combinational active-high output*, S_1S_0 is made 11. This selects the inverted output of the OR gate as the input to the inverting three-state buffer that drives the I/O pin.

For a *combinational active-low output*, S_1S_0 is made 10. Under this condition, the output of the OR gate is directly input to the inverting output buffer.

In both the previous cases S_1 is 1, so the two-input multiplexer at the lower left of the macrocell feeds the signal at the I/O pin back, in both its normal and complement forms, as inputs to the AND array.

Registered Macrocell Output

A *registered output* is one derived from an OLMC's flip-flop. For a *registered active-high output*, S_1S_0 is made 01 and the $\overline{Q}$ output of the D flip-flop is input to the inverting output buffer. For a *registered active-low output*, S_1S_0 is made 00 and the Q output of the D flip-flop is directly input to the inverting output buffer. For either registered output, S_1 is 0, and the $\overline{Q}$ output of the flip-flop is fed back, in its normal and complement forms, as inputs to the AND array.

Each output logic macrocell is independently programmable, providing substantial flexibility. Like the reprogrammable interconnects in the AND gate array, the configuration bits S_1 and S_0 are also reprogrammable.

Programmable Output Polarity

It is clear from the previous discussion that an OLMC provides versatility, but how is this versatility used and why is it advantageous?

When implementing a combinational function, the OLMC allows either the sum-of-products function or its complement to appear at its output pin. Some Boolean sum-of-product functions are most efficiently implemented in their normal (true) form. Others are more efficiently implemented in their complement form. That is, either the normal or complement form of a particular function might require fewer product terms. A function that is a wide OR function is most efficiently implemented in a SPLD-type architecture as the complement of the complement of the function.

For example, the complement of a large sum of individual terms may be expressed as a single product term. Consider a function that is the OR of 18 inputs. To implement this function directly in the programmable AND–fixed OR architecture of a PAL requires 18 AND gates, each with only one of its inputs used. In other words, each input to the OR gate requires a product term consisting of a single literal. The output OR gate sums these 18 product terms to realize the function. Assume that you had a SPLD where the most product terms that can be ORed together as input to an OLMC is 16. This function requiring 18 AND gates cannot be directly implemented in this SPLD using a single output.

However, by application of DeMorgan's Law, the OR of 18 inputs is equivalent to the complement of the AND of the complements of each of the 18 inputs. Since each input is available in its complement form in the AND array of a SPLD, all we need is a single AND gate with at least 18 inputs. This gives the complement of the desired function at the output of the OR gate. The output of the OR gate is then complemented in the OLMC to obtain the desired function.

With programmable output polarity, synthesis and place-and-route software can perform logic optimization on an expression and its complement and use the form that requires the fewest logic resources.

Popular PAL-type SPLDs

There are a number of popular PAL-type SPLDs that find widespread use. Part numbers for these devices typically start by designating the total number of I/Os. This is followed by a letter indicating the type of flip-flop or logic in the OLMC. Next is a number indicating the maximum number of outputs. A hyphen then precedes a number indicating the SPLD's speed. Other letters following the speed may indicate power level, package type, and temperature grade.

The most popular SPLDs are the 16V8, 18V10, 20V8, 22V10, and 26V12. The most popular SPLD from this group is the 22V10. The "V" stands for versatile, meaning that the device contains OLMCs allowing a variety of I/O configurations. Versions of the previous devices are available from manufacturers such as Atmel, Cypress, and Lattice Semiconductor. Data sheets and application notes providing details on these devices are available from the manufacturers' web sites.

The 22V10 SPLD

A block diagram of a 22V10 is shown in Figure 1.8.2. A 22V10 has 12 dedicated input pins, shown on the left of the diagram, and 10 input/output (I/O) pins, shown on the right. I/O pins are pins that can be configured as either inputs or outputs. One manufacturer rates this device as having a logic capacity of 500 gates.

The "22" in the designation 22V10 indicates that the device provides up to 22 inputs. All 22 inputs are available only when all 10 I/O pins are configured as inputs and used along with the 12 dedicated input pins. The "10" in the designation indicates that as many as 10 outputs are available. This situation exists when all of the I/O pins are configured as outputs. Practically speaking, only 21 inputs are available, since in any useful application at least one I/O pin needs to be configured as an output.

The 22V10's logic diagram is shown in Figure 1.8.3. The pin numbers are those for a 28-pin PLCC package. The 12 dedicated input pins are shown on the left side and bottom right side of this diagram. The input at pin 2 can be used as either the global clock input to the D flip-flops or as an input to the AND array. Each input is buffered. The buffered normal and complement forms of each input are connected to separate columns, generating 24 columns. For example, the buffered normal and complement forms of the input at pin 3 are available on columns 4 and 5, respectively.

The other 20 columns are the normal and complement forms of signals either fed back from the 10 macrocells on the right side of the diagram or input from I/O pins configured as inputs. Together these 44 columns provide the AND array the normal and complement forms of the 22 possible inputs.

The rows of the AND array are the AND gate product lines. Each AND gate has 44 inputs. For each output, one AND gate, the *output enable product term,* is used to control that output's inverting three-state buffer. The remaining AND gates associated with an output provide the inputs to the OR gate that feeds the output's OLMC.

Two other AND gates are shared by all the OLMCs. One controls the asynchronous resets to all the flip-flops in the OLMCs and the other controls their synchronous

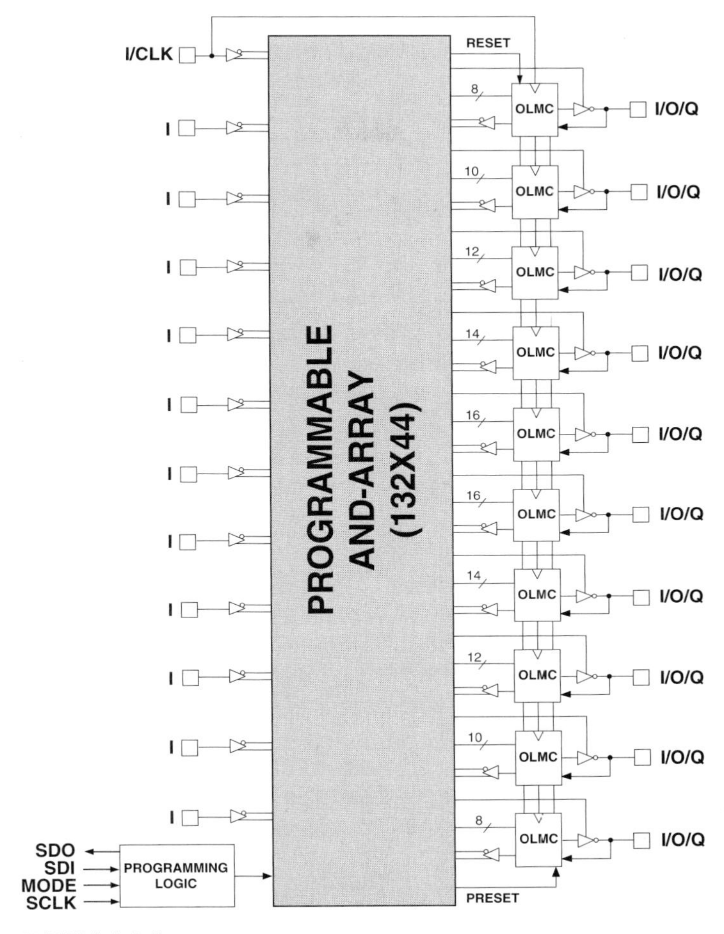

FIGURE 1.8.2

Functional block diagram of a 22V10 SPLD. *Courtesy of Lattice Semiconductor Corporation. All rights reserved.*

presets. All total, there are 132 AND gates, each with 44 inputs. Accordingly, the AND array is designated as 132×44 in Figure 1.8.2.

22V10 Product Term Distribution

Not all of the 22V10's OR gates have the same number of inputs (AND gate outputs). This is seen clearly in the block diagram. This type of arrangement, called *varied product term distribution,* is based on the fact that in a typical application some logic expressions require fewer product terms than others. If each OR gate summed 16 product terms, for some outputs many product terms would go unused.

For example, the OR gate providing the output at pin 27 has 8 inputs and the OR gate providing the output at pin 23 has 16 inputs (Figures 1.8.2 and 1.8.3). As a result, the output at pin 27 can implement a sum-of-products expression that is the sum of as many

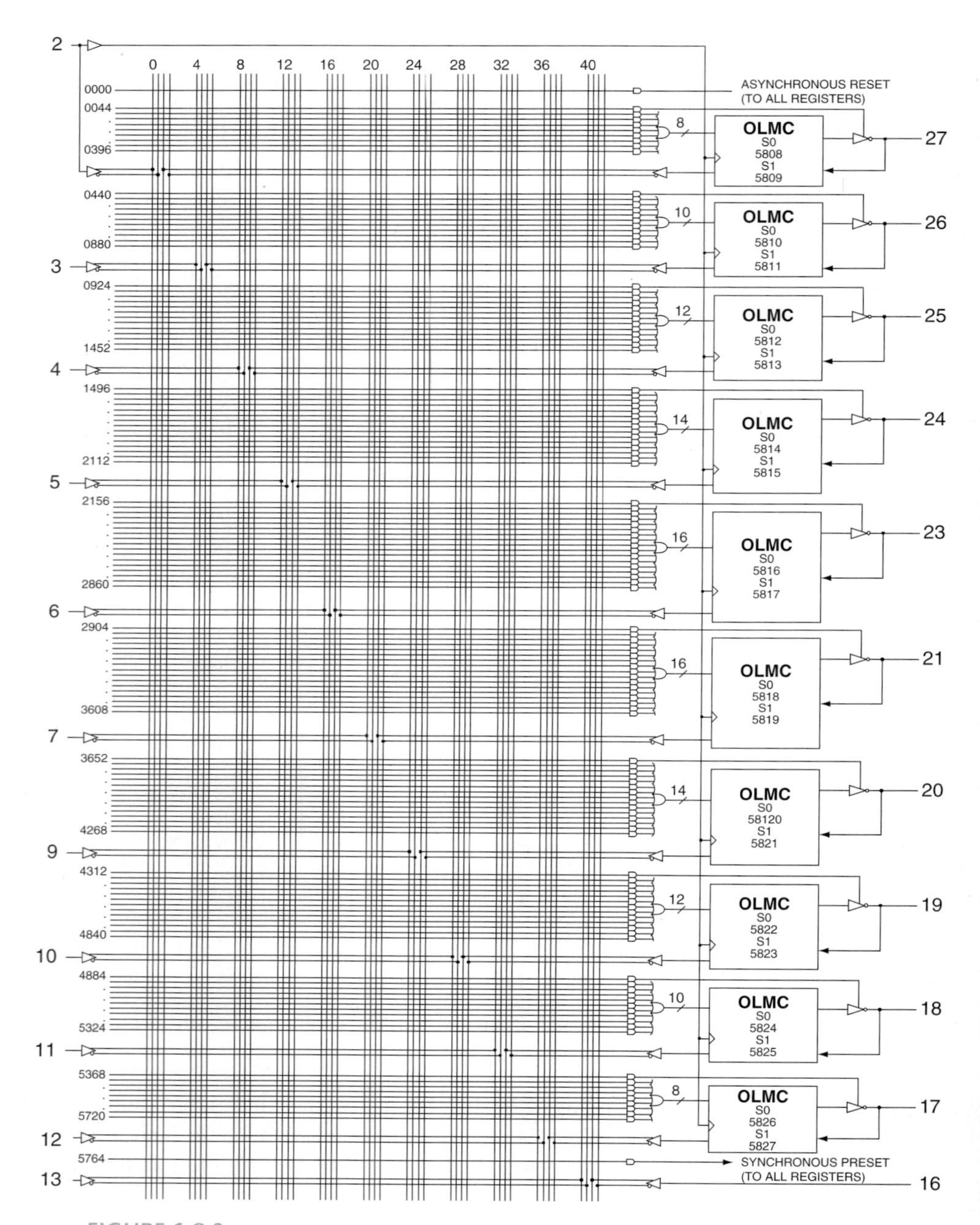

FIGURE 1.8.3
22V10 SPLD logic diagram. *Courtesy of Lattice Semiconductor Corporation.*

as 8 product terms, each of 22 variables, and the output at pin 23 can implement a sum-of-products expression that is the sum of as many as 16 product terms, each of 22 variables.

22V10 Output Logic Macrocells

The 22V10's OLMCs are similar to that shown in Figure 1.8.1. It can be configured to provide a combinational active-low, combinational active-high, registered active-low, or registered active-high output. The flip-flops in all 10 OLMCs share a common clock, asynchronous reset, and synchronous preset. The asynchronous reset and synchronous preset are each derived from product terms in the AND array. Therefore, they can be an AND function of any of the inputs.

While the 22V10 has considerable logic capacity (500 gate equivalents) and versatility (provided by its OLMCs), it is still a SPLD, and as such is among the simplest and least capable PLDs.

1.9 LOGIC SYNTHESIS FOR THE TARGET PLD

The next group of phases is performed to synthesize logic for the target PLD. These phases are:

- Select PLD
- Synthesize logic
- Post-synthesis simulation

This portion of the design flow starts with selection of the target PLD, as shown in Figure 1.9.1.

Selecting the Target PLD

A PLD must be selected that has sufficient logic capacity to implement the logic to be synthesized. The selected PLD is called the *target PLD*. The target PLD must also be fast enough to meet the system's timing requirements. To optimally make this selection, we must have a good knowledge of PLDs and their architectures.

For complex designs, a PLD is typically selected such that no more than approximately 80 to 90% of its resources are expected to be used. This provides extra resources to ensure a successful routing and possibly allow additional features to be added later. For our half-adder design, very little logic capacity is required, so a 22V10 SPLD is a more than adequate choice.

Logic Synthesis

Once the target PLD has been selected, a synthesizer is used to synthesize the logic described by the design description. *Synthesis* is the process of automatically translating a VHDL design description to logic. The synthesized logic is optimized in terms of *area* (number of gates) and/or *speed* (number of gate levels).

It is important to keep in mind that while a synthesizer optimizes the logic it produces, it can only translate the given VHDL description to hardware. If what is described is an inferior solution to the original problem, the synthesized results will be less than optimal.

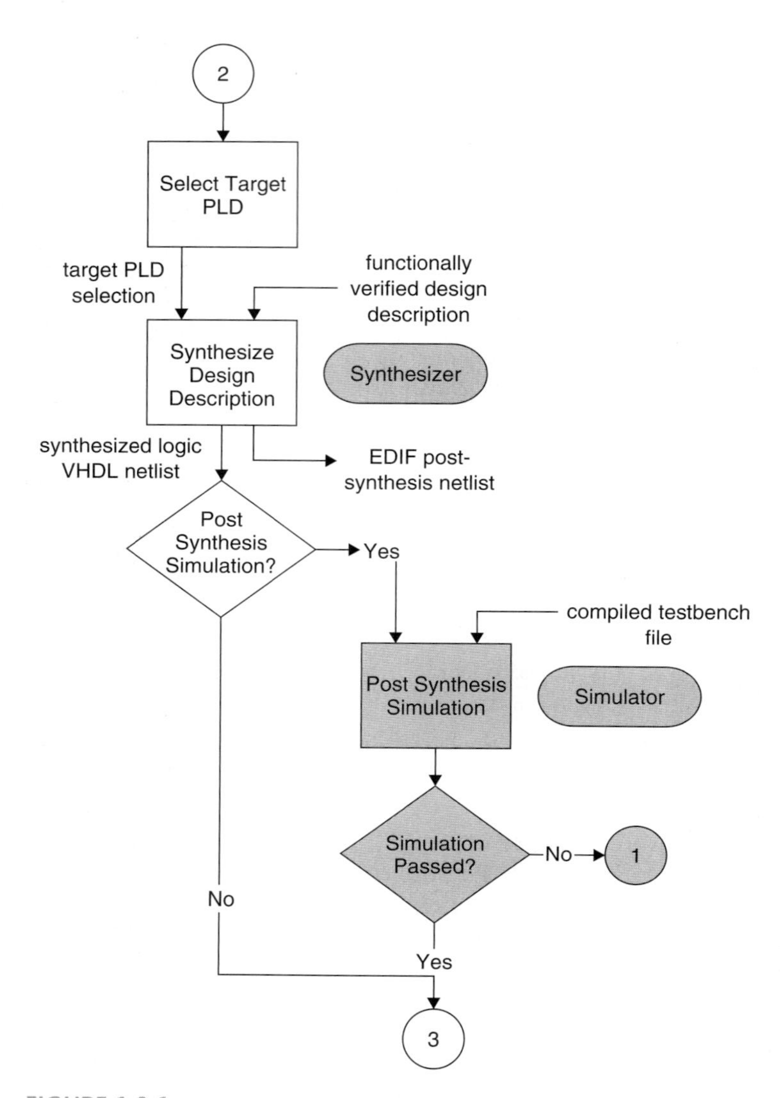

FIGURE 1.9.1
Detailed view of post-synthesis simulation portion of VHDL/PLD design flow.

A synthesizer requires two inputs: the design description file and the specification of the target PLD.

Synthesizer

The objective of a *synthesizer* is to ***synthesize logic that behaves identically to the simulated behavior of the design description.*** In effect, a synthesizer translates a design description into a functionally equivalent gate-level logic implementation.

A synthesizer generates the required Boolean equations. It automatically carries out logic and state minimization and stores the simplified Boolean equations in an intermediate form. The output of the synthesizer is a gate-level logic description.

Gate-Level Logic

A *gate-level logic implementation* is sometimes referred to as a *register transfer level (RTL)* implementation. This level describes the logic in terms of registers and the Boolean equations for the combinational logic between the registers. Of course, for a combinational system there are no registers and the RTL logic consists only of combinational logic. Register transfer logic is discussed in greater detail in Chapters 10, 11, and 15.

A synthesizer may include its own compiler to compile a design description into the intermediate form used by the synthesizer. This compiler can highlight synthesis errors in the source code. Not all VHDL language constructs can be synthesized. Some *synthesis errors* result from the use of constructs that can't be synthesized.

Typically, a synthesizer performs three steps during synthesis:

1. *Language synthesis*: the design description is transformed into a representation based on Boolean equations. These equations represent the interconnection of generic logic elements or functional blocks.
2. *Optimization*: algorithms apply the rules of Boolean algebra to optimize the logic for area and/or speed. These optimizations are independent of the technology of the target PLD and produce a technology independent gate-level netlist.
3. *Technology mapping*: the logic is mapped to the target PLD using techniques specific to the target PLD's architecture and further optimizations are performed. This step corresponds to transforming the technology independent netlist to a technology dependent netlist.

For our simple half adder, the logic that results from step 1 of the synthesis process is given in Figure 1.9.2.

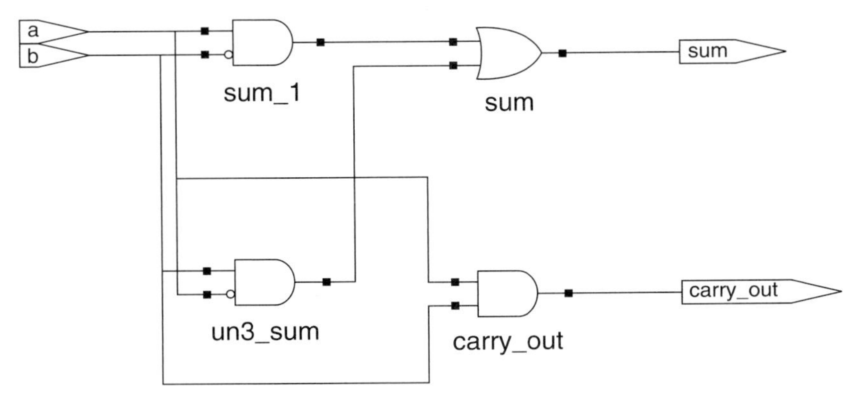

FIGURE 1.9.2
RTL view of half-adder synthesized logic from synthesis step 1.

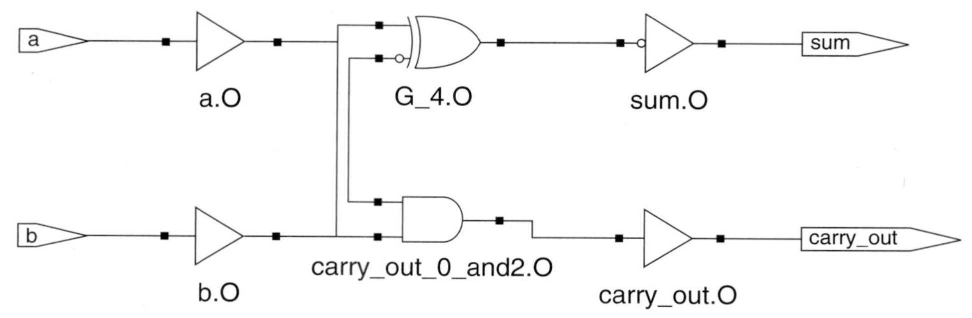

FIGURE 1.9.3
Technology dependent view of half-adder synthesized logic from synthesis step 3.

The logic from step 3 is represented in Figure 1.9.3. This diagram includes input and output buffers that are part of the target 22V10 SPLD architecture.

Synthesizer Output

The synthesizer produces two output files: a VHDL netlist and a technology dependent gate-level netlist.

A *netlist* is a textual representation of the interconnection of logic elements.

VHDL Netlist

A *VHDL netlist* is a design file that describes, in VHDL structural style, the connectivity of the optimized logic implemented using the target PLD's primitives. The VHDL netlist is used as the UUT model in a post-synthesis simulation.

Post-Synthesis (Gate-Level) Simulation

To verify the synthesized logic, the VHDL netlist is used as the UUT in the testbench. Simulation of the VHDL netlist is called post-synthesis simulation or gate-level simulation, since it simulates the structural interconnection of the gates and flip-flops synthesized for a design. It verifies the functional operation of the synthesized logic.

The results from this simulation are compared with the results from the functional simulation of the design description. These results should be the same.

For post-synthesis simulation of our half adder, its design description is replaced as the UUT model in the testbench by the VHDL netlist. This replacement is accomplished by replacing the file that contains the design description of the half adder with the file generated by the synthesizer that contains the VHDL netlist of the synthesized logic.

If the synthesized logic is functionally correct, the post-synthesis simulation waveforms will be the same as the functional simulation waveforms in Figure 1.6.2. Simulators usually have a feature that allows waveforms from different simulations to be compared. This makes it easy to compare waveform results from a post-synthesis simulation with those from the corresponding functional simulation.

Skipping Post-Synthesis Simulation

If a timing simulation is to be performed, post-synthesis simulation may be skipped. A timing simulation can simultaneously verify both the function and timing of the synthesized logic, making post-synthesis simulation unnecessary.

Accordingly, the activities related to post-synthesis simulation are shaded in Figure 1.9.1.

Technology Dependent Netlist

A *technology dependent netlist* describes the connectivity of the optimized logic using the target PLD's logic primitives. These *primitives* are the logic elements available in the target PLD's architecture. The technology dependent netlist is in a format readable by the PLD vendor's place-and-route tool. Typically, the EDIF netlist format is used.

EDIF Netlist Format

EDIF (Electronic Data Interchange Format) is a format issued by the Electronic Industries Association, as EIA-548. Its purpose is to provide a standard format for transferring design information between EDA tools.

1.10 PLACE-AND-ROUTE AND TIMING SIMULATION

The next phases are:

- Place and route logic to PLD
- Timing simulation

These phases are included in the last portion of the detailed VHDL/PLD design flow shown in Figure 1.10.1.

Place-and-Route Operations

A *place-and-route* (or *fitter*) tool is software used to automatically map or fit synthesized logic to a target PLD's architecture. The *place operation* selects and configures specific logic primitives in the PLD's architecture for each logic primitive in the technology dependent netlist.

The *route operation* determines the path for each connection between two logic primitives and each connection between a logic primitive and a pin of the PLD. The place-and-route tool is usually obtained from the PLD vendor.

Place-and-Route Tool

A place-and-route tool takes as its inputs a technology dependent gate-level EDIF netlist and constraint information such as pin assignments for port signals and timing constraints.

The place-and-route tool maps the EDIF netlist into the PLD's architecture by determining exactly which instance of each primitive logic element in the PLD's architecture to use for each primitive in the synthesized logic. It also takes into account such things as desired pin assignments for port signals and timing goals.

Pin Assignments

In our half-adder design example we did not concern ourselves with the assignment of input and output signals to specific pins of the PLD. The place-and-route tool can automatically make these assignments. However, sometimes we want to make these assignments ourselves, rather than depending on the place-and-route tool's default

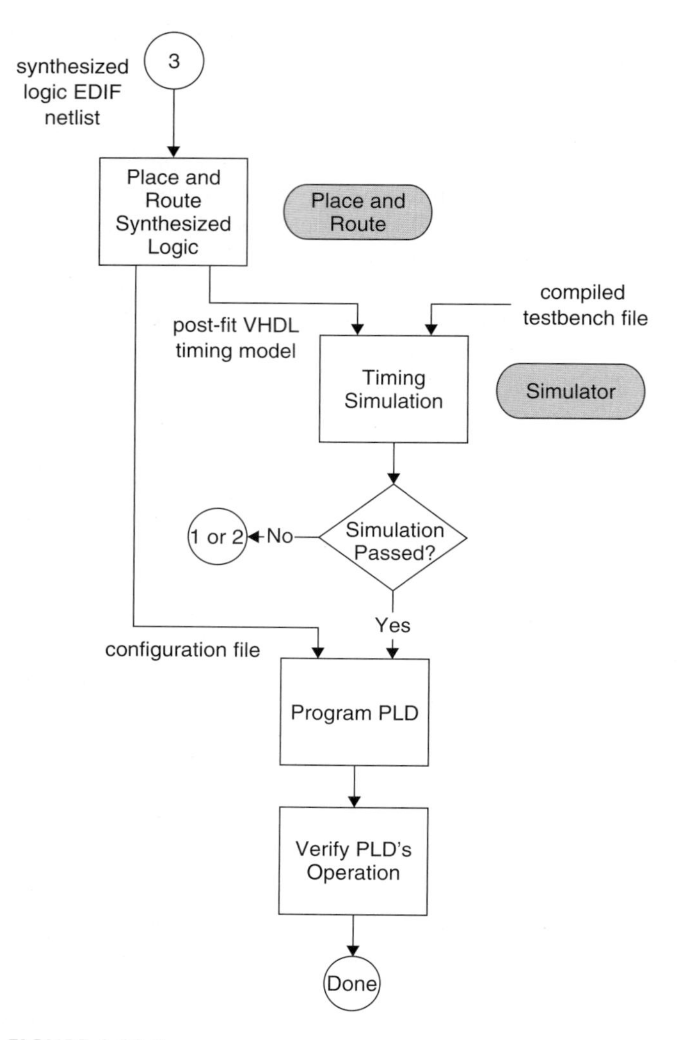

FIGURE 1.10.1
Detailed view of timing simulation portion of VHDL/PLD design flow.

assignments. For example, if we modify some of the logic of a design and reprocess the modified description, the place-and-route tool may assign the signals to different pins. If the printed circuit board for the PLD has been laid out based on the previous pin assignments, we will have a problem.

A place-and-route tool allows us to specify pin assignments. The ability of the place-and-route tool to achieve specified pin assignments and to maintain these assignments after a design's logic has been modified depends on the routing resources of the PLD.

User-Defined Attributes

How pin assignments are specified depends on the particular place-and-route tool. One approach is to use VHDL's user-defined attribute feature. In addition to predefined attributes, VHDL allows the user to define attributes. *User-defined attributes* are VHDL statements that we can include in our design description file. A user-defined attribute does not affect the semantics of the VHDL code. It is recognized only by the EDA tool that defines it. These attributes are used to pass information to EDA tools used in the design flow.

Synthesis and place-and-route tool vendors create their own "user"-defined attributes to allow us to specify information about how a design should be synthesized or placed and routed. Such attributes are described in each tool's documentation.

For example, attributes defined by Lattice Semiconductor to specify pin assignments for their SPLDs are placed in the entity declaration as shown in Listing 1.10.1.

These attributes are ignored by the compiler and synthesizer. However, they are passed along to the place-and-route tool in the EDIF file generated by the synthesizer. Our design description examples usually do not include specification of pin assignments.

Place-and-Route Tool Outputs

A place-and-route tool generally produces three outputs: a chip report, a configuration file, and a VHDL timing model.

Chip Report

The chip report documents which port signals are assigned to which PLD pins and how much of the PLD's logic capacity is used.

LISTING 1.10.1
Use of attributes to specify assignment of ports to device pins.

```
library ieee;
use ieee.std_logic_1164.all;

entity half_adder is
   port (a, b : in std_logic; sum, carry_out: out std_logic);

   attribute loc: string; -- pin assignment attribute
   attribute loc of a : signal is "P3"; -- pin assignments
   attribute loc of b : signal is "P4";
   attribute loc of sum : signal is "P25";
   attribute loc of carry_out : signal is "P24";

end half_adder;

architecture dataflow of half_adder is
   begin
   sum <= (not a and  b) or (a and not b);
   carry_out <= a and b;
end dataflow;
```

For example, the 22V10 SPLD is available in several packages including a 24-pin dual-in-line package (DIP). As previously discussed, this chip has 12 dedicated inputs and 10 I/O pins that can be configured as either inputs or outputs. The other two pins are power and ground. Our half adder requires only two input pins and two I/O pins configured as outputs. Slightly less than 20% of the 22V10's pins and logic capacity are required. We could put four additional half adders in the same PLD, or use the unused logic for other functions.

Configuration File

The *configuration file* (programming file) contains the interconnection and configuration data necessary to program the PLD. This file specifies exactly which programmable interconnects in the target PLD are to be programmed as connections and which are not.

VHDL Timing Model

The VHDL timing model is a file containing a structural-style VHDL program that describes the logic and timing of the synthesized logic mapped to the target PLD. This model includes information detailing the propagation delays of signals through the PLD.

Timing Parameter Values

The VHDL timing model uses a set of timing parameters to specify the propagation delays of the PLD's primitive elements. The timing parameter values are usually passed to the model using constants called *generics*. Actual generic values are either defined directly in the timing model or specified in a separate file that is generated by the place-and-route tool. If a separate file is used it usually specifies the generic timing values in a format called *standard delay format (SDF)*.

A PLD's timing is a function of both the propagation delays of its primitive elements and the specific delay paths that result when the synthesized logic is routed for the target PLD.

Timing (Post-Route) Simulation

A synthesized circuit's operation may be functionally correct, but it may not meet one or more timing constraints given in its specification. A *timing simulation* allows us to determine the speed at which the synthesized logic will operate when later programmed into the target PLD. Timing simulations are also referred to as *post-route, post-fit,* or *post-implementation simulations,* since the simulation is performed after the synthesized logic has been placed and routed for the target PLD.

The VHDL timing model is simulated to provide the timing verification. When we simulate using the VHDL timing model as the UUT, we obtain output waveforms that show not only the functional operation of the design, but also the effects of signal propagation delays through the target PLD.

Half-Adder Timing Simulation

For example, after place and route, the VHDL timing model of the half adder is used as the UUT in the same testbench (Listing 1.5.1) used for the functional simulation. The VHDL timing model entity declaration has the same name and includes all the ports of the original design description in Listing 1.3.1. However, it also includes ports for all the unused PLD pins.

The architecture body of the half-adder VHDL timing model is significantly different from the half-adder design description architecture body. The timing model's architecture is a structural style architecture that represents the half-adder logic mapped to the target PLD's architecture.

The target 22V10 SPLD in our half-adder example has a 10 ns propagation delay. After applying a stimulus value, the testbench waits for a time equal to the constant `period` before verifying the output values. In Listing 1.5.1, `period` is equal to 20 ns. Since any change in the output occurs within 10 ns of the associated input change, the outputs are stable for 10 ns before they are verified.

The timing simulation waveforms for the half adder mapped to a 22V10 SPLD with a 10 ns delay are given in Figure 1.10.2. Note that at $t = 0\ ns$, the testbench assigns both of the UUT's inputs the value 0. The outputs of the UUT are initially undefined for the first 10 ns. This is how long it takes for the effect of the new input values to propagate through the PLD. In the timing waveforms, this undefined signal state is represented by dashed lines halfway between the logic levels.

The value of the constant `period` in the testbench can be changed to reflect a different timing constraint value. The appropriate value would be determined from timing constraints included in the half-adder's specification.

Point Tools versus an IDE

The previously discussed EDA tools may each be provided as a separate program. In such cases, each tool is referred to as a *point tool,* because it is used at a specific point in the design flow.

Alternatively, several, or all, of the tools may be integrated together in a single EDA tool providing an *integrated design environment (IDE).* In an IDE, a design is usually set up as a *project* that keeps all of the files associated with the design together in one directory. An IDE may be simpler to use because the interface between the various tools is seamless and the IDE provides a single consistent graphical user interface for all of the tools. PLD vendors often provide such an IDE. However, a PLD vendor's IDE has the limitation that it only supports design for that vendor's PLDs. Another limitation is that the simulator and synthesizer integrated into a PLD vendor's IDE may not be as powerful as available third-party point tools.

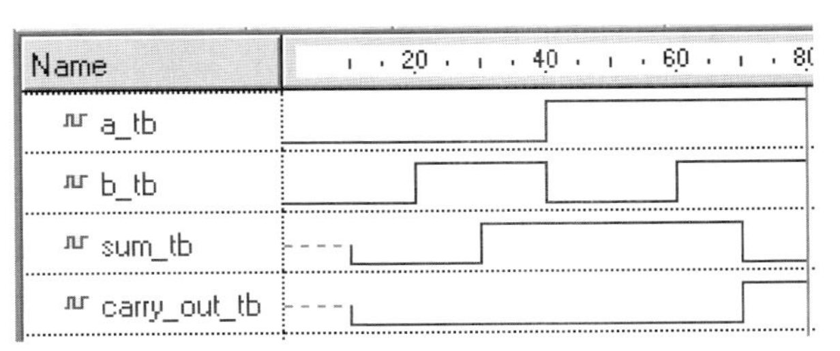

FIGURE 1.10.2
Timing simulation output for half adder.

Design Flow Manager

Some simulators include a design flow manager that makes it easy to use point tools from different EDA vendors and PLD vendors.

For example, *Aldec's Active-HDL* simulator contains a design flow manager that allows us to use our choice of external synthesizer and place-and-route tools. For a particular design, we configure the design flow through menus to specify the target PLD and which synthesis and place-and-route tools to use. Based on these menu selections, Active-HDL automatically creates a graphical design flowchart consisting of button icons for each phase of the design flow. Simply clicking on a button icon launches the appropriate external tool and causes that phase of the flow to be executed.

Aldec's design flow manager provides the ease of use of an IDE along with the flexibility of using a wide choice of target PLDs, synthesizers, and place-and-route tools.

The Aldec Active-HDL design flow for the half adder using a Lattice Semiconductor ispGAL22V10 as the target PLD, Synplicity's Synplify as the synthesizer, and Lattice Semiconductor's ispLEVER for place and route is shown in Figure 1.10.3.

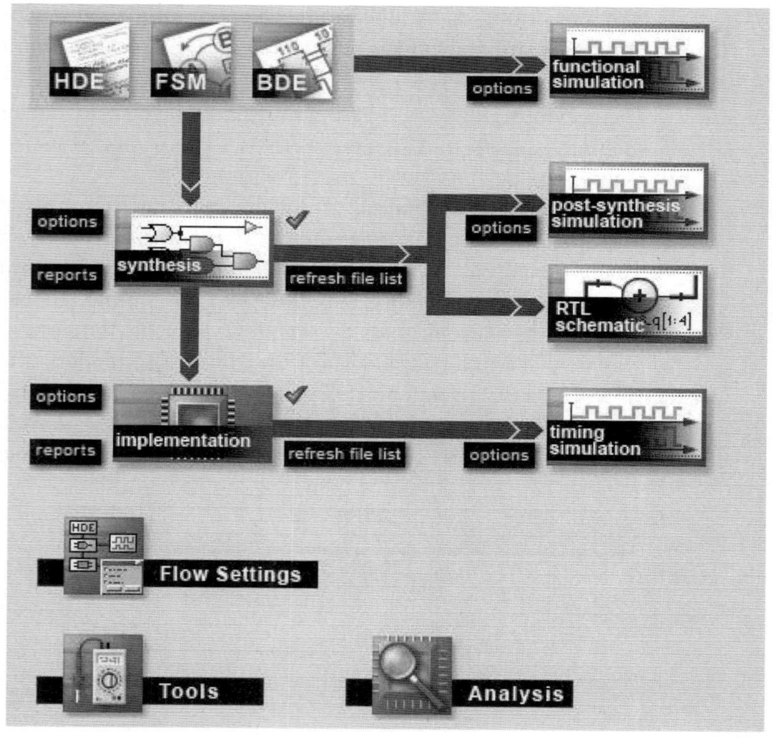

FIGURE 1.10.3
Aldec Active-HDL Design Flow Manager.

1.11 PROGRAMMING AND VERIFYING A TARGET PLD

The last two phases of the design flow are:

- Program PLD
- Verify PLD's operation

Programming a PLD

The configuration file generated by the place-and-route tool is also called a *programming file, fuse map,* or *JEDEC file.*

Device Programmer

We program the programmable interconnects in a PLD by loading configuration data into the PLD. For some PLDs, this configuration data is loaded prior to the PLD being placed on the system's printed circuit board. An electronic instrument called a *device programmer* is used to load the configuration data into such a PLD.

The device programmer is connected to a host computer through a communications port. The device programmer's application software, executing on the host computer, transfers the configuration file from the host computer to the device programmer. If necessary, the device programmer erases the PLD and then programs the programmable interconnects as specified by the configuration file. Finally, the device programmer confirms that the PLD was successfully programmed.

For our half-adder design, a Lattice Semiconductor ispGAL22V10 is placed in a socket on a device programmer and programmed. We now have a physical implementation of our half adder.

In-System Programming

In addition to being programmed by a device programmer, some PLDs can be programmed after they have been placed on a printed circuit board. These types of PLDs are called *in-system programmable (ISP).* In these devices, logic to program the PLD is contained on the PLD chip itself. The serial connection to transfer the configuration data is made directly to the PLD, through a connector and traces on the printed circuit board on which the PLD is mounted. Use of an ISP PLD leads to reduced system production costs.

The "isp" prefix in the ispGAL22V10 part number indicates that this SPLD can be in-system programmed.

Verifying a Programmed PLD

As the last phase in the design methodology, we want to verify the function and timing of an actual programmed PLD. This is the final verification that the design programmed into the target PLD actually meets its specifications. This requires appropriate electronic test equipment for generating input sequences and verifying output sequences. Alternatively, we may be able to adequately verify the operation of the PLD by placing it in the system for which it was designed. However, it is important to understand that verifying a programmed PLD must not be used as a substitute for design verification using simulation.

Production Testing of PLDs

Verification of a single programmed PLD at the end of the design flow is quite different from *production testing* (manufacturing verification) of each PLD after it is programmed during a production run. The primary purpose of a production test is to determine that the configuration data has been properly programmed into each PLD and that the PLD is not defective. The purpose of production testing is not to verify the accuracy and performance of the design.

1.12 VHDL/PLD DESIGN METHODOLOGY ADVANTAGES

Some advantages that accrue from using the VHDL/PLD design methodology are obvious from our simple half-adder example; many others become evident after more complex design examples are introduced.

Management of Complex Designs

Using VHDL allows us to succinctly describe complex systems. It also allows us to manage designs of enormous complexity. Complex systems can be described behaviorally at a high level of abstraction, so that the description is independent of any particular logic implementation. Alternatively, a system can be described structurally, as an interconnection of logic components or subsystems.

A Single Comprehensive Design Language

VHDL can be used for specification, description, simulation, and synthesis. This eliminates the need for us to learn different languages for different phases in the design flow.

A Nonproprietary Language

Since VHDL is a nonproprietary standard, descriptions are portable to different vendors' tools or to different vendors' PLDs. Previously verified descriptions can be reused in later designs. VHDL descriptions for some complex systems can be purchased as *intellectual property (IP)* from IP vendors and included as components in our designs.

Wide Selection of Tools

Because of its wide acceptance as a standard, many EDA tool vendors provide compilers, synthesizers, and simulators for VHDL. Competition between tool vendors leads to the availability of more powerful and lower cost tools. PLD vendors often provide low-cost or free EDA tools that include a place-and-route tool that maps designs only to that vendor's PLDs. Because VHDL is a standard, we can use whichever VHDL tools we prefer.

Device Independent Design

VHDL descriptions can be written to be device independent. We can functionally verify our designs using a simulator, before specifying a target PLD (or other means) to implement them.

Rapid Prototyping

The VHDL/PLD design methodology can be used for proof-of-concept to rapidly prototype and implement a system in small quantities to prove its viability. If the system is successful, the same kind of PLDs can be used for large production runs. When a design is well proven and needs to be produced in very large quantities, it may be economical to create an application-specific integrated circuit (ASIC) instead of using PLDs. The same VHDL description used to implement a design targeted to a PLD can be used as the design description from which an ASIC is created.

1.13 VHDL'S DEVELOPMENT

VHDL is an acronym for *VHSIC (Very High Speed Integrated Circuit)* Hardware Description Language. VHDL has its origins in a research project funded in the early 1980s by the U.S. Department of Defense (DoD).

IEEE Std 1076

In 1985 the DoD released VHDL for public use. To encourage widespread adoption, a proposal was made to the IEEE to make VHDL a standard. A modified version of the original VHDL became IEEE Std 1076-1987. "Std" is the IEEE's abbreviation for "Standard." The standard's number is 1076. The number following the hyphen is the date that the standard was issued or revised.

The document that defines the precise syntax and simulation semantics of VHDL is the *IEEE Standard VHDL Language Reference Manual (LRM)*. Revised versions of this standard are dated 1993, 2000, and 2002.

1.14 VHDL FOR SYNTHESIS VERSUS VHDL FOR SIMULATION

VHDL was originally designed to be used for description, simulation, and documentation of digital systems, but not for synthesis. It was later adopted as a design description language for synthesis.

Learning VHDL for the purpose of synthesis involves more than just IEEE Std 1076. Related IEEE standards, such as IEEE Std 1164, IEEE Std 1076.3, and IEEE Std 1076.6, also need to be utilized so that we can make our VHDL designs as portable and "standard" as possible. These supporting standards are also discussed in this book. The Bibliography lists applicable VHDL standards.

VHDL for Synthesis

As previously stated, some VHDL constructs are not synthesizable. Therefore, it is possible to write a VHDL description that can be simulated, but not synthesized.

For a design description to be synthesizable, we must use only those constructs that are acceptable to our synthesis tool. A synthesis tool infers the logic it synthesizes from the context in which specific constructs appear in a design description. If the VHDL code is physically meaningless or too far removed from the hardware it attempts to describe, it may not be synthesizable.

Use of the term *design description* in this book is restricted to synthesizable VHDL programs. For brevity, design description will often be shortened to either *design* or *description*.

IEEE Std 1076.6

Prior to 1999 there was no standard that specified which VHDL language constructs a synthesis tool must be able to synthesize. As a consequence, the subsets of VHDL constructs that could be synthesized differed from one synthesis tool to the next. This limited the portability of VHDL descriptions that were to be synthesized.

In 1999, the IEEE issued *IEEE Std 1076.6-1999, IEEE Standard for VHDL Register Transfer Level (RTL) Synthesis.* This standard described a subset of IEEE Std 1076 suitable for RTL synthesis. It also described the syntax and semantics of this subset with regard to synthesis. The purpose of this standard was to have all compliant RTL synthesis tools produce functionally equivalent synthesis results, just as IEEE Std 1076 ensures that all compliant simulation tools produce equivalent simulation results. A revision of this standard was issued in 2004.

Example design descriptions in this book are written to be compliant with IEEE Std 1076.6-2004. These descriptions have been simulated and synthesized and should be synthesizable by most recent synthesizer tools, whether IEEE 1076.6 compliant or not.

VHDL for Simulation

Testbenches are written to apply stimulus to a UUT and to verify the UUT's output values during simulation. Testbenches are not synthesized. Accordingly, there are no constraints on which VHDL constructs we can use in a testbench. All testbenches in this book have been simulated using Aldec's Active-HDL simulator. They should be capable of being simulated on any IEEE Std 1076-1993 or later compliant VHDL simulator.

1.15 THIS BOOK'S PRIMARY OBJECTIVE

Providing you with the necessary fundamentals to use VHDL for description, verification, and synthesis of digital systems that are to be implemented using PLDs is this book's primary objective. Therefore, using VHDL for synthesis and using VHDL for simulation are both focuses of this book.

In this book, the presentation of fundamental concepts is ordered so that you can accomplish complete designs of simple systems as early as possible, while still in the process of learning VHDL.

PROBLEMS

1.1 For each phase in the VHDL/PLD design methodology design flow in Figure 1.1.1 give a brief description (one or two sentences) of what is accomplished.

1.2 What important descriptive feature of HDLs is not common to conventional high-level programming languages?

1.3 In terms of a signal, what is an event?

1.4 What happens to the traditional design tasks of minimizing the number of gates and flip-flops when the VHDL/PLD design methodology is used?

1.5 Describe some advantages of using a PLD as opposed to fixed-function ICs to implement a digital system.

1.6 What is the advantage, if any, of having the specification for a system originate in VHDL form rather than in a natural language form?

1.7 If you have a system specification that originated in VHDL form, can you simply synthesize it directly?

1.8 Write a brief requirements definition and specification for a 1-bit half subtractor.

1.9 What is a concurrent signal assignment statement and when is it executed? Does the textual order of concurrent signal assignment statements in an architecture body make any difference?

1.10 If you have a function expressed as a Boolean equation, should you attempt to simplify it before using it in a signal assignment statement in a VHDL description? Explain your answer.

1.11 Design a half adder using the traditional approach. From the truth table in Figure 1.3.1 create a Karnaugh map for each output. From the Karnaugh maps, determine the simplified sum-of-products expression for each output. Draw a logic diagram for two implementations of the circuit, one using AND, OR, and NOT gates and the other using only NAND gates.

1.12 Draw a flowchart showing the design flow for the traditional method of designing a half adder followed in Problem 1.11. Show each phase involved. Compare your flowchart with the one in Figure 1.1.1 for the VHDL/PLD design methodology. In particular, for corresponding phases of the two methodologies, comment on the relative time each approach would take. Also, discuss how this time might increase at each phase for the application of each approach to the design of a much more complex system.

1.13 Compare the traditional implementation of a half-adder physical prototype using SSI NAND gates (74HC00) with that of a 22V10 PLD implementation using the VHDL/PLD methodology. Qualitatively compare component cost, construction time, printed circuit board space, power consumption, reliability, and ease of modification if an error were made in the logical design.

1.14 Using the half-adder description in Listing 1.3.1 as a template, write a description of a half subtractor that subtracts `b` from `a` and produces a difference (`a - b`) output named `diff` and a borrow output named `borrow`. It might be helpful to start by developing a truth table for your half subtractor.

1.15 Explain the difference between a syntax error and a semantic error.

1.16 What is the difference between a static semantic error and a dynamic semantic error? Which tool is used to detect each type of error?

1.17 Using the half-adder description in Listing 1.3.1 as a starting point, write a description of a full adder. The full adder has inputs `a`, `b`, and `carry_in` and produces outputs `sum` and `carry_out`.

1.18 What kind of simulation is used to logically debug a design and why?

1.19 If you needed to verify that your synthesizer is correctly generating logic for a particular design, what type of simulation must you perform? What kind of VHDL model must you simulate and what would be the source of that model? If you found that the logic synthesized is not correct, how would you remedy this problem?

1.20 If you needed to verify that a particular design would be fast enough when targeted to a specific PLD, what type of simulation must you perform? What kind of VHDL model must you simulate and what would be the source of this model? If you found that the design was too slow when targeted to the specified PLD, how might you remedy this problem?

1.21 If you developed an algorithm to perform the logical operations required in a system's specification, what type of simulation should you perform to verify the algorithm? What kind of VHDL model would you simulate and what would be the source of the model? If you found an error in the algorithm, how would you remedy this problem?

1.22 What is a testbench and what is the relationship of a design entity to its testbench? Can the same testbench be used for each of the three kinds of simulations?

1.23 What is a UUT?

1.24 What are the primary advantages of using a testbench for simulation rather than preforming an interactive simulation or a command–line–driven simulation?

1.25 Why does a testbench not require inputs or outputs?

1.26 Modify the half-adder testbench in Listing 1.5.1 to create a testbench for the half subtractor of Problem 1.14.

1.27 Examine the testbench in Listing 1.5.1. Create a table listing, in order, the stimulus values applied and the expected output values for each stimulus value.

1.28 Draw the waveform that you would expect from the functional simulation of the half subtractor of Problem 1.14 assuming that the stimulus values are the same as in Listing 1.5.1.

1.29 What is a programmable logic device? What are the three major classes of programmable logic devices?

1.30 Make a copy of Figure 1.7.1(a) and mark it to represent the programming of this device so that output 0_1 is the equality function.

1.31 What is a product line? What is a sum line?

1.32 Using a copy of the PLD logic diagram in Figure 1.7.1(a), mark it to represent the programming of this device to implement the half subtractor of Problem 1.14.

1.33 What is an erasable PLD?

1.34 What is an output logic macrocell (OLMC)?

1.35 Why choose a PLD with slightly more logic capacity than is expected to be required for a design? Why not choose a PLD with substantially more logic capacity than what is expected to be required?

1.36 What is the primary purpose of post-synthesis simulation and under what conditions might it be skipped?

1.37 What are the basic inputs to and outputs from a synthesizer?

1.38 What are the three steps performed by a synthesizer during synthesis?

1.39 Explain what a place-and-route tool does.

1.40 What are the basic inputs to and outputs from a place-and-route tool?

1.41 How can port signals be assigned to specific pins of a target PLD?

1.42 What is the advantage of an erasable PLD over one that is not?

1.43 What is the advantage of an in-circuit programmable PLD over one that is not?

1.44 List six advantages of the VHDL/PLD methodology over traditional digital design.

1.45 For which current use was the VHDL language not originally intended?

1.46 Can every VHDL program that can be simulated be synthesized? If not, explain why?

Chapter 2

Entities, Architectures, and Coding Styles

The simplest VHDL design description is one for a single design entity that consists of an entity declaration and an associated architecture body together in one design file.

Since there are many ways to implement the same function, we can write alternative architecture bodies for the same entity declaration. An alternative architecture body normally describes a different way of accomplishing the same function.

An architecture body can be written in one of three basic coding styles: dataflow, behavioral, or structural. These styles differ in their degree of abstraction—how closely they directly relate to their hardware implementations. More commonly, an architecture body is written in a combination of these styles.

A single design entity may comprise an entire system, or it may be used as component in a larger system. When using hierarchical design, a top-level design entity is structured from interconnected lower-level design entities. Each of these design entities can, in turn, be constructed from even lower-level design entities.

In this chapter we examine design entities and their entity declarations and architecture bodies in greater detail. An overview of coding styles, which are discussed in detail in later chapters, is provided. Hierarchical design is also introduced.

2.1 DESIGN UNITS, LIBRARY UNITS, AND DESIGN ENTITIES

Design Units

A *design unit* is a VHDL construct that can be separately compiled and stored in a design library. Design units provide modularity for design management of complex systems.

A design unit consists of a context clause followed by a library unit. Compilation of a design unit defines the corresponding library unit, which is stored in a design library.

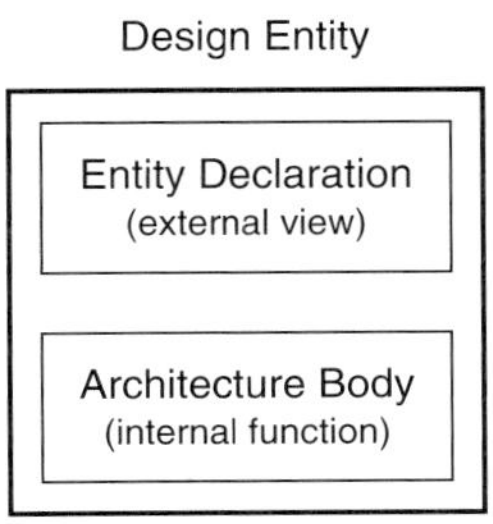

Design Entity

Entity Declaration
(external view)

Architecture Body
(internal function)

FIGURE 2.1.1
A design entity is comprised of an entity declaration and an associated architecture body.

Library Units

There are five kinds of *library units*:

- Entity declaration
- Architecture body
- Package declaration
- Package body
- Configuration declaration

Entity declarations and architecture bodies are discussed in detail in this chapter. The other library units are discussed in later chapters.

Design Entity

A *design entity* is the basic unit of a hardware design. Its inputs, outputs, and the function it performs are completely defined. A single design entity may represent the entirety of the system being designed or it may only represent a component of a larger system. For example, a design entity can represent a single gate, a subsystem, the entire system, or any portion of hardware in between. Multiple copies of the same design entity may appear in a larger design.

Entity Declaration and Architecture Body

A design entity consists of an entity declaration and an associated architecture body, Figure 2.1.1. Separating a design entity in this manner makes it easier to experiment with alternative implementations. Also, in a team design effort, this separation allows the entity declaration to be written by one person and the architecture body by another.

2.2 ENTITY DECLARATION

An *entity declaration* gives a design entity its name and defines its interface to the environment in which it is used. It names the entity's inputs and outputs and defines the type of information they carry. It provides all of the information needed to physically connect the design entity to the outside world or to connect it as a component in a larger system. An entity declaration describes only the external view of a design entity; it does not tell us how the design entity performs its function.

The entity declaration for the half adder from Chapter 1 is:

```
entity half_adder is
      port (a, b : in std_logic;
      sum, carry_out: out std_logic);
end half_adder;
```

An entity declaration starts with the keyword **entity** followed by the entity's name. The name chosen for this entity is half_adder. An entity declaration is named so that it can be referenced in other design units. For example, to associate an architecture body with the entity declaration or to use the resulting design entity as a component in a larger system.

Following the keyword **port** is a list of the entity's input and output ports. Each port's name, direction, and the type of data it transfers is listed. In the preceding example, a and b are inputs, as denoted by keyword **in**. Their data type is declared as std_logic. Keyword **out** denotes that sum and carry_out are outputs. These outputs are also type std_logic.

Entity Declaration Syntax

A simplified syntax for an entity declaration is given in Figure 2.2.1. Our examination of this syntax definition in the following section will also be used to explain the notation used. In this book, some syntax definition figures consist of multiple boxes. In such cases, the top box gives the syntax of the construct being defined. The lower box gives the syntax of selected items used in the definition.

entity_declaration ::=
entity identifier **is**
[**port** (*port*_interface_list) ;]
end [**entity**] [*entity*_simple_name] ;

*port*_interface_list ::=
[**signal**] identifier_list : [mode] subtype_indication [:= *static*_expression]
{; [**signal**] identifier_list : [mode] subtype_indication [:= *static*_expression] }

subtype_indication ::=
[*resolution_function*_name] type_mark [constraint]

type_mark ::=
*type*_name | *subtype*_name

FIGURE 2.2.1
Simplified syntax for an entity declaration.

2.3 VHDL SYNTAX DEFINITIONS

Syntax refers to the pattern or structure of the word order in a phrase. *Syntax definitions* specify how we may combine the lexical elements of the language to form valid constructs. Many syntax definitions in this book are simplified for instructional purposes. This means that not all of the possible variations of a particular syntactic category are given.

A simplified syntax definition provided at a given point in this book is sufficient for the topics being covered at that point. If needed at a later time, an expanded version of the syntax is then given.

Table 2.3.1 summarizes the notation used in the syntax definitions.

::= Symbol

The symbol ::= is read as "can be replaced by" in syntax definitions. The text on the left-hand side of the symbol can be replaced by the text on the right-hand side of the symbol. For example, the syntax for a port interface list is defined in Figure 2.2.1. So, a port interface list can be replaced by (is equivalent to) the optional keyword `signal` followed by an identifier list (a list of port signal names) followed by an optional mode specification and a required subtype indication. In its simplest form, a *subtype indication* is the name of a type or subtype. The mode specification and subtype indication apply to all the ports in the identifier list.

The term *port*_interface_list in the syntax definition has the italicized prefix *port_*. Italicized prefixes provide semantic information in syntax definitions. In this case, *port_* indicates that the interface list must be a list of ports. Ports can be listed in any order.

Default Port Mode

A port's mode specifies its direction of information transfer. If the mode specification is omitted, it defaults to `in`. So, we could omit specifying mode `in` for input

Table 2.3.1
Notation used in syntax definitions.

Symbol	Meaning
::=	The text to the left of the symbol can be replaced by the text to the right of the symbol in a production.
[ ]	Square brackets enclose optional items.
{ }	Braces enclose repeated items. The items may be repeated zero or more times.
|	A vertical bar separates alternative items, unless it occurs immediately after an opening brace, in which case it stands for itself.
text	Underlined syntax text is ignored by IEEE Std 1076.6 compliant synthesizers.
~~text~~	Strikethrough syntax text is not supported by IEEE Std 1076.6 compliant synthesizers.

ports, but cannot omit specifying mode **out** for output ports. The port clause for the entity half_adder could be written as:

```
port (a, b : std_logic;
sum, carry_out: out std_logic);
```

However, to make our code more readable, we will always specify all port modes.

Square Brackets

Square brackets ([ ]) indicate that the items they enclose are optional. The square brackets do not actually appear in the VHDL source code. So, from Figure 2.2.1, it is clear that the port clause in an entity declaration is optional. We saw an example of this in Chapter 1 for the design entity testbench, Listing 1.5.1:

```
entity testbench is
end testbench;
```

Entity testbench has no ports. Entity declarations for testbenches are usually completely self-contained and are the only entity declarations that we will encounter that don't have a port clause.

As another example, from the bottom of Figure 2.2.1, we see that the *subtype indication* in a port interface list can be as simple as a type mark. And, that a *type mark* is simply the name of a type or subtype.

Braces

Braces ({ }) indicate that the items they enclose can be repeated zero or more times. The braces do not actually appear in the VHDL source code. For example, in the syntax for a port interface list we see inside the braces that if we have additional signals of a different mode or type, they are separated from the initial list by a semicolon.

Vertical Bar

A vertical bar (|) is used to separate alternative choices, unless it occurs immediately after an opening brace, in which case it stands for itself.

Port Default Value

A default value can be assigned to a port signal. The default value is used if the port is left unconnected. If the port is connected, the default value is ignored. ***Default port values are used by a simulator, but are ignored by an IEEE 1076.6 compliant synthesizer.***

IEEE Std 1076.6 Compliant Synthesizers

A synthesis tool is said to *accept* a VHDL construct if it allows the construct to be a legal input. In contrast, a synthesis tool is said to *interpret* a construct if it synthesizes hardware that represents the construct. An IEEE Std 1076.6 compliant synthesizer is not required to interpret every construct it accepts, but is required to interpret those constructs specified in IEEE Std 1076.6.

Under IEEE Std 1076.6, a VHDL construct can be categorized as either:

- *Supported:* the synthesizer interprets the construct (maps the construct to an equivalent hardware representation).
- *Ignored:* the synthesizer ignores the construct and produces a warning. When the synthesizer encounters the construct, the synthesis does not fail. However, synthesis results may not match simulation results.
- *Not supported:* the synthesizer does not support the construct. The synthesizer does not expect to encounter the construct and may fail. However, failure is not required by the standard and a particular synthesizer may treat the construct as ignored.

IEEE Std 1076.6 Compliant Design Descriptions

A design description that is compliant with IEEE Std 1076.6 will, when synthesized, produce logic whose functional characteristics are independent of which vendor's IEEE Std 1076.6 compliant synthesizer is used.

A design description is compliant with IEEE Std 1076.6 if it:

- Uses only constructs that are supported or ignored by the standard
- Adheres to the semantics of the standard
- Does not describe logic with an oscillatory behavior

Syntax Underlines and Strike-Throughs

The syntax specifications in this book indicate which parts of a construct are ignored and which are not supported by an IEEE Std 1076.6 compliant synthesizer. An underline indicates that an item is ignored. A strike-through indicates that an item is not supported.

For example, in Figure 2.2.1, the underline under *static*_expression (:= *static*_expression) indicates that initial values for port signals are ignored by an IEEE Std 1076.6 compliant synthesizer.

Keywords

Keywords (reserved words) have a special meaning in a programming language and must be used only for their predefined purposes. VHDL's keywords are listed in Figure 2.3.1.

VHDL Identifiers

The descriptive name we choose for an entity declaration is an example of an identifier. A VHDL *identifier* is used to identify various VHDL objects: design entities, signals, procedures, functions, processes, and so on. An identifier is made up of alphabetic, numeric, and underscore characters. The first character must be a letter. Underscores are significant and can be embedded in an identifier. However, two underscores in succession are not allowed and the last character cannot be an underscore.

When choosing names (identifiers) for an entity declaration and its ports, we must not use any VHDL keywords.

```
abs            exit       not         severity
access         file       null        shared
after          for        of          signal
alias          function   on          sla
all            generate   open        sll
and            generic    or          sra
architecture   group      others      srl
array          guarded    out         subtype
assert         if         package     then
attribute      impure     port        to
begin          in         postponed   transport
block          inertial   procedure   type
body           inout      process     unaffected
buffer         is         proctected  units
bus            label      pure        until
case           library    range       use
component      linkage    record      variable
configuration  literal    register    wait
constant       loop       reject      when
disconnect     map        rem         while
downto         mod        report      with
else           nand       return      xnor
elsif          new        rol         xor
end            next       ror         variable
entity         nor        select
```

FIGURE 2.3.1
VHDL keywords.

Case Sensitivity

VHDL is not case sensitive. Therefore, `Half_Adder` and `half_adder` are treated by VHDL as the same identifier. However, some place and route, as well as other EDA tools, may be case sensitive. So, in practice it is best not to use names that differ only in case to represent the same identifier.

Ports as Data Objects

Ports are external signals and all signals are data objects. Data objects transfer or hold values of specified types. An entity declaration must specify each port's type. There are many possible types for ports. The only types we have encountered, so far, are bit and std_logic. Data objects and types are discussed in greater detail in Chapter 3.

2.4 PORT MODES

As previously stated, a port's *mode* specifies its direction of information transfer. There are five modes in VHDL: **`in`**, **`out`**, **`inout`**, **`buffer`**, and **`linkage`** (Figure 2.4.1). If no mode is specified, the default is **`in`**. Mode information is used by the compiler

```
mode ::=
in | out | inout | buffer | linkage
```

FIGURE 2.4.1
Port mode syntax.

to determine whether a design entity is being properly connected when used as a component in a larger system.

Mode Linkage

Mode `linkage` has primarily been used to link VHDL models with models written in other languages or to interface with non-VHDL simulators. This mode is not supported for synthesis in IEEE Std 1076.6 and is being considered for removal from future versions of IEEE Std 1076. As a result, mode linkage is not considered further in this book.

Reading and Writing Signals

To read a signal means to access its value. For example, in a signal assignment statement, any signal that appears on the right-hand side of the assignment symbol is read. To write a signal means to assign a value to the signal. In a signal assignment statement, the signal that appears on the left-hand side of the assignment symbol is written.

Signal Sources

A *source* is a contributor to the value of a signal. In VHDL there are two kinds of sources, one is a port of mode out, inout, or buffer and the other is a driver. A *driver* is created by a signal assignment statement that assigns a value to a signal. For example, in our half-adder implementation, input `a` is a port that might be driven by an output port of some other design entity. In contrast, output port `carry_out` is driven by the driver associated with the assignment statement:

```
carry_out <= a and b;
```

Usually, a signal has a single driver. However, in a design with bused connections, a bus signal has multiple drivers. How drivers are created from signal assignment statements is explained in Chapter 6.

Port Modes Conceptual Representations

Port modes `in`, `out`, `inout`, and `buffer` are conceptually represented in Figure 2.4.2. The triangle symbols represent sources, except for the triangle symbol with a B inside, which represents a buffer. A source symbol shown inside a design entity represents a driver created by an assignment statement in the entity's architecture body. A source symbol outside a design entity represents a port created by an assignment statement in the architecture body of some other design entity.

In this discussion of port modes, "value read" indicates the value obtained when a port is read by an assignment statement in the architecture body of the design entity.

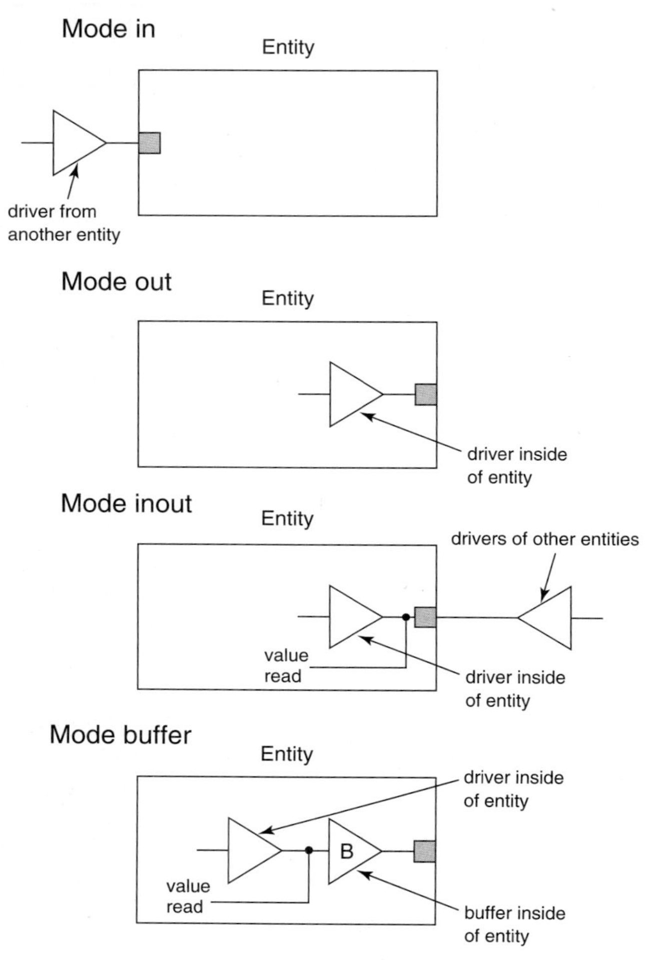

FIGURE 2.4.2
Input, output, and inout modes for a port.

"Value written" indicates the value assigned to a port by an assignment statement in the architecture body of the design entity.

A port cannot be used in a manner that violates its defined direction of information transfer.

Mode In

A mode `in` port is an input to a design entity. Statements in the design entity's architecture body can only read a value from the port; they are not allowed to assign a value to it. Accordingly, a mode in port's name can only appear on the right-hand side of an assignment statement.

Mode Out

A mode **out** port is an output from a design entity. Statements in the design entity's architecture body can only write (assign a value to) the port signal; they are not allowed to read a value from it. Accordingly, a mode out port's name can only appear on the left-hand side of a signal assignment statement.

Mode Inout

A mode **inout** port is bidirectional. Statements in the design entity's architecture body can read and write the port. Therefore, a mode inout port's name can appear on either side of a signal assignment statement.

For a mode inout port that is driving an external signal that is also driven by other ports, the value that is written to the port is the value driven by the port. The value that is read from the port is determined from the value driven by the port and the values driven by the other ports connected to the common external signal (see Section 3.4).

Mode Buffer

A mode **buffer** port is an output port that can also be read from inside the architecture body. The signal is modeled as if the driver is connected to the port through a buffer. Statements in the entity's architecture body can read and write the port. Therefore, the port's name can appear on either side of a signal assignment statement.

For a port with mode buffer that is driving an external signal that is also driven by other ports, the value that is written to the port is the value driven by the port. However, unlike mode inout, with mode buffer the value obtained when the port is read is the same as the value driven by the port. This value is not necessarily the same as the value of the external signal connected to the buffer port. The value of the external signal is determined from the value driven by the port and the values driven by other ports connected to the external signal.

A buffer port is useful when a signal value that is being output from an entity needs to be read within the entity's architecture body. This is a feedback situation that is useful in sequential systems. The use of mode buffer ports is discussed in detail in Section 9.2.

For combinational systems, which will be our initial focus, only modes in and out are needed.

2.5 ARCHITECTURE BODY

An *architecture body* defines either the behavior or structure of a design entity—how it accomplishes its function or how it is constructed. As such, it provides the internal view of a design entity.

Entity/Architecture Pair

An architecture body must be associated by name with a specific entity declaration to form an *entity/architecture pair* (a design entity). An architecture body cannot exist without an associated entity declaration. We can have different architectures associated with the same entity declaration, creating different design entities. This typically occurs when alternative implementations of a design entity are developed.

The architecture body for the half adder from Chapter 1 is:

```
architecture dataflow of half_adder is
begin
     sum <= (not a and b) or (a and not b);
     carry_out <= a and b;
end dataflow;
```

An architecture body starts with the keyword **architecture**. The name of this architecture body is `dataflow` and its corresponding entity declaration is `half_adder`. The entity name referenced in the architecture body must be the same name used in the entity declaration to which the architecture is associated.

Following the keyword **begin** are two concurrent signal assignment statements. The first computes output `sum`. The second computes output `carry_out`.

Logical operators **and**, **or**, and **not**, along with the other logical operators, are predefined in VHDL only for data objects of types bit and boolean. The package STD_LOGIC_1164 overloads (extends) these operator definitions to data objects of type std_logic.

A simplified syntax definition for an architecture body is given in Figure 2.5.1.

As shown in Figure 2.5.1, an architecture body is divided into two parts, the declarative part and the statement part.

Architecture Body Declarative Part

The declarative part starts immediately following the keyword **is** and ends at keyword **begin**. The declarative part is used to declare signals and other items used in the architecture. In general, there are numerous kinds of items that can be declared. However, in our half-adder architecture, there are no items to declare. A structural style example that has signals and components declared in its architecture body's declarative part is given in Section 2.6.

Architecture Body Statement Part

The statement part of an architecture body starts immediately following keyword **begin** and ends at keyword **end**. The statement part consists of one or more

```
architecture_body ::=
architecture identifier of entity_name is
     { block_declarative_item}   -- declarative part
begin
     { concurrent_statement }   -- statement part
end [ architecture ] [ architecture_simple_name ] ;
```

FIGURE 2.5.1
Simplified syntax for an architecture body.

concurrent statements. The concurrent statements that we will initially encounter are classified as either:

- Concurrent signal assignment statement
- Process statement
- Component instantiation statement

Concurrent statements are common in hardware description languages. As previously stated, the order in which they appear in the statement part of the architecture body is not relevant. For example, in the half-adder architecture body there are two concurrent signal assignment statements:

```
sum <= (not a and b) or (a and not b);
carry_out <= a and b;
```

If the order of these two statements is reversed:

```
carry_out <= a and b;
sum <= (not a and b) or (a and not b);
```

the operation of the system remains the same.

Conceptually, concurrent statements are able to perform their operations simultaneously (concurrently). When simulated, these statements can execute in parallel in simulated time. Concurrent statements model portions of hardware that can execute in parallel in actual systems.

Informal Use of the Terms Entity and Architecture

The terminology of the LRM is precise. The terms design entity, entity declaration, and architecture body are clearly distinguished and have the previously described meanings. However, informally, entity declaration is often shortened to entity and architecture body is shortened to architecture. Thus, the term entity/architecture pair refers to an entity declaration and an associated architecture body.

Unfortunately, the term design entity is also often informally shortened to entity. So, the term entity alone might stand for either design entity or entity declaration. In this book, the term entity is sometimes used alone, when its meaning is clear from the context in which it is used.

2.6 CODING STYLES

An architecture can be written in one of three basic coding styles: dataflow, behavioral, or structural. The distinction between these styles is based on the type of concurrent statements used:

- A dataflow architecture uses only concurrent signal assignment statements.
- A behavioral architecture uses only process statements.
- A structural architecture uses only component instantiation statements.

Instead of writing an architecture exclusively in one of these styles, we can mix two or more, resulting in a mixed style.

In theory, any description can be written using any of these styles. However, depending on the system being designed, one style may be more intuitive than the others. For example, for a design entity comprised of several lower-level design entities, a structural style is appropriate. For a system whose function is algorithmic, or which we wish to describe in a very abstract fashion, behavioral style is more intuitive. We may choose to write in a particular style depending on the level of abstraction desired, or simply based on our personal preference.

This section briefly introduces the different styles by describing a half adder in each style. Subsequent chapters cover the details of these styles and introduce various VHDL constructs using them, first to describe and synthesize combinational systems, then to describe and synthesize sequential systems.

Dataflow Style

Dataflow style describes a system in terms of how data flows through the system. Data dependencies in the description match those in a typical hardware implementation. A dataflow description directly implies a corresponding gate-level implementation.

Dataflow descriptions consist of one or more concurrent signal assignment statements. A variation of the previous dataflow style description of a half adder is shown in Listing 2.6.1.

This description uses two concurrent signal assignment statements:

```
sum <= a xor b;
carry_out <= a and b;
```

Here, advantage has been taken of the fact that the sum-of-products expression for output `sum`, in the earlier half-adder description, is equivalent to the exclusive-OR (XOR) function of `a` and `b`.

LISTING 2.6.1
Dataflow style half-adder description.

```
library  ieee;
use ieee.std_logic_1164.all;

entity half_adder is
   port (a, b : in std_logic;
      sum, carry_out: out std_logic);
end half_adder;

architecture dataflow2 of half_adder is
   begin
   sum <= a xor b;
   carry_out <= a and b;
end dataflow2;
```

The first assignment statement describes how input data flows from inputs a and b through an XOR function to create sum. The second assignment statement describes how input data flows through an AND function to produce carry_out. Anytime there is an event on either input, the statements concurrently compute an updated value for each output.

The concurrent signal assignment statements in this description directly imply a hardware implementation consisting of an XOR gate and an AND gate.

Behavioral Style

Behavioral style is the most abstract style. A behavioral description describes a system's behavior or function in an algorithmic fashion. The description is abstract in the sense that it does not directly imply a particular gate-level implementation.

Behavioral style consists of one or more process statements. Each process statement is a single concurrent statement that itself contains one or more sequential statements. *Sequential statements* are executed sequentially by a simulator, the same as the execution of sequential statements in a conventional programming language.

A behavioral style description of a half adder is given in Listing 2.6.2. The entity declaration is the same as for the dataflow architecture. However, the architecture body is quite different. This architecture consists of a single process statement. Process statements are discussed in detail in Chapter 5. The process statement in Listing 2.6.2 is briefly discussed here to provide an overview of the behavioral style.

LISTING 2.6.2
Behavioral style half-adder description.

```
library ieee;
use ieee.std_logic_1164.all;

entity half_adder is
   port (a, b : in std_logic;
   sum, carry_out: out std_logic);
end half_adder;

architecture behavior of half_adder is
begin
   ha: process (a, b)
   begin
      if a = '1' then
         sum <= not b;
         carry_out <= b;
      else
         sum <= b;
         carry_out <= '0';
      end if;
   end process ha;

end behavior;
```

The process statement in Listing 2.6.2 starts with the label `ha` followed by the keyword **`process`**. A label on a process is optional, but is useful to differentiate processes in designs that contain multiple processes. Accordingly, we will follow the practice of labeling all processes.

Sensitivity List

Following the keyword **`process`** is a list of signals in parentheses, called a *sensitivity list.* A sensitivity list enumerates exactly which signals cause the process to be executed. Whenever there is an event on a signal in a process's sensitivity list, the process is executed.

Between the second **`begin`** keyword and the keywords **`end process`** is a sequential if statement. This if statement is executed whenever the process executes. The VHDL if statement is similar to that of conventional high-level programming languages. For this half adder, each branch of the if statement contains two assignment statements. Since all statements contained within a process are sequential, these are sequential signal assignment statements. The distinctions between concurrent and sequential signal assignment statements will be made clearer in later chapters.

The if statement describes the half adder behaviorally. A quick look at the truth table for a half adder, Table 1.3.1, confirms that when input `a` is a 1, output `sum` is the complement of input `b`, and output `carry_out` is equal to `b`. In contrast, when `a` is 0, output `sum` is equal to `b` and output `carry_out` is 0.

Looking closely at the behavioral style description, it becomes apparent that it correctly describes the function of a half adder. However, this description is abstract with respect to how the half adder is implemented at the gate level. It is not immediately obvious that this description is equivalent to an XOR gate and an AND gate.

Structural Style

Structural style describes a system as a hierarchical interconnection of design entities. The *top-level design entity's* architecture describes the interconnection of lower-level design entities. Each lower-level design entity can, in turn, be described as an interconnection of design entities at the next-lower level, and so on.

Structural style is most useful and efficient when a complex system is described as an interconnection of moderately complex design entities. This approach allows each design entity to be independently designed and verified before being used in the higher-level description.

When structural style is used to describe a very simple circuit, such as a half adder, the description is overly long when compared to either a dataflow or behavioral style description.

Listing 2.6.3 is a structural style half-adder description. The half adder is described as an interconnection of an XOR gate design entity and an AND gate design entity. A total of three design entities are described in the design file: the XOR gate (`xor_2`), the AND gate (`and_2`), and, at the end of the file, the top-level entity `half_adder`. Design entity `half_adder` describes how the XOR gate and the AND gate are connected to implement a half adder. It is this top-level entity that has a structural style description.

In Listing 2.6.3, the lower-level design entities appear at the beginning of the file. Alternatively, and more commonly, lower-level design entities are described in separate design files that are separately compiled into a library. This later approach allows for better management of large designs and for reuse of design entities as components in other designs.

LISTING 2.6.3

Structural style half adder using direct entity instantiation.

```
----------------------- XOR gate component ------------------------
library ieee;
use ieee.std_logic_1164.all;

entity xor_2 is  -- Entity declaration for 2 input XOR gate
   port (i1, i2 : in std_logic;
   o1: out std_logic);
end xor_2;

architecture dataflow of xor_2 is  -- Architecture body for 2 input XOR
begin
   o1 <= i1 xor i2;
end dataflow;

------------------------ AND gate component ------------------------
library ieee;
use ieee.std_logic_1164.all;

entity and_2 is  -- Entity declaration for 2 input AND gate
   port (i1, i2 : in std_logic;
   o1: out std_logic);
end and_2;

architecture dataflow of and_2 is  -- Architecture body for 2 input AND
begin
   o1 <= i1 and i2;
end dataflow;

------------------------ HALF-ADDER structure -------------------------
library ieee;
use ieee.std_logic_1164.all;

entity half_adder is  -- Entity declaration for half adder
   port (a, b : in std_logic;
   sum, carry_out: out std_logic);
end half_adder;
```

(Cont.)

LISTING 2.6.3 *(Cont.)*

```
architecture structure of half_adder is  -- Architecture body for half adder

begin
      u1: entity xor_2 port map (i1 => a, i2 => b, o1 => sum);
      u2: entity and_2 port map (i1 => a, i2 => b, o1 => carry_out);

end structure;
```

Each design entity in Listing 2.6.3 consists of an entity declaration and its associated architecture body. The first design entity, `xor_2`, represents a two-input XOR gate. The entity declaration describes this gate as having inputs `i1` and `i2` and output `o1`. The architecture body for the XOR gate entity uses a single concurrent signal assignment statement to describe its function. Thus, the `xor_2` architecture is written in the dataflow style. AND gate design entity, `and_2`, is described in a similar fashion. These two design entities are graphically illustrated in Figure 2.6.1.

The last design entity in Listing 2.6.3 is the half adder itself. It is the half-adder design entity whose architecture is structural.

Component Instantiation

In the statement part of the half-adder architecture are two component instantiation statements. Each one creates an *instance* (copy) of a design entity. Component instantiation statements require unique labels.

Component Instantiation Syntax

Two different forms of component instantiation syntax are given in Figure 2.6.2.

Direct Instantiation of a Design Entity

The first form of component instantiation in Figure 2.6.2, called *direct design entity instantiation,* is more concise and is the form used in Listing 2.6.3. This form directly specifies the design entity to be used and, optionally, its architecture. If an architecture body is not specified, the most recently compiled architecture associated with the entity declaration is used by default.

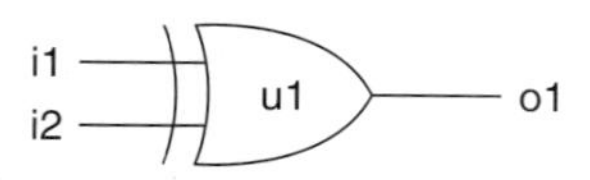

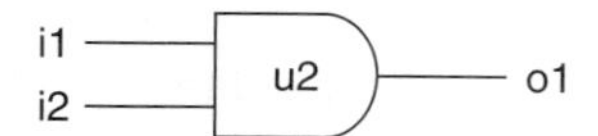

FIGURE 2.6.1
XOR and AND gate design entities used in the half adder described in Listing 2.6.3.

component_instantiation ::=
instantiation_label: **entity** entity_name [(architecture_identifier)] **port map** (port_association_list)] ;
| instantiation_label: [**component**] component_name **port map** (port_association_list)] ;

port_association_list ::=
[*port*_name =>] *signal*_name | expression | **open** {, [*port*_name =>] *signal*_name | expression | **open**}

FIGURE 2.6.2
Simplified syntax for a component instantiation statement.

The first component instantiation in Listing 2.6.3, with label `u1`, specifies an instance of an `xor_2` design entity:

```
u1: entity xor_2 port map (i1 => a, i2 => b,
        o1 => sum);
```

Note that this statement is called a component instantiation statement even though it is a design entity that is actually being instantiated.

Port Map

A port map tells how a design entity is connected in the enclosing architecture. Each port of the design entity is associated with either a signal in the enclosing architecture, an expression, or the keyword **open**.

Signals in the enclosing architecture include both the ports of the enclosing architecture as well as any signals declared in the declarative part of the enclosing architecture.

If a port is associated with an expression, the value of the expression must be a constant. For example, the expression cannot involve other signals.

Keyword Open

Associating a port with the keyword **open** leaves the port unassociated (unconnected). Any unused output ports of a design entity may be left **open**, but input ports may be left **open** only if a default value has been specified for the port in its entity declaration.

The previous *port map* specifies how the `xor_2` design entity is connected in the enclosing half adder. The `xor_2`'s `i1` input is connected to port `a`, its `i2` input is connected to port `b`, and its `o1` output is connected to `sum`. The interconnection specified by the port map is shown in Figure 2.6.3. Each symbol in the figure has a unique instance identifier (u1 and u2) used to identify the symbol.

Association Symbol

In a port map, the *arrow symbol* (=>) indicates the association (connection) of a design entity's port, on the symbol's left-hand side, to a signal or port of the enclosing architecture, on the symbol's right-hand side. The arrow symbol is read as "connected to." This method of association is called *named association*.

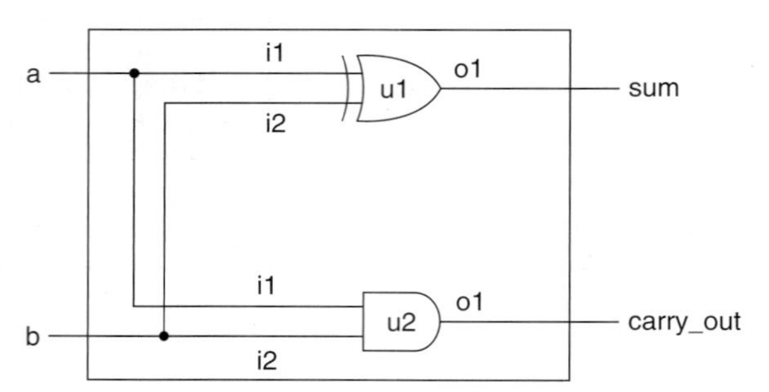

FIGURE 2.6.3
Half-adder structural implementation using XOR and AND design entities.

Named Association

Using named association, each association is explicit and the listing order of the associations in the port map is of no concern. For example, a design entity instantiated with a port map equivalent to the previous one is:

```
u1: entity xor_2 port map (i2 => b, o1 => sum,
     i1 => a);
```

Positional Association

Alternatively, we can use positional association to specify how a design entity is connected. Using *positional association,* signals are associated based on their relative positions as specified in the entity's declaration. The corresponding component instantiation statement using positional association is:

```
u1: entity xor_2 port map (a, b, sum);
```

While positional association is more concise, named association is preferable because it is more readable and less error prone. It eliminates the need to refer back to the design entity's declaration to determine to which of its ports a signal is connected.

The second component instantiation statement, with the label u2, describes how the and_2 design entity is connected:

```
u2: entity and_2 port map (i1 => a, i2 => b,
     o1 => carry_out);
```

Since component instantiation statements are concurrent statements, an event on any signal associated with a design entity's inputs (in its port map) causes the design entity to "execute." In the structural half-adder example, an event on either input a or b causes each design entity to simultaneously update its output value.

Components in VHDL

The form of component instantiation statement that directly instantiates a design entity, as used in Listing 2.6.3, did not become a part of VHDL until 1993. Prior to that, structural VHDL descriptions specified the interconnection of design entities indirectly through the use of components.

In VHDL, a *component* is actually a placeholder for a design entity. A structural design that uses components simply specifies the interconnection of the components. Ultimately, the actual design entity that each component represents must be established. This action is said to *bind* each component to a design entity. This binding can be specified explicitly or can occur by default.

Since a VHDL component is a virtual design entity, informally, the terms design entity and component are often used interchangeably. However, they represent distinct concepts.

A structural architecture that uses components must declare each component. This declaration can be done in the declarative part of the architecture. The component instantiation statements then describe the interconnection of the declared components. Finally, the binding of each component to a design entity must be established.

The syntax of a component instantiation statement that instantiates components (as opposed to design entities) corresponds to the second form in Figure 2.6.2. This form is for *indirect design entity instantiation* and is illustrated by the top-level half-adder design entity in Listing 2.6.4.

LISTING 2.6.4
Structural style half adder using indirect design entity instantiation (components).

```
------------------------- HALF-ADDER structure ----------------------------
library ieee;
use ieee.std_logic_1164.all;

entity half_adder is -- Entity declaration for half adder
   port (a, b : in std_logic;
   sum, carry_out: out std_logic);
end half_adder;

architecture structure of half_adder is -- Architecture body for half adder

   component xor_2                       -- xor_2 component declaration
      port (i1, i2 : in std_logic;
      o1: out std_logic);
   end component;

   component and_2                          -- and_2 component declaration
      port (i1, i2 : in std_logic;
      o1: out std_logic);
   end component;

begin
      u1: xor_2 port map (i1 => a, i2 => b, o1 => sum);
      u2: and_2 port map (i1 => a, i2 => b, o1 => carry_out);

end structure;
```

Component Declarations

As previously stated, when components are used, each must be declared. A component declaration is similar to an entity declaration in that it provides a listing of the component's name and its ports.

Component declarations start with the keyword **component**. In the declarative part of the `half_adder` architecture in Listing 2.6.4, components `xor_2` and `and_2` are declared. The component declaration for the XOR gate is:

```
component xor_2
     port (i1, i2 : in std_logic;
     o1: out std_logic);
end component;
```

Component Declaration Syntax

A component declaration has the simplified syntax shown in Figure 2.6.4.

Component Instantiation

The first component instantiation in Listing 2.6.4, with label `u1`, specifies an instance of an `xor_2` component:

```
u1: xor_2 port map (i1 => a, i2 => b, o1 => sum);
```

The `xor_2` component in this instantiation statement must be bound to a design entity. To take full advantage of the power and flexibility of components, this binding could be specified explicitly, using a configuration declaration. Configuration declarations are discussed in detail in Chapter 15. However, for simple situations a default configuration is used.

Default Configuration Rules

In Listing 2.6.4, we take advantage of VHDL's *default configuration rules*. These rules implicitly bind a component to a design entity having the same name and same port interface. Port interfaces are the same if the ports have identical names and types.

Since direct instantiation of design entities is more concise, it is used extensively in examples in this book until Chapter 15. In Chapter 15 we examine components in greater detail and explore situations in which their use is advantageous.

component_declaration ::=
component identifier [**is**]
  [**port** (*local_port*_interface_list);]
end component [component_simple_name] ;

*local_port*_interface_list ::=
[**signal**] identifier_list : [mode] subtype_indication [:= *static* expression]
{; [**signal**] identifier_list : [mode] subtype_indication [:= *static* expression] }

FIGURE 2.6.4
Simplified syntax for a component declaration.

Mixed Style

There is no restriction against using different kinds of concurrent statements in the same architecture body. Doing so results in a *mixed style*. In fact, most designs use a mixed coding style.

A mixed style half adder is given in Listing 2.6.5. Output `sum` is described using a dataflow concurrent statement. Output `carry_out` is described using a process concurrent statement labeled `co`.

Behavior versus Structure

In describing architectural coding styles, we have made a distinction between dataflow and behavioral styles. Both of these styles actually describe behavior. Examination of either of these two kinds of descriptions tells us the behavior of the design entity. In the dataflow style, behavior is described indirectly by concurrent signal assignment statements, from which the corresponding behavior can be inferred. In the behavioral style, behavior is described algorithmically using process statements.

We will see later that a concurrent signal assignment statement used in dataflow style architecture is easily converted to an equivalent process statement that can be used in a behavioral style architecture. In fact, a concurrent signal assignment statement is simply a shorthand representation of a process.

LISTING 2.6.5
Mixed style half-adder description.

```
library ieee;
use ieee.std_logic_1164.all;

entity half_adder is
   port (a, b : in std_logic;
      sum, carry_out: out std_logic);
   end half_adder;

architecture mixed of half_adder is
   begin

   sum <= a xor b;    -- dataflow concurrent statement

   co: process (a,b) -- start of process concurrent statement
      begin
      if a = '1' then
         carry_out <= b;
      else
         carry_out <= '0';
      end if;
   end process co; -- end of process concurrent statement

end mixed;
```

In contrast, in a structural style description the top-level entity simply describes the interconnection of components; it provides no behavioral information. Ultimately, the lowest level components in a structural design must each be described behaviorally in their respective architectures. Otherwise, the top-level description cannot be compiled, simulated, or synthesized.

2.7 SYNTHESIS RESULTS VERSUS CODING STYLE

When the same system is described using different coding styles, but synthesized to the same target PLD, we might expect that the resulting synthesized logic at the gate level is identical. This is usually the case for simple designs. However, for complex designs, the gate level logic, while functionally equivalent, may not be identical. This can occur because when starting from different functionally equivalent descriptions of the same system, the synthesizer optimization processes may not produce the same gate level results. Still, the results are functionally equivalent.

Diagrams Representing Synthesized Logic

Some synthesizers provide a *RTL hierarchical view* of the synthesized logic as well as a *technology dependent flattened to gates view.* The RTL level hierarchical view is what results from step 1 of the synthesis process (see page 29) and depends on the style of the description. However, it is the technology dependent flattened to gates view (from synthesis step 3) of different descriptions of the same system that are likely to be the same. Our half-adder descriptions are simple enough that this is the case.

Half-Adder Dataflow Description Hierarchical Representation

The RTL hierarchical representation of the hardware synthesized from the dataflow style half-adder description in Listing 2.6.1 is shown in Figure 2.7.1. As we might have expected, an AND gate and an XOR gate are shown.

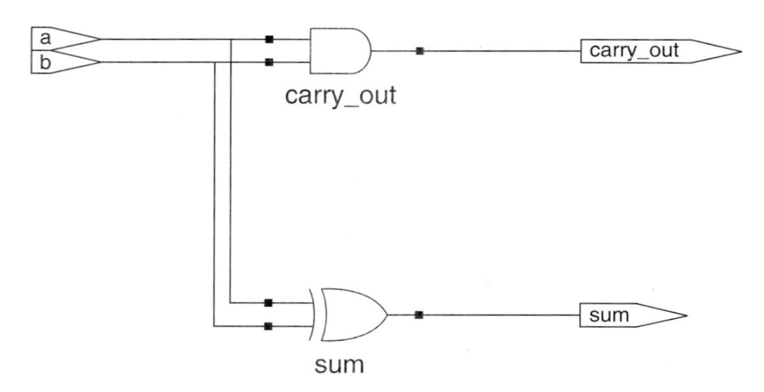

FIGURE 2.7.1
RTL hierarchical view of synthesized logic from the dataflow style description of a half adder.

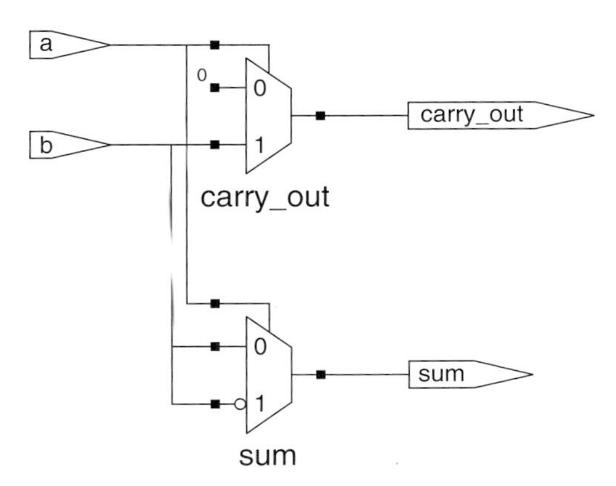

FIGURE 2.7.2
RTL hierarchical view of synthesized logic from the behavioral style description of a half adder.

Half-Adder Behavioral Description Hierarchical Representation

The RTL hierarchical representation of the hardware synthesized from the behavioral style half-adder description in Listing 2.6.2 is given in Figure 2.7.2. Perhaps unexpectedly, this representation shows two 2-input multiplexers.

The if statement from Listing 2.6.2 is initially synthesized as two multiplexers. The correlation between the code and the synthesized logic is straightforward. Input `a` is the select input to both multiplexers. If `a = '1'`, the bottom multiplexer routes the complement of `b` to its output `sum` and the top multiplexer routes `b` to its output `carry_out`.

The if statement's else branch corresponds to the situation where `a = '0'`. For this condition, the bottom multiplexer routes `b` to `sum` and the top multiplexer routes `'0'` to `carry_out`.

Half-Adder Structural Description Hierarchical Representation

The RTL hierarchical representation of the hardware synthesized from the structural style half-adder description in Listing 2.6.3 is given in Figure 2.7.3. The two component instantiation statements cause two design entities to be instantiated. These design entities are represented as rectangles, each labeled with the design entity's name. It is only because descriptive names were chosen for the instantiated design entities that we might infer their functions. Distinctive shape logic symbols are not used because in a structural style design what is being represented is the interconnection of design entities, not their functions.

Technology Dependent Flattened to Gates View

When the synthesis process is completed for the half-adder description written in each style, the results are all the same, as shown in Figure 1.9.3, which is repeated here as Figure 2.7.4. This is the final technology dependent flattened to gates view. In addition to the logic gates, this representation shows buffers associated with the input and output pins of the target PLD.

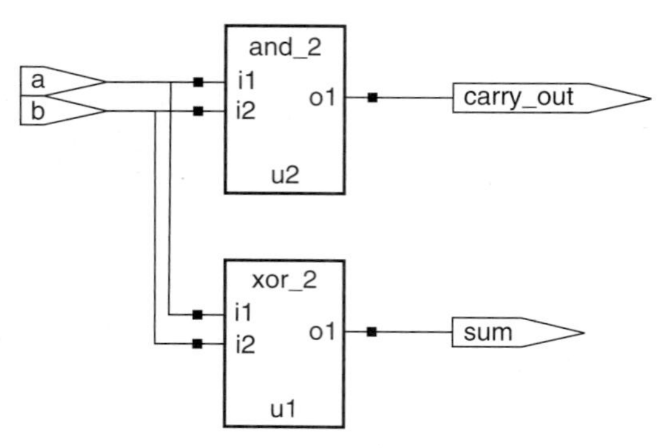

FIGURE 2.7.3
RTL hierarchical view of synthesized logic from the structural style description of a half adder.

Synthesis Constraints

Another way of synthesizing functionally equivalent logic that differs at the gate level is through the use of synthesis constraints. *Synthesis constraints* are design goals that we can specify for a synthesizer's optimizer to try to achieve. These constraints are not part of the VHDL language and are only discussed here in general terms.

Different synthesis constraints may cause different optimized circuits to be synthesized from the same architecture. All the optimized circuits are functionally identical.

Area Constraints

The most common synthesis constraints are for area and speed. *Area constraints* limit the area of the circuit in terms of the number of target technology primitives used. Area is often expressed in terms of the number of equivalent gates.

Timing Constraints

Timing constraints are expressed in terms of when signals arrive at the circuit's inputs and when they need to arrive at the circuit's outputs. Sometimes the implementation with the smallest area is also the fastest. Yet, these two goals often represent a trade-off,

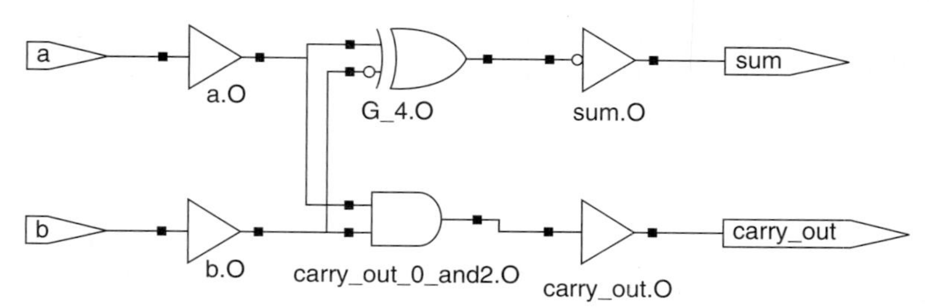

FIGURE 2.7.4
Technology dependent flattened-to-gates representation of synthesized half adder.

with timing being minimized by carrying out logic operations in parallel, thus increasing the area.

Some synthesizers allow us to specify constraints in a graphical user interface. Others use command lines or a separate constraint file. VHDL's user attribute feature even allows constraints to be specified in the design description. However, this latter approach may limit a design description's portability.

2.8 LEVELS OF ABSTRACTION AND SYNTHESIS

The *level of abstraction* used in describing a system refers to the amount of implementation or structural detail in a description. *Behavior* refers to the function of an intended system independent of the level of abstraction used in its description. Use of a high level of abstraction allows us to describe a complex system's behavior without being overwhelmed by the large amount of detail and complexity required to describe the system at the gate level.

Table 2.8.1 is a listing of hierarchically related levels of abstraction for a system, from most abstract at the top to least abstract (most detailed) at the bottom. This characterization also indicates the "components" used to represent the system at each level.

The levels of abstraction in Table 2.8.1 are commonly used in traditional design methodologies to describe or model a design. In a traditional top-down design methodology, a design is conceived at the behavioral level. The design is then developed as a stepwise refinement from one level of abstraction to the next lower level. When either the gate or transistor level is reached, depending on the technology to be used for implementation, all the detail necessary to implement the design is available.

Table 2.8.1
Levels of abstraction.

Level	Components	Representation
Behavioral	Algorithms, FSMs	Algorithmic representation of system functions
System (architectural)	Processors, memories, functional subsystems	Interconnection of major functional units
Register Transfer (dataflow)	Registers, adders, comparators, ALUs, counters, controllers, bus	Data movement and transformation and required sequential control
Gate (logic level)	Gates, flip-flops	Implementation of register level components using gates and flip-flops
Transistor	Transistors, resistors, capacitors	Implementation of gates and flip-flops using transistors, capacitors, and resistors

Using VHDL, we can describe a system at any level of abstraction. We can describe a system at the highest level of abstraction in a purely behavioral or algorithmic fashion, without any consideration of its structure or logic implementation. Such a description represents the system as a black box describing its behavior as a function of its input values over time.

Alternatively, we can describe a system structurally at a high level of abstraction as an interconnection of subsystems, without any detail concerning each subsystem's behavior. In contrast, we can also describe a system structurally at (what is for our purposes) its most detailed level, gate level.

RTL Synthesizers versus Behavioral Synthesizers

Given a natural language behavioral specification for a simple system, we can write a corresponding high-level VHDL behavioral style description of the system. As we have seen with the half-adder example, a properly written behavioral description of a simple system can be directly synthesized.

Given a natural language behavioral specification for a complex system, we can also write a corresponding high-level VHDL behavioral style description. We may not be able to directly synthesize this description. This is particularly true for a description that involves complex algorithms. However, this description can be used to verify the desired functionality.

For synthesis in this latter situation, we may have to translate the high-level behavioral description into a high-level structural description of an architecture that can implement the algorithm. For complex arithmetic designs, issues such as scheduling, hardware allocation, resource sharing, and memory inferencing are a part of creating such an optimal architecture.

Register Transfer Level Synthesizers

In our VHDL/PLD design methodology, translation of a high-level behavioral description involving complex algorithms to a high-level structural description is done manually. The resulting structural description is called a register transfer level (RTL) description. A properly written RTL VHDL description can be directly synthesized to gate-level logic by a RTL VHDL synthesizer. The vast majority of VHDL synthesizers are RTL synthesizers. The designs in this book are all synthesizable by a RTL VHDL synthesizer. IEEE Std 1076.6 is for RTL VHDL synthesizers.

Behavioral Synthesizers

For complex behavioral descriptions, a relatively new type of synthesizer called a behavioral synthesizer is available. A behavioral synthesizer can often translate a complex behavioral VHDL description to a RTL description and from there to a gate-level implementation. Behavioral synthesis has advantages for implementation of complex algorithmic systems. However, behavioral synthesizers are expensive and their use is beyond the scope of this book.

2.9 DESIGN HIERARCHY AND STRUCTURAL STYLE

As seen in the half-adder example in Section 2.6, use of structural style is relatively inefficient when describing a system at the gate level. The advantages of structural style become more evident as the complexity of the components used increases, requiring that the system be designed in a hierarchical fashion.

Hierarchical Design

The hierarchical design of a complex system begins with a *top-down* successive decomposition of the system's specification into specifications for simpler component design entities. Each design entity at a given level may be further decomposed into constituent design entities that comprise the next lower level of the decomposition. Eventually, a point is reached where each design entity is simple enough to be behaviorally described in VHDL.

Each design entity is individually coded and then verified using a specifically written testbench. The complete system is then implemented from the *bottom up* by interconnecting its constituent design entities. This approach describes and synthesizes a complex system as a *hierarchy* of design entities. Finally, a testbench is written to verify the complete system.

A particular design entity can be instantiated (used) multiple times in an architecture, in the same way that multiple instances of the same IC component are used in traditional hardware design.

Top-Level Design Entity

The design entity that represents the entire system is the *top-level design entity*. The architecture of the top-level design entity defines the system in terms of the interconnection of component design entities.

Component design entities are interconnected by signals. *Signals* are analogous to wires or printed circuit board traces in a corresponding hardware implementation.

External and Local Signals

From the viewpoint of the top-level design entity, signals are divided into two kinds: external and local. *External signals* are accessed from outside the top-level design entity. External signals allow input to and output from the top-level design entity. Such signals are *port signals* (or simply *ports*).

Component design entities are interconnected by *local* (*internal*) *signals* to form the top-level design entity. A local signal cannot be accessed from outside of its enclosing architecture body. Local signals are simply referred to as *signals*.

Full-Adder Structural Description

A simple example of hierarchical design using structural style is given here for a full adder. Hierarchical design is discussed in detail in Chapter 15.

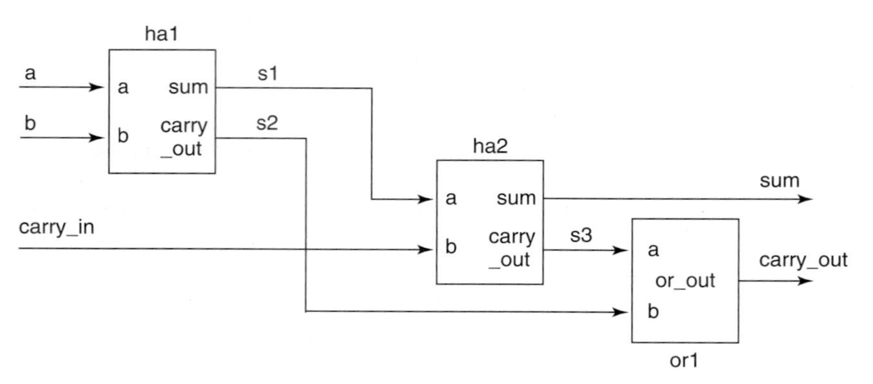

FIGURE 2.9.1
Block diagram of a full adder composed of two half adders and an OR gate.

The design file for the half adder in Listing 1.3.1 contains a single design entity. If our objective is simply to design a half adder, then this design entity comprises our entire target system. However, if our objective is to design a full adder, then we might design the half adder as a component for use in the full adder. Figure 2.9.1 shows a block diagram of the structure of a full adder composed of two half adders and an OR gate.

Single Design File Description

A structural description of the full adder using a single design file is given in Listing 2.9.1. The design entities for the half-adder and OR gate components appear first in the file. The last design entity in the file is the top-level design entity.

When the file in Listing 2.9.1 is compiled, each design entity is processed in the order of its appearance. Thus, the component design entities are compiled before the top-level design entity. As a result, when these components are referenced in the top-level design entity, they are already known to the compiler.

LISTING 2.9.1
Structural style full-adder description in a single design file.

```
------------------ Half-adder design entity ----------------------------------

library ieee;
use ieee.std_logic_1164.all;

entity half_adder is
   port (a, b : in std_logic;
      sum, carry_out: out std_logic);
end half_adder;
```

```
architecture dataflow of half_adder is
begin
   sum <= a xor b;
   carry_out <= a and b;
end dataflow;

------------------- Two input OR gate design entity ----------------------------

library ieee;
use ieee.std_logic_1164.all;

entity or_2 is
   port (a, b : in std_logic;
      or_out : out std_logic);
end or_2;

architecture dataflow of or_2 is
begin
   or_out <= a or b;
end dataflow;

------------------- Full-adder design entity ------------------------------------

library ieee;
use ieee.std_logic_1164.all;

entity full_adder is
   port (a, b, carry_in : in std_logic;
      sum, carry_out: out std_logic);
end full_adder;

architecture structure of full_adder is

   signal s1, s2, s3 : std_logic; -- Signals to interconnect components

begin
   -- Each component instantion below is a concurrent statement

   ha1: entity half_adder port map (a => a, b => b, sum =>s1, carry_out => s2);
   ha2: entity half_adder port map (a => s1, b => carry_in, sum => sum,
         carry_out => s3);
   or1: entity or_2 port map (a => s3, b => s2, or_out => carry_out);

end structure;
```

Declaring (Local) Signals

Signals s1, s2, and s3, needed to connect the design entities, are declared in the declarative part of the full-adder's architecture for use in its statement part.

In the architecture's statement part there are three component instantiation statements. Each describes an instance of a design entity and how it is interconnected in the larger structure. The half-adder design entity is instantiated twice. Signals s1, s2, and s3 are used in the port maps to interconnect the design entities.

Multiple Design File Description

The same structural full-adder design can be described using three separate files, each containing only one design entity. The text in the separate files for the half adder and the OR gate would be the same as for their respective design entities in Listing 2.9.1.

The separate design file for the top-level design entity appears in Listing 2.9.2. This file assumes that the half-adder and OR gate design entities have been compiled to a user-created library called parts. The second library statement makes components in the library parts available. Otherwise, the description of the top-level design entity is the same as in the single design file version.

LISTING 2.9.2
Design file containing top-level design entity for a multiple design file full adder.

```
------------------ Full-adder design entity ----------------------------------

library ieee;
use ieee.std_logic_1164.all;
library parts;
use parts.all;

entity full_adder is
   port (a, b, carry_in : in std_logic;
   sum, carry_out: out std_logic);
end full_adder;

architecture structure of full_adder is

   signal s1, s2, s3 : std_logic; -- Signals to interconnect components

begin

   ha1: entity half_adder port map (a => a, b => b, sum => s1, carry_out => s2);
   ha2: entity half_adder port map (a => s1, b => carry_in, sum => sum,
       carry_out => s3);
   or1: entity or_2 port map (a => s2, b => s3, or_out => carry_out);

end structure;
```

For a complex design, having each design entity in a separate file allows us to design and verify each one separately. This simplifies design management, allowing us to focus on one thing at a time. As a further step, placing design entities into a library also allows us to easily reuse them in other designs.

PROBLEMS

2.1 What is a design unit?

2.2 List the five kinds of library units.

2.3 Why is a design entity separated into an entity declaration and an architecture body?

2.4 What information does an entity declaration provide about a design entity?

2.5 Explain the meanings of the following symbols when used in syntax definitions: ::=, |, [ ], and { }.

2.6 What does it mean for an IEEE Std 1076.6 compliant synthesizer to accept a construct? What does it mean for an IEEE Std 1076.6 synthesizer to interpret a construct?

2.7 What are the meanings of the terms supported, ignored, and not supported in the context of classifying VHDL constructs for an IEEE Std 1076.6 compliant synthesizer? How are parts of a construct that are ignored by an IEEE Std 1076.6 compliant synthesizer indicated in a syntax definition? How are parts of a construct that are not supported indicated?

2.8 Which of the following can be used as a valid identifier in a VHDL program and which cannot? For those that cannot, explain why.

(a) `adder`
(b) result_
(c) `units`
(d) 4_to_1_mux
(e) top_level
(f) `port`

2.9 For each port mode, list the mode and the possible direction(s) of information transfer. Specify these directions as in, out, or in and out. Which port mode is not acceptable for synthesis?

2.10 What is a signal source and what are the different kinds of signal sources?

2.11 In terms of their mode, which ports can be written by a signal assignment statement in an architecture body? Which ports can be read by a signal assignment in an architecture body?

2.12 In each of the following concurrent signal assignment statements, indicate which signals are written, which are read, and to which signals each statement is sensitive.

(a) `f <= a and b and c;`
(b) `g <= not (w and x and y);`
(c) `eq <=(not x and not y) or (x and y);`
(d) `eq <= (x_bar and y_bar) or (x and y);`
(e) `alarm <= armed and (d1 or d2 or d3) or panic;`

2.13 Given the following architecture body, write the associated entity declaration:

```
architecture csa of and_or is
    signal s1, s2 : std_logic ;
begin
   s1 <= a and b;
   s2 <= a and c;
   f <= s1 or s2;
end csa;
```

2.14 What is the difference in operation of a port of mode inout and one of mode buffer?

2.15 What information about a design entity does an architecture body provide?

2.16 Into what two parts is an architecture body divided? What is the purpose of each part?

2.17 What are the three different pure coding styles for an architecture body and how is each distinguished?

2.18 A two-to-one multiplexer has two data inputs, `a` and `b`, and one select control input, `s`. The multiplexer has a single output `c`. If `s = 0`, `c` is equal to `a`. If `s = 1`, `c` is equal to `b`.

(a) Write an entity declaration for the multiplexer named `mux`. Use only the data type std_logic.
(b) Write an architecture for `mux` that uses Boolean equations in a dataflow style architecture.
(c) What other statements must be included in the design file containing the multiplexer's entity and architecture in order to compile the design entity? Why must these statements be included?

2.19 Given the following truth table:

a	b	c	x	y
0	0	0	0	1
0	0	1	1	0
0	1	0	1	0
0	1	1	0	0
1	0	0	1	0
1	0	1	0	1
1	1	0	0	1
1	1	1	1	0

(a) Write a canonical sum-of-products Boolean equation for each output.
(b) Write a complete VHDL design description of a design entity that accomplishes the function defined by the truth table. Use a simple dataflow architecture where the signal assignment statements are Boolean equations.

2.20 Write a design description that accomplishes the function defined by the following truth table. Use a dataflow style architecture consisting of Boolean equations.

a	b	c	x	y
0	0	0	0	1
0	0	1	1	0
0	1	0	0	1
0	1	1	0	0
1	0	0	1	0
1	0	1	0	0
1	1	0	0	0
1	1	1	0	1

2.21 A 74HC10 IC contains three 3-input NAND gates.

(a) Write an entity declaration for the design entity `triple_3` that is functionally equivalent to a 74HC10.

(b) Write an architecture in the dataflow style for the design entity `triple_3`. Use only signal assignment statements with Boolean equations.

2.22 What is a component in VHDL?

2.23 What does binding mean in terms of components and design entities? What is the default binding rule for components?

2.24 Using only the design entities that follow, write the `top_level` design entity to complete the description of a circuit that is functionally equivalent to the Boolean equation:

$$y = abc + \bar{c}$$

Use direct design entity instantiation.

```
------------------------- OR gate component -------------------------
library ieee; -- Load the IEEE 1164 library
use ieee.std_logic_1164.all; -- Make package visible to the next entity

entity or_2 is -- Entity declaration for 2 input or
   port (i1, i2 : in std_logic;
   o1: out std_logic);
end or_2;

architecture dataflow of or_2 is -- Architecture body for 2 input or
begin
   o1 <= i1 or i2;
end dataflow;
```

(Cont.)

(Cont.)

```
------------------------- AND gate component -------------------------
library ieee; -- Load the IEEE 1164 library
use ieee.std_logic_1164.all; -- Make package visible to the next entity

entity and_2 is -- Entity declaration for 2 input and
   port (i1, i2 : in std_logic;
   o1: out std_logic);
end and_2;

architecture dataflow of and_2 is -- Architecture body for 2 input and
begin
   o1 <= i1 and i2;
end dataflow;


------------------------- NOT gate component -------------------------
library ieee;-- Load the IEEE 1164 library
use ieee.std_logic_1164.all; -- Make package visible to the next entity

entity not_1 is -- Entity declaration for 2 input and
   port (i1 : in std_logic;
   o1: out std_logic);
end not_1;

architecture dataflow of not_1 is -- Architecture body for 2 input and
begin
   o1 <= not i1;
end dataflow;


------------------------- TOPLEVEL structure -------------------------
-- put your top-level design unit here
```

2.25 Given the following entity declaration and architecture body:

```
entity nand_2 is
      port (i1, i2 : in std_logic;
      o : out std_logic);
end nand_2;

architecture dataflow of nand_2 is
      begin
            o <= i1 nand i2;
end dataflow;
```

Write a design description of a two-input OR gate design entity that has inputs `a` and `b` and output `c`. Use a structural architecture consisting of only `nand_2` components.

2.26 The figure below gives the pin assignments and logic diagram of a Fairchild NC7SP19 1-of-2 Decoder/Demultiplexer IC.

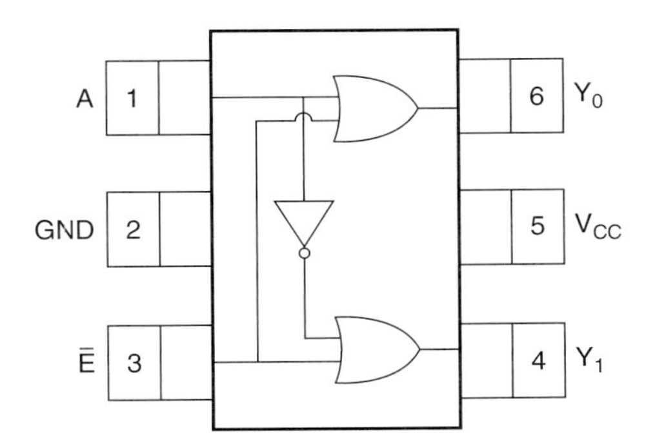

(a) Write an entity declaration named `nc7sp19` for an equivalent function. Label the inputs and outputs to correspond as closely as possible to the names used in the diagram. Power and ground pins are not included as ports in entity declarations. Also, ignore the pin numbers.

(b) Write a complete design file for the entity `nc7sp19` that includes a top-level structural description. Use the OR gate component `or_2` and the inverter component `not_1` from Problem 2.24. Label the components in the diagram `u1` to `u3` from top to bottom. Use direct design entity instantiation in the top-level structural architecture. Use named association in the port map.

2.27 As a hot-shot VHDL designer, you have been hired by Chapter 11 Electronics to salvage their biggest project, which is in trouble because their chief engineer left the company. All that remains is a logic diagram covered with coffee stains and a partially completed VHDL design file. The partially completed file is the same as that given in Problem 2.24. To save the company, you must finish the design by writing the description of the top-level design entity!

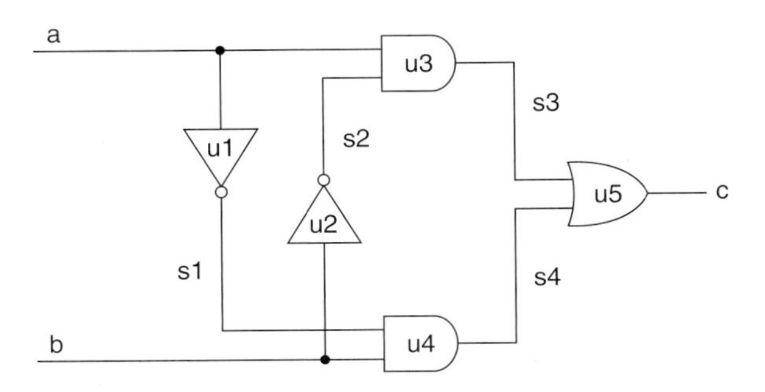

2.28 For the logic diagram shown below:

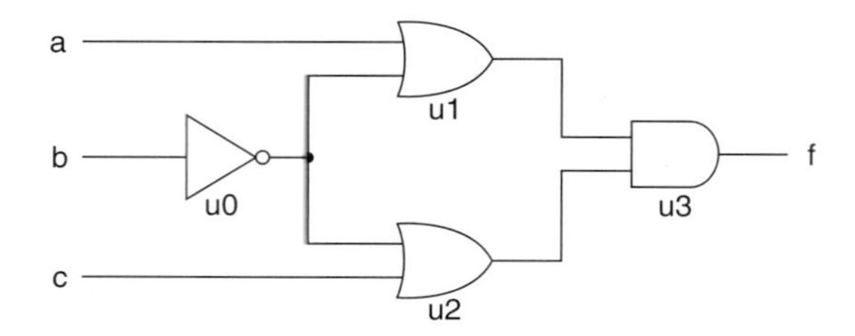

Write a complete structural style design description consisting of a single design file. This description should correspond as closely as possible to the diagram. Use direct design entity instantiation. Also, use named association in the port maps.

2.29 For the logic diagram shown below:

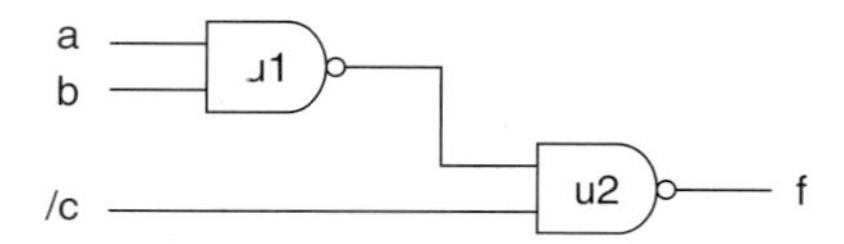

(a) Write a complete structural style design description consisting of a single source file. Name the architecture `structural`. This description should have a one-to-one correspondence with the diagram. Use direct instantiation. Also use only named association in port maps.

(b) Write an alternative structural architecture for part (a) that uses indirect instantiation (components).

(c) Write a complete dataflow style design that uses a single concurrent signal assignment consisting of a Boolean expression that uses only NAND operators.

2.30 For the design description that follows, draw the corresponding logic diagram.

```
library ieee;
use ieee.std_logic_1164.all;

entity gate_ckt is
   port (a, b, c : in std_logic; f: out std_logic);
end gate_ckt;

architecture structure of gate_ckt is
   signal s1, s2: std_logic;
   begin
   u0 : entity nand_2 port map (i1 => a, i2 => b, o1 => s1);
   u1 : entity nand_2 port map (i1 => s1, i2 => s2, o1 => f);
   u2 : entity invert port map (i1 => c, o1 => s2)
end structure;
```

Assume that the nand_2 entity is a two-input NAND gate and the invert entity is a NOT gate. Port names starting with an i indicate inputs and port names starting with an o indicate outputs.

2.31 Of the three typical steps carried out by a logic synthesizer (see page 29) to the result of which step does the RTL hierarchical view of the synthesized logic correspond? To the result of which synthesis step does the technology dependent flattened-to-gates view correspond?

2.32 For descriptions of the same system written in different coding styles and synthesized for the same target PLD, under what conditions would we expect the gate-level synthesized logic to be identical? Under what conditions would we expect the gate-level synthesized logic to not be identical? Should the gate-level synthesized logic be functionally the same under all conditions?

2.33 Why is the RTL hierarchical view of the logic synthesized for a design described in structural style always represented as an interconnection of rectangular blocks?

2.34 What does "level of abstraction" mean in terms of the description of a system?

2.35 What is the advantage of using a description with a high level of abstraction when describing a complex system?

Chapter 3
Signals and Data Types

In digital systems, the logic signals that transfer information from one component to another, and from one system to another, are voltages that vary in a discrete fashion with respect to time and are carried by wires. In VHDL, signals play a similar and equally important role. Local signals connect components within an architecture to form a larger system. Port signals provide a system's interface to its environment.

All signals in VHDL have a declared type. A signal's type determines the set of values it can take and the operations that can be performed on those values.

Every signal must be declared before it is used. Where a signal is declared determines where in the source code text it can be used—its scope or visibility. A signal can only be referenced by statements that appear within the scope of its declaration.

Constants are also introduced in this chapter. A constant is a named object that is assigned a value when it is declared. Once a constant is assigned a value, its value cannot be changed. Use of constants make a description more readable and easier to modify.

3.1 OBJECT CLASSES AND OBJECT TYPES

Signals belong to a class of VHDL language elements called objects. An *object* is a named item that has a value of a specified type. To place the study of signals and constants in context, this section presents an overview of objects and types.

Object Classes

An object's *class* represents the nature of the object and how it is used, for example, whether the object's value can be modified and whether it has a time element associated with it. Conventional programming languages typically have three classes of objects: constants, variables, and files. In contrast, VHDL has four:

- *Signal:* an object with a current value and projected (future) values. The projected values can be changed, as many times as desired, using signal assignment statements.
- *Constant:* an object whose value cannot be changed after it is initially specified.
- *Variable:* an object with only a current value. A variable's value can be changed, as many times as desired, using variable assignment statements.
- *File:* an object that consists of a sequence of values of a specified type.

Signals, constants, and variables are synthesizable and are used in design descriptions. Files are not synthesizable; they are used mostly in testbenches.

Object Types

Each VHDL object must be of some type. An object's *type* defines the set of values the object can assume and the set of operations that can be performed on those values.

There are five types:

- *Scalar type:* has a single indivisible value, which can be either numeric or enumerated.
- *Composite type:* consists of a collection of elements each with a value. There are two kinds of composite types: arrays and records. All elements in an array are of the same type. Elements in a record can be of different types.
- *Access type:* provides access to objects of a given type, similar to a pointer in conventional programming languages.
- *File type:* provides access to objects containing a sequence of values of a given type (such as disk files). The value of a file type is the sequence of values contained in the host system file.
- *Protected type:* provides atomic and exclusive access to variables accessible to multiple processes (global variables).

Only scalar and composite types are synthesizable.

Strong Typing

VHDL is a *strongly typed* language with strictly enforced type rules. If we mix different types in an expression or exceed a type's range of values, the compiler or simulator generates an error message. For example, the integer value `0`, the real number `0.0`, and the bit value `'0'` are not the same type, and, therefore, are not the same. VHDL's strong typing makes it easier for a compiler to detect errors.

Object Declarations

All objects must be declared before being used. A *declaration* introduces the name of an object, defines its type, and, optionally, assigns an initial value.

This chapter focuses on signal and constant objects of scalar and composite types. Neither signals nor constants are allowed to be of access type.

3.2 SIGNAL OBJECTS

As seen in Chapter 2, the ports of a design entity are signals. From the viewpoint of a design entity, its ports can be thought of as *external signals*, which are simply referred to as ports. Signals within a design entity can be thought of as *local* or *internal signals,* which are simply referred to as signals.

Scope and Visibility

For each signal declared, VHDL language rules define a certain region of text, called *scope* or *namespace,* where the signal's name can be used. A signal is said to be *visible* (accessible) within its scope. Outside of its scope, a signal is unknown. Any attempt to use a signal outside of its scope causes a compilation error.

A signal's scope is determined by where it is declared. A signal can be declared in the declarative part of an entity declaration, the declarative part of an architecture body, or in a package declaration.

Port Visibility

A port's name is visible in the entity declaration in which it is declared and in any architecture bodies associated with that entity declaration. A component declaration in the declarative part of an architecture makes that component's port signals visible in that architecture.

Signal Visibility

A signal declared in the declarative part of an architecture body can only be used within that architecture; it is not visible outside of that architecture.

Information Hiding

Once a design entity's functionality has been verified, the details of its internal operation are not subsequently of concern to a user of the design entity. This is consistent with the concept of *information hiding.* Therefore, it is appropriate that a signal declared in the declarative part of an architecture body not be visible outside of that architecture.

Port Declarations

The term port is used to refer only to signals declared by port declarations. Ports are declared in the port clause of an entity declaration, as was shown in Figure 2.1.1, which is repeated here as Figure 3.2.1.

```
entity_declaration ::=
    entity identifier is
            [port ( port_interface_list ) ;]
    end [ entity ] [ entity_simple_name ] ;

port_interface_list ::=
[signal] identifier_list : [ mode ] subtype_indication [ := static_expression ]
{ [signal] identifier_list : [ mode ] subtype_indication [ := static_expression ] }
```

FIGURE 3.2.1
Simplified syntax for an entity declaration.

The port clause contains a port interface list that consists of one or more interface signal declarations. Each interface signal declaration has the form:

```
[signal] identifier_list : [ mode ]
      subtype_indication [ := static_expression ]
```

Each signal name in the identifier list creates a new signal. The mode specifies the direction of the port's information transfer. If the mode is not specified, it defaults to mode `in`. The subtype_indication specifies the port's type.

Port Declarations Example

For example, consider the design of a comparator that compares two inputs `x` and `y` and makes its output `eq` a `'1'` if its inputs are equal.

A truth table and list of minterms for the comparator are given in Table 3.2.1.

A dataflow description of this comparator is given in Listing 3.2.1.

Three signals, `x`, `y`, and `eq`, are declared in the port clause:

```
port (x, y: in std_logic; eq: out std_logic);
```

Table 3.2.1
Truth table and minterms for a comparator.

x	y	eq	minterms
0	0	1	$\bar{x}\,\bar{y}$
0	1	0	
1	0	0	
1	1	1	$x\,y$

LISTING 3.2.1
Dataflow style comparator.

```
library ieee;
use ieee.std_logic_1164.all;

entity compare is
   port (x, y: in std_logic; eq: out std_logic);
end compare;

architecture dataflow of compare is
begin
  eq <= (not x and not y) or (x and y);
end dataflow;
```

A multiple object declaration, like the preceding, is equivalent to a sequence of the corresponding number of single-object declarations. For example:

```
port (x: in std_logic;
      y: in std_logic;
      eq: out std_logic);
```

Single-object declarations are more readable, particularly when each is followed by an appropriate comment that states the purpose of the signal, but they require more space.

For port declarations, the convention is to leave out the optional keyword **signal**. So instead of writing:

```
signal x: in std_logic;
```

we simply write:

```
x: in std_logic;
```

Since `x`, `y`, and `eq` are declared in the port clause for `compare`, these signals are visible (can be used) in any architecture associated with `compare`. Accordingly, the architecture `dataflow` of `compare` can (and does) use these port signals.

The architecture consists of a single concurrent signal assignment statement:

```
eq <= (not x and not y) or (x and y);
```

This is the canonical sum-of-products expression determined from the truth table in Table 3.2.1.

Signal Declarations

Signals are declared in the declarative part of an architecture using a signal declaration statement. The syntax for such a statement is given in Figure 3.2.2.

The syntax for a signal declaration looks just like that of an interface signal declaration, except that the keyword **`signal`** is required and no mode is specified. Unlike ports, signals do not have a mode. A signal can transfer information in either direction.

:= Symbol

The symbol used for assignment of an initial value to a signal in its declaration is `:=`. This is in contrast to the symbol `<=` used for signal assignment. *Initial signal values* are useful for simulation, but not for synthesis. ***For synthesis, any initial or default value assigned to a signal is ignored by an IEEE 1076.6 compliant synthesizer.*** This is because the synthesizer cannot assume that the target hardware has the capability to be powered up with its signals in the specified states.

Note that there are some synthesizers that attempt to synthesize initial values (with varying degrees of success). However, for a design description to be IEEE Std 1076.6 compliant, initial values must not be assigned to signals or variables.

For purposes of simulation, in the absence of an explicit initial value expression, an implicit default value is assigned to a signal. This default value is equal to the leftmost value in the range of values for the specified type.

Using Signals to Connect Concurrent Statements

We have previously used signals in structural style architectures, where it was clear that their purpose was to connect design entities, just as wires connect components in an electronic system. Signals are also used to connect concurrent signal assignment statements in dataflow style architectures and processes in behavioral style architectures. In effect, signals allow concurrent statements of any kind to communicate with each other.

Connecting Concurrent Signal Assignment Statements

The dataflow architecture for design entity `compare` in Listing 3.2.1 did not use any signals. A second dataflow style architecture for `compare` is given in Listing 3.2.2.

Signals `x_bar` and `y_bar` are declared in the declarative part of architecture `dataflow2`. These signals are visible only within this architecture. This architecture is named `dataflow2` to distinguish it from the previous architecture (`dataflow`) for the same entity declaration.

```
signal_declaration ::=
signal identifier_list : subtype_indication [:= static_expression] ;
```

FIGURE 3.2.2
Simplified syntax for a signal declaration.

LISTING 3.2.2
Dataflow style comparator using local signals.

```
library ieee;
use ieee.std_logic_1164.all;

entity compare is
   port (x, y: in std_logic; eq: out std_logic);
end compare;

architecture dataflow2 of compare is
signal x_bar, y_bar: std_logic;
begin
   x_bar <= not x;
   y_bar <= not y;
   eq <= (x_bar and y_bar) or (x and y);
end dataflow2;
```

Use of signals `x_bar` and `y_bar` in Listing 3.2.2 demonstrates that signals can connect concurrent signal assignment statements in an architecture. In a more complex dataflow architecture, it might be advantageous to break a single complex concurrent signal assignment statement into multiple constituent statements to improve readability.

3.3 SCALAR TYPES

The signals we have encountered in previous programs were all scalar types. As previously stated, a *scalar type* represents a single indivisible value. Scalar types are classified as either enumeration, integer, physical, or floating point (Figure 3.3.1).

Discrete Types and Numeric Types

Enumeration types and integer types are discrete types. *Discrete types* represent sets of distinct values. Integer types are also classified as *numeric types,* as are physical types and floating point types. All numeric types can be specified with a range that constrains the set of possible values.

Type Declarations

Just as all signals must be declared with a type before they are used, all types must also be declared before use. Types are declared using a type declaration (Figure 3.3.2). The type definition portion of a type declaration defines the set of values an object of that type can have.

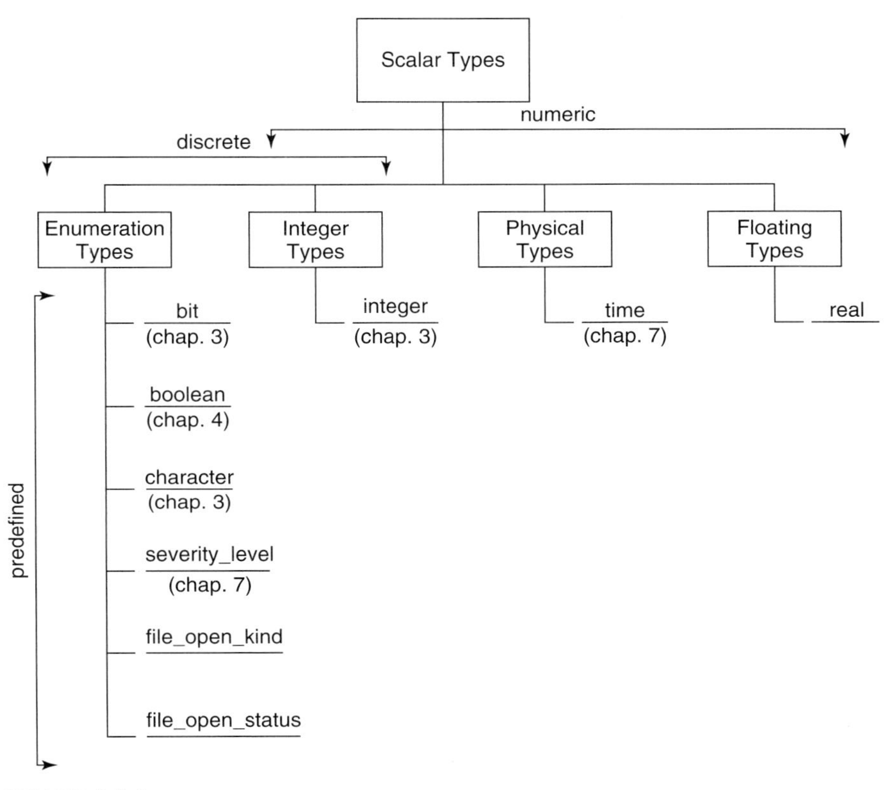

FIGURE 3.3.1
Predefined scalar types.

Types used for ports must be declared in a package, so that the type is visible to entity declarations. A package must be compiled before any entity declaration that uses a type declared in that package is compiled.

Predefined Types

Predefined (*built-in*) *types* are those defined in packages STANDARD and TEXTIO in the library STD. This library is provided with all VHDL compilers. Package TEXTIO is provided to support formatted human-readable I/O.

```
type_declaration ::=
type identifier is type_definition ;
```

FIGURE 3.3.2
Simplified syntax for a type declaration.

```
enumeration_type_declaration ::=
type identifier is ( enumeration_literal { , enumeraton_literal } )

enumeration_literal ::= identifier | character_literal
```

FIGURE 3.3.3
Simplified syntax for an enumeration type declaration.

The predefined scalar types in Figure 3.3.1 are defined in package STANDARD. Of these, only bit, boolean, character, and integer are synthesizable. Types character, bit, and integer are discussed in this chapter. The other predefined scaler types are discussed in later chapters.

To use a type that is declared in a package, we must precede each entity declaration that uses that type with an appropriate context clause. Because every design unit in a description is automatically preceded by the *implicit context clause*:

```
library std, work;
use std.standard.all;
```

the predefined types in package STANDARD are directly visible in a design without explicitly using any library or use clauses.

Enumeration Types

An *enumeration type* is declared by simply listing (enumerating) all possible values for the type. There must be at least one value in the list. The syntax for declaring an enumeration type is given in Figure 3.3.3.

Identifier is the name given to the type. The type definition portion of an enumeration type declaration consists of the list of enumeration literals (values) for the type. An *enumeration literal* can be an identifier or a character literal.

Like all other VHDL identifiers, an identifier used as an enumeration literal is not case sensitive. In contrast, character literals are case sensitive. A character literal is a single character enclosed in single quotes. For example, the single characters `'a'` and `'A'` are distinct. An enumeration type is said to be a *character type* if at least one of its enumeration literals is a character literal.

Each enumeration literal must be unique within the type being declared. However, different enumeration types may use the same literal. In such cases, the literal is said to be *overloaded.*

All scalar types are ordered. All relational operators (>, <, =, and so on) are predefined for their values. Enumeration literals are ordered, in ascending order, by their textual position in the list of literals defining the type.

Type Character

Type character is an enumeration type, predefined in package STANDARD, that corresponds to the *International Organization for Standardization (ISO)* 8-bit encoded character set ISO 8859-1:1987. This character set includes graphic characters and format effectors. The definition of type character contains a mixture of character literals and identifiers as elements:

```
type character is (NUL, SOH, STX, ETX, EOT, ENQ, ACK, BEL,
                   BS,   HT,   LF,   VT,   FF,   CR,   SO,   SI,
                   DLE,  DC1,  DC2,  DC3,  DC4,  NAK,  SYN,  ETB,
                   CAN,  EM,   SUB,  ESC,  FSP,  GSP,  RSP,  USP,
                   ' ',  '!',  '"',  '#',  '$',  '%',  '&',  ''',
                   '(',  ')',  '*',  '+',  ',',  '-',  '.',  '/',
                   '0',  '1',  '2',  '3',  '4',  '5',  '6',  '7',
                   '8',  '9',  ':',  ';',  '<',  '=',  '>',  '?',
                   '@',  'A',  'B',  'C',  'D',  'E',  'F',  'G',
                   'H',  'I',  'J',  'K',  'L',  'M',  'N',  'O',
                   'P',  'Q',  'R',  'S',  'T',  'U',  'V',  'W',
                   'X',  'Y',  'Z',  '[',  '\',  ']',  '^',  '_',
                   '`',  'a',  'b',  'c',  'd',  'e',  'f',  'g',
                   'h',  'i',  'j',  'k',  'l',  'm',  'n',  'o',
                   'p',  'q',  'r',  's',  't',  'u',  'v',  'w',
                   'x',  'y',  'z',  '{',  '|',  '}',  '~',  DEL,
              C128, C129, C130, C131, C132, C133, C134, C135,
              C136, C137, C138, C139, C140, C141, C142, C143,
              C144, C145, C146, C147, C148, C149, C150, C151,
              C152, C153, C154, C155, C156, C157, C158, C159,
                   ' ',  '¡',  '¢',  '£',  '‡',  '¥',  '¦',  '§',
                   '¨',  '©',  'ª',  '«',  '¬',  '-',  '®',  '¯',
                   '°',  '±',  '²',  '³',  '´',  'µ',  '¶',  '·',
                   '¸',  '¹',  'º',  '»',  '¼',  '½',  '¾',  '¿',
                   'À',  'Á',  'Â',  'Ã',  'Ä',  'Å',  'Æ',  'Ç',
                   'È',  'É',  'Ê',  'Ë',  'Ì',  'Í',  'Î',  'Ï',
                   'Ð',  'Ñ',  'Ò',  'Ó',  'Ô',  'Õ',  'Ö',  '×',
                   'Ø',  'Ù',  'Ú',  'Û',  'Ü',  'Ý',  'Þ',  'ß',
                   'à',  'á',  'â',  'ã',  'ä',  'å',  'æ',  'ç',
                   'è',  'é',  'ê',  'ë',  'ì',  'í',  'î',  'ï',
                   'ð',  'ñ',  'ò',  'ó',  'ô',  'õ',  'ö',  '÷',
                   'ø',  'ù',  'ú',  'û',  'ü',  'ý',  'þ',  'ÿ');
```

The first 128 characters in this enumeration are the ASCII characters. The identifiers from NUL to USP and DLE are the nonprintable ASCII control characters. Characters C128 to C159 in the ISO character set do not have names. In VHDL, these characters are given names based on their positions in the enumeration.

The character in position 160 is the nonbreaking space character, in contrast to the ordinary space character in position 32. The character at position 173 is the soft hyphen, in contrast to the ordinary hyphen in position 45. The soft hyphen is used to represent a line break within a word.

Type Bit

Type bit is an enumeration type used as a simple representation for a logic signal. The type declaration for bit appears in package STANDARD as:

```
type bit is ('0','1');
```

```
subtype_declaration ::=
subtype identifier is [ resolution_function_name ] type_mark [ constraint ] ;

type_mark ::= type_name | subtype_name
constraint ::= range_constraint | index_constraint
range_constraint ::= range range
index_constraint ::= ( discrete_range { , discrete_range } )
```

FIGURE 3.3.4
Simplified syntax for a subtype declaration.

Here character literals `'0'` and `'1'` are used to represent logic 0 and logic 1. Hence, type bit is also a character type. Because of their order in the enumeration list, `'0'` is less than `'1'`.

Subtype Declarations

A type derived from another type is a *subtype*. The type from which the subtype is derived is its *base type*. Creating and using a subtype allows us to make it clear which values from the base type are valid for an object.

The simplified syntax for a subtype declaration is given in Figure 3.3.4. From the syntax, we see that a subtype is simply a type together with either a resolution function or a constraint, or both.

Inclusion in a Subtype

A value belongs to a subtype of a given type if it belongs to the base type and satisfies the constraint, if any, of the subtype declaration. A constraint is a restriction on the set of possible values for the subtype from the values allowed for the base type. A type is a subtype of itself. Such a subtype is unconstrained because it corresponds to a situation that imposes no restrictions.

A subtype declaration does not define a new type. The type of a subtype is the same as its base type. Accordingly, the set of operations allowed on a subtype is the same as the set allowed on its base type.

An example of a subtype declaration using a resolution function is given in the next section. Examples of subtype declarations using constraints are given in Section 3.10.

Type Compatibility in Assignments

When a signal assignment is made, the objects on both sides of the assignment symbol must be of the same type. Otherwise, the compiler gives a type conflict error. This ensures that an object is only assigned values of its own type and avoids mixing values that represent different things.

Since a subtype declaration does not define a new type, a value of a subtype can be assigned to an object of its base type.

Conversely, a value of a base type can also be assigned to an object of its subtype without producing a syntax error. However, if, during simulation, an assignment of a value of a given type is made to an object of its subtype and the value assigned does not meet the subtype constraint, a dynamic semantic error occurs.

To assign a value of one type to an object of a different type requires a type conversion. Type conversions are either built-in or must be provided as functions. Type conversions are introduced in "Types Unsigned and Signed" on page 107. Writing your own functions for type conversion is discussed in Chapter 12, "Type Conversions," on page 557.

3.4 TYPE STD_LOGIC

While VHDL has several predefined types, it does not have one that provides more than two values to represent a logic signal. Type bit, with its values `'0'` and `'1'`, is not sufficient to represent a logic signal when more detailed information about the signal's logical and/or electrical properties is needed for simulation or synthesis. In addition to the conventional logic levels, we may need to be able to represent different signal strengths and conditions, such as a high impedance or an unknown logic level.

Multivalued Logic System

A logic system that has more than two logic states (0 and 1) is a *multivalued logic system*. VHDL allows user-defined types, so we can easily create any multivalued logic system that we might desire. However, creating and using a nonstandard multi-valued logic system reduces the interoperability of code that uses that logic system. To resolve this issue, the IEEE created a standard multivalued logic system that it felt would adequately meet the need for specifying detailed information on the properties of a logic signal.

Package STD_LOGIC_1164

The IEEE Library, included with VHDL compilers, includes package STD_LOGIC_1164. This package is defined in the standard *IEEE Std 1164 — IEEE Standard Multi-value Logic System for VHDL Model Interoperability (Std_logic_*1164).

Nine Value Logic System

Package STD_LOGIC_1164 defines a nine value logic system. Type std_ulogic and its subtype std_logic are defined in this package. They are enumerated types. Also defined in this package are operators and functions for operands of these types.

Std_ulogic

Type std_ulogic is declared in package STD_LOGIC_1164 as:

```
type std_ulogic is (
      'U' -- Uninitialized
      'X' -- Forcing Unknown
      '0' -- Forcing Low
      '1' -- Forcing High
      'Z' -- High Impedance
      'W' -- Weak Unknown
      'L' -- Weak Low
      'H' -- Weak High
      '-' -- Don't Care);
```

Since the character literals used in this type declaration are uppercase, lowercase versions of these characters are not valid representations of std_ulogic values.

There are two ways of categorizing the nine std_logic values. One way is in terms of whether the value represents a logical value or a metalogical value. The other way is in terms of a value's properties of state and strength.

Logical Values versus Metalogical Values

Values `'1'`, `'H'`, `'0'`, and `'L'` from std_ulogic are *logical values* that represent only two logic levels. Each logic level typically corresponds to one of two distinct voltage ranges in a circuit. In contrast, values `'U'`, `'X'`, `'W'`, and `'-'` are metalogical values. *Metalogical values* are used to represent the behavior of the VHDL model itself, rather than the behavior of the hardware being modeled or synthesized. The value `'Z'` is not classified as either a logical value or a metalogical value.

State and Strength Properties

Std_ulogic values can be categorized in terms of their state and strength. The *state* of a std_ulogic value denotes its logic level. The *strength* of a std_ulogic value denotes the electrical characteristics of the source that drives the signal. Not all of the std_ulogic values have both of these properties. Each of the std_ulogic values and its state and strength properties is listed in Table 3.4.1.

Driving Strengths

Three *driving strengths* are represented by std_ulogic: forcing, weak, and high impedance. These driving strengths represent the electrical characteristics of a signal's driver.

Uninitialized Value

The *uninitialized* value (`'U'`) is the default value given to all std_ulogic signals before the start of a simulation. At that point no values have been assigned to signals

Table 3.4.1
State and strength properties of std_ulogic values.

Value	State	Strength
U	uninitialized	none
X	unknown	forcing
0	0	forcing
1	1	forcing
Z	none	high impedance
W	unknown	weak
L	0	weak
H	1	weak
-	don't care	none

by the program being simulated. The uninitialized value has no strength property since there is no source driving an uninitialized signal. During simulation, if a signal still has the value `'U'`, it means that the signal has not been assigned a value during execution of the program. This typically represents a design error. Without the `'U'` value it would be difficult to determine whether a signal had not been assigned a value or had only been assigned values equal to its initial value.

Forcing Strength

Signals driven by active output drivers are referred to as *forcing strength* signals. Three such values are provided: forcing a low (`'0'`), forcing a high (`'1'`), and forcing an unknown value (`'X'`).

"Normal" Logic 0 and 1

Forcing low (`'0'`) and *forcing high* (`'1'`) are the normal logic 0 and logic 1 values, respectively. These values represent the signal from an active output driver, such as that of a CMOS circuit.

Unknown Value

The *unknown* value (`'X'`) is used to represent a signal that is being driven, but whose value cannot be determined to be a 0 or a 1. For example, a signal whose voltage is in the "unallowed" voltage range between the valid voltage ranges for the target technology's logic levels. This value often results from bus contention that occurs when a multiply driven signal has one of its drivers driving a `'0'` and another driving a `'1'`.

Weak Strength

Weak strength signals—weak low (`'L'`), weak high (`'H'`), and weak unknown (`'W'`)—are used to represent signals from resistive drivers, such as pull-up resistor or pull-down resistor outputs, or pass transistors.

High Impedance

High impedance (`'Z'`) represents the output of a three-state buffer when it is not enabled. A high impedance output makes no effective contribution to the resolved value of a multiply driven signal to which it is connected.

Don't Care

The interpretation of *don't care* (`'-'`) is that for certain input conditions the description does not specify a choice for the output value. Don't care (`'-'`) is interpreted by a synthesis tool as representing the common don't care condition used in logic design. Its use allows a synthesizer to decide whether to assign a `'0'` or `'1'` value to an output for each input combination where the output is specified as `'-'`. The synthesizer makes these assignments in a way that results in minimal synthesized logic.

Position Number

In an enumeration type, each enumeration literal has a specific *position number*. This number is determined by the textual order (position) in which the literal appears in the definition of the type. For example, in std_ulogic the first literal (`'U'`) has the position number 0, the next literal ('X') has the position number 1, and so on. One use of position numbers is to order the enumeration literals. The literals in an enumeration are ordered in ascending textual order. For example, in terms of ordering, 'U' is less than '0', because the position number for 'U' (0) is less than (left of) the position number for '0' (2).

Unresolved Type

Type std_ulogic is an *unresolved type.* The 'u' in std_ulogic stands for unresolved. By default, types, whether predefined or user defined, are unresolved. It is illegal for two (or more) sources to drive the same signal if that signal is of an unresolved type (Figure 3.4.1). If the signals of the circuit we are describing do not have multiple sources, then we can use std_ulogic.

Compiler errors are generated when there are multiple sources for a signal of an unresolved type. These multiple sources could be the result of having the same signal assigned values in two different processes. So, using std_ulogic has an advantage that if our design unintentionally creates two sources for a signal, we can catch this error during compilation.

Resolved Type

A *resolved type* is a type declared with a resolution function. A *resolution function* is a function that defines, for all possible combinations of one or more source values, the resulting (resolved) value of a signal.

If we are describing a circuit with an output that drives a signal that is also driven by one or more other circuits, such as a circuit with three-state outputs used in a bus interface, we need to use a resolved data type. This is a situation where we intend for a signal to have multiple sources (Figure 3.4.1).

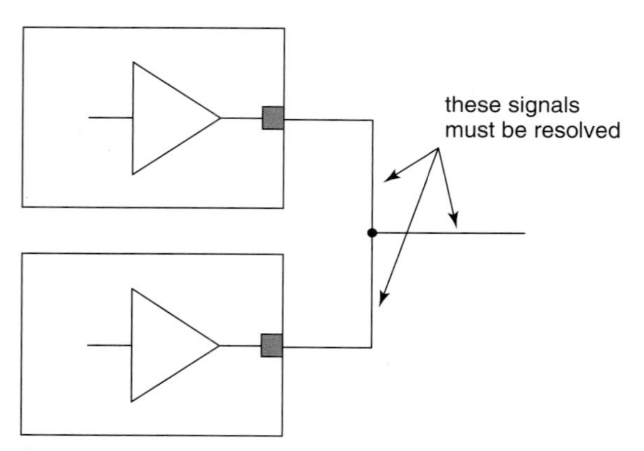

FIGURE 3.4.1
Two sources driving the same signal.

Std_logic as a Resolved Subtype

Std_logic is a subtype of std_ulogic and is declared in package STD_LOGIC_1164 as:

```
subtype std_logic is resolved std_ulogic;
```

This subtype declaration specifies a resolution function named `resolved`, but it does not specify a constraint. Thus, std_logic consists of the same set of nine values as std_ulogic. As a subtype, all operations and functions defined for std_ulogic apply to std_logic.

Std_logic's resolution function `resolved` is defined in package STD_LOGIC_ 1164. In the case of multiple sources for a signal, the effect of the multiple source values is resolved to a single value.

The resultant value for two sources driving the same signal is indicated in Table 3.4.2. Function `resolved` uses a VHDL representation of this table along with a table lookup operation to determine the result. This resolution function is discussed in detail in Chapter 13, "Resolution Function for Std_logic" on page 510. Function `resolved` is automatically invoked by the simulator whenever a std_logic signal has two or more sources.

Forcing Strength versus Weak Strength

Table 3.4.2 clarifies the effect of forcing strength values (`'X'`, `'0'`, `'1'`) and weak strength values (`'W'`, `'L'`, `'H'`) on a common signal. If one source drives a forcing strength value and the other drives a weak strength value, the forcing strength value dominates. If both sources drive different values of the same strength, the result is the unknown value for that strength, either `'X'` or `'W'`.

Table 3.4.2
Resolution table for std_logic.

	U	X	0	1	Z	W	L	H	-
U	U	U	U	U	U	U	U	U	U
X	U	X	X	X	X	X	X	X	X
0	U	X	0	X	0	0	0	0	X
1	U	X	X	1	1	1	1	1	X
Z	U	X	0	1	Z	W	L	H	X
W	U	X	0	1	W	W	W	W	X
L	U	X	0	1	L	W	L	W	X
H	U	X	0	1	H	W	W	H	X
-	U	X	X	X	X	X	X	X	X

Bused Std_logic Signals

When one source's value is high-impedance ('Z') and the other source is either a forcing strength value or a weak strength value, the resulting value is the forcing strength value or weak strength value, respectively. This represents the case where three-state outputs have been used to connect signals to a common bus. When designing bused circuits, we must carefully design the circuit so that only one output driving a common connection is enabled (forcing or weak strength value) at a time and all other outputs driving that common connection are in their high-impedance states. For example, if by design error, two enabled forcing strength sources try to force different values on the same signal during simulation, the result is the unknown value 'X'.

During a simulation, observation of a signal having the value 'X' or 'W' is usually an indication of a design error.

The value of a std_logic signal with more than two sources is the resolved value of all the source values driving the signal. This is computed by the repeated application of the resolution function to the individual source values.

Value Read from a Multiply Driven Inout Port

For a mode inout port driving a signal that is also driven by other ports, the value that is assigned to the port is the value driven by the port. The value read from the port is the result of the resolution of the value assigned to the port and the values driven by all other sources that connect to the port. In contrast, for a mode buffer port, the value read by the port is the same as the value driven by the port.

Use of Std_logic versus Std_ulogic

A disadvantage of using std_logic instead of std_ulogic is that signals that are unintentionally multiply driven will not be detected as an error during compilation.

However, IEEE Std 1164 recommends that std_logic be used instead of std_ulogic, even if a signal has only a single source. The reason is that the standard expects simulator vendors to design their simulators to optimize the simulation of models using the resolved subtype, but they need not optimize the simulation of models using unresolved types.

Additional reasons for using std_logic are discussed in the section "Port Types for Synthesis" on page 116.

Our Use of Std_logic Values

We are interested in writing design descriptions that will be synthesized and then implemented using PLDs. In our descriptions we will assign only the values '0', '1', or '-' to std_logic signals. For ports of the top-level design entity that are mode out, buffer, or inout, we add 'Z' to the previous list of values. Assigning values 'H' and 'L' to signals is not compatible with the device technology normally used in PLDs.

For testbenches we typically assign only the values '0' and '1'as inputs to the UUT. However, for some special timing simulations we may have the testbench assign the value 'X' to a UUT input. During simulation we may observe the value 'U' and sometimes the value 'X'. Since we will not assign values 'H' and 'L' to signals, we don't expect to observe the value 'W'.

3.5 SCALAR LITERALS AND SCALAR CONSTANTS

A *literal* is a value that is expressed as itself. A literal can be directly assigned to a signal or used in an expression that determines the value assigned to a signal. Only character, enumeration, string, and numeric literals can be used in synthesizable descriptions.

Literals don't have an explicit type. The type of an object being assigned a literal value or the type implied by the expression in which a literal appears determines whether use of a given literal is valid.

Character Literals

As previously mentioned, a *character literal* is formed by enclosing one of the 191 graphic (printable) characters (including the space and nonbreaking space characters) from the ISO 8859-1:1987 character set between two single quote characters (e.g., `'Z'`).

Enumeration Literal

Also as previously mentioned, an *enumeration literal* is a character literal or an identifier. Enumeration literals are used in the definition of an enumeration type and as literal values to assign to an object of that type.

Each item in the definition of type character is an enumeration literal. The previous definition of std_ulogic is an example of an enumeration of character literals. Use of character literals in multivalued logic enumerations makes it clear that a logic value (`'0'` or `'1'`) is being represented, as opposed to an integer value (`0` or `1`).

Scalar Constants

A *constant* is an object that is assigned a value only once, normally when it is declared. The value of a constant cannot be changed in a program. We can think of a constant as a literal value with a name and type. Constants are used to make programs more readable and to make it easier to change a value throughout a program, by changing the value only in the constant's declaration.

Constants must be declared before they are used. Constants are declared using the syntax in Figure 3.5.1.

If the assignment symbol `":="` followed by an expression is not present in the declaration, then the declaration declares a *deferred constant.* Deferred constants may only appear in declarations in a package. The full constant declaration would then appear in the package body.

```
constant_declaration ::=
constant identifier_list : subtype_indication := expression ;
```

FIGURE 3.5.1
Simplified syntax for a constant declaration.

An example of a scalar constant declaration is:

```
constant high_impedance : std_logic := 'Z';
```

Constant Visibility The visibility rules for constants are the same as for signals.

3.6 COMPOSITE TYPES

A *composite type* consists of a collection of related elements that form either an array or a record. Each element has a value. The elements of an array are all of the same type. The elements of a record may be of different types. A composite type may contain elements that are scalar, composite, or access type. However, file types are not allowed in a composite type. VHDL's composite types are shown in Figure 3.6.1.

A composite object can be operated on as a single object or its constituent elements can be operated on individually. An element of an array is selected by the array's name and the element's index value. An element of a record is selected by the record's name and element's name.

Arrays are discussed in detail in this chapter. Records are discussed in Chapter 7.

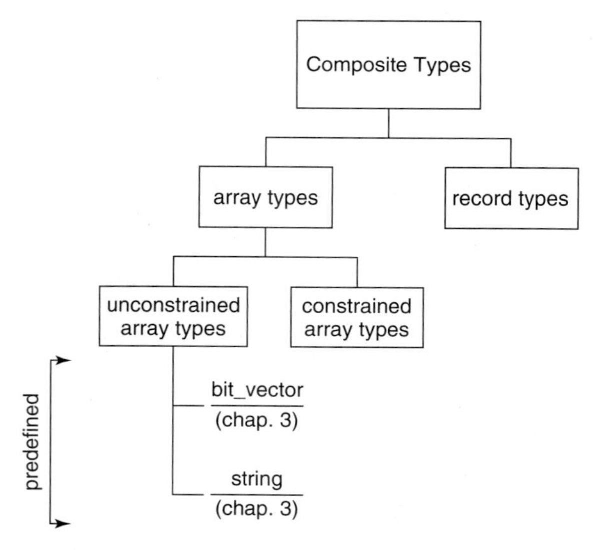

FIGURE 3.6.1
Composite types.

3.7 ARRAYS

Array types group elements of the same type together as a single object. Arrays can be one dimensional or multidimensional. A one-dimensional array is also called a *vector*. It has an index whose value selects an element in the array by its position. An element of a multidimensional array is selected using its indices.

Synthesizable Array

IEEE Std 1076.6 compliant synthesizers only support one-dimensional arrays for synthesis. A one-dimensional array is commonly used to represent a bus in a digital system; for example, the address bus, data bus, or control bus of a computer.

Even though a two-dimensional array is not supported for synthesis, an array of arrays is supported. So, the equivalent of a synthesizable two-dimensional array can be created.

Unconstrained Array

Before an array object can be declared, its array type must be declared using an array type declaration. One way to declare an array type is as an unconstrained array. An *unconstrained array* (unbounded array) is an array whose dimensions and indices types are specified when the array type is declared, but the bounds for each dimension are left unspecified.

Use of an unconstrained array allows multiple subtypes to share a common base type. Unconstrained arrays are also often used to allow subprograms to be written to handle arrays of any length. The simplified syntax of an unconstrained array type declaration is given in Figure 3.7.1.

Box Symbol

The symbol "<>", called "box," is a placeholder for the index range. When an unconstrained type is used in the declaration of an array object, the object's index bounds must then be specified. The type_mark specifies the type of the index. The element_subtype_indication specifies the type of the array's elements.

Type Bit_vector

For example, type bit_vector is a predefined one-dimensional unconstrained array of elements of type bit, defined in package STANDARD as:

```
type bit_vector is array ( natural range <> ) of bit;
```

unconstrained_array_type_declaration ::=
type identifier **is array** (type_mark **range** <> ~~{ , type_mark **range** <> }~~) **of** element_subtype_indication;

FIGURE 3.7.1
Simplified syntax for an unconstrained array type declaration.

The type of the index range is specified as `natural`, which is a type that consists of all integer values from 0 up to the highest value that can be represented in the particular VHDL implementation, usually 2,147,483,647 (32 bits).

Type Std_logic_vector

Std_logic_vector is a one-dimensional unconstrained array of elements of type std_logic that is defined in package STD_LOGIC_1164 as:

```
type std_logic_vector is array
      ( natural range <> ) of std_logic;
```

Declaring an Object of an Unconstrained Array Type

When we declare an object of an unconstrained array type, we must specify a constraint for the index bounds. One way of doing this, assuming the index type is integer, is to specify the index range as either an *ascending* (integer **to** integer) or *descending* (integer **downto** integer) *integer range*. For example:

```
signal databyte: std_logic_vector (7 downto 0);
```

declares an array named `databyte`. This array's type is std_logic_vector and its index is an integer that ranges from 7 down to 0.

Another example of declaring array objects of an unconstrained array type is the declaration of input ports for the 4-bit binary comparator in Listing 3.7.1.

LISTING 3.7.1
A 4-bit comparator using std_logic_vector inputs.

```
library ieee;
use ieee.std_logic_1164.all;

entity compare_vect is
   port (
      a: in std_logic_vector (3 downto 0);
      b: in std_logic_vector (3 downto 0);
      equal: out std_logic);
end compare_vect;

architecture behavioral of compare_vect is
begin
   process (a, b)
   begin
      if a = b then
         equal <= '1';
      else
         equal <= '0';
      end if;
   end process;
end behavioral;
```

The inputs to the comparator are two 4-element std_logic_vectors. Port a is declared in the port interface list as:

```
a: in std_logic_vector (3 downto 0);
```

The discrete range specifies both the width of vector a and the order of its elements. The most significant element of a is element 3 and the least significant is element 0. If the declaration for a were written as:

```
a: in std_logic_vector (0 to 3);
```

Then the most significant element of a would be element 0, and the least significant element would be element 3.

Declaring an Array Object of a Constrained Subtype

Another way to declare an array object is to declare a constrained subtype of an unconstrained type and then declare the array object using that subtype. For example:

```
subtype byte is std_logic_vector (7 downto 0);
signal databyte: byte;
```

Unconstrained arrays allow multiple subtypes to share a common base type.

Predefined Unconstrained Array Types

The predefined unconstrained array types are bit_vector and string. The declaration of bit_vector in package STANDARD makes use of the subtype natural, which is also declared in that package. The declaration for bit_vector was given previously as an example of an unconstrained one-dimensional array.

Type String

The definition of type *string* makes use of the subtype positive, which is also declared in package STANDARD. Subtype positive consists of all the integers from 1 up to the highest positive integer that can be represented in a particular VHDL implementation. The declaration for string is:

```
type string is array (positive range <>)
     of character;
```

This declaration defines type string as an unconstrained one-dimensional array of characters.

Declaring a Constrained Array Type

Another way to declare an array type is as a *constrained array* (bounded array) type that has its index range constrained in its type declaration. The simplified syntax for the declaration of a constrained array is given in Figure 3.7.2.

Discrete_range is a subrange of an enumeration or integer type. The type of each array element is represented by *element*_subtype_indication.

constrained_array_type_declaration ::=
type identifier **is array** (discrete_range ~~{ , discrete_range }~~) **of** element_subtype_indication ;

FIGURE 3.7.2
Simplified syntax for a constrained array declaration.

For example:

```
type word is array (15 downto 0) of std_logic;
signal dataword: word;
```

Referencing Arrays and Array Elements

Either an entire array, a single element of an array, or a slice (consecutive elements) of an array can be used in an expression.

Treating an Array as a Single Object

We can treat an array as a single composite object. Assignment of the value of one array to another requires that the arrays be of the same length. For example, consider the following signal declaration:

```
signal x, y : std_logic_vector (3 downto 0);
```

To make the value of all the elements of array x equal to the matching element values of array y, we simply write:

```
x <= y;
```

Selecting Array Elements

Sometimes we want to deal with a single array element or a subset of consecutive array elements. We can assign a value to a single array element by using the element's index. The assignment statement:

```
x(3) <= '0';
```

assigns the value '0' to the leftmost element of x, while leaving the other three element values unchanged.

Array Slices

A *slice* denotes a one-dimensional array composed of a sequence of consecutive elements from another one-dimensional array. Slices are often used to split a single bus into two or more buses.

A slice of a signal is a signal with the same base type as the array from which the slice is taken. For example, if we declare signals:

```
signal address : std_logic_vector (15 downto 0);
signal low_address: std_logic_vector (7 downto 0);
```

then we can make an assignment of the low byte of `address` to `low_address` as:

```
low_address <= address(7 downto 0);
```

This statement assigns a slice consisting of the rightmost eight elements of `address` to the eight elements of `low_address`.

Aggregates

An *aggregate* combines one or more values to form a composite. This is one approach used to combine scalars into an array or to combine multiple buses into one. The simplified syntax for an aggregate is given in Figure 3.7.3.

For example, to assign the value 1011 to the signal `x`, an aggregate can be used to form character literal values into an array and then this array is assigned to `x`.

```
x <= ('1','0','1','1');
```

Positional Association

If element associations are implied by the position of the expressions in the aggregate, the association is called *positional association*. This is the form of an aggregate where the optional choice and arrow symbol part of the syntax are not used and the items in the parentheses form a list of expressions. In the previous example, each expression is a single character literal. Each element of array `x` is associated with a value in the aggregate by its position. As a result, the previous assignment statement is equivalent to the four assignment statements:

```
x(3) <= '1';
x(2) <= '0';
x(1) <= '1';
x(0) <= '1';
```

Named Association

If element associations are specified explicitly, then the association is called *named association.* Using named association, elements can be listed in any order. ***For array elements, named association uses the element's index.*** The previous assignment could be written using named association as:

```
x <= (3 =>'1', 1 => '1', 0 => '1', 2 => '0');
```

```
aggregate ::=
( [ choice { | choice } => ] expression {, [ choice { | choice } => ] expression} )

choice ::= simple_expression | discrete_range | others
```

FIGURE 3.7.3
Simplified syntax for an aggregate.

In each of these named associations, an integer specifying the index value appears on the left-hand side of the arrow symbol and a character literal appears on the right-hand side specifying the element's value.

Mixing Positional and Named Association

Positional associations and named associations may be mixed in a single aggregate. The positional associations must come before the named associations. Once a named association is given, all the remaining associations must be named.

Others Keyword

If the keyword **others** is used in an aggregate, it can only appear at the end of an association list. All remaining elements are assigned the value specified by **others**. For example:

```
x <= (2 => '0', others => '1');
```

This assignment is equivalent to the previous assignments.

The symbol | can be used to separate a list of index values for which all elements have the same value. For example:

```
x <= (3|1|0 =>'1', 2 => '0');
```

If we wanted to make the value of all the elements in the vector equal to `'0'` we can simply write:

```
x <= (others => '0');
```

Using a Discrete Range in an Aggregate

A discrete range can be used to assign a sequence of elements the same value. For example, if `d` is declared as:

```
signal d : std_logic_vector(7 downto 0);
```

we can make each of the most significant four elements a `'1'` and the least significant four elements a `'0'` by writing:

```
d <= (7 downto 4 => '1', 3 downto 0 => '0');
```

Using an Aggregate to Combine Scalar Signals into a Vector

An aggregate can also be used to combine scalar signals together so that their values can be assigned to an array and treated as a vector within an architecture's body. Given the signal declarations:

```
signal rd_bar, wr_bar, s2_bar : std_logic;
signal cntrl : std_logic_vector(2 downto 0);
```

The assignment statement:

```
cntrl <= (rd_bar, wr_bar, s2_bar);
```

assigns these three scalar signals to the `cntrl` bus (array).

Aggregates as Assignment Targets

An aggregate can also be the target of an assignment statement. This allows a composite value to be split into a number of scalar signals. For example, the assignment statement:

```
(rd_bar, wr_bar, s2_bar) <= "010";
```

allows the three scalar signals to be assigned values using a single character string literal.

Concatenation Operator

Another way to combine scalars into an array or to combine arrays is to use concatenation. The *concatenation operator* `'&'` is predefined for any one-dimensional array type. It is used to join two one-dimensional arrays end to end. For example:

```
d <= "1011" & "0101";
```

joins the two 4-bit strings to form an 8-bit string that is assigned to `y`.

A scalar can also be concatenated to a bit string to produce a longer bit string. For example, to assign a 7-bit string to an 8-bit vector, a `'0'` can be appended to the 7-bit string:

```
d <= '0' & "1100101";
```

3.8 TYPES UNSIGNED AND SIGNED

Type std_logic_vector is not defined as a numeric representation. It is simply an array of elements of type std_logic. No arithmetic operators are defined for it in package STD_LOGIC_1164.

To use std_logic_vector to represent a numeric value, arithmetic operators such as addition, subtraction, and inequality would have to be defined, and their functions written. Some of these functions would have to differ depending on whether a std_logic_vector was to be interpreted as an unsigned or signed value.

Package Numeric_std

To avoid confusion as to how a std_logic_vector is to be interpreted, separate types were created for numeric representation. IEEE Std 1076.3, *IEEE Standard VHDL Synthesis Packages*, provides package NUMERIC_STD, which defines two vector types whose elements are type std_logic. These types are named *unsigned* and *signed* because they are intended to represent unsigned and signed numeric values. The related arithmetic operators are also defined in this package. IEEE Std. 1076.6 states that types unsigned and signed from package NUMERIC_STD (or from package NUMERIC_BIT if type bit is used) are the only array types that shall be used to represent unsigned or signed numbers.

Type Unsigned

Type *unsigned* is a one-dimensional unconstrained array of elements of type std_logic. This type is specifically defined to represent unsigned binary numeric values.

Type unsigned is declared in package NUMERIC_STD as:

```
type unsigned is array ( natural range <> ) of
      std_logic;
```

With the exception of the type name, this declaration is identical to that of std_logic_vector. The difference in these two types is in how they are interpreted and the operators that are available for them. Type unsigned is interpreted as an unsigned binary number with the leftmost element as the most significant bit.

Type Signed

Type *signed* is a one-dimensional unconstrained array of type std_logic elements. This type is specifically defined to represent signed binary numeric values.

Type signed is declared in package NUMERIC_STD as:

```
type signed is array ( natural range <> ) of
      std_logic;
```

Type signed is interpreted as a signed binary number in 2's complement form. The leftmost element is the sign bit.

Context Clause to Use Types Unsigned or Signed

To use either type unsigned or signed a design unit must be preceded by the following context clause:

```
library ieee;
use ieee.std_logic_1164.all;
use ieee.numeric_std.all;
```

The first use clause makes the declaration of type std_logic visible, which is needed to define the elements of either type unsigned or type signed. The second use clause makes the declarations of types unsigned and signed and their operators visible.

Interpretation of Unsigned and Signed Operators

The VHDL arithmetic operators **abs**, *, /, **mod**, **rem**, +, and – are overloaded in package NUMERIC_STD for types unsigned and signed. For these overloaded operators the operands can be both unsigned, both signed, unsigned and natural, or unsigned and integer.

The relational operators are overloaded to handle operands that are both unsigned, both signed, unsigned and natural, or unsigned and integer. When both operands are unsigned or both or signed they are treated as the numerical values they represent and the relation is evaluated accordingly.

Conversion between Std_logic_vector, Unsigned, and Signed

Although types std_logic_vector, unsigned, and signed all have elements that are type std_logic, they are different types. As a result, a value of one of these types cannot be directly assigned to one of the other types.

To make an assignment of the value of one type to one of the others, the type of the value being assigned must be converted to the target type. This conversion is easy to accomplish because in VHDL these types are considered *closely related*. Type conversion between closely related types is accomplished by simply using the name of the target type as if it were a function. That is, the name of the type we are converting to is used as the function name and the signal we are converting is placed inside parentheses as the function's argument. For example, if signal x is declared as type std_logic_vector and signal y is declared as type unsigned, and they are of equal length, each of the following assignments is illegal:

```
x <= y;    -- illegal assignment, type conflict
y <= x:    -- illegal assignment, type conflict
```

However, appropriate type conversions allow the following assignments to be made:

```
x <= std_logic_vector(y);    -- valid assignment
y <= unsigned(x):            -- valid assignment
```

"n-bit" Vectors with Std_logic Elements

Types std_logic_vector, unsigned, and signed are all vectors whose elements are type std_logic. Signals of these types having n-elements are often referred to as n-bit vectors. While this terminology is a misnomer, since the elements of the vector are not actually type bit but are instead type std_logic, it is commonly used. This terminology reflects the actuality of the hardware being described.

Functions for Conversion to and from Type Integer

Package NUMERIC_STD defines several functions that are useful in both nonarithmetic and arithmetic applications. Several of these functions are introduced as they are needed throughout the text. Others are discussed in Chapter 13 where package NUMERIC_STD and several other standard and de facto standard packages are the focus.

Three of these functions—*to_unsigned*, *to_signed*, and *to_interger*—are introduced here in Table 3.8.1. These functions allow conversion between types unsigned or signed and type integer. Since type integer is not similar to either type unsigned or type signed, these functions are needed to make such conversions.

For example, to convert type integer to type unsigned, the function `to_unsigned` is used. This function's first parameter is the non-negative integer to be converted. Its second parameter is the number of elements in the unsigned vector it is being

Table 3.8.1
Functions to convert between types unsigned and signed and integer.

Function Name	1st Parameter (value to convert)	2nd Parameter (width of target)	Result Type
to_unsigned	integer	integer	unsigned
to_signed	integer	integer	signed
to_integer	unsigned or signed	none	integer

converted to. If i is an integer and unsigned signal y has eight elements, the following statement assigns the value of integer i to unsigned signal y:

```
y <= to_unsigned(i,8);
```

To assign integer i to std_logic_vector signal x, the previous two techniques are combined:

```
x <= std_logic_vector(to_unsigned(i,8));
```

3.9 COMPOSITE LITERALS AND COMPOSITE CONSTANTS

A *composite literal* is an array or record of literal values. An appropriate composite literal can be assigned to an array or used within an expression that determines the value assigned to an array.

String Literals

A *string literal* is a sequence of graphic characters bounded by quotation marks (double quotes " "). If a quotation mark is to be represented in a string, it must be represented by a pair of adjacent quotation marks.

A string literal has a value that is the sequence of characters in the string, excluding the bounding quotation marks. The length of a string literal is the number of character values in the sequence, excluding the bounding quotation marks. Each pair of adjacent quotation marks in a string is counted as a single character in determining the string's length.

A string literal must fit on a single line. Longer string literals can be created by using the concatenation operator (&). A long string literal can be separated into substrings on different lines joined by concatenation operators.

The type of a string literal is determined from the context of its use. There are two kinds of string literals: character string and bit string.

Character String Literals

A *character string literal* is a string literal consisting of a sequence of characters. We will often use them in testbenches to display text on the console.

For example:

```
'test failed"     -- string literal of length 11
```

A character string can be assigned to a std_logic_vector as long as each character in the string is a character in the std_ulogic enumeration. For example:

```
outvector <= "ZZZZZZZZ";
```

Bit String Literals

A *bit string literal* is a special form of string literal used to represent a sequence of bit values as either binary, octal, or hexadecimal numeric data. A bit string literal can contain any of the characters from `0` through `9`, and `a` through `f`, or `A` through `F`.

A bit string literal must be prefixed by a `B`, `O`, or `X` to indicate binary, octal, or hexadecimal base, respectively. The base may be indicated in upper- or lowercase. Characters used in the bit string must be appropriate to the base. For binary, only 0 and 1 are allowed. For octal, only 0 through 7 are allowed. And, for hexadecimal, 0 through 9 and A through F are allowed. Since VHDL is case insensitive, a through f can also be used.

Regardless of the base used, a bit string literal has a value that is a string literal consisting of the character literals `'0'` and `'1'`. If the base specifier is `B`, the value of the bit string is the sequence of 0s and 1s given explicitly by the bit string itself. If the base specifier is `O`, the value of the bit string is the character string obtained by replacing each digit in the bit string by the equivalent sequence of three `'0'` and `'1'` character literals. For a base specifier of `X`, each character in the bit string is replaced by a sequence of four `'0'` and `'1'` characters.

The following two signal assignment statements all assign the value "01111001" to the vector `low_address`:

```
low_address <= B"01111001";
low_address <= X"79";
```

However, the following assignment is invalid:

```
low_address <= O"171";   -- invalid assignment
```

because the bit string literal has a length of 9 and the target's length is 8.

Underscores in Bit String Literals

One or more underscore characters `'_'` may be used between characters in a bit string literal to improve its readability, with the exception that two successive underscores cannot be used. Underscores are ignored in determining a bit string literal's value.

For example, the first of the previous assignments can be written as:

```
low_address <= B"0111_1001";
```

In contrast, the following statement is invalid:

```
low_address <= "0111_1001";   -- invalid assignment
```

because the composite literal is a string literal, not a bit string literal. In a string literal an underscore is a character in the string. So, the latter string literal has a length of 9, which is incompatible with the length of the target. In addition, the underscore is not a valid element value for the target type.

Since a literal does not have an explicit type, a composite literal's interpretation is based on its use. For example, given the following signal declarations:

```
signal x : bit_vector (7 downto 0);
signal y : std_logic_vector (7 downto 0);
```

and the following signal assignment statements:

```
x <= B"1011_0101";
y <= B"1011_0101";
```

The composite literal `B"1011_0101"` assigned to `x` is interpreted as a bit vector and the composite literal `B"1011_0101"` assigned to `y` is interpreted as a std_logic_vector.

Composite Constants

A composite constant is a constant that is an array or record. When a constant of an unconstrained array type is declared, its array bounds can be inferred from the expression used to initialize the constant. For example, in the constant declaration:

```
constant pattern : std_logic_vector := "11010101";
```

`pattern` is inferred to have a range from 0 to 7 since the index type for std_logic_vector is defined to be natural and natural is defined to be an ascending range starting at 0.

3.10 INTEGER TYPES

An *integer type* has values that are whole numbers in a specified range. The syntax for an integer type declaration is given in Figure 3.10.1.

An integer type declaration defines both a type and a subtype. The type is an anonymous type. *Anonymous types* cannot be referred to directly because they have no names. The range of an anonymous type is dependent on the implementation, but must include the range given in the integer type declaration.

The subtype is a named subtype of the anonymous base type. The name of the subtype is given by the corresponding type declaration and the range of the subtype

integer_type_declaration ::=
type identifier **is range** range;

range::= *simple_expression* **to** *simple_expression*
| *simple_expression* **downto** *simple_expression*

FIGURE 3.10.1
Simplified integer type declaration syntax.

is the given range. The range includes all of the contiguous integers between and including those specified by the two bounding expressions. These expressions must evaluate to integers.

For example, we can declare two integer types:

```
type ten is range 0 to 9;
type digit is range 0 to 9;
```

We could then declare the following signals

```
signal x, y: ten;
signal f, g: digit;
```

In the architecture we could have the statement:

```
x <= y;
```

because these signals are the same type. However, we could not have the statement:

```
x <= g;
```

because these signals are of different types. This is true even though the two types have the same set of integer values.

Predefined Type Integer

The only predefined integer type is the type *integer*. The range of type integer is implementation dependent, but must be at least –2,147,483,647 ($-2^{31}+1$) to +2,147,483,647 ($+2^{31}-1$), inclusive. The range is defined to be symmetric around 0. Type integer includes all of the whole numbers representable on a particular host computer.

Type integer is declared in package STANDARD as:

```
type integer is range implementation_defined;
```

We can restrict the range of type integer with a range constraint.

For example, we can declare two type integer subtypes:

```
subtype tenn is integer range 0 to 9;
subtype digitt is integer range 0 to 9;
```

we could then declare the following signals

```
signal x, y: tenn;
signal f, g: digitt;
```

In the architecture we could have the statement:

```
x <= y;
```

because these signals are the same type. However, this time we can also have the statement:

```
x <= g;
```

because both of these subtypes have type integer as their base type. For type checking of subtypes, it is the base type that is checked.

Predefined Integer Subtypes

There are two predefined type integer subtypes: natural and positive. These subtypes have been used previously in this chapter. They are defined in package STANDARD. The declaration of the integer subtype *natural* is:

```
subtype natural is integer range 0 to integer'high;
```

This declaration defines the type natural as a subtype of the predefined type integer constrained to the integers from 0 to the highest integer that can be represented in the particular VHDL implementation (`integer'high`). This constraint excludes all negative integers from this subtype.

The declaration of the type integer subtype *positive* is:

```
subtype positive is integer range 1 to integer'high;
```

When a design requires that an integer number not be negative, it is good design practice to use either the natural or positive subtype rather than integer.

Integer Literals

All integer literals are literals of an anonymous predefined type called *universal_integer*. For each integer type there exists an implicit conversion that converts a value of type universal_integer to the corresponding value (if any) of the integer type. Integer literals can be represented as decimal literals or based literals.

Decimal Literals

A *decimal literal* representing an integer can be simply a sequence of digits or exponential notation can be used. The implicit base is 10. An underline character can be inserted between adjacent digits for clarity without affecting the value of the literal. For example, the value 100 might be represented as:

100 or 1E2 or 1e2

An exponent for an integer literal must not have a minus sign.

Based Literals

A *based literal* uses an explicitly specified base between 2 and 16. The number is enclosed in number signs (#) and has the base as a prefix and an optional exponent as the suffix. An underline character can be inserted between adjacent digits of the literal for clarity without affecting its value. The base and exponent, if any, are in decimal notation. For example, the value 100 might be represented as:

10#100# or 2#1100100# or 2#0110_0100# or 16#64#

Integer Signals

We will encounter a number of uses for integers in this book. Some of these uses will be quite ordinary, others may at first be unexpected. Since a signal has been likened to a wire in a circuit, it may seem odd that a signal could be of type integer.

To illustrate this, consider the comparator in Listing 3.10.1.

Input ports `a` and `b` in this example are type integer and the output port is type boolean. As will be seen in the next section, neither integer nor boolean types are normally used for ports in designs that are to be synthesized, but they are sometimes used for local signals. When an integer signal is synthesized, it is translated by the synthesizer into a std_logic_vector. The length of the vector is just large enough to encode the numbers in the integer's range.

Synthesis of an Unconstrained Integer

An unconstrained integer is synthesized as 32 bits. For this reason, when used for synthesis, an integer signal should always be constrained to the smallest range of values required. A synthesizer will map constrained integers to the minimum number of bits required to represent the specified range. This reduces the amount of hardware synthesized.

LISTING 3.10.1
A 4-bit comparator with integer inputs and a boolean output.

```
entity compare_int is
   port (
      a: in integer range 0 to 15;
      b: in integer range 0 to 15;
      equal: out boolean);
end compare_int;

architecture behavioral of compare_int is
begin
   process (a, b)
   begin
      if a = b then
         equal <= true;
      else
         equal <= false;
      end if;
   end process;
end behavioral;
```

When assignments are made between integers of different sizes, the synthesis tool will automatically perform any necessary padding.

3.11 PORT TYPES FOR SYNTHESIS

When writing a VHDL program, we must choose a type for each port and signal. Sometimes use of abstract types can simplify coding and maintenance. However, since a design description will be synthesized, particular care must be taken in choosing its types, especially for ports.

A synthesizer must translate all types used for signals into types that can represent wires. Typically, a synthesizer converts all types to either std_logic or std_logic_vector. For example, when the comparator of Listing 3.10.1 is synthesized, the synthesizer translates the entity declaration from:

```
entity compare_int is
      port (
            a: in integer range 0 to 15;
            b: in integer range 0 to 15;
            equal: out boolean);
end compare_int;
```

to:

```
entity compare_int is
      port(
             a : in std_logic_vector(3 downto 0);
             b : in std_logic_vector(3 downto 0);
             equal :  out std_logic);
end compare_int;
```

in the post-synthesis model.

The integer input ports were translated to std_logic_vectors. Since the integer input ports have a range of 0 to 15, they can be represented by a four-element (bit) std_logic_vector. The boolean output port was translated to std_logic.

Nonabstract and Abstract Data Types

For purposes of the current discussion, it is helpful to categorize data types as either nonabstract or abstract, based on how they are treated by the synthesizer.

A *nonabstract type* is a type that the synthesizer directly uses to represent a wire. This category includes std_logic and std_logic vector.

An *abstract type* is a type that the synthesizer must convert to a nonabstract type to represent wires in the gate-level netlist that it produces. In the previous example, abstract ports of type integer were converted to nonabstract std_logic_vector ports and the abstract port of type boolean was converted to a nonabstract std_logic port.

Table 3.11.1
Type translations made by a synthesizer.

Pre-synthesis type	Post-synthesis type
bit	std_logic
bit_vector	std_logic_vector
boolean	std_logic
integer	std_logic_vector
enumeration	std_logic_vector
record	(record elements are split)
2-D array	(array elements are split)

Table 3.11.1 lists a number of abstract data types that can be used in models and the types they are translated to by a synthesizer.

Since the post-synthesis model has std_logic (std_logic or std_logic_vector) ports, it follows that the timing model derived from the post-synthesis model will also have std_logic ports.

Limiting Port Types to Std_logic or Std_logic_vector

Ideally, we want to use the same testbench for functional, post-synthesis, and timing simulations, with only the instantiated UUT model changed.

Since, as a result of the synthesis, the post-synthesis model and timing model will have std_logic ports, it is preferable that the original design description also have std_logic ports. This way a testbench that uses std_logic signals to connect the UUT can connect to any of the three UUT simulation models. Otherwise, two versions of the testbench are required, one with signals compatible to the abstract signals used in the design description and another with std_logic signals compatible with the post-synthesis and timing models.

For example, it might seem simpler to have used type bit for the ports of our half adder. If we had done so, our testbench for functional simulation would have to be written using type bit signals to connect to the half adder. After functional simulation is completed, the half adder would be synthesized. The post-synthesis half-adder model would have std_logic ports. This model could not be instantiated into the existing testbench as the UUT, because of the type conflicts between the testbench's type bit signals and post-synthesis models type std_logic ports. We would have to change the type bit signals in the testbench to std_logic, to connect to the post-synthesis model of the UUT.

Use of std_logic for ports provides other benefits as well. It allows our design descriptions to be easily used as components in larger designs. Std_logic is also used for ports in design entities produced by third-party intellectual property (IP) vendors. We can purchase these design entities and use them as components in our designs. For all of these reasons, std_logic or std_logic_vector are used for all ports in the design descriptions and testbenches that follow in this book.

If there are advantages in using abstract type signals in a design description, those signals are declared inside the description. If any of these abstract signals connect to ports, the description has to include statements to convert the nonabstract port types to the abstract types, and vice versa.

Limiting Port Index Order to Descending

As a practical matter, it is recommended that for ports that are vectors the port index ordering be specified as descending (`downto`) and that the lower bound be 0. Some synthesis and place-and-route tools will automatically reorder the index of a port specified as ascending in the design description to descending in the VHDL post-synthesis or timing model produced. When this model is instantiated as the UUT for a post-synthesis or timing simulation, results may be incorrect because the port elements are incorrectly referenced.

3.12 OPERATORS AND EXPRESSIONS

In the beginning of this chapter it was stated that an object's type defines the set of values it can assume and the set of operations that can be performed on those values. An *expression* is a formula that defines the computation of a value. Expressions are formed from object operands and operators.

Operator Precedence

VHDL's operators are listed in Figure 3.12.1 from higher to lower precedence. Operators on the same line are of equal precedence. Thus, the exponentiation (**), absolute value (`abs`), and complement (`not`) operators have the highest precedence. Operators of higher precedence are associated with their operands before operators of lower precedence. Where a sequence of operators are allowed, operators having the same precedence in an expression are associated with their operands in textual order from left to right. Parentheses may be used to control the order of association of operators and operands.

The predefined operators are declared in package STANDARD.

```
miscellaneous_operator ::= ** | abs | not
multiplying_operator ::= * | / | mod | rem
sign ::= + | -
adding_operator ::= + | - | &
shift_operator ::= sll | srl | sla | sra | rol | ror
relational_operator ::= = | /= | < | <= | > | >=
logical_operator ::= and | or | nand | nor | xor | xnor
```

FIGURE 3.12.1
VHDL operators listed from higher to lower precedence.

Not all of the operators in Figure 3.12.1 can be used with every data type. We will start to see which operators can be used with which types in the next chapter.

Locally or Globally Static Expressions

A *locally static expression* is an expression whose value can be completely determined when the design unit in which it appears is compiled. The value of a locally static expression depends only on declarations that are local to the design unit or to any packages used by the design unit.

In contrast, a *globally static expression* is an expression whose value can be determined only when the design hierarchy in which it appears is elaborated prior to simulation.

Mod Operator

Most of the operators in Figure 3.12.1 are familiar. However, operators **mod** and **rem**, which apply to integers, are described here in more detail. Further details regarding the other operators are given in later chapters when they are used with various types.

The modulus operator **mod** must satisfy the following relation:

$$A = B \times N + (A \text{ mod } B)$$

where A mod B has the sign of B and an absolute value less than the absolute value of B.

Rem Operator

The remainder operator **rem** is defined by the following relation:

$$A = (A/B) \times B + (A \text{ rem } B)$$

where (A rem B) has the sign of A and an absolute value less than the absolute value of B.

When the sign of A equals the sign of B, A mod B and A rem B are equal. When the signs of A and B differ, A mod B may not be equal to A rem B.

PROBLEMS

3.1 List and define the four classes of VHDL objects. Which of these classes are not common to conventional programming languages? Which classes are synthesizable and which are not? How are the classes that are not synthesizable used?

3.2 List and define the five kinds of VHDL types. Which types are used in design descriptions for synthesis? Where might the other types be used?

3.3 What is meant by strong typing in VHDL and why is it useful?

3.4 What name does VHDL use for a design entity's external signals? What name does VHDL use for a design entity's internal signals?

3.5 Where are a design entity's ports declared and what is their scope? Where are a design entity's signals declared and what is their scope?

3.6 A two-to-one multiplexer has two data inputs, `a` and `b`, and one control input, `s`. The multiplexer has a single output, `c`. Write an entity declaration, `mux`, for the multiplexer. The inputs and output are all type std_logic.

3.7 Write an entity declaration for a combinational circuit named `bcd2xs3` that converts a BCD input code for a decimal digit to an excess three (XS3) output code for a decimal digit. The input code is represented by four scalar signals of type std_logic. These signals have the identifiers `D`, `C`, `B`, and `A`, where `D` is the most significant bit. The output is represented by four scalar signals of type std_logic. The identifiers for the outputs are `XS3`, `XS2`, `XS1`, and `XS0`.

3.8 Write an entity declaration for a 4-to-2 priority encoder named `encode_4_to_2`. The inputs are four scalar std_logic signals named `x3_bar`, `x2_bar`, `x1_bar`, and `x0_bar`. The outputs are two std_logic signals named `y1` and `y0`.

3.9 Can a signal be assigned an initial value when it is declared? When simulated, will the signal start out with this initial value? If a signal is assigned an initial value in a design description, will that description be IEEE 1076.6 compliant? Can a synthesizer synthesize an initial value given a signal?

3.10 What is a scalar type? List the predefined scalar types. Where are these predefined types declared?

3.11 What is a discrete type? List the predefined discrete types. Which of these are synthesizable?

3.12 Write the implicit context clause that automatically precedes every design unit.

3.13 What is a subtype? What is a base type? Can a signal that is of a subtype be assigned to a signal that is of the base type of the subtype? Can a signal that is of a base type be assigned to a signal that is of a subtype of the base type? Under what condition will these kinds of assignments cause an error during a simulation?

3.14 What is the purpose of the type std_ulogic?

3.15 List the enumeration literals in type std_ulogic. For each enumeration literal indicate its state and strength.

3.16 What is the relationship between std_ulogic and std_logic? Under what condition(s) would each be preferred as the type of a signal in a design description?

3.17 Assume that a std_logic signal has two sources. For each of the following combinations of source values, what is the resolved value of the signal?

(a) `'0'` and `'Z'`
(b) `'1'` and `'X'`
(c) `'H'` and `'0'`
(d) `'U'` and `'Z'`
(e) `'0'` and `'1'`

3.18 Assume that a std_logic signal has three sources. For each of the following combinations of source values, what is the resolved value of the signal?

(a) `'0'`, `'Z'`, and `'Z'`
(b) `'0'`, `'1'`, and `'Z'`
(c) `'H'`, `'Z'`, and `'0'`
(d) `'1'`, `'1'`, and `'L'`
(e) `'0'`, `'U'`, and `'1'`

3.19 You are designing a system using std_logic signals, some of which represent bused signals. During a functional simulation which std_logic values would you expect to see and which values would be cause for concern?

3.20 Write a type declaration to create a new type named `mv3` that consists of the values `'0'`, `'1'`, and `'Z'` and has the resolution function `resolvmv3`. Can a concurrent signal assignment statement assign the value of a signal of type `mv3` to a signal of type std_logic?

3.21 Which of the following statements are true and which are false? Justify your answers.

(a) `'1'> '0'`(type std_logic)
(b) `'1'> '0'`(type character)
(c) `'H'> '1'`(type std_logic)
(d) `'A'> 'a'`(type character)
(e) `'1'> '0'`(type bit)

3.22 Write a statement or statements to declare constants `vcc` and `gnd`. Constant `vcc` has a value of `'1'` and constant `gnd` has a value of `'0'`. Both are type std_logic.

3.23 What are the two composite types in VHDL and what are their similarities and differences?

3.24 Write declarations for the three signals named `address_bus`, `data_bus`, and `control_bus`. Each signal is a std_logic_vector with a descending range ending in `'0'`. The number of elements in the vectors are 16, 8, and 6, respectively.

3.25 Write declarations for three constrained subtypes of std_logic_vector with descending ranges ending in `'0'`. Subtype `word` has 16 elements, subtype `dword` has 32 elements, and subtype `qword` has 64 elements. Write declarations for the signal `address` as type `dword` and the signal `data` as type `word`.

3.26 Write an entity declaration for the design entity `memory` represented by the following symbol. Its inputs and outputs are all type std_logic or std_logic_vector.

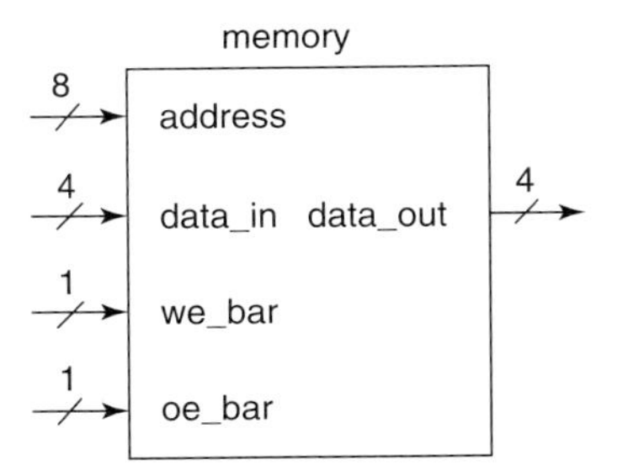

3.27 Given the following declarations:

```
signal a, b : std_logic_vector (3 downto 0);
signal p, q : std_logic_vector (7 downto 0);
```

Give the value of the signal after the following assignments:

```
a <= (2 => '1', 1 => '0', 0 => '0', 3 => '1');
p <= (7|5|3 => '1', others => '0');
b <= (others => '0');
q <= (6 downto 4 => '0', others => '1');
```

3.28 Given the following declarations:

```
constant m : unsigned := "1101";
constant p : signed := "1101";
constant i : integer := 145;
constant r : integer := -31;
constant q : integer := 19;
```

For the following assignment statements give the type, number of elements, and value of the target signal:

```
f <= to_unsigned(i, 8);
g <= to_signed(r,8);
h <= std_logic_vector(to_unsigned(q, 5));
j <= to_integer(m);
k <= to_integer(p);
```

3.29 Signals a, b, c, d, and e are std_logic_vectors with an index from 7 down to 0. What is the value, expressed as a string literal consisting of only 0s and 1s, assigned to each target signal? If an assignment is invalid, explain why.

```
a <= X"8_2";
b <= b"01101101";
c <= "00" & o"46";
d <= "1101_0011";
e <= x"AC";
```

3.30 Write each of the following integer literals as a simple decimal integer.

(a) `1E3`
(b) `3#120#`
(c) `2#0110_1101#`
(d) `16#33#`
(e) `16#af#`

3.31 If the following integers are synthesized, what will be the lengths of the std_logic_vectors used to implement them?

(a) `signal u : integer range 0 to 7;`
(b) `signal v : integer range 0 to 31;`
(c) `signal w : integer range 0 to 1;`
(d) `signal x : integer range 0 to 6;`
(e) `signal yu : integer range 0 to 18;`

Chapter 4

Dataflow Style Combinational Design

Combinational systems have no memory. A combinational system's outputs are a function of only its present input values.

In traditional combinational design, the task of logic minimization constitutes a substantial amount of the effort involved. When using VHDL, logic minimization is performed automatically by the synthesizer and place-and-route tools. This results in a significant reduction in design effort.

VHDL provides a full set of logical operators that can be used on scalar or array objects. Using these operators, we can easily write descriptions for combinational systems.

As previously shown, combinational system descriptions can be written in dataflow, behavioral, or structural styles. This chapter discusses in greater detail writing dataflow style descriptions. Selected and conditional concurrent signal assignment statements, used in dataflow style descriptions, are introduced in this chapter.

4.1 LOGICAL OPERATORS

VHDL provides the logical operators **`and`**, **`or`**, **`nand`**, **`nor`**, **`xor`**, and **`xnor`** (Figure 4.1.1). The logical operator **`not`** is also provided, but in VHDL it is classified as a miscellaneous operator. When we refer to logical operators in this book, we generally include the **`not`** operator.

The logical operators, which are predefined in package STANDARD for types bit and boolean, are listed in Table 4.1.1 for type bit. They have the same functional interpretation as the correspondingly named logic gates.

logical_operator ::=
and | **or** | **nand** | **nor** | **xor** | **xnor**

FIGURE 4.1.1
VHDL logical operators.

Overloaded Logical Operators

Package STD_LOGIC_1164 provides functions to define these same logical operators for type std_ulogic. These functions *overload* (extend) the definitions of the logical operators to operands of type std_ulogic and, by extension, to std_logic. When a VHDL compiler encounters a logical operator, it determines which version of that operator to use by the operands' types.

As defined for std_ulogic operands, the logical operators can return any one of the values `'U'`, `'X'`, `'0'`, or `'1'`. However, if operand values are limited to `'0'` and `'1'`, the result is also limited to `'0'` or `'1'`. With this constraint, Table 4.1.1 also represents the logical operations for std_logic. Thus, we can use these operators on std_logic signals to design combinational systems.

The inputs and output of the two input OR gate in Listing 2.9.1 from Chapter 2 are declared as std_logic:

```
entity or_2 is
      port (a, b : in std_logic;
      or_out : out std_logic);
end or_2;

architecture dataflow of or_2 is
begin
      or_out <= a or b;
end dataflow;
```

Operands of the logical operators in Figure 4.1.1, as defined in packages STANDARD and STD_LOGIC_1164, must be of the same type. Because `a` and `b` are std_logic, the OR operation in the signal assignment statement:

```
or_out <= a or b;
```

Table 4.1.1
Logical operators defined for type bit.

A	B	not A	A and B	A nand B	A or B	A nor B	A xor B	A xnor B
0	0	1	0	1	0	1	0	1
0	1	1	0	1	1	0	1	0
1	0	0	0	1	1	0	1	0
1	1	0	1	0	1	0	0	1

is the overloaded OR operation defined in package STD_LOGIC_1164. If the input values to entity `or_2` are limited to `'0'` and `'1'`, its output is either `'0'` or `'1'`.

Operator Precedence

Unlike in Boolean algebra and most high-level programming languages, all the operators in Table 4.1.1, except for **not**, have the same level of precedence. As previously stated, in VHDL **not** is classified as a miscellaneous operator. Miscellaneous operators have the highest precedence. In Figure 3.12.1, all the VHDL operators are listed, from highest precedence to lowest. Operators on the same row have the same precedence. As shown in Figure 3.12.1, the logical operators have the lowest precedence of all VHDL operators.

Order of Evaluation

In the absence of parentheses, operators at the same precedence level are associated with their operands in textual order from left to right. Parentheses must be used in an expression to force any other order of evaluation. Expressions within parentheses are evaluated first.

For example, if we want to form the XOR of `a` and `b` using only AND, OR, and NOT operations and assign the result to `c`, the following statement is sufficient in many high-level languages:

```
c = a and not b or not a and b
```

However, in VHDL, since the **and** operator does not have higher precedence than the **or** operator, the previous expression is interpreted as:

```
c = ((a and (not b)) or (not a)) and b
```

We must use parentheses to ensure that our expressions are interpreted as we intend. Accordingly, the previous assignment statement must be written in VHDL as:

```
c <= (a and not b) or (not a and b);
```

Associative Logical Operators

Because the predefined versions of the operators **and**, **or**, **xor**, and **xnor** are defined as *associative,* a sequence of any one of these operators (whether predefined or user-defined) is allowed. Thus, to create a three-input AND using the **and** operator, we can write:

```
f <= a and b and c;
```

Nonassociative Logical Operators

In contrast, a sequence of either **nand** or **nor** operators (whether predefined or user-defined) is not allowed, because the predefined versions of these operators are nonassociative. Accordingly, the following attempt to produce a three-input NAND using the **nand** operator creates a compiler error:

```
g <= a nand b nand c;   -- invalid
```

Instead, we could write:

```
g <= not (a and b and c);   -- valid
```

which is acceptable.

Logical Operations on Array Elements

The logical operators can be used to operate on elements of arrays by indexing specific elements. For example, the following design entity accomplishes the AND of the corresponding elements from two input vectors:

```
entity and_vector1 is
      port (x, y : in std_logic_vector(3 downto 0);
            f : out std_logic_vector(3 downto 0));
end and_vector1;

architecture dataflow1 of and_vector1 is
begin
      f(3) <= x(3) and y(3);
      f(2) <= x(2) and y(2);
      f(1) <= x(1) and y(1);
      f(0) <= x(0) and y(0);
end dataflow1;
```

Logical Operations on Entire Arrays

When dealing with two arrays of the same length and type, the logical operators can be applied to the entire arrays. The result is an array of the same length. Each element of the resultant array is obtained by applying the logical operation to the matching (corresponding) elements in the operand arrays. Taking advantage of this, the previous assignments can be accomplished with a single signal assignment statement:

```
architecture dataflow2 of and_vector1 is
begin
      f <= x and y;
end dataflow2;
```

Array Element Matching for Logical Operations

For logical operations, the matching of operand array elements is positional, starting with the leftmost position, and is not based on the index. Assume that the index range for input vector `x` is defined as descending and the index range for an input vector `y` is defined as ascending, as in the following entity:

```
entity and_vector2 is
      port (x : in std_logic_vector(3 downto 0);
            y : in std_logic_vector(0 to 3);
            f : out std_logic_vector(3 downto 0));
end and_vector2;
```

then the architecture

```
architecture dataflow1 of and_vector2 is
begin
      f <= x and y;
end dataflow1;
```

computes a result that is equivalent to the architecture:

```
architecture dataflow2 of and_vector2 is
begin
      f(3) <= x(3) and y(0);
      f(2) <= x(2) and y(1);
      f(1) <= x(1) and y(2);
      f(0) <= x(0) and y(3);
end dataflow2;
```

Element `f(3)` is the AND of `x(3)` and `y(0)`, because each of these is the leftmost element of its array. It follows that `f(0)` is the AND of `x(0)` and `y(3)` because each of these is the rightmost element of its array.

4.2 SIGNAL ASSIGNMENTS IN DATAFLOW STYLE ARCHITECTURES

In previous examples, we used signal assignment statements that assigned a scalar or composite value to a signal.

The full syntax for a signal assignment statement is rather complex. However, it is substantially simplified for assignment statements that are to be synthesized. The syntax for synthesizable assignment statements is given in Figure 4.2.1.

A synthesizable signal assignment statement has the form:

```
target <= value_expression;
```

where the target can be a signal name or an aggregate.

We discuss more complex forms of the signal assignment statement, useful for timing models and testbenches, in Chapters 6 and 7.

```
signal_assignment_statement ::=
target <= value_expression ;

target ::= name | aggregate
```

FIGURE 4.2.1
Syntax for synthesizable signal assignment statement.

A signal assignment statement placed in the statement part of an architecture is a concurrent statement. We have previously used a form of concurrent signal assignment statement where the value expression is a Boolean expression.

There are actually two forms of concurrent signal assignment statements: selected and conditional. A Boolean expression concurrent signal assignment statement is a simplified form of a conditional signal assignment statement. However, for instructional purposes we treat the Boolean expression form separately. As a result, three kinds of concurrent signal assignment statements are covered in this chapter:

- Concurrent signal assignment statement using a Boolean expression
- Selected signal assignment statement
- Conditional signal assignment statement

Signal Assignments Using Boolean Expressions

As we have seen, some combinational systems are easily described using Boolean expressions. A concurrent signal assignment statement using a Boolean expression assigns the value of the expression to a signal. For example:

```
f <= a and b;
```

A Multiplexer Using Boolean Expressions

Figure 4.2.2 is a logic symbol for a 4-to-1 multiplexer (functionally equivalent to one-half of a 74HC153). When the enable input `g_bar` is `'0'`, select inputs `b` and `a` select the value at one of the data inputs—`c3`, `c2`, `c1`, or `c0`—to be the value of output `y`. Otherwise, `y` is `'0'`.

A description of this multiplexer using a single Boolean expression concurrent signal assignment statement is shown in Listing 4.2.1.

If we chose to treat inputs `c3`, `c2`, `c1`, and `c0` as a single vector, we could rewrite the multiplexer design entity as shown in Listing 4.2.2.

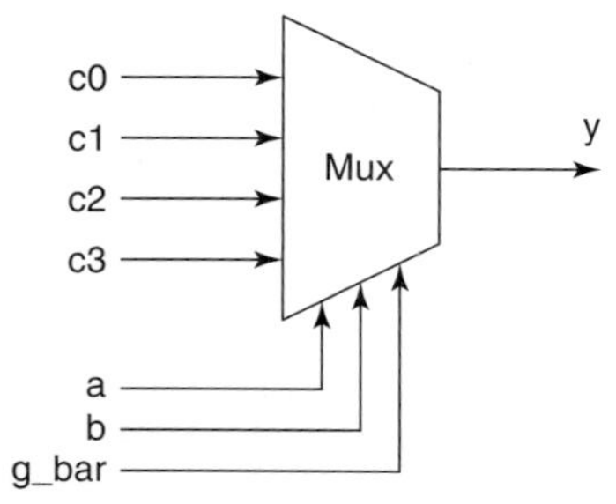

FIGURE 4.2.2
Logic symbol for a 4-to-1 multiplexer.

LISTING 4.2.1
A 4-to-1 multiplexer using a Boolean expression.

```
library ieee;
use ieee.std_logic_1164.all;

entity mux4to1 is
   port (c3, c2, c1, c0: in std_logic;
      g_bar, b, a: in std_logic;
      y: out std_logic);
end mux4to1;

architecture dataflow of mux4to1 is
   begin

   y <= not g_bar and (
      (not b and not a and c0)
      or (not b and a and c1)
      or (b and not a and c2)
      or (b and a and c3));

end dataflow;
```

LISTING 4.2.2
A 4-to-1 multiplexer with data inputs treated as a vector.

```
library ieee;
use ieee.std_logic_1164.all;

entity mux4to1 is
   port (c: in std_logic_vector(3 downto 0);
      g_bar, b, a: in std_logic;
      y: out std_logic);
end mux4to1;

architecture dataflow2 of mux4to1 is
   begin

   y <= not g_bar and (
      (not b and not a and c(0))
      or (not b and a and c(1))
      or (b and not a and c(2))
      or (b and a and c(3)));

end dataflow2;
```

Note the differences in notation used to represent inputs as elements of a vector in the signal assignment statement in Listing 4.2.2 and as scalars in Listing 4.2.1.

4.3 SELECTED SIGNAL ASSIGNMENT

A *selected signal assignment* statement, also called a *with-select-when* statement, allows one of several possible values to be assigned to a signal based on a *select expression*. A simplified form of the selected signal assignment syntax is given in Figure 4.3.1.

Select Expression

The select expression must be a discrete type or a one-dimensional character array type. Accordingly, there will be only a finite number of possible values for the select expression.

Selected Signal Assignment Choices

The choices specified must be *locally static* values of the select expression. This means that their values must be determinable when the architecture containing the selected signal assignment statement is compiled.

One and only one of the choices in a selected signal assignment statement must match the value of the select expression. Accordingly, the choices listed must be mutually exclusive.

Listing Choices

A list of choices can be associated with a single *value*_expression by separating each choice in the list using the vertical bar symbol (|).

Describing Combinational Logic Using a Selected Signal Assignment

To describe combinational logic, the choices listed in a selected signal assignment must be all inclusive. That is, they must include every possible value of the select expression. Exactly one choice must appear for each possible value of the select expression. This ensures that each time the assignment statement is executed, its target is assigned a value.

```
selected_signal_assignment ::=
with expression select
  target <= { value_expression when choices , }
            value_expression  when choices ;
```

FIGURE 4.3.1
Simplified syntax for selected signal assignment.

The form of selected signal assignment statement that we will most often use is:

```
with expression select
target <=  value_expression1 when choices1,
           value_expression2 when choices2,
                ...
           value_expressionn when others;
```

A selected signal assignment statement is evaluated by first determining the value of the select expression. This value is then simultaneously compared with the values of the choices. The value of one and only one choice must equal the value of the select expression. The value expression associated with this choice is assigned to the target.

Keyword Others as a Choice

When using std_logic to synthesize a combinational circuit, the last when clause must use the keyword **others** to assign a default value to the target for all values of the select expression that are not explicitly listed in the previous when clauses.

For example, the 4-to-1 multiplexer is described using a selected signal assignment in Listing 4.3.1.

LISTING 4.3.1
Using an aggregate in a select expression.

```
library ieee;
use ieee.std_logic_1164.all;

entity mux4to1 is
   port (c3, c2, c1, c0: in std_logic;
      g_bar, b, a: in std_logic;
      y: out std_logic);
end mux4to1;

architecture selected of mux4to1 is
begin

   with std_logic_vector'(g_bar, b, a) select
   y <= c0 when "000",
        c1 when "001",
        c2 when "010",
        c3 when "011",
        '0' when others;     -- default assignment

end selected;
```

Type Qualification

In Listing 4.3.1, we want to use an aggregate formed from the enable and select input signals as the select expression:

```
(g_bar, b, a);
```

However, if we use this aggregate alone, the compiler cannot tell the aggregate's type from the context. The compiler does not simply assume that an aggregate of std_logic elements is type std_logic_vector, since there are other array types, such as unsigned, that have std_logic elements.

We must explicitly specify the aggregate's type. This is accomplished using a type qualification. A *type qualification* consists of writing the intended type name followed by a single quote character preceding an expression enclosed in parentheses. The single quote (`'`) is sometimes referred to as a *tick*. For example:

```
std_logic_vector'(g_bar, b, a)
```

specifies that the aggregate is type std_logic_vector.

If the multiplexer is enabled (`g_bar` = `'0'`), and inputs `b`, `a` have the values `'1'` and `'0'`, respectively, the select expression is the std_logic_vector `"010"`, and the value of `c2` is assigned to `y`.

Because every possible value of the select expression must be accounted for in the listed choices, the last clause of the selected signal assignment statement uses the keyword **`others`** to assign the value of `'0'` to `y` for the 725 values of the select expression not explicitly listed. While only 8 of the input combinations make sense to a synthesis tool, for the code to be strictly VHDL compliant it must account for all possible values.

4.4 TYPE BOOLEAN AND THE RELATIONAL OPERATORS

Conditional signal assignments, discussed in the next section, require evaluation of one or more conditions. A *condition* is an expression involving objects and relational operators. The result from evaluating a condition is type boolean.

Type boolean is an enumeration type predefined in package STANDARD as:

```
type boolean is (false, true);
```

Type boolean is an abstract type and is not used to directly represent signals in synthesizable descriptions. In VHDL, boolean and bit are different types, so boolean values `true` and `false` are not equivalent to bit values `'1'` and `'0'`.

The logical operators also apply to type boolean. Table 4.1.1 can be used to define the logical operators for type boolean by replacing the 1s and 0s with true and false, respectively.

Table 4.4.1
VHDL relational operators.

Operator	Operation	Left Operand	Right Operand	Result
=	equality	any except file	same as left	boolean
/=	inequality			
<	less than	scalar or	same as left	boolean
<=	less than or equal	composite		
>	greater than	discrete type		
>=	greater than or equal			

Relational Operators

Package STANDARD also implicitly defines a set of *relational operators,* as shown in Table 4.4.1.

Relational operators are used to test the equality, inequality, or ordering of two scalar types or composites of discrete types. Operand types must be the same. The result of a relational operation is always type boolean.

The equality and inequality operators are defined for all types except file. In contrast, the other relational operators are only defined for scalars or composites of any discrete type. Since package STD_LOGIC_1164 does not overload any of the relational operators, the predefined interpretations of these operators apply to type std_logic.

Relational Operations on Enumerations

For scalars of enumeration types the interpretation of equality and inequality is straightforward. For example, the equality operator returns `true` if the two operands are the same and returns `false` otherwise.

Interpretations for the operators less than and greater than are based on the operands' position numbers in the enumeration type definition. The leftmost literal in an enumeration listing is in position 0. The position of each additional literal is one higher than the literal to its left. As a result, each enumeration type definition defines an ascending range of position numbers. When comparing two scalar enumeration literals, the one with a higher position number in the type definition is greater than the other. Thus, in std_logic, `'Z'` (with position number 4) is greater than `'1'` (with position number 3).

Relational Operations on Arrays

The interpretation of relational operators on arrays is a little more complex. However, we are interested only in the interpretation as it applies to one-dimensional arrays of equal length. For equality, two arrays are equal if each pair of matching elements in the two arrays is equal. If any pair of matching elements is not equal, then the two arrays are not equal.

Ordering of One-dimensional Arrays

Evaluation of the ordering operators applied to one-dimensional arrays of equal length is done starting with the leftmost elements of the arrays and preceding to the right until the determination can be made. For example, given two signal vectors:

```
signal a, b : std_logic_vectors (3 downto 0);
```

To determine if `a < b`, element `a(3)` is first compared to `b(3)`. If `a(3)` is less than `b(3)`, then `a < b`.

However, if `a(3) = b(3)`, then `a(2)` is compared with `b(2)`. If `a(2) < b(2)`, then `a < b`. This process is continued, comparing the next corresponding elements to the right until the determination can be made.

Less Than or Equal Operator (<=)

The evaluation of `a <= b` is `true` if either `a < b` or `a = b` is `true`.

Note that the interpretation of the operator `<=` as either the "less than or equal" operator or as the "assignment" operator is based on the context in which it is used.

A description of a 4-bit magnitude comparator using relational operators acting on vectors (one-dimensional arrays) was given in Listing 3.7.1.

4.5 CONDITIONAL SIGNAL ASSIGNMENT

A *conditional signal assignment* allows a signal to be assigned a value based on a set of conditions. The conditions are expressions involving relational operators. Conditional assignment is accomplished using a conditional signal assignment statement, also called a *when-else* statement. A simplified form of the conditional signal assignment syntax is given in Figure 4.5.1.

When there are zero occurrences of the "*value*_expression **when** condition **else**" clause and no final "**when** condition" clause, this syntax reduces to:

```
target <= value_expression;
```

which is the syntax of the simple Boolean signal assignment statement.

conditional_signal_assignment ::=
target <= { *value*_expression **when** condition **else** }
    *value*_expression [**when** condition];

FIGURE 4.5.1
Simplified syntax for conditional signal assignment.

Describing Combinational Logic Using a Conditional Signal Assignment

To describe combinational logic using a conditional signal assignment statement, the last value_expression must not have an associated when condition. This ensures that every time the conditional signal assignment statement is executed its target is assigned a value.

The form of conditional signal assignment that we will most often use is:

```
target <=   value_expression1 when condition1 else
      value_expression2 when condition2 else
      ...
      value_expressionn-1 when conditionn-1 else
      value_expressionn;
```

A conditional signal assignment statement is evaluated by first evaluating its first condition. If this condition evaluates `true`, the associated value expression is assigned to the target and the execution is complete. If the first condition evaluates `false`, the next condition is evaluated. This sequence of evaluations is continued until a condition evaluates `true`, at which point its associated value expression is assigned to the target and execution is complete. If none of the conditions evaluates `true`, and the last value expression does not have a when condition clause, this value is assigned to the target.

For example, Listing 4.5.1 illustrates the use of three separate conditional signal assignment statements, each containing one occurrence of the "*value*`_expression` **`when`** condition **`else`**" clause, to describe a 4-bit comparator that has three outputs.

LISTING 4.5.1
A 4-bit comparator using multiple conditional signal assignment statements.

```
library ieee;
use ieee.std_logic_1164.all;

entity mag_comp is
   port ( p, q : in std_logic_vector (3 downto 0);
      p_gt_q_bar, p_eq_q_bar, p_lt_q_bar : out std_logic);
end mag_comp;

architecture conditional of mag_comp is
begin

   p_gt_q_bar <= '0' when p > q else '1';
   p_eq_q_bar <= '0' when p = q else '1';
   p_lt_q_bar <= '0' when p < q else '1';

end conditional;
```

This comparator uses relational operators to compare the 4-bit std_logic_vectors `p` and `q`. These std_logic_vectors are interpreted as representing unsigned binary values. The comparator has three outputs, each asserted low. Output `p_gt_q_bar` is asserted if `p` is greater than `q`. Output `p_lt_q_bar` is asserted if `p` is less than `q`. Finally, output `p_eq_q_bar` is asserted if `p` is equal to `q`.

As another example, the 4-to-1 multiplexer is described using a single conditional signal assignment statement with multiple conditions in Listing 4.5.2.

A signal `tmp` is declared that is a three-element std_logic_vector. Port signals `g_bar`, `b`, and `c` are aggregated and assigned to `tmp`.

The conditional signal assignment statement compares aggregate `tmp` to different literal vector values, as shown in Listing 4.5.2. For example, the third condition is true if `tmp = "010"`, which indicates that `g_bar = '0'`, `b = '1'`, and `a = '0'`.

In this example, using the signal `tmp` is more concise than repeatedly using the aggregate with a std_logic_vector type qualifier:

```
std_logic_vector'(g_bar, b, a)
```

Avoiding Implied Latches

Since our goal is to synthesize a combinational system, it is critical that the last value expression provides a default value for the situation where the input value does not cause any of the specified conditions to be true. This means that the last *value*_expression must not have an associated when condition clause.

LISTING 4.5.2

A 4-to-1 multiplexer using an aggregate for control and select inputs.

```
library ieee;
use ieee.std_logic_1164.all;

entity mux4to1 is
   port (c0, c1, c2, c3: in std_logic;
      g_bar, b, a: in std_logic;
      y: out std_logic);
end mux4to1;

architecture conditional of mux4to1 is
   signal tmp : std_logic_vector(2 downto 0);
   begin
   tmp <= (g_bar, b, a);
   y <= c0 when tmp = "000" else
      c1 when tmp = "001" else
      c2 when tmp = "010" else
      c3 when tmp = "011" else
      '0';     -- default assignment

end conditional;
```

For example, if the conditional signal assignment is written as:

```
y <= c0 when tmp = "000" else
     c1 when tmp = "001" else
     c2 when tmp = "010" else
     c3 when tmp = "011";
```

The VHDL interpretation of this construct is that, for the four specified values of `tmp`, a new value should be assigned to `y`. However, for all other values of `tmp` there should be no new value assigned to `y`. This implies that the `y` should retain its previous value. Thus, a synthesizer's interpretation is that the circuit needs to remember the previous value of `y`. As a result, the synthesizer generates a latch at the output of the combinational logic to store the value of `y`.

Figure 4.5.2 shows the logic synthesized for a modified version of Listing 4.5.2 that uses the previous conditional signal assignment, where the last value expression has a when clause.

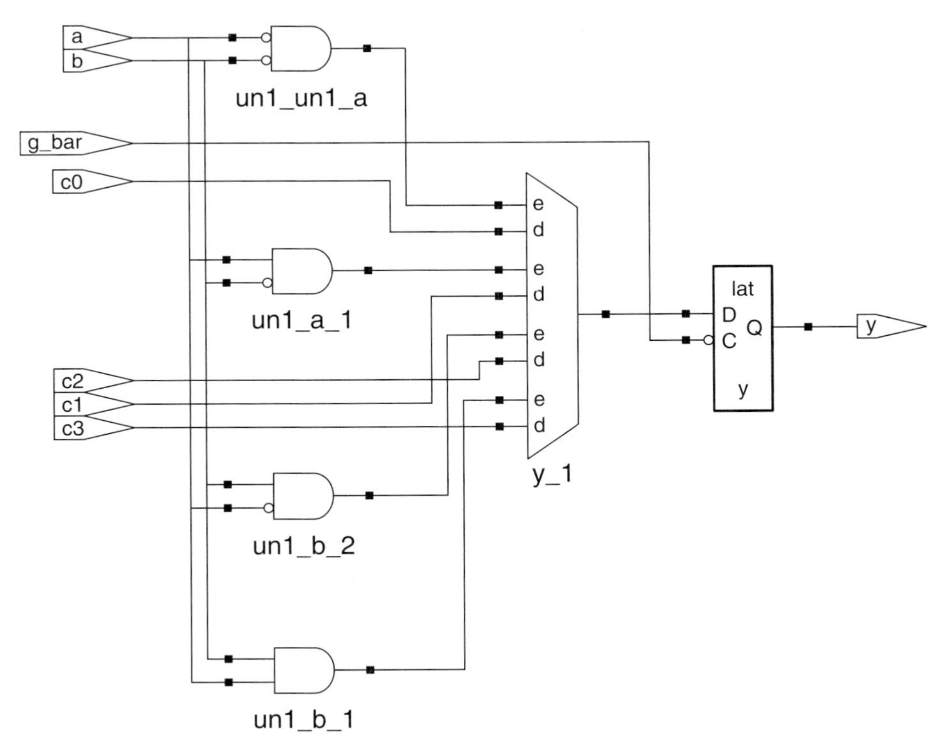

FIGURE 4.5.2
Synthesized multiplexer logic with undesired implied latch.

LISTING 4.5.3
Two-input AND using conditional signal assignment without a default condition.

```
library ieee;
use ieee.std_logic_1164.all;

entity and_2_bad is
   port (
      a: in std_logic;
      b: in std_logic;
      c: out std_logic);
end and_2_bad;


architecture and_2_bad of and_2_bad is
   signal tmp : std_logic_vector(1 downto 0);
   begin
   tmp <= (a, b);
    c <= '0' when tmp = "00" else
       '0' when tmp = "01" else
       '0' when tmp = "10" else
       '1' when tmp = "11";
end and_2_bad;
```

The form of multiplexer symbol used in Figure 4.5.2 has each data input (d) paired with an enable input (e). Only one enable input is asserted at any time. The data at the data input whose associated enable is asserted appears at the multiplexer's output.

The memory element synthesized when a default assignment is omitted is a latch. The synthesized circuit still works properly when `g_bar` is asserted (because then the latch is transparent). However, instead of producing `y = '0'` when `g_bar` is unasserted, the circuit holds `y` at the value it had at the time `g_bar` was unasserted. Of course, the unwanted latch also makes the circuit slower.

The All Possible Input Combinations Trap

Consider the description in Listing 4.5.3 for a two-input AND gate. This description uses a conditional signal assignment statement that has a final when condition clause (Figure 4.5.1). There is a subtle trap here for the unwary.

At first glance, it may appear that all the possible input value combinations for a two-input gate have been specified as conditions in the conditional signal assignment statement. However, this design may also cause memory to be synthesized.

The catch is that all input combinations have been specified only if the gate's inputs are of a type (such as bit) that has only the values `'0'` and `'1'`. However, this gate's inputs are type std_logic. Each std_logic input has nine possible values. So, for a two-input gate there are $9^2 = 81$ different input combinations. Only four have been listed. Some synthesizers may add a latch to remember the previous value of `y` for those cases where the input is one of the 77 other possible values. Other synthesizers may correctly interpret the intent of what was written and synthesize only an AND gate.

4.6 PRIORITY ENCODERS

Three forms of concurrent signal assignment statements have been introduced. For a combinational design, any of these constructs might be used. However, for some functions, one may be more concise or intuitive than the others. For a particular function, the circuits synthesized from all of these constructs are functionally equivalent.

Inherent Priority of Conditional Signal Assignment

Because a conditional signal assignment statement assigns the value associated with its first condition that is true, it inherently establishes a priority. This makes a conditional signal assignment statement ideal for describing priority circuits.

Consider a 4-to-2 priority encoder that has four inputs `i3`, `i2`, `i1`, and `i0` and outputs `a1`, `a0`. The outputs are the complement of the binary encoding of the suffix of the highest priority input asserted. Inputs are asserted low. Functionally, this encoder is a smaller and more limited version of the 74HC148 eight-line to three-line priority encoder.

Input `i3` has the highest priority and input `i0` has the lowest. For example, if inputs `i3` and `i2` are simultaneously asserted, the output is `"00"`. This is so because `"00"` is the complement of `"11"`, which is the binary encoding of the numeric suffix `3` of `i3`. The VHDL code for this circuit is given in Listing 4.6.1.

LISTING 4.6.1
A 4-to-2 priority encoder using a conditional signal assignment.

```
library ieee;
use ieee.std_logic_1164.all;

entity encoder is
   port ( i3, i2, i1, i0 : in std_logic;
      a : out std_logic_vector(1 downto 0));
end;

architecture csa_cond of encoder is

   --signal outputs: std_logic_vector(1 downto 0);
   begin

   a <= "00" when i3 = '0' else
      "01" when i2 = '0' else
      "10" when i1 = '0' else
      "11";
end csa_cond;
```

The conditional signal assignment statement first tests whether `i3 = '0'`. If this is true, the output is assigned the value `"00"`. Only if `i3` does not equal `'0'` is the second condition tested. The second condition tests whether `i2 = '0'`. If `i2 = '0'`, then the output is assigned the value `"01"`. However, since the second condition is only tested after the first condition fails, the output is only assigned the value `"01"` when `i3 = '1'` and `i2 = '0'`. This property of the conditional signal assignment statement provides inherent priority.

4.7 DON'T CARE INPUTS AND OUTPUTS

As previously discussed, std_logic values `'1'`, `'H'`, `'0'`, and `'L'` are interpreted as representing one of two logic levels. In contrast, std_logic metalogic values `'U'`, `'X'`, `'W'`, and `'-'` are interpreted as defining the behavior of a model itself, rather than the behavior of the hardware being synthesized.

Don't Care Value

In traditional logic design, a "don't care" is used to specify a value in a situation where we don't care whether the value is a `'0'` or a `'1'`. In VHDL, `'-'` represents the "don't care" value. There are some important differences in interpretation of a "don't care" in VHDL compared to traditional logic design. Importantly, the "don't care" value is treated differently by a simulator than by a synthesizer.

Simulator Interpretation of a `'-'`

Simulators treat a `'-'` input or output as a literal value. So, to a simulator, `'-'` literally means the character `'-'`, rather than meaning that the character `'-'` can be replaced by either a `'0'` or `'1'`. When `'-'` appears as an output in a description, a simulator's interpretation is that this symbol is being used to indicate that a choice for that output's state has not been made.

Synthesizer Interpretation of a `'-'`

Normally, a synthesizer is supposed to synthesize logic that behaves the same way as does the design description's simulation. A synthesizer also treats a `'-'` input as a literal value.

In contrast, a synthesizer treats a `'-'` output as a true "don't care" condition for logic minimization purposes. This allows the synthesizer to take advantage of any don't care output assignments to minimize the logic it synthesizes.

IEEE Std 1076.3 Interpretation of Metalogical Values

The IEEE Standard Synthesis Package (IEEE Std 1076.3) requires that, if a metalogical value occurs as a choice in a statement interpreted by a synthesizer, the synthesizer shall interpret the choice as one that will never occur. For a metalogical value in an assignment statement, the synthesizer shall accept the value, but is not required to provide any particular interpretation.

Don't Care Inputs

Listing 4.7.1 shows an attempt to use "don't cares" to describe a 4-to-2 priority encoder.

A simulator compiles this design without any syntax errors. However, when we simulate this description by applying all input combinations where each input is either a `'0'` or a `'1'`, the output is always `"11"`. This result occurs because the literal input values (`"0---"` or `"10--"` or `"110-"`) that can cause an output other than `"11"` are never applied as inputs during the simulation.

When this design is synthesized, the synthesizer's compiler also generates no syntax errors. However, because the synthesizer interprets conditions involving a `'-'` input value (such as `"0---"`) as conditions that can never occur in a physical circuit, it generates warnings that the inputs are unused and synthesizes a circuit with its outputs hardwired to `"11"`.

Std_match Function

But what if, for the purpose of making comparisons of std_logic_vectors, we want `'-'` inputs to be treated by both the simulator and synthesizer as "don't cares" in the traditional sense? A function is available in package NUMERIC_STD that can be used to accomplish this. The function is named `std_match`. To use this function, a use clause specifying package NUMERIC_STD must be included in the description. Package NUMERIC_STD is discussed further in Chapter 13.

LISTING 4.7.1
A 4-to-2 priority encoder—an incorrect approach using "don't care" inputs.

```
library ieee;
use ieee.std_logic_1164.all;

entity encoder is
   port ( i3, i2, i1, i0 : in std_logic;
      a : out std_logic_vector(1 downto 0));
end;

architecture csa_cond of encoder is

signal tempi: std_logic_vector(3 downto 0);
begin

   tempi <= (i3, i2, i1, i0);

   a <= "00" when tempi = "0---" else
         "01" when tempi = "10--" else
         "10" when tempi = "110-" else
         "11";

end csa_cond;
```

The `std_match` function can be used to compare two scalars, which are type std_ulogic, or two vectors whose elements are type std_ulogic. Thus, since std_logic is a subtype of std_ulogic, this function can be used to compare std_logic vectors.

If the vectors match, a boolean `true` is returned. Element values in corresponding positions in a vector are considered a match if they are not metalogical and are equal, or if one is not metalogical and the other is `'-'`, or if both element values are `'-'`s. In effect, `'-'` acts as a "wild card" for the comparison.

Vectors match if their element values in all corresponding positions match. For example, the value `"10--"` is considered by the `std_match` function to match all the following values: `"101-"`, `"1010"`, and `"1001"`.

We can't use the `std_match` function to represent choices in a selected signal assignment, because these choices must be statically deterministic. A description of a 4-to-2 priority encoder using the `std_match` functions in a conditional signal assignment is given in Listing 4.7.2.

While the code in Listing 4.7.2 is valid, it is not an improvement in terms of conciseness over that in Listing 4.6.1.

LISTING 4.7.2
A 4-to-2 priority encoder using the std_match function to interpret "don't care" inputs, '-'.

```
library ieee;
use ieee.std_logic_1164.all;
use ieee.numeric_std.all;

entity encoder is
   port ( i3, i2, i1, i0 : in std_logic;
      a : out std_logic_vector(1 downto 0));
end entity;

architecture cond_match of encoder is

   signal tempi: std_logic_vector(3 downto 0);
   begin

   tempi <= (i3, i2, i1, i0);

   a <= "00" when std_match (tempi, "0---") else
   "01" when std_match (tempi, "10--") else
   "10" when std_match (tempi, "110-") else
   "11";

end cond_match;
```

Don't Care Outputs Assignments of "don't care" values to outputs are very useful because they allow a synthesizer to take advantage of the "don't cares" to minimize the logic it synthesizes. This is equivalent to the use of don't cares to minimize a Boolean function represented by a Karnaugh map.

Consider the design of a system that takes a BCD input and generates a `'1'` output only when the BCD input is valid and represents a digit greater than 5. In Listing 4.7.3, a description that does not take advantage of don't care conditions is written.

When this design is synthesized, the equation that the resulting gate-level logic implements is:

$$y = \bar{d}cb + d\bar{c}\bar{b}$$

Taking advantage of "don't cares" allows the synthesizer to produce a simpler circuit. If it is known that only valid BCD codes will be input to the system, then the six binary codes corresponding to `"1010"` through `"1111"` will never occur and the outputs for these input combinations can be specified as "don't cares." We also don't care what the output is for all the other std_logic input combinations that contain an input value that is not a `'0'` or `'1'`. These input combinations are all covered in the description by using the **`others`** keyword.

LISTING 4.7.3
BCD digit greater than 5 design that does not use "don't care" outputs.

```
library ieee;
use ieee.std_logic_1164.all;

entity bcdgt5 is
   port ( d, c, b, a : in std_logic;
      y : out std_logic);
end;

architecture csa_sel of bcdgt5 is
   signal tmp : std_logic_vector(3 downto 0);

begin

   tmp <= ( d, c, b, a );   -- Create an aggregate for the BCD inputs

   with tmp select

   y <= '1' when "0110" | "0111" | "1000" | "1001",   -- Inputs where fcn. is 1
      '0' when others;   -- All other std_logic input values

end csa_sel;
```

LISTING 4.7.4
BCD digit greater than 5 design that uses "don't care" outputs.

```
library ieee;
use ieee.std_logic_1164.all;

entity bcdgt5 is
   port ( d, c, b, a : in std_logic;
      y : out std_logic);
end;

architecture csa_sel of bcdgt5 is
   signal tmp : std_logic_vector(3 downto 0);
begin
   tmp <= ( d, c, b, a );
-- Outputs for the ten inputs that can actually
      -- occur in the circuit are specified
   with tmp select
   y <= '0' when "0000" | "0001" | "0010" | "0011"
               | "0100" | "0101",
      '1' when "0110" | "0111" | "1000" | "1001",
      '-' when others; -- Don't care  for input conditons that
                          -- cannot occur
end csa_sel;
```

The description in Listing 4.7.4 specifies a don't care output value `'-'` for the **`others`** choice.

Listing 4.7.4 is longer than Listing 4.7.3, because the cases where the output must be `'0'` must be distinguished from the "don't care" cases. However, the gate-level logic synthesized from Listing 4.7.4 realizes the simpler equation:

$$y = d + cb$$

As this equation indicates, the output from the synthesized hardware never actually has the value `'-'`. For each input condition where the output is specified as don't care, the minimization algorithm has chosen a value of either `'0'` or `'1'` for the output, whichever results in the simplest function.

'X' as a "Don't Care" for Outputs

Some synthesizers also accept use of the std_logic value `'X'` to specify a "don't care" output. X is also traditionally used on IC data sheets to indicate a "don't care" value, as opposed to its primary use in VHDL to represent a forcing unknown.

4.8 DECODERS

Like encoders, decoders are readily described using dataflow style architectures. Two example decoder designs are given in this section.

74HC138 Decoder Equivalent

Consider a decoder function like that of the 74HC138 three-line to eight-line decoder. It has three enable inputs: g1, g2a_bar, g2b_bar. When g1, g2a_bar, g2b_bar equal "100", the decoder is enabled. For any other combination of the enable inputs, the decoder is disabled.

When disabled, the decoder's eight outputs (y7 down to y0) are all '1's. When enabled, one of its outputs is '0' and all the others are '1's. The output that is '0' is determined by the binary value of the three select inputs: c, b, and a. For example, if the decoder is enabled and c, b, a equals "110", then output y6 is a logic '0'. A description of this decoder is given in Listing 4.8.1.

The select and the enable inputs are declared as scalars in the entity declaration. In the architecture declarative part, two signals, enables and cba, are each declared as three-element std_logic_vectors. In the architecture statement part, two aggregates

LISTING 4.8.1
A 3-to-8 decoder description.

```
library ieee;
use ieee.std_logic_1164.all;

entity decoder_3to8 is
   port (
      c, b, a: in std_logic;
      g1, g2a_bar, g2b_bar: in std_logic;
      y: out std_logic_vector (7 downto 0));
end decoder_3to8;


architecture csa_cond of decoder_3to8 is
   signal enables, cba : std_logic_vector(2 downto 0);

   begin

   enables <= (g1, g2a_bar, g2b_bar);
   cba <= (c, b, a);

   y <= "11111110" when enables = "100" and cba = "000" else
   "11111101" when enables = "100" and cba = "001" else
   "11111011" when enables = "100" and cba = "010" else
   "11110111" when enables = "100" and cba = "011" else
   "11101111" when enables = "100" and cba = "100" else
   "11011111" when enables = "100" and cba = "101" else
   "10111111" when enables = "100" and cba = "110" else
   "01111111" when enables = "100" and cba = "111" else
   "11111111";

end  csa_cond;
```

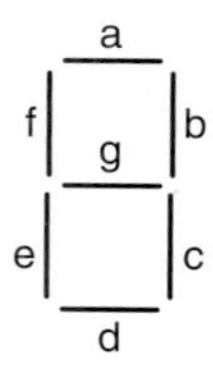

FIGURE 4.8.1
Segment pattern for a seven-segment display.

are used. One aggregates the enable inputs into the std_logic_vector `enables`. The other aggregates the select inputs into the std_logic_vector `cba`.

A single conditional signal assignment statement is then used to describe the decoder function. Each of the eight conditions, corresponding to the assertion of one of the decoder's outputs, is the logical AND of a condition for the enable inputs and a condition for the select inputs. The default assignment handles the case when the decoder is not enabled.

A single aggregate could be used to aggregate all six inputs, but the functionality of the design would probably be less clear when reading the code.

BCD to Seven-Segment Decoder

An often used decoder function is a BCD to seven-segment decoder. This decoder takes as its inputs four bits that represent a binary coded decimal (BCD) value and generates a 7-bit output to drive a seven-segment LED. The segments of a seven-segment LED are shown in Figure 4.8.1.

A description of a BCD to seven-segment decoder is given in Listing 4.8.2.

The decoder's BCD input is the std_logic_vector `dcba`. There are two other inputs to the decoder. When asserted, blanking input `bi_bar` forces all the decoder's outputs to `'1'`s. A `'1'` output turns OFF its associated LED segment; a `'0'` output

LISTING 4.8.2
Description of a BCD to seven-segment decoder.

```
library ieee;
use ieee.std_logic_1164.all;

entity bcd_7seg is
   port (
      bi_bar : in std_logic; -- blanking input
      lt_bar : in std_logic; -- lamp test input
      dcba : in std_logic_vector (3 downto 0); --BCD input
      seg: out std_logic_vector(6 downto 0) -- segments in order from a to g
      );
end bcd_7seg;

architecture conditional of bcd_7seg is
```

```
begin
   seg <= "1111111" when bi_bar = '0' else
   "0000000" when lt_bar = '0' else
   "0000001" when (dcba = "0000") else
   "1001111" when (dcba = "0001") else
   "0010010" when (dcba = "0010") else
   "0000110" when (dcba = "0011") else
   "1001100" when (dcba = "0100") else
   "0100100" when (dcba = "0101") else
   "1100000" when (dcba = "0110") else
   "0001111" when (dcba = "0111") else
   "0000000" when (dcba = "1000") else
   "0001100" when (dcba = "1001") else
   "-------";

end conditional;
```

turns ON its associated LED segment. If the blanking input is not asserted and lamp test input `lt_bar` is asserted, all the outputs are `'0'`s, forcing all the LED segments to be ON. If neither the blanking input nor the lamp test input is asserted, the output is determined by the value of the BCD input. If the BCD input has a value from 0 to 9 (`"0000"` to `"1001"`), the output consists of the values needed to display the corresponding decimal digit. If the BCD input is greater than 9, all the outputs are don't cares, since these input values should not occur.

The priority characteristic of a conditional signal assignment is particularly helpful for this design since the `bi_bar` input has priority over the `lt_bar` input, which has priority over the `dcba` input.

4.9 TABLE LOOKUP

A simple way to describe a combinational system is to use a table lookup. For a system with a single output the table is represented as a constant vector. For a system with multiple outputs an array of constant vectors is used. For any input combination we can then determine the output by simply looking it up in the table.

Consider the combinational system represented by the truth table in Table 4.9.1.

This system has a single output that is specified to be a don't care value for two of its input combinations. A description of this system is given in Listing 4.9.1.

A constant vector named `output` with a range from 0 to 7 is declared. The value of this vector, `"01-01-10"`, corresponds to the values of output `y` for each possible combination of the inputs. These values are simply the values of the column labeled `y` taken from top to bottom. That is, from row 0 to row 7.

Table 4.9.1
Truth table for a system with a single output.

a	b	c	y
0	0	0	0
0	0	1	1
0	1	0	-
0	1	1	0
1	0	0	1
1	0	1	-
1	1	0	1
1	1	1	0

If the binary input value is converted to an integer, this integer can be used as an index into the table to select the table entry that is the desired output. The architecture body consists of a single statement:

```
y <= output(to_integer(unsigned'(a, b, c)));
```

This statement first aggregates the inputs `a`, `b`, `c` and type qualifies them as an unsigned vector, so that the `to_integer` function can be used to convert the input values to an integer. This integer is then used as the index into the constant vector to look up the appropriate value for the output.

LISTING 4.9.1
Table lookup for a system with a single output.

```
library ieee;
use ieee.std_logic_1164.all;
use ieee.numeric_std.all;

entity table_lookup_vector is
    port( a, b, c : in std_logic;
        y : out std_logic );
end table_lookup_vector;

architecture lookup of table_lookup_vector is
constant output : std_logic_vector(0 to 7) := "01-01-10";
begin

    y <= output(to_integer(unsigned'(a, b, c)));

end lookup;
```

Table 4.9.2
Truth table for a 2-bit binary to reflected code conversion.

input	output
00	00
01	01
10	11
11	10

For a system with multiple outputs, an array of constant vectors can be used to define the table that describes the system. Consider the truth table representation of the combinational function that converts a 2-bit binary value to a 2-bit reflected code, Table 4.9.2.

A description of this table lookup is given in Listing 4.9.2.

LISTING 4.9.2
A 2-bit binary to reflected code conversion described as a table lookup.

```
library ieee;
use ieee.std_logic_1164.all;
use ieee.numeric_std.all;

entity bin2reflected is
   port (input: in std_logic_vector (1 downto 0);
      output: out std_logic_vector (1 downto 0));
end bin2reflected;

architecture look_up of bin2reflected is

type bin2refl is array (0 to 3) of std_logic_vector(1 downto 0);

-- table of values
constant reflected_tt : bin2refl := (
"00",
"01",
"11",
"10");

begin

  output <= reflected_tt(to_integer(unsigned(input)));

end look_up;
```

In the architecture's declarative part an array type named `bin2refl` is declared.

```
type bin2refl is array (0 to 3) of std_logic_vector(1 downto 0);
```

Type `bin2refl` is an array of vectors. This array type consists of four elements representing rows (0 to 3) of a truth table. Each element (row) in this array is a two-element std_logic_vector, representing the outputs for that row of the truth table.

A constant named `reflected_tt` is then declared.

```
constant reflected_tt : bin2refl := (
"00",
"01",
"11",
"10");
```

This constant is of the previously declared type `bin2refl` and its value is specified by four 2-character strings.

The architecture's statement part consists of a single signal assignment statement:

```
output <= reflected_tt(to_integer(unsigned(input)));
```

To_integer Function

In this assignment statement, the binary input vector `input` is first converted from type std_logic to type unsigned by a type conversion. Because these two types are closely related, this type conversion can be accomplished by simply using the name of the target type as if it were a function.

```
unsigned(input)
```

The unsigned result of this operation is then converted to an integer using the function `to_integer`. Function `to_integer` is defined in package NUMERIC_STD. The integer result returned by the function is used as the index to select the element (row vector) of constant array `reflected_tt`. This row vector is assigned to `output`. For example, if `input` equals `"10"` then the function

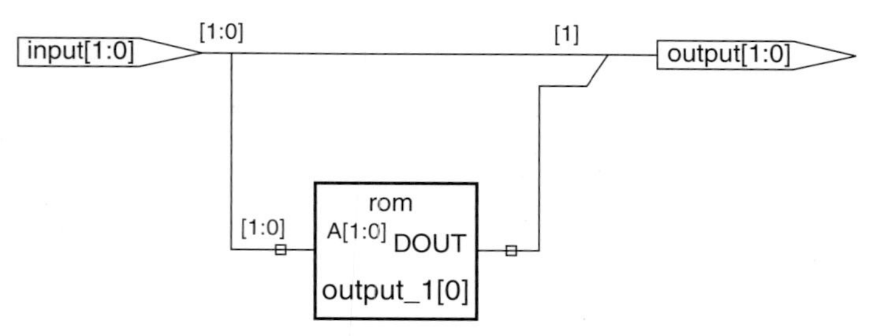

FIGURE 4.9.1

RTL hierarchical view of table lookup description of binary to reflected code description.

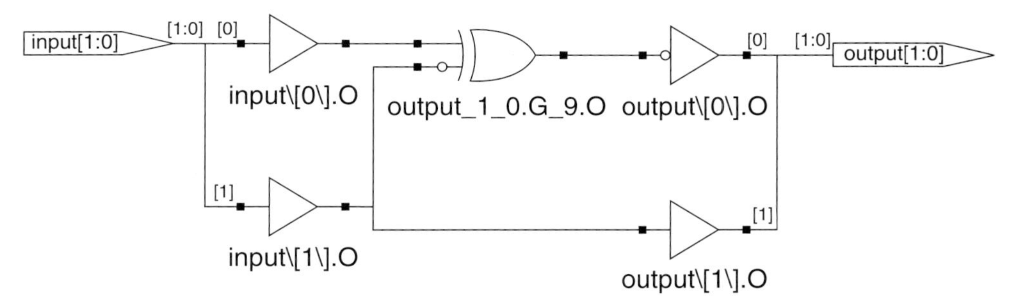

FIGURE 4.9.2
Technology dependent flattened-to-gates view of table lookup description of binary to reflected code description.

`to_integer` returns the integer 2. For this input value, the assignment statement is equivalent to:

```
output <= reflected_tt(2);
```

or

```
output <= "11";
```

One way to view the hardware implementation of the table lookup process is as read-only memory (ROM). The system's inputs are the ROM's address inputs and the ROM's contents are the table's output values. For a particular input value (address) the ROM outputs the data stored at that address.

A synthesis tool may also view a table lookup in terms of ROM. A hierarchical view of the description in Listing 4.9.2 is given in Figure 4.9.1. However, when target of the synthesis is a PLD, the synthesizer does not synthesize the circuit as ROM. Instead, it implements the inferred logic in its reduced logic form.

The reduced logic created by the synthesizer is given in Figure 4.9.2.

The reduced synthesized logic is:

$$\text{output}(1) = \text{input}(1)$$
$$\text{output}(0) = \overline{\text{input}(1)} \cdot \text{input}(0) + \text{input}(1) \cdot \overline{\text{input}(0)}$$

4.10 THREE-STATE BUFFERS

A *three-state buffer* (*tristate buffer*) has a data input and an enable input. Its enable input controls whether the three-state buffer is OFF and its output is high impedance (`'Z'`), or whether it is ON and its output is driven by its data input. Thus, the output of a three-state buffer can be either `'0'`, `'1'`, or `'Z'`.

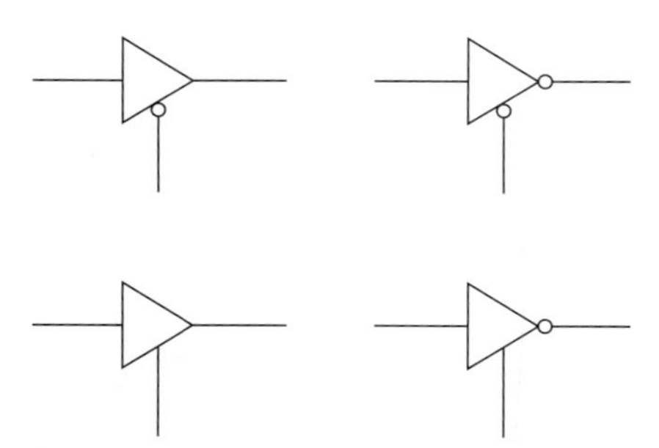

FIGURE 4.10.1
Logic symbols representing noninverting and inverting three-state buffers.

A noninverting three-state buffer is represented by a triangle symbol. An inversion circle is added at the symbol's output for an inverting three-state buffer (Figure 4.10.1).

The enable input is shown connected to the side of the triangle symbol. As shown in Figure 4.10.1, some three-state buffers are enabled when their enable input is `'0'` and others when it is `'1'`.

Describing Three-state Output Ports in VHDL

In VHDL, the fact that an output port has three-state capability is not indicated by its mode. Instead, this capability is implied in the architecture by statements that assign values to the output. If, for one or more conditions, an output is assigned the value `'Z'`, then the output is a three-state output.

A VHDL description of a non-inverting three-state buffer is given in Listing 4.10.1. If the enable input `en_bar` is asserted, the buffer's output is the same as its input. If the enable input is not asserted, the buffer's output is `'Z'`.

LISTING 4.10.1
Three-state buffer description.

```
library ieee;
use ieee.std_logic_1164.all;

entity three_state_buffer is
   port (
      d_in, en_bar : in std_logic;
      d_out: out std_logic
   );
end three_state_buffer;

architecture dataflow of three_state_buffer is
begin

d_out <= d_in when en_bar = '0' else 'Z';

end dataflow;
```

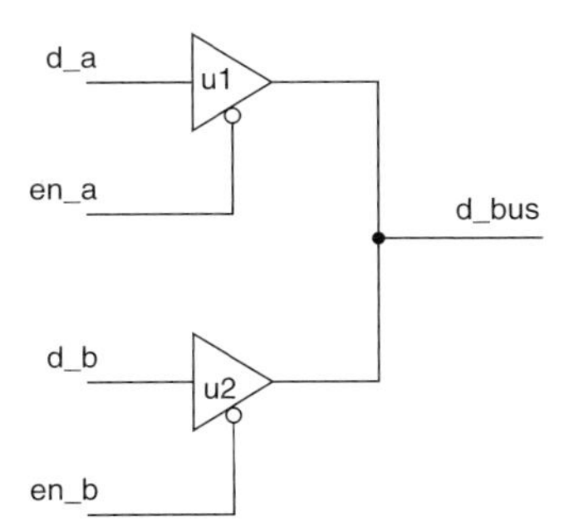

FIGURE 4.10.2
Connecting the outputs of two three-state buffers to accomplish multiplexing.

Multiplexing Two Data Sources

Multiplexing signals from two data sources using three-state buffers is achieved by connecting the two three-state buffers' outputs together to drive the shared signal, Figure 4.10.2. While this figure represents the three-state buffers as separate logic elements, they could be integrated into the outputs of two more functionally complex systems.

When the type of connection in Figure 4.10.2 is made, the designer must add logic that controls the three-state buffers' enable inputs so that only one or none of the buffers is enabled at a time. If only one buffer is enabled, that output determines the logic level for the shared signal. If none of the buffers is enabled, both outputs are in their high impedance states and the shared signal they drive has the high impedance value.

Bus Contention

The prohibited situation, where two or more three-state outputs connected to a shared signal are simultaneously enabled, creates a serious problem if one of the enabled outputs is trying to drive the shared signal to `'1'` and another is trying to drive it to `'0'`. The resulting voltage level is in the forbidden region. This situation is referred to as *bus contention.* In std_logic terms, the shared signal's value is `'X'` (forcing unknown). In a hardware implementation, excessive current may flow between the two enabled three-state outputs under this condition.

The resolution function for std_logic properly models three-state outputs driving a shared signal. In Table 3.4.2, if both outputs (drivers) are `'Z'`, the shared output is `'Z'`. If one output is `'Z'` and the other is `'1'`, the result is `'1'`. If one output is `'Z'` and the other is `'0'`, the result is `'0'`. If both outputs are enabled and drive different values, `'0'` and `'1'` or `'1'` and `'0'`, the output is `'X'`.

A structural description of the circuit in Figure 4.10.2 is given in Listing 4.10.2. The description assumes that the three-state buffer entity from Listing 4.10.1 has been compiled to the library `three_state_buffer`.

Waveforms from the simulation of this description are presented in Figure 4.10.3. The input stimulus just counts through all possible binary input combinations. The first input combination has the data input to each buffer equal to `'0'` and both

LISTING 4.10.2
Two three-state buffers with their outputs connected.

```
library ieee;
use ieee.std_logic_1164.all;

library three_state_buffer;  -- library containing three-state buffer
use three_state_buffer.all;

entity three_state_bus is
   port (
      d_a: in std_logic;        -- data input buffer a
      en_a_bar: in std_logic;   -- enable input buffer a
      d_b: in std_logic;        -- data input buffer b
      en_b_bar: in std_logic;   -- enable input buffer b
      d_bus: out std_logic      -- bused data output
      );
end three_state_bus;

architecture three_state_bus of three_state_bus is
begin
   u1: entity three_state_buffer port map (d_a, en_a_bar, d_bus);
   u2: entity three_state_buffer port map (d_b, en_b_bar, d_bus);
end three_state_bus;
```

enables are asserted. This corresponds to an invalid situation because the two buffers should not be enabled simultaneously. However, the output is a valid `'0'` because both buffers' data inputs are `'0'`.

For the second input combination, indicated by the cursor, the input data at buffer `u1` is a `'1'` and at buffer `u2` is `'0'`, and both buffers are enabled. The output for this case is `'X'`, because one buffer is trying to drive the output to a `'1'` and the other is trying to drive it to a `'0'`. On the output waveform, the times during which the output is forcing an unknown value (`'X'`) are represented by two lines, one at the `'1'` logic level and the other at the `'0'` logic level.

Name	Value
d_a	1
en_a_bar	0
d_b	0
en_b_bar	0
d_bus	X

FIGURE 4.10.3
Timing waveform for two three-state buffers with their outputs connected.

For input conditions corresponding to one buffer enabled and the other not enabled, the output is the same as the enabled buffer's input value. When both buffers are not enabled, the output is `'Z'`. This value is represented on the waveform by a line halfway between the `'0'` and `'1'` logic levels.

Three-state Output Capability of a PLD

If some of the outputs of a PLD need to be multiplexed with outputs from other ICs, these outputs can be described in VHDL as three-state outputs. Since almost all PLDs have output pins with three-state capability, the described functionality is easily synthesized. One of the inputs to the PLD must be an enable signal to control the three-state outputs. Note that ports of a PLD described as mode **out** or **inout** (bidirectional) can be described as three-state outputs.

We will see a number of examples later in this book where the system being described in VHDL and implemented using a PLD connects to the system bus of a microprocessor. Those signals that connect to the data bus portion of the system bus must be three-state.

Internal Three-state Buffers in a PLD

If we need to implement a multiplexing operation within a design description, we might choose to describe this operation using three-state buffers. However, while PLDs have three-state buffers at their output pins, many PLDs do not have internal three-state buffers. When such a description is synthesized, the synthesizer attempts to replace the functionality implemented by the three-state buffers with a multiplexer or equivalent gate logic. Even so, some PLD manufacturers recommend that internal three-state buffers not be used in a description.

Furthermore, if we are writing descriptions of entities to be used as components in a hierarchical system, we should refrain from using three-state buffers on their outputs, unless the three-state outputs of the component entity will connect directly to outputs of the top-level entity. This is because three-state outputs from components that do not connect directly to outputs of the top-level entry become internal three-state buffers at the top level. In conclusion, it is best to use three-state buffers only when they are driving top-level bidirectional or output pins.

4.11 AVOIDING COMBINATIONAL LOOPS

When synthesizing a description of a combinational system, the synthesizer may issue a warning that the description contains a combinational loop. A *combinational loop* (Figure 4.11.1) is an output looped back (fed back) to one of the inputs. If we are attempting to describe a combinational system, our description must not contain any combinational loops. The existence of such a loop can lead to circuit instability.

Figure 4.11.1(a) shows the general case of a combinational loop. In the description of a complex system, it may not be obvious that such a loop exists. Fortunately, the synthesizer will detect the loop's existence and issue a warning message.

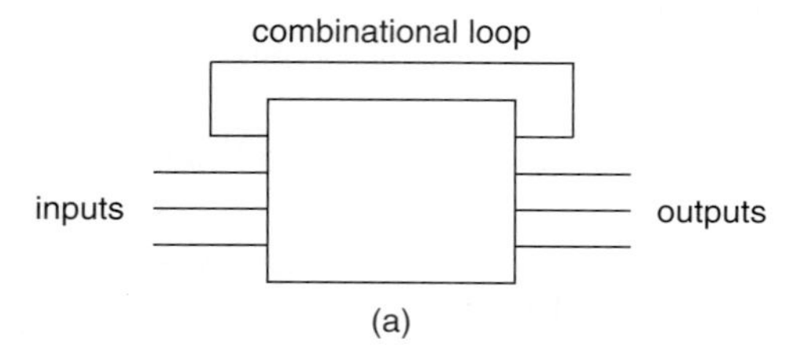

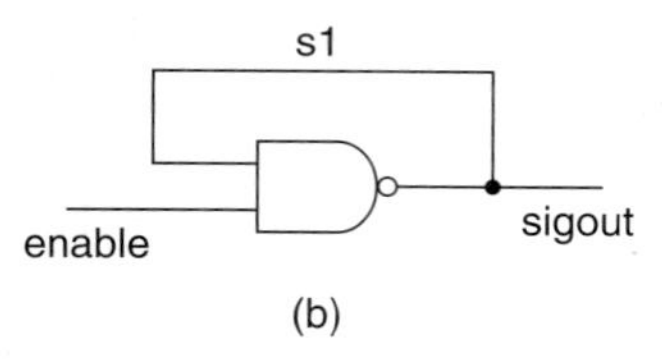

FIGURE 4.11.1
Combinational loops.

In Listing 4.11.1 of the circuit in Figure 4.11.1(b), existence of a combinational loop is readily apparent.

Since port `sigout` is declared as mode out, it cannot be read in the architecture. Instead, signal `s1` is declared and used to feed the NAND gate's output back to one of its inputs. Signal `s1` can then be assigned to `sigout`.

LISTING 4.11.1
Dataflow description of a NAND gate circuit containing a combinational loop.

```
library ieee;
use ieee.std_logic_1164.all;

entity nandwfb is
    port(
       enable : in std_logic;
       sigout : out std_logic
        );
end nandwfb;

architecture dataflow of nandwfb is
signal s1: std_logic;
begin

s1 <= enable nand s1;
sigout <= s1;

end dataflow;
```

The concurrent statement:

```
s1 <= enable nand s1;
```

creates the combinational loop. In general, whenever a signal appears on both the left-hand and right-hand sides of an assignment statement, a combinational loop results. However, in a more complex description, the loop may not be so direct. That is, an output signal may be assigned to another signal, or chain of signals, where the last signal is assigned to an input.

Issues regarding simulation and instability of circuits with a combinational loop are discussed in detail in "Delta Delays and Combinational Feedback" on page 235 in Chapter 6.

PROBLEMS

4.1 List the logical operators available for std_logic in order of their precedence. Include the **not** operator in your list. Where are these operators defined? Are they overloaded? Which of these operators are associative and which are not?

4.2 Write a signal assignment statement corresponding to each of the following Boolean equations:

(a) $y = \bar{a} \cdot b \cdot c + a \cdot \bar{b} \cdot \bar{c} + a \cdot \bar{b} \cdot c$

(b) $x = (a + b + c) \cdot (a + \bar{b} + \bar{c})$

(c) $p = a + \bar{b} \cdot c$

(d) $q = a \cdot b + \bar{a} \cdot \bar{b}$

4.3 Write a signal assignment statement that assigns to signal x the NOR of inputs a, b, c, and d.

4.4 If a logical operation is performed on two arrays, how are the elements of the arrays matched to perform the operation? What restrictions apply to the lengths of the arrays?

4.5 Given the following entity and architecture:

```
entity xor_vector is
  port (x : in std_logic_vector(0 to 2);
    y : in std_logic_vector(2 downto 0);
    f : out std_logic_vector(2 downto 0));
end xor_vector;

architecture dataflow of xor_vector is
begin
  f <= x xor y;
end dataflow;
```

write a functionally equivalent architecture, dataflow2, that separately computes each element of the output.

4.6 What kinds of statements make up a dataflow style architecture?

4.7 Name the basic forms of concurrent signal assignment statements. Another form of signal assignment statement is a simplified version of one of these basic forms. Name this derived form and the basic form from which it is derived.

4.8 For the design entity described below:

```
library ieee;
use ieee.std_logic_1164.all;

entity myckt is
   port (a, b, c : in std_logic;
   w, y: out std_logic);
end myckt;

architecture dataflow of myckt is
begin
   w <= not a or (b and c);
   y <= ((a and b) or c) or (not a and not b);

end dataflow;
```

(a) Draw a block diagram of the system with inputs and outputs labeled.
(b) Directly translate the architecture to a gate level logic diagram of the circuit (without minimization).

4.9 The truth table that follows represents an arbitrary three-input three-output combinational function. Note that outputs f2 and f3 are actually the same function.

a	b	c	f1	f2	f3
0	0	0	1	0	0
0	0	1	1	1	1
0	1	0	0	1	1
0	1	1	1	0	0
1	0	0	0	1	1
1	0	1	1	1	1
1	1	0	0	0	0
1	1	1	0	1	1

Write the CSOP functions for f1 and f2. For f3, which is functionally equivalent to f2, write a CSOP expression for the complement of the function and then complement the entire expression. This results in an equivalent expression that is the complement of a CSOP expression that has only three minterms.

4.10 For the truth table in Problem 2.19, write a complete VHDL design description of a design entity that accomplishes the function defined by the truth table. Inputs and outputs are type std_logic. Use selected signal assignment statements.

4.11 A 74HC10 contains three independent three-input NAND gates.

(a) Write an entity declaration for the entity `triple_3`, which is equivalent to a 74HC10.

(b) Write an architecture body in the dataflow style for the entity `triple_3`. Use only selected signal assignment statements.

4.12 A 4-bit binary value named `fuel` is input to a combinational system named `fuel_encoder`. This input value represents the quantity of fuel in a tank to a resolution of 1/16 of the tank's maximum capacity. Four LEDs are used to display the quantity of fuel in the tank to a resolution of 1/4 of the tank's maximum capacity. The LEDs form a vertical column. The system must output a 4-bit vector name `leds` to drive the LEDs. The most significant element of vector `leds` drives the top LED. The next most significant element drives the next LED down in the column, and so on. A 0 value for an element of `leds` turns its associated LED ON.

Write a description of the system `fuel_encoder`. The architecture body must be named `selected` and use only selected signal assignment statements.

4.13 If x and y are type std_logic, give the results of the evaluation of the following relationships. For each case explain how the result is determined:

(a) x < y, where x = '0' and y = 'Z'

(b) x >= y, where x = '0' and y = '1'

(c) x > y, where x = 'Z' and y = '1'

4.14 Assuming that a and b are equal length std_logic_vectors, explain how the relationship a >= b is evaluated.

4.15 For the truth table in Problem 2.19, write a complete VHDL design description of a design entity that accomplishes the function defined by the truth table. Inputs and outputs are type std_logic. Use conditional signal assignment statements.

4.16 Rewrite Listing 4.5.2 to implement the 4-to-1 multiplexer using only a single conditional signal assignment statement. That is, do not assign an aggregate to a local signal.

4.17 A comparator has two 4-bit inputs `a` and `b` and two outputs. Output `equal` is asserted only if the two inputs are equal. Output `cmpl` is asserted only if the two inputs are the complement of each other. Write an entity declaration for the comparator. Write an architecture named `selected` that implements each output with a separate selected signal assignment statement. Write an alternative architecture named `conditional` that implements each output with a separate conditional signal assignment statement.

4.18 A demultiplexer has a 4-bit data input named `datain`. It also has four 4-bit data outputs named `route0`, `route1`, `route2`, and `route3`. The demultiplexer has two select inputs `s1` and `s0`. The value of `s1,s0` determines on which one of the four outputs the input data will appear. For example, if `s1,s0` is 1,0, the input data will appear on `route2`. The three 4-bit data outputs that are not selected at any given time must be placed in their high impedance states (three-stated).

(a) Write the entity declaration for the demultiplexer. Name the entity `mux`.

(b) Write an alternative architecture named `condsa` using one or more conditional signal assignment statements.

(c) Write an architecture name `selectsa` using one or more selected signal assignment statements.

4.19 A system named `ones_count` has an input that is a 3-bit vector named `inp`. The system's output is a 4-bit vector named `num_ones`. The system's output indicates the number of 1s in the input vector. That is, the element in the output vector that is a 1 indicates the number of 1s in the input vector. Output num_ones(0) is a 1 if the number of ones in the inp is 0, otherwise it is 0. Output num_ones(1) is a 1 if the number of ones in the inp is 1, otherwise it is 0, and so on for the other outputs.

Write an entity declaration for ones_count. Write an architecture for ones_count named mixed. This architecture must use a separate concurrent signal assignment statement to compute each output. Output num_ones(0) must be computed by a Boolean expression assignment statement. Output num_ones(1) must be computed by a selected signal assignment statement. Output num_ones(2) must be computed by a conditional signal assignment statement. Output num_ones(3) may be computed by any of the three previous techniques.

4.20 In the description of the 4-to-2 priority encoder in Listing 4.6.1, the output is '11' when only input i0 is asserted and when no inputs are asserted. Rewrite the description to add an output named asserted, which allows these two cases to be distinguished. This output is a '1' only if one or more of the inputs is asserted.

4.21 Write a description of the 4-to-2 priority encoder in Listing 4.6.1 using a selected signal assignment. Take advantage of the vertical bar symbol (|) to put input combinations that produce the same output into the same list of choices. Compare your description to the one in Listing 4.6.1 in terms of conciseness and readability.

4.22 Write a description of the 4-to-2 priority encoder in Listing 4.6.1 using Boolean expression signal assignment statements. Compare your description to the one in Listing 4.6.1 in terms of conciseness and readability.

4.23 If you had to describe a priority encoder using a dataflow architecture and any kind of concurrent signal assignment statement, what choices of statements would be available to you? Which kind of concurrent signal assignment statement would be preferable and why?

4.24 Given the following truth table:

a	b	c	g
0	0	0	1
0	0	1	0
0	1	0	1
0	1	1	0
1	0	0	-
1	0	1	-
1	1	0	1
1	1	1	0

(a) Write a complete design description of the function in the truth table using a conditional signal assignment statement. Call the entity function_g. Treat the inputs as scalars in the entity declaration. Write the description so that the synthesizer can take advantage of the don't care cases to simplify the function.

(b) Write an alternative architecture for entity function_g that uses a selected signal assignment.

4.25 Given the following truth table:

a	b	c	x
0	0	0	1
0	0	1	1
0	1	0	0
0	1	1	1
1	0	0	0
1	0	1	-
1	1	0	-
1	1	1	1

Write a complete VHDL design description of a design entity that accomplishes the function defined by the truth table. The inputs and outputs are std_logic type. Use a conditional signal assignment architecture that allows the synthesizer to take advantage of the don't cares for logic reduction.

4.26 An n-bit thermometer code has $n + 1$ valid code words. The first code word has all its bits equal to 0s. The other n code words in sequence correspond to all the possible bit combinations with one or more consecutive bits equal to 1 starting with the least significant bit. All other bit combinations are invalid in this code. This code is called a thermometer code because if the code words are viewed as a vertical stack of bits, with the most significant bit at the top, the level of 1s rises like the mercury in a thermometer rises with temperature.

The analog comparators in a flash analog-to-digital converter (ADC) generate a thermometer code. However, it is required that the output of the ADC be a binary code. This is accomplished by including a thermometer to binary encoder on the ADC IC. The entity declaration for a thermometer to binary encoder for a 3-bit (output) ADC has a 7-bit input. Its entity declaration is:

```
entity therm2bin is
    port(
       therm : in std_logic_vector(6 downto 0); -- thermometer code
       bin : out std_logic_vector(2 downto 0)   -- binary code
        );
end therm2bin;
```

The output binary code simply indicates how many 1s there are in a valid input code. Write an architecture body named `selected` that uses a selected signal assignment with don't cares to describe the system. Write an alternative architecture named `conditional` that uses a conditional signal assignment with don't cares to describe the system.

4.27 A BCD-to-bargraph decoder is to be designed. Its entity declaration is:

```
entity bcd_2_bar is
   port ( bcd : in std_logic_vector (3 downto 0);
   bar_graph : out std_logic_vector (8 downto 0));
end bcd_2_bar;
```

The inputs represent a binary value between 0 and 9. There are nine outputs. Each output drives an LED. When an output is 0, its associated LED is ON. When it is 1, its associated LED is OFF. The LEDs are stacked in a vertical bar. The top LED is driven by `bargraph(8)` and the bottom LED is driven by `bargraph(0)`.

When the input combination is 0000, all the LEDs are OFF. As the input is stepped from 0001 to 1001, the next higher LED turns ON in sequence, from the bottom to the top. When the input is 1001 all the LEDs are ON. As a result, the number of LEDs that are ON indicate the value of the BCD input.

Input values from 1010 to 1111 are invalid. If it were possible in an application that these numbers could actually occur as input values, you could simply leave all the LEDs ON for these input combinations. If it was known that none of these inputs would occur, the decoder's outputs could be specified as don't cares for these input combinations. Thus, the two different designs.

(a) Write a description of the converter using a selected signal assignment. Have all outputs 1s for the six invalid input codes.
(b) Write a description of the converter using selected signal assignments. Assign all outputs don't care values for the six invalid input codes.
(c) Write a description of the converter using a conditional signal assignment. Have all outputs 1s for the six invalid input codes.
(d) Write a description of the converter using conditional signal assignments. Assign all outputs don't care values for the five invalid input codes.
(e) Synthesize the designs and compare the amount of logic required for each implementation.

4.28 A 74HC139 contains two 1-out-of-4 decoders in a single MSI circuit. The truth table for a single 1-out-of 4 decoder is given below.

Inputs			Outputs			
	Select					
$\overline{G}$	B	A	Y0	Y1	Y2	Y3
1	X	X	1	1	1	1
0	0	0	0	1	1	1
0	0	1	1	0	1	1
0	1	0	1	1	0	1
0	1	1	1	1	1	0

As can be seen from the truth table, the 1-out-of-4 decoder has three inputs and four outputs. Inputs B and A select which output is asserted when the decoder is enabled. The decoder is enabled when the enable input $\overline{G}$ is a 0. Since VHDL does not allow overbars in identifiers, a common way of naming a signal that is active-low is to attach the suffix_bar to the signal name. Accordingly, you will use `g_bar` in your VHDL code to represent $\overline{G}$.

(a) Write an entity declaration for the entity `decode`, which is a single 1-out-of-4 decoder whose truth table is given above.
(b) Write an architecture in the dataflow style for the entity `decode`. Use only conditional signal assignment statements.

4.29 Modify the BCD to seven-segment decoder in Listing 4.8.2, to create a hexadecimal to seven-segment decoder named `hex_7seg`. For this decoder, input `dcba` is interpreted as a 4-bit hexadecimal code. Output `seg` displays the value of the hexadecimal input on the seven-segment LED display.

4.30 Given the entity declaration for a BCD to seven-segment decoder in Listing 4.8.2, where each output that drives a segment is a `'0'` to turn the segment ON and a `'1'` to turn the segment OFF.

(a) Write a design description whose architecture uses a selected signal assignment to implement the conversion function. Specify the outputs as don't cares for non-BCD input values, so that the synthesizer can produce the simplest logic.

(b) Write a design description whose architecture uses a table lookup to implement the conversion function. Specify the outputs as don't cares for non-BCD input values so that the synthesizer can produce the simplest logic.

4.31 For the function specified by the truth table in Problem 4.25, write a complete description that implements the function using a table lookup. The architecture must allow the synthesizer to take advantage of the don't care conditions in implementing the function.

4.32 Write an alternative architecture that uses table lookup to describe the BCD to seven-segment decoder of Listing 4.8.2. The architecture must allow the synthesizer to take advantage of the don't care conditions in implementing the function.

4.33 Given the following truth table:

a	b	c	x	y
0	0	0	1	0
0	0	1	0	-
0	1	0	1	0
0	1	1	0	-
1	0	0	0	0
1	0	1	-	1
1	1	0	0	0
1	1	1	1	1

(a) Write a complete design description for a design entity named `two_out_tt` that implements the function described by the truth table above. The architecture body must use only a table lookup to implement the function. The name of the table is `outputs` and its type is `table`.

(b) Draw Karnaugh maps and determine how the synthesizer would assign 0s or 1s to the don't care output values to synthesize minimal SOP logic for the outputs.

4.34 Design a 2421 code to BCD converter. The input to the converter is the std_logic_vector `c2421` (3 downto 0). The input vector represents a 2421 code for a decimal digit between 0 and 9. The converter's output is the std_logic_vector `BCD` (3 downto 0). The output vector represents the BCD code for the decimal digit indicated by the 2421 code.

Decimal Digit	2421 Code	BCD Code
0	0000	0000
1	0001	0001
2	0010	0010
3	0011	0011
4	0100	0100
5	1011	0101
6	1100	0110
7	1101	0111
8	1110	1000
9	1111	1001

Input values that are not valid 2421 code words are to be treated as don't cares for invalid input combinations.

4.35 A 74LS251 is an 8-to-1 multiplexer with three-state outputs. It has eight data inputs D7 down to D0 and an enable input $\overline{G}$. When it is enabled ($\overline{G} = 0$), the data at the data input selected by its select inputs C, B, A appears at its Y output and the complement of this data appears at its W output. When it is not enabled, both of its outputs are in their high impedance states.

(a) Write a design description of an entity `ic74LS251` that uses selected signal assignment statement(s) and is functionally equivalent to the 74LS251.

(b) Write an alternate architecture for the entity `ic74LS251` that uses conditional signal assignment statement(s).

4.36 A 74HC541 is an octal three-state buffer. It has eight inputs and eight corresponding three-state outputs. The outputs are enabled only when the 74HC541's two enable inputs `oe1_bar` and `oe2_bar` are simultaneously asserted. Each of these enable inputs is active low. Write a description of a functionally equivalent buffer. Name the entity `ic74HC541`. Name the input vector `a` and the output vector `b`.

Chapter 5

Behavioral Style Combinational Design

A behavioral style architecture uses algorithms in the form of sequential programs to describe a system. These sequential programs are called processes. Unlike dataflow or structural styles, a behavioral style description does not imply a particular hardware implementation. It simply tells how the system is to behave.

The statement part of a behavioral style architecture consists of one or more process statements. Each process statement is, in its entirety, a concurrent statement. Accordingly, processes can execute simultaneously.

Processes communicate with each other, and with other concurrent statements, using signals. This communication is analogous to the way components in a structural description communicate.

Statements inside a process are sequential statements. They are executed in sequence and their order is critical to their effect. Sequential statements are introduced in this chapter and used to describe combinational systems. Starting in Chapter 8, sequential statements are also used to describe sequential systems.

Variables are also introduced in this chapter. They are used to store local data inside a process. A variable is declared inside a process or subprogram and is only visible in that process or subprogram. Variables in VHDL are similar to those in conventional programming languages. A variable's value is changed using a variable assignment statement.

5.1 BEHAVIORAL STYLE ARCHITECTURE

Recall that the statement part of an architecture body can contain only concurrent statements. A behavioral style architecture contains one or more processes. Each process is a concurrent statement, so processes are able to execute simultaneously. A two-process version of the half adder is shown in Listing 5.1.1.

This half adder consists of two processes, labeled `p1` and `p2`. Each process starts with the keyword **`process`**. Following keyword **`process`** is a sensitivity list (in parentheses). Process `p1` has two signals, `a` and `b`, in its sensitivity list. This process executes whenever there is an event on either of these signals. Thus, this process is sensitive to `a` and `b`.

Process `p1` contains a single signal assignment statement:

```
sum <= a xor b;
```

Sequential Signal Assignment Statements

This same statement appeared in some previous dataflow versions of the half adder. However, because it then appeared by itself in the statement part of an architecture body, it was a concurrent signal assignment statement. In Listing 5.1.1, because the same signal assignment statement appears within a process, it is a sequential signal assignment statement. Thus, whether a signal assignment statement is concurrent or sequential depends on its location (context).

LISTING 5.1.1
Half adder using two processes.

```
library ieee;
use ieee.std_logic_1164.all;

entity half_adder is
   port (a, b : in std_logic;
   sum, carry_out : out std_logic);
end half_adder;

architecture two_processes of half_adder is
   begin

   p1: process (a,b)
   begin
   sum <= a xor b;          -- sequential signal assignment
   end process p1;

   p2: process (a,b)
   begin
   carry_out <= a and b; -- sequential signal assignment
   end process p2;

end two_processes;
```

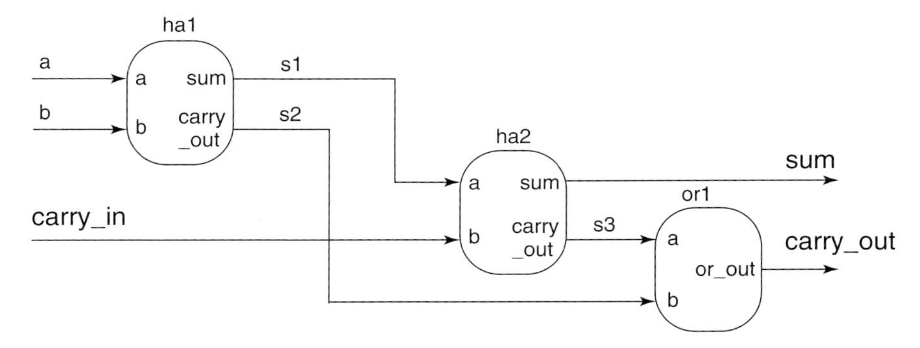

FIGURE 5.1.1
Three processes communicating, using signals, to implement a full adder.

In this example, processes `p1` and `p2` are independent of each other because each is sensitive only to input port signals and not to any signal assigned a value by the other process. Because they are independent, they do not need to communicate.

Process Communication

A block diagram of a full adder described by three separate communicating processes is shown in Figure 5.1.1. These processes communicate using signals `s1`, `s2`, and `s3`. Process communication using signals is similar to the way components in a structural description communicate (are connected). The interprocess communication in this example is identical to that of the half-adder and OR components in the structural full-adder architecture in Figure 2.9.1.

A behavioral description of this full adder appears in Listing 5.1.2.

LISTING 5.1.2
Full-adder behavioral description using communicating processes.

```
library ieee;
use ieee.std_logic_1164.all;

entity full_adder is
   port (a, b, carry_in : in std_logic;
   sum, carry_out: out std_logic);
end full_adder;

architecture processes of full_adder is

   signal s1, s2, s3 : std_logic;  -- Signals to interconnect processes
```

(Cont.)

LISTING 5.1.2 *(Cont.)*

```
begin
   -- Each process is a concurrent statement

   ha1: process (a, b) -- ha1 process
   begin
      if a = '1' then
         s1 <= not b;
         s2 <= b;
      else
         s1 <= b;
         s2 <= '0';
      end if;
   end process ha1;

   ha2: process (s1, carry_in) -- ha2 process
   begin
      if s1 = '1' then
         sum <= not carry_in;
         s3 <= carry_in;
      else
         sum <= carry_in;
         s3 <= '0';
      end if;
   end process ha2;

   or1: process (s3, s2) -- or1 process
   begin
      if (s3 = '1') or (s2 = '1') then
         carry_out <= '1';
      else
         carry_out <= '0';
      end if;
   end process or1;

end processes
```

Processes `ha1` and `ha2` are algorithms for half-adder functions. In fact, they are the identical algorithm operating on different sets of signals. Process `or1` is an algorithm for the OR function. Signals `s1`, `s2`, and `s3` are declared in the architecture body's declarative part.

5.2 PROCESS STATEMENT

A process statement may represent the behavior of all or some portion of a design. A *process statement* is a concurrent statement that itself is comprised of sequential statements. Included in these sequential statements can be statements that call subprograms. The simplified syntax for a process statement is given in Figure 5.2.1.

Following the keyword **`process`** is an optional sensitivity list, which is followed by the optional keyword **`is`**.

The declarative part of a process consists of zero or more declarations. Among other items, variables can be declared here. ***No signals can be declared in a process.***

The statement part of a process follows the keyword **`begin`** and consists of zero or more sequential statements. Sequential statements are executed in sequence and their order is critical. The order of the sequential statements affects what logic is synthesized. The sequential statement part of a process is terminated by the keywords **`end process`**.

Sensitivity List

Since the sensitivity list is optional, there are two kinds of processes: those with a sensitivity list and those without. Processes used to synthesize logic typically have sensitivity lists.

A *sensitivity list* is a list of signals to which a process is sensitive. Signals in the sensitivity list may be signals declared in the enclosing architecture and/or ports of the associated entity declaration. The order of signals in a sensitivity list is not important.

Recall that an event was defined in Chapter 1 as a change in a signal's value. Whenever there is an event on a signal in its sensitivity list, the process is executed. If a signal name that denotes a composite type signal appears in a sensitivity list, an event on any element of the composite is an event on the composite signal, causing the process to execute.

```
process_statement ;;=
[ process_label : ]
   process [ ( sensitivity_list ) ] [ is ]
            { process_declarative_item } -- process declarative part
   begin
            { sequential_statement } -- process statement part
   end process [ process_label ] ;
```

FIGURE 5.2.1
Simplified syntax for a process statement.

A process with a sensitivity list must not contain any wait statements. A process without a sensitivity list must contain at least one wait statement, or its simulation will never end.

Process Execution

A process can be viewed as an infinite loop whose execution is repeatedly suspended and resumed. The sequential statements in a process are executed from the first to the last, then back to the first. A process continues to execute its sequential statements until the process is suspended.

Suspending a Process That Has a Sensitivity List

A process that has a sensitivity list automatically suspends after the last statement in the process is executed. When there is a subsequent event on a signal in its sensitivity list, the process resumes execution, starting with the first statement, and executes to its last statement. Since a process with a sensitivity list cannot contain any wait statements, any procedure that it calls cannot contain a wait statement.

Suspending a Process That Has No Sensitivity List

A process that does not contain a sensitivity list suspends when it encounters a wait statement. The process resumes execution when the wait statement's conditions are satisfied. For now our focus is solely on processes that have sensitivity lists. Wait statements and processes containing them are discussed in detail in Chapter 7.

Using a Process to Describe Combinational Logic

Processes can be used to describe combinational or sequential systems. Two requirements for a process to synthesize to a combinational system are:

1. The process's sensitivity list must contain all signals read in the process.
2. A signal or variable assigned a value in a process must be assigned a value in all possible executions of the process.

5.3 SEQUENTIAL STATEMENTS

Sequential statements are executed one after another. They can appear only inside of a process or a subprogram and are used to describe algorithms. In this chapter, the discussion is limited to use of sequential statements in processes. Sequential statements in subprograms are discussed in Chapter 12. VHDL's sequential statements are listed in Table 5.3.1.

Sequential signal assignment statements were introduced earlier in this chapter. A more detailed discussion of signal assignment statements is presented in Chapter 6, where we look at additional variations of the signal assignment statement and examine in detail just when a new value is assigned to a signal.

Variables are introduced in Section 5.7, near the end of this chapter. Variable assignments statements are used to assign values to variables and are discussed in that section.

Table 5.3.1
VHDL's sequential statements.

Sequential Statements
signal assignment
variable assignment
wait
assertion
report
procedure call
return
case
null
if
loop
next
exit

Wait, assertion, and report statements are considered in Chapter 7 in terms of their use in testbenches. Wait statements can also be used to synthesize storage elements as described in Chapter 8.

Procedure call statements and return statements are discussed along with subprograms in Chapter 12.

Sequential Control Statements

The remaining sequential statements in Table 5.3.1 are control statements. *Control statements* select between alternative statement execution paths. Control statements are introduced in this chapter, along with examples of their use in describing combinational systems.

5.4 CASE STATEMENT

The case statement and the if statement both allow us to select one sequence of statements for execution from a number of alternative sequences.

Use of a *case statement* is appropriate when the selection can be based on the value of a single expression.

The case statement's simplified syntax is given in Figure 5.4.1.

When used to assign a value to a single signal, a sequential case statement is essentially equivalent to a concurrent selected signal assignment statement. The selected signal assignment statement is often referred to as a shorthand version of such a case statement.

```
case_statement ::=
case expression is
when choices => sequence_of_statements
{ when choices => sequence_of_statements }
end case [ case_label ] ;

choices ::= choice { | choice }
choice ::= simple_expression | discrete_range | element_simple_name | others
```

FIGURE 5.4.1
Simplified syntax for a case statement.

Case Expression

The expression following the keyword `case` is called the case expression (or selector expression). This expression must be of a discrete type or of a one-dimensional character array type.

Case Choices

The choices specify for which values of the case expression a particular branch is executed. Each branch contains an alternative sequence of statements. The choices must be mutually exclusive locally static values of the case expression. Locally static means that the choice values must be determinable when the architecture containing the case statement is compiled. This restriction rules out the use of a variable in specifying a choice.

One and only one choice in a case statement must match the value of the case expression. Accordingly, in addition to the choices being mutually exclusive, exactly one choice must appear for every possible value of the case expression. That is, the choices must be exhaustive.

Listing Choices

A list of choices can be associated with a single branch by separating each choice in the list using the vertical bar (|) symbol.

Others Choice in a Case Statement

If the set of choices does not enumerate every possible value of the case expression type, then the last when choices clause must be an `others` choice.

The `others` choice handles all possible values of the case expression not explicitly listed in previous choices. Only one branch may have an `others` choice and, if it is included, it must be the only choice of the last branch in the case statement.

A branch with an `others` choice can still be included, even if all possible values of the case expression have been covered by previous choices. In this situation, the branch associated with the `others` choice will never be selected.

LISTING 5.4.1
A two-input XOR using a case statement.

```
library ieee;
use ieee.std_logic_1164.all;

entity xor_2 is
   port ( a, b : in std_logic;
      c : out std_logic);
end xor_2;

architecture bhvxor2c of xor_2 is
begin
   process (a, b)
   begin
      case std_logic_vector'(a, b) is
         when "01" => c <= '1';
         when "10" => c <= '1';
         when others => c <= '0';
      end case;
   end process;
end bhvxor2c;
```

Others Choice and Don't Care Output Values

For a combinational design that has input combinations for which we wish to assign don't care output values, the **others** choice can be used to select a sequence of statements that make the don't care output assignments.

Case Statement Examples

In Listing 5.4.1 a two-input exclusive-OR function is described using a case statement.

In this description, the case expression is a std_logic_vector aggregate of the entity's inputs. The first two branches specify choices that are values of the inputs that cause the XOR function to be a `'1'`. The **others** choice handles all the other possible values of the aggregate. For these values, the output is assigned a value of `'0'`.

By taking advantage of the fact that we can specify more than one choice in a choice list by using the | symbol, the case statement in Listing 5.4.1 can be written more concisely as:

```
case tmp is
     when "01" | "10" => c <= '1';
     when others => c <= '0';
end case;
```

A 2-to-4 decoder is described using a case statement in Listing 5.4.2.

LISTING 5.4.2
A 2-to-4 decoder using a case statement.

```
library ieee;
use ieee.std_logic_1164.all;

entity decode is
   port (g_bar, b, a : in std_logic;
      y0, y1, y2, y3 : out std_logic);
end;

architecture behavior of decode is
begin
   casey: process (g_bar, b, a)
   begin
      -- default value assigned to outputs
      y0 <= '1'; y1 <= '1'; y2 <= '1'; y3 <= '1';

      case std_logic_vector'(g_bar, b, a) is
         when "000" => y0 <= '0';
         when "001" => y1 <= '0';
         when "010" => y2 <= '0';
         when "011" => y3 <= '0';
         when others => null;
      end case;

   end process;
end behavior;
```

When this decoder is enabled, only one of its outputs should be `'0'`. This output is determined by the values of `b` and `a`. Each of the first four case branches assigns a value to only one of the four outputs. If any of these branches is taken, the other three outputs are not assigned a value by the case statement. Recall that, for a process to describe a combinational system, one of the requirements is that in every execution of the process all of the outputs must be assigned a value. To ensure this, default assignment statements assign values to all of the outputs at the beginning of the process. This prevents latches from being inferred.

The fifth case branch results in the execution of a null statement. Execution of a null statement causes no action to take place. So, if this branch is taken, all of the outputs are left with the default values assigned at the beginning of the process.

Null Statement

A *null statement* performs no action. Its only function is to pass control on to the next statement.

The simplified syntax for a null statement is given in Figure 5.4.2.

A null statement is used to explicitly state to a reader that no action is to be taken. It is often used in a case statement to allow all possible values of the case expression to be covered. For certain choices it may be that no action is required, as in Listing

```
null_statement ::=
null ;
```

FIGURE 5.4.2
Simplified syntax for a null statement.

5.4.2. Use of the null statement makes our intentions clearer than simply leaving that branch void of any statements.

Example of a 74F539 Decoder Equivalent

A 74F539 is a dual 2-to-4 decoder IC with three-state outputs. The truth table for this decoder is given in Table 5.4.1.

The 74F539 has three more control inputs than the previously considered 2-to-4 decoders. If the output enable input $\overline{OE}$ is a 1, all outputs are in their high-impedance states. If enable input $\overline{E}$ is a 1, no output is selected and all outputs have the value of the polarity input P. When both $\overline{OE}$ and $\overline{E}$ are 0s, the output selected by inputs A_1 and A_0 has the value $\overline{P}$, and all other outputs have the value P.

Since the 74F539 is an obsolete part, we could design our own equivalent and implement it in a PLD. While this decoder's functionality is relatively complex, its VHDL description is very concisely written using a case statement, Listing 5.4.3.

Inputs A_1 and A_0 are declared as the two-element std_logic_vector `a`. Outputs O_3 to O_0 are declared as the four-element vector `o`. The process assigns a default value of `p` to all outputs. This is accomplished by creating an aggregate of four copies of `p`. This aggregate does not need to be type qualified, because the compiler can determine from the context (the aggregate is assigned to std_logic_vector `o`) that it is type std_logic_vector. A case statement is then used to assign a value to the output vector based on the values of `oe_bar` and `e_bar` and, in the case of the first branch, vector `a`.

Table 5.4.1
Truth table for one half of a 74F539 dual 2-to-4 decoder.

Function	Inputs				Outputs			
	$\overline{OE}$	$\overline{E}$	A_1	A_0	O_0	O_1	O_2	O_3
High impedance	1	–	–	–	Z	Z	Z	Z
Disable	0	1	–	–	$O_n = P$			
Active High Output (P = 0)	0	0	0	0	1	0	0	0
	0	0	0	1	0	1	0	0
	0	0	1	0	0	0	1	0
	0	0	1	1	0	0	0	1
Active Low Output (P = 1)	0	0	0	0	0	1	1	1
	0	0	0	1	1	0	1	1
	0	0	1	0	1	1	0	1
	0	0	1	1	1	1	1	0

LISTING 5.4.3
Description of one half of a 74F359 dual 2-to-4 decoder.

```
library ieee;
use ieee.std_logic_1164.all;
use ieee.numeric_std.all;

entity ic74f539 is
   port (
      a : in std_logic_vector(1 downto 0);      -- address (select) inputs
      p :  in std_logic;                        -- polarity input
      e_bar :  in std_logic;                    -- enable input
      oe_bar : in std_logic;                    -- output enable input
      o : out std_logic_vector(3 downto 0)      -- outputs
      );
end ic74f539;

architecture behav_case of ic74f539 is
begin
   process(a, p, oe_bar, e_bar)
   begin
      o <= (p,p,p,p);
      case std_logic_vector'(oe_bar, e_bar) is
         when "00" => o( to_integer(unsigned(a))) <= not p;
         when "01" => null;
         when others => o <= "ZZZZ";
      end case;
   end process;
end behav_case;
```

5.5 IF STATEMENT

An *if statement* selects one or none of its alternative sequences of statements for execution, depending on the value of each branch's condition. It establishes a priority in the determination of which sequence of statements is executed. An if statement has the simplified syntax given in Figure 5.5.1.

When used to assign a value to a single signal, a sequential if statement is essentially equivalent to a concurrent conditional signal assignment statement. The conditional signal assignment statement is often referred to as a shorthand version of such an if statement.

If Statement Evaluation

If the first condition in an if statement evaluates true, the sequence of statements following the first **then** is executed and the construct is terminated. If the first condition is not true, and an elsif clause is included, the condition associated with the

```
if_statement ::=
    if condition then
        sequence_of_statements
  { elsif condition then
        sequence_of_statements }
  [ else
        sequence_of_statements ]
  end if [ if_label ] ;
```

FIGURE 5.5.1
Simplified syntax for an if statement.

first **`elsif`** is tested. If this condition is true, the sequence of statements following the first **`elsif`** is executed and the construct is terminated.

Since the elsif clause in the syntax definition is enclosed in curly brackets, zero, one, or a multiple number of elsif clauses may appear in the construct. The condition associated with each elsif clause is evaluated in succession until one evaluates true. The sequence of statements associated with the first elsif clause that evaluates true is then executed and the construct is terminated.

If none of the elsif clauses evaluate true and an else clause is included, the sequence of statements following the **`else`** is executed and the construct is terminated.

If none of the conditions in the if statement evaluate true and there is no else clause, then none of the included sequences of sequential statements is executed.

Common errors made when first writing if statements are to write **`elsif`** as two words (else if) or to write **`end if`** as a single word (endif).

If Statement Examples

A two-input XOR gate is described in Listing 5.5.1 using an if statement. The condition associated with the if clause is the condition that causes the output to be `'1'`. Accordingly, the sequential statement following **`then`** assigns `'1'` to `c`. The else clause is included to handle all other input combinations. It assigns the value of `'0'` to `c`.

LISTING 5.5.1
XOR gate described using an if statement with an else clause.

```
library ieee;
use ieee.std_logic_1164.all;

entity xor_2 is
   port ( a, b : in std_logic;
      c : out std_logic);
end xor_2;
```

(Cont.)

LISTING 5.5.1 *(Cont.)*

```
architecture bhvxor2i of xor_2 is
begin
   process ( a, b)
   begin
      if a /= b then
         c <= '1';
      else
         c <= '0'; -- default
      end if;
   end process;
end bhvxor2i;
```

Adding default assignments at the beginning of a process eliminates the need for an else clause when describing combinational circuits using an if statement. In Listing 5.5.2 the default assignment to c is made at the beginning of the process and the else clause is eliminated.

Since there is no limitation on the kind of sequential statements in an if statement, if statements can be nested. In Listing 5.5.3, a 2-to-4 decoder is described using nested if statements.

LISTING 5.5.2
XOR gate described using a default signal assignment and an if without an else.

```
library ieee;
use ieee.std_logic_1164.all;

entity xor_2 is
   port ( a, b : in std_logic;
      c : out std_logic);
end xor_2;

architecture bhvxor2id of xor_2 is
   begin
   process ( a, b)
      begin

      c <= '0'; -- default

      if a /= b then
         c <= '1';

      end if;
   end process;
end bhvxor2id;
```

LISTING 5.5.3
A 2-to-4 decoder using nested if statements.

```
library ieee;
use ieee.std_logic_1164.all;

entity decode is
   port (g_bar, b, a : in std_logic;
   y0, y1, y2, y3 : out std_logic);
end;

architecture behavior of decode is

begin
   iffy: process (g_bar, b, a)
      begin
         y0 <= '1';-- default output values
         y1 <= '1';
         y2 <= '1';
         y3 <= '1';

         if (g_bar ='0') then
            if ((b ='0') and (a = '0')) then
               y0 <= '0';
            elsif ((b ='0') and (a = '1')) then
               y1 <= '0';
            elsif ((b ='1') and (a = '0')) then
               y2 <= '0';
            elsif ((b ='1') and (a = '1')) then
               y3 <= '0';
            end if;
         end if;
      end process;
end behavior;
```

An example of a behavioral description of a 4-bit magnitude comparator described using multiple processes, each consisting of a single if statement, is shown in Listing 5.5.4. This comparator is functionally equivalent to the one in Listing 4.5.1.

LISTING 5.5.4
A 4-bit magnitude comparator using processes containing if statements.

```
library ieee;
use ieee.std_logic_1164.all;

entity mag_comp is
   port ( p, q : in std_logic_vector (3 downto 0);
```

(Cont.)

LISTING 5.5.4 *(Cont.)*

```
      p_gt_q_bar, p_eq_q_bar, p_lt_q_bar : out std_logic);
end mag_comp;

architecture behavior of mag_comp is
begin
   gt: process (p, q)
   begin
      if p > q then
            p_gt_q_bar <= '0';
      else
            p_gt_q_bar <= '1';
      end if;
   end process gt;

   lt: process (p, q)
   begin
      if p < q then
            p_lt_q_bar <= '0';
      else
            p_lt_q_bar <= '1';
      end if;
   end process lt;

   eq: process (p, q)
   begin
      if p = q then
            p_eq_q_bar <= '0';
      else
            p_eq_q_bar <= '1';
      end if;
   end process eq;

end behavior;
```

This architecture has three independent processes. Each process computes one of the three outputs. The condition in each if statement uses a relational operator to make the comparison.

Case Statement versus If Statement

Case and if statements are similar in that they both allow us to select between execution of alternative sequences of sequential statements. Their difference is in how the selection is made.

A case statement evaluates a single case expression and executes the sequence of statements associated with the choice that matches the value of the case expression.

An if statement evaluates one or more conditions in succession and executes the sequence of statements associated with the first condition that evaluates true. An if statement provides inherent priority.

Since an if statement is capable of evaluating multiple arbitrarily complex conditions to make a selection, it is more flexible than a case statement.

5.6 LOOP STATEMENT

A *loop statement* contains a sequence of sequential statements that can be repeatedly executed, zero or more times. It has the simplified syntax shown in Figure 5.6.1.

Execution of a loop continues until the loop is terminated as a result of any of the following:

- Completion of its iteration scheme
- Execution of an exit statement
- Execution of a next statement that specifies a label outside of the loop
- Execution of a return statement

A return statement can appear inside a loop only if the loop is in a subprogram.

Infinite Loop

As indicated in Figure 5.6.1, the iteration scheme is optional. A loop statement without an iteration scheme is an infinite loop, one whose sequence of statements is executed endlessly. An infinite loop cannot be synthesized.

Iteration Schemes

There are two possible iteration schemes, resulting in two kinds of loops: for loops and while loops. For loops are synthesizable, but while loops are not. Although not synthesizable, while loops are useful in testbenches.

```
loop_statement ::=
[ loop_label:]
[ iteration_scheme ] loop
sequence_of_statements
end loop [ loop_label ] ;

iteration_scheme ::= for identifier in discrete_range | while condition
discrete_range ::= discrete_subtype_indication | range
```

FIGURE 5.6.1
Simplified syntax for a loop statement.

For Loop

Substituting the for iteration scheme into the syntax of Figure 5.6.1 gives:

```
[loop_label:] for identifier in range loop
      sequence_of_statements
end loop [loop_label];
```

In a for loop, the identifier following keyword **for** serves as an implicit declaration of a *loop parameter* with the specified name. Thus, we do not need to explicitly declare a loop parameter. The loop parameter's type is the base type of the discrete range. The loop parameter only exists while the loop is being executed, and is not visible outside of the loop.

Within the sequence of sequential statements, the loop parameter is treated as a constant. Thus, it can be read by statements within the loop, but it cannot be written.

For Loop Execution

The discrete range specifies how many iterations of the loop are to occur. To execute a for loop, the discrete range is first evaluated. If the discrete range is null, the loop's execution is complete and the sequence of statements in the loop is not executed.

If the discrete range is not null, the instructions in the loop are normally executed once for each value of the discrete range. However, if a next, exit, or return statement appears in the loop, it can cause fewer iterations of the loop to occur before it is terminated.

Prior to each loop iteration, the corresponding value of the discrete range is assigned to the loop parameter. Values from the discrete range are assigned to the loop parameter in left-to-right order. After the loop has been executed the specified number of times, the iteration scheme is complete and the loop is terminated.

While Loop

Substituting the while iteration scheme into the syntax of Figure 5.6.1 gives:

```
[loop_label:] while condition loop
      sequence_of_statements
end loop [loop_label];
```

The condition of a while loop is evaluated ***before*** each iteration of the loop, including the first iteration. If the condition is true, the sequence of statements in the loop is executed. If false, the iteration scheme is complete and the loop is terminated. The condition is not tested during the execution of the sequence of statements in the loop.

With a while loop, we must ensure that statements inside the loop will eventually cause the loop's condition to evaluate false or cause an exit statement or return statement to be executed to terminate the loop. Otherwise, the loop is infinite.

Next and Exit Statements

Next and exit statements are used only inside of loops. They are used to terminate the execution of a single iteration of a loop or to terminate an entire loop, respectively.

```
next_statement ::=
[ label : ] next [ loop_label ] [ when condition ] ;
```

FIGURE 5.6.2
Simplified syntax for a next statement.

Next Statement

A *next statement* is used to terminate the current iteration of an enclosing loop statement and go to the next iteration. If the next statement includes a condition, completion of the current loop iteration is conditional. The simplified syntax is given in Figure 5.6.2.

A next statement that has a loop label is only allowed within the loop that has that label, and applies to only that loop. A next statement without a loop label is only allowed within a loop, and applies only to the innermost enclosing loop (whether the loop is labeled or not).

For the execution of a next statement, the condition, if present, is first evaluated. The current iteration of the loop is terminated if the condition is true or if there is no condition. If the next statement has no label or if the label is that of the innermost loop containing the next statement, the number of complete iterations of the loop can be less than specified by the discrete range.

Execution of a next statement can cause a loop to be terminated, if the loop is nested inside an outer loop and the next statement has a loop label that denotes the outer loop. In this situation, the current iteration of the outer loop is terminated, causing the termination of the inner loop.

Exit Statement

An *exit statement* is used to terminate the execution of an enclosing loop statement. If the exit statement includes a condition, termination of the loop is conditional. The simplified syntax for an exit statement is given in Figure 5.6.3.

An exit statement with a loop label is only allowed within the loop that has that label, and applies to that loop. An exit statement without a loop label is only allowed within a loop, and applies only to the innermost enclosing loop (whether labeled or not).

For the execution of an exit statement, the condition, if present, is first evaluated. The loop is terminated if the value of the condition is true or if there is no condition.

Description of a 4-bit Comparator Using a Loop

A description of a 4-bit magnitude comparator using a loop is given in Listing 5.6.1. While this certainly is not the most concise way to describe a comparator, it serves as a simple example of the use of a for loop with exit statements.

```
exit_statement ::=
[ label : ] exit [ loop_label ] [ when condition ] ;
```

FIGURE 5.6.3
Simplified syntax for an exit statement.

LISTING 5.6.1
A 4-bit magnitude comparator described using a for loop.

```
library ieee;
use ieee.std_logic_1164.all;

entity mag_comp is
   port ( p, q : in std_logic_vector (3 downto 0);
      p_gt_q_bar, p_eq_q_bar, p_lt_q_bar : out std_logic);
end mag_comp;

architecture behavior of mag_comp is
begin
   comp: process (p, q)
   begin

      p_gt_q_bar <= '1';  --defaults
      p_eq_q_bar <= '1';
      p_lt_q_bar <= '1';

      for i in 3 downto 0 loop
         if ((p(i) = '1') and (q(i) = '0')) then
            p_gt_q_bar <= '0';
            exit;
         elsif((p(i) = '0') and (q(i) = '1')) then
            p_lt_q_bar <= '0';
            exit;
         elsif i = 0 then
            p_eq_q_bar <= '0';
         end if;
      end loop;

   end process comp;
end behavior;
```

The process first sets all the outputs to `'1'`s, then uses a for loop to determine which output to make a `'0'`.

The range of the loop parameter `i` is from 3 down to 0. During each loop iteration, the if statement compares two corresponding elements of `p` and `q` to determine which vector is greater. As soon as the determination can be made, the loop is exited.

In each pass through the loop, the if clause determines whether `p(i)` is a `'1'` and `q(i)` is a `'0'`. If this condition is true, output `p_gt_q_bar` is assigned `'0'`, and the exit statement terminates the loop.

If the previous condition does not evaluate true, the first elsif clause determines whether `p(i)` is a `'0'` and `q(i)` is a `'1'`. If this condition is true, output `p_lt_q_bar` is assigned `'0'` and the exit statement terminates the loop.

The last elsif determines whether loop parameter `i` equals `0`. This condition can be true only on the fourth pass through the loop. If this condition is true, output `p_eq_q_bar` is assigned `'0'`, because all of the corresponding elements of the two vectors must have been equal to reach this point.

5.7 VARIABLES

The four kinds of VHDL objects were introduced in Chapter 3: signals, constants, variables, and files. Signals and constants have been used extensively in previous examples. We now examine variables and use them in descriptions.

Normal Variables and Shared Variables

There are two kinds of variables in VHDL, normal variables and shared variables. *Shared variables* can be accessed from multiple processes and are not synthesizable. Normal variables can only be accessed from a single process or subprogram and are synthesizable. The subsequent discussions of variables in this book are limited to normal variables.

Storing Intermediate Values in a Process or Subprogram

Variables are used to store intermediate values in a process or subprogram. Variables must be declared in the declarative part of the process or subprogram in which they are used. A variable is local to the process or subprogram in which it is declared (visible only within that process or subprogram).

Variables in a process preserve their values between executions of the process. That is, between its suspension and reactivation. In contrast, variables in a subprogram do not preserve their values between executions of the subprogram. Variables in subprograms are discussed further in Chapter 12.

The syntax for a variable declaration is given in Figure 5.7.1.

Variable Initial Value

A variable can optionally be assigned an initial value when it is declared. An initial value is assigned by using the variable assignment symbol := . For example:

```
variable count : std_logic_vector (3 downto 0)
          := "0000";
```

This statement declares variable `count` and gives it an initial value of `"0000"`. ***Importantly, IEEE 1076.6 compliant synthesizers ignore initial values assigned to***

> variable_declaration ::=
> [~~shared~~] **variable** identifier_list : subtype_indication [<u>:= expression</u>] ;

FIGURE 5.7.1
Syntax for a variable declaration.

variables. Accordingly, in a description, if a variable requires an initial value, it should be assigned by a separate variable assignment statement in the process or subprogram that uses the variable.

Variable Assignment Statement

Variables are assigned values by using a variable assignment statement. To distinguish assignments to variables from assignments to signals, two different assignment symbols are used. The symbol := is the *variable assignment symbol*. For example:

```
count := "0000";
```

assigns the value "0000" to the variable count.

Variable Assignments Take Effect Immediately

When a variable is assigned a value, it takes that value immediately. This action is the same as for a variable in a conventional programming language. However, it is different from the situation for a signal. When a signal is assigned a value, the assigned value does not take effect immediately. When a signal is assigned a value in a process, the signal does not take the new value until after the process has suspended. This characteristic of a signal has not been previously emphasized. It is considered in detail in Chapter 6.

It will also become more clear in Chapter 6 when use of a variable in a process is preferable to use of a signal. However, as a general guideline, a variable is used to store data within a process when the new value assigned to the variable is to be used (read) in the same execution of the process.

Variable and Signal Differences

At this point, some of the obvious differences between a variable and a signal in a process can be stated.

- A variable used in a process must be declared inside the process. A signal used in a process must be declared outside of the process (in the declarative part of the architecture or as a port in the associated entity declaration).
- A variable is only visible in the process in which it is declared. A signal is visible to all processes in the architecture in which it is declared.

As an example of the use of variables, the magnitude comparator is described using variables in Listing 5.7.1.

LISTING 5.7.1
A 4-bit magnitude comparator description using variables.

```
library ieee;
use ieee.std_logic_1164.all;

entity mag_comp is
   port ( p, q : in std_logic_vector (3 downto 0);
      p_gt_q_bar, p_eq_q_bar, p_lt_q_bar : out std_logic);
end mag_comp;
```

```
architecture vari of mag_comp is
begin
   comp: process (p, q)

   variable  p_gt_q_bar_v, p_lt_q_bar_v, p_eq_q_bar_v : std_logic;

   begin

      p_gt_q_bar_v  := '1';   -- default values
      p_lt_q_bar_v  := '1';
      p_eq_q_bar_v  := '1';

      for i in 3 downto 0 loop
         if ((p(i) = '1') and (q(i) = '0')) then
            p_gt_q_bar_v := '0';
            exit;
         elsif ((p(i) = '0') and (q(i) = '1')) then
            p_lt_q_bar_v := '0';
            exit;
         end if;
      end loop;

      if ((p_gt_q_bar_v = '1') and (p_lt_q_bar_v = '1')) then
         p_eq_q_bar_v := '0';
      end if;

      p_gt_q_bar <= p_gt_q_bar_v;   -- assign variable values to ports
      p_lt_q_bar <= p_lt_q_bar_v;
      p_eq_q_bar <= p_eq_q_bar_v;

   end process comp;
end vari;
```

In this description, the determination as to whether to assert `p_eq_q_bar` is made based on the values of `p_gt_q_bar` and `p_lt_q_bar` determined during the same execution of the process. Variables are used to store intermediate values.

Variables `p_gt_q_bar_v`, `p_lt_q_bar_v`, and `p_eq_q_bar_v` are declared in the declarative part of the process. In the statement part of the process these variables are all given a default value of `'1'`. A loop is used to determine if either `p_gt_q_bar_v` or `p_lt_q_bar_v` should be asserted. If either of these variables is to be asserted, it is assigned a `'0'`. This assignment takes effect immediately.

After the loop is terminated, an if statement is used to determine whether variable `p_eq_q_bar_v` should be asserted. If either `p_gt_q_bar_v` or `p_lt_q_bar_v` were assigned a value of `'0'` in an execution of the loop statement, that `'0'` is detected in the execution of the if statement.

Before the process ends, the values of the variables are assigned to the corresponding ports. This must be done inside the process because variables are not visible outside of a process.

5.8 PARITY DETECTOR EXAMPLE

A few examples of designs for a parity detector are given in this section. These examples clarify some important concepts concerning variables and signals. Design entity `parity` takes a four-element std_logic_vector named `din` as its input and has a single std_logic output `oddp`. Output `oddp` is asserted when the number of 1s in the input vector is odd, otherwise it is unasserted.

Dataflow Description of Parity Detector

If we are aware that a four-input XOR gate detects odd parity for a 4-bit vector, the simplest way to describe the parity detector is using a single concurrent signal assignment statement. A dataflow style description using this approach is given in Listing 5.8.1.

A simulation of this design description results in the waveforms in Figure 5.8.1.

The testbench sequences through all possible binary input combinations. Input vector `din` is displayed as both a composite signal, whose value is expressed in hexadecimal, and in terms of its separate elements. The waveforms indicate that the description is functionally correct.

Synthesis of the design description in Listing 5.8.1 produces the hierarchical representation of the logic in Figure 5.8.2. The notation [3:0] represents a 4-bit bus. The notation [3] indicates bit 3 of the bus. As expected, the logic is a four-input XOR gate.

LISTING 5.8.1
Dataflow design description of parity detector.

```
library ieee;
use ieee.std_logic_1164.all;

entity parity is
   port (
      din: in std_logic_vector (3 downto 0);  -- 4-bit input vector
      oddp: out std_logic                     -- asserted for odd parity
      );
end parity;

architecture dataflow of parity is
begin
   oddp <= din(3) xor din(2) xor din(1) xor din(0);
end dataflow;
```

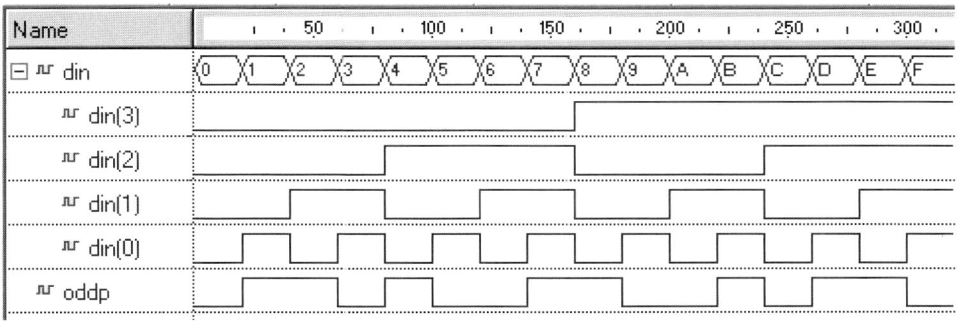

FIGURE 5.8.1
Simulation of dataflow style parity detector.

Loop and Signal Description of Parity Detector

In contrast to the dataflow approach, we might envision the function of the parity detector in terms of an algorithm that examines each bit of the input vector one at a time and keeps track of whether the parity of all bits previously examined is odd. After the last bit has been examined, the parity of the entire vector is known.

An architecture using this iterative approach is given in Listing 5.8.2. The entity declaration for the parity detector remains the same.

In Listing 5.8.2, a loop is used to examine each bit of the input vector in turn, starting with the leftmost bit. A signal named `odd` is used to keep track of the computed parity. This signal is declared in the declarative part of the architecture.

In the process, signal `odd` is first assigned the value `'0'`. The loop is then entered. Each time through the loop, `odd` is assigned the XOR of the current value of `odd` and the element of the input vector selected by the loop index. When the loop is complete, the value of signal `odd` is assigned to output port `oddp`.

When the description in Listing 5.8.2 is simulated using the same testbench as before, the waveforms in Figure 5.8.3 result.

It is obvious from the waveforms that the output is incorrect. The value of `oddp` remains `'U'` for the entire simulation. The value `'U'` is indicated in the waveform by the dotted line halfway between the `'0'` and `'1'` logic levels.

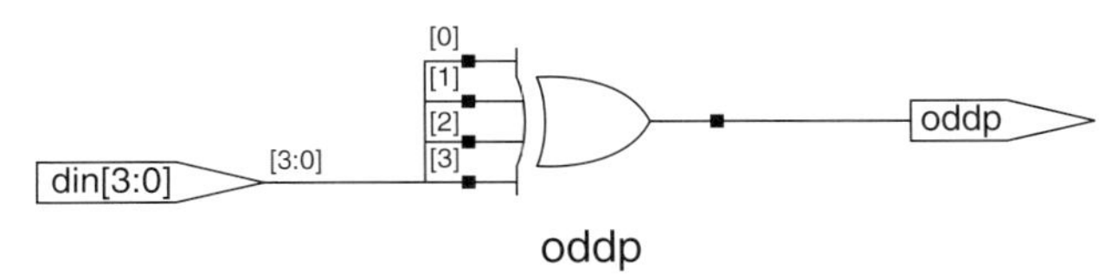

FIGURE 5.8.2
Logic synthesized from the dataflow style parity detector description in Listing 5.8.1.

LISTING 5.8.2
Architecture for parity detector written using a loop and a signal.

```
architecture behav_sig of parity is
signal odd : std_logic;
begin
   po: process (din)
   begin
      odd <= '0';
      for index in 3 downto 0 loop
         odd <= odd xor din(index);
      end loop;
   end process;
      oddp <= odd;
end behav_sig;
```

Before examining why this description failed, lets see what would have happened if, because the design seemed so simple, we skipped simulation and went directly to synthesis. Synthesis using the architecture in Listing 5.8.2 produces the logic in Figure 5.8.4.

The diagram shows element `din(0)` of the four-element input vector `din` as an input to a two-input XOR gate. The other input to the XOR gate is the gate's output fed back. When the description is synthesized, the synthesizer gives a warning that the design contains a "combinational loop."

Unrolling a Loop

Insight into what is wrong with the description in Listing 5.8.2 can be gained by considering how a RTL synthesizer synthesizes a loop. The synthesizer "unrolls" the loop to create the sequence of statements that corresponds to the complete execution of the loop. One copy of the sequence of statements inside the loop is created for each loop iteration. In each copy, the loop parameter is replaced by its value for that loop iteration. The loop is replaced by the statements created by unrolling the loop. Unrolling a loop does not change its behavior.

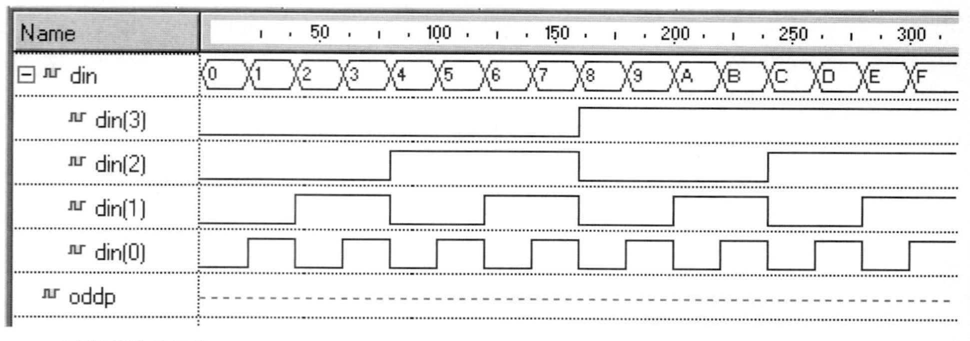

FIGURE 5.8.3
Waveforms from simulation of parity detector design in Listing 5.8.2.

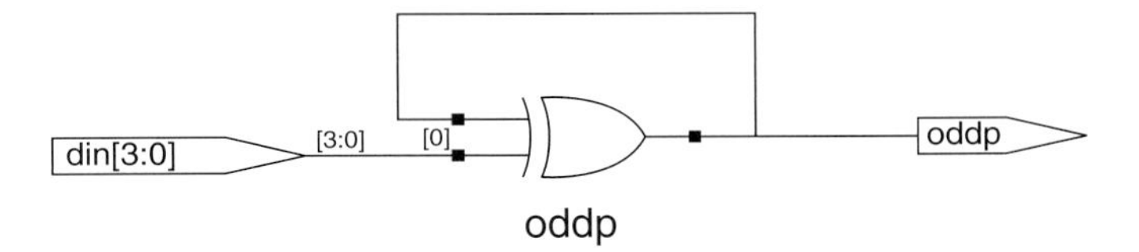

FIGURE 5.8.4
Logic synthesized from the loop and signal description of the parity detector in Listing 5.8.2.

In this description, unrolling the loop produces four signal assignment statements, each of which assigns a value to signal `odd`. When the assignment to `odd` at the beginning of the process is included, the process contains a total of five sequential assignments to signal `odd`:

```
odd <= '0';
odd <= odd xor din(3);
odd <= odd xor din(2);
odd <= odd xor din(1);
odd <= odd xor din(0);
```

Since `odd` is a signal, the assignment of a value to it does not take effect until after the process suspends. Each of these sequential signal assignment statements simply schedules a value to be assigned to `odd` after the process suspends. As a result, only the last assignment to `odd` before the process suspends is meaningful. Accordingly, the entire sequence of five sequential assignments to `odd` reduces to the last assignment, which is:

```
odd <= odd xor din(0);
```

The synthesizer synthesizes logic that corresponds to only this last assignment. Thus, it synthesizes the logic in Figure 5.8.4.

It is now clear from this last assignment statement why the simulation resulted in `oddp` always having the value `'U'`. At initialization, `odd` is given the value `'U'`. After the process suspends `odd` is assigned `'U' xor din(0)`, which evaluates to `'U'`. Thus, the process always computes the value `'U'` for `odd`. The concurrent signal assignment statement following the process assigns `odd` to `oddp`.

Loop and Variable Description of Parity Detector

The solution to the problem in Listing 5.8.2 is to realize that, for our algorithm to work, we need each assignment to `odd` to take effect immediately. Thus, we need to use a variable to store the value of `odd`. The design description in Listing 5.8.2 is modified by replacing the signal declaration for `odd` in the declarative part of the architecture body with a variable declaration for `odd` in the declarative part of the process. Since we are now assigning values to a variable, the assignment operator must also be changed. The architecture of the resulting description is given in Listing 5.8.3.

LISTING 5.8.3
Architecture for parity detector written using a loop and variable.

```
architecture behav_var of parity is
begin
   po: process (din)
   variable odd : std_logic;   -- declare a variable
   begin
      odd := '0';
      for index in 3 downto 0 loop
         odd := odd xor din(index);
      end loop;
      oddp <= odd;

   end process;
end behav_var;
```

Simulation using the architecture in Listing 5.8.3 gives the correct outputs, producing waveforms identical to those in Figure 5.8.1.

The hierarchical representation of the logic produced by synthesis of the description in Listing 5.8.3 is shown in Figure 5.8.5.

The sequence of sequential assignments made to variable odd, including those from unrolling the loop, are:

```
odd := '0';
odd := odd xor din(3);
odd := odd xor din(2);
odd := odd xor din(1);
odd := odd xor din(0);
```

Since odd is now a variable and assignments to variables take effect immediately, the algorithm works as desired.

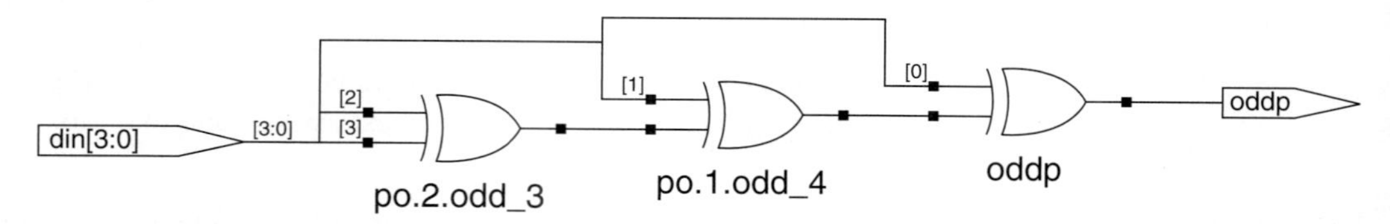

FIGURE 5.8.5
Logic resulting from synthesis using the architecture of Listing 5.8.3.

5.9 SYNTHESIS OF PROCESSES DESCRIBING COMBINATIONAL SYSTEMS

When a process is used to describe a combinational system, what is important is the final result computed by the sequence of operations that comprises the algorithm, not the sequence of operations itself. The synthesizer's objective is to synthesize logic that computes the same final result as the algorithm.

This fact is illustrated by the architecture description in Listing 5.8.3 for the odd parity detector. The hierarchical representation of the synthesized logic in Figure 5.8.5 reflects the order of these assignments corresponding to the unrolled loop:

```
odd := '0';
odd := odd xor din(3);
odd := odd xor din(2);
odd := odd xor din(1);
odd := odd xor din(0);
```

However, note that while the description is sequential, the synthesized logic is combinational.

If the target PLD's library has no XOR primitives larger than two-input XOR gates, the hierarchical representation of the synthesized logic translates to the technology dependent flattened-to-gates representation in Figure 5.9.1. Here the four-input XOR function is realized by a two-level combination of three two-input XORs. The hierarchical representations of both the dataflow architecture in Listing 5.8.1 (Figure 5.8.2) and the process architecture using a loop and variable in Listing 5.8.3 (Figure 5.8.5) produce the same flattened-to-gates representation in Figure 5.9.1 when targeted to the same PLD.

The 4-bit magnitude comparator in Listing 5.7.1 also illustrates the fact that what is synthesized for a process is functionally equivalent to the overall effect of the process, rather than its sequential nature.

Synthesizing Loops

As previously discussed, to synthesize a loop, a RTL synthesizer must unroll the loop. This requires that the number of loop iterations be known during synthesis. With respect to a for loop, this determination can be made if the loop's range bounds are statically deterministic. Thus, a for loop with constant bounds can be synthesized.

In contrast, if a variable were used in specifying the loop's range bounds, the synthesizer could not statically determine the number of loop iterations. Also, a synthesizer cannot statically determine the number of loop iterations of a while loop,

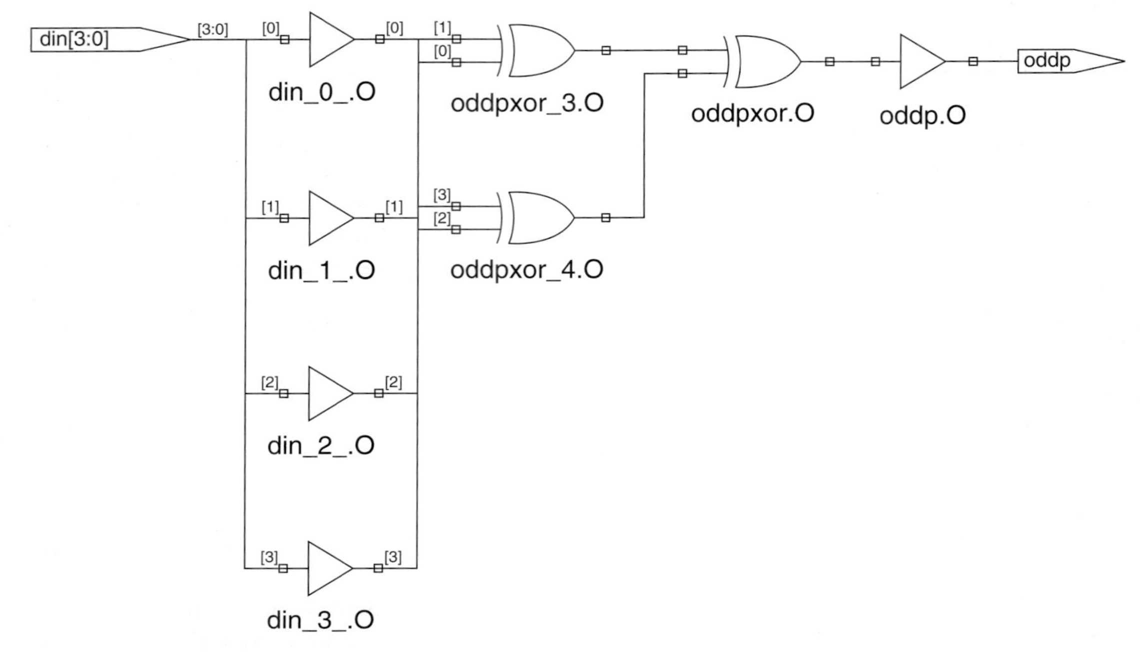

FIGURE 5.9.1
Flattened-to-gates representation of four-input odd parity detector.

since completion of a while loop is dependent on data generated during the loop's execution. Consequently, for loops with nonconstant bounds, all while loops, and infinite loops cannot be synthesized.

Avoiding Latches To ensure that the logic synthesized from a process is combinational, all signals and variables assigned a value in the process must be assigned a value each time the process executes. If this requirement is not met, latches may be inferred. To ensure this requirement is met it might be helpful to assign a default value at the beginning of the process to each signal or variable that is assigned values later in the process.

PROBLEMS

5.1 If a changes from `'1'` to `'0'` in Listing 5.1.1, what is the order of process execution and why?

5.2 If `carry_in` changes from `'0'` to `'1'` in Listing 5.1.2, what is the order of process execution and why?

5.3 What is the effect on the design of changing the textual order of the processes in the architecture body in Listing 5.1.2 so that process `ha2` is first, followed by `or1` and then `ha1`? Explain your answer.

5.4 What causes a process with a sensitivity list to suspend its execution? What causes a process without a sensitivity list to suspend its execution?

5.5 What is a control statement? List VHDL's control statements.

5.6 Write a behavioral style description of the logic circuit that follows. The description must have a separate process corresponding to each gate in the logic diagram. Each process must consist of a single sequential signal assignment statement.

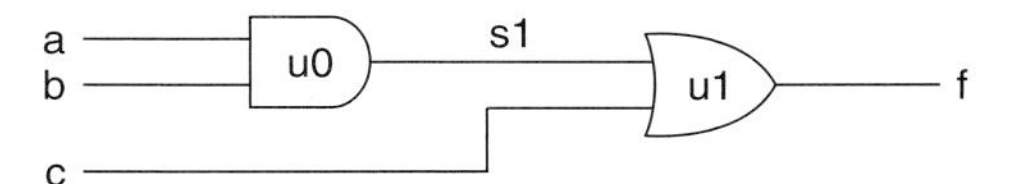

5.7 For the logic diagram shown below:

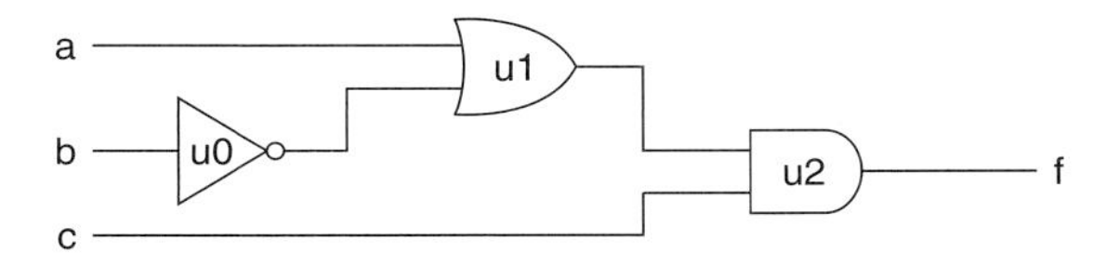

(a) Write an entity declaration for the circuit.
(b) Write a behavioral style architecture that is functionally equivalent to the logic diagram. This architecture must have a separate process for each gate. Label each process using the same label as used in the logic diagram. The statement part of each process must contain only signal assignment statements.
(c) Write another behavioral style architecture that is functionally equivalent to the logic diagram. This architecture must consist of only a single process.

5.8 Write a description of a combinational system that converts a BCD input code for a decimal digit to the excess three (XS3) code for the same decimal digit. The input code is the std_logic_vector `bcd`. The output code is the std_logic_vector `xs3`. Use a case statement to describe the logic. Write the case statement so that the synthesizer can synthesize the least amount of logic necessary to implement the converter.

5.9 The following truth table describes a function that must be implemented. In addition to this truth table the specification for the system states that input combinations 110 and 111 will never occur in the application:

a	b	c	w	y
0	0	0	1	0
0	0	1	0	1
0	1	0	1	1
0	1	1	1	0
1	0	0	1	1
1	0	1	0	0

Using a case statement, write a description that implements the function.

5.10 Repeat Problem 4.12 using a case statement to describe the system. Name the architecture `behav_case`.

5.11 Repeat Problem 4.36 using an if statement in the architecture body.

5.12 Given the following truth table:

(a) Write an entity declaration for the design.
(b) Write an architecture body that uses a process that contains an if statement.
(c) Write an alternative architecture body that uses a process that contains a case statement.

a	b	c	x
0	0	0	1
0	0	1	0
0	1	0	1
0	1	1	1
1	0	0	1
1	0	1	0
1	1	0	1
1	1	1	1

5.13 An excess-3 to decimal decoder is to be designed. The decoder's four inputs represent a decimal digit encoded in excess-3 code. There are 10 outputs, `y0` to `y9`. The single output corresponding to the digit represented by the excess-3 input code must be asserted high, all other outputs must be low. Write a behavioral style description of the decoder. Inputs and outputs must be represented as std_logic_vectors. The decoder's outputs for nonvalid excess-3 codes must be assigned don't care values.

w	x	y	z	digit
0	0	1	1	0
0	1	0	0	1
0	1	0	1	2
0	1	1	0	3
0	1	1	1	4
1	0	0	0	5
1	0	0	1	6
1	0	1	0	7
1	0	1	1	8
1	1	0	0	9

5.14 A two-to-one multiplexer has two data inputs, `a` and `b`, and one select input `s`. The multiplexer has a single output `c`. If `s` = `'0'`, `c` is equal to `a`. If `s` = `'1'`, `c` is equal to `b`.

(a) Write an entity declaration, named `mux`, for the multiplexer. Use only the data type std_logic.

(b) Write an architecture for `mux`, named `if_bod`, that uses a process statement containing an if statement to describe the architecture.

(c) Write an architecture for `mux`, named `case_bod`, that uses a process statement containing a case statement to describe the architecture.

5.15 Given the following function:

$$f = \bar{a}b\bar{c} + a\bar{b}\bar{c} + ab\bar{c}$$

(a) Write an entity declaration.

(b) Write an architecture body named `fcn_ife` that uses an if statement to implement the function.

(c) Write an alternative architecture body name `fcn_case` that uses a case statement to implement the function.

5.16 For the function specified by the truth table in Problem 4.24:

(a) Write a complete design description of the function in the truth table using a case statement. Call the entity `function_g`. Treat the inputs as scalars in the entity declaration. Write the description so that the synthesizer can take advantage of the don't care cases to simplify the function.

(b) Write an alternative architecture for entity `function_g` that uses an if statement.

5.17 A Fairchild 74LCX257 IC is a quad two-input multiplexer with three-state outputs. When its enable input `oe_bar` is asserted, it selects 4 bits of data from one of two sources `i0` or `i1`, selected by a common select input `s`, to appear on its 4-bit output. When `s` is 0 it selects `i0`, when it is 1 it selects `i1`. When `oe_bar` is not asserted, the outputs are in their high impedance state. Write an entity declaration named `lcx257` for an equivalent system. Represent the data inputs as std_logic_vectors `i0` and `i1`. Name the output vector `o`. Write an architecture body named `behavioral` for design entity `lcx257` that uses a process containing nested if statements to describe the system.

5.18 Repeat the design of the thermometer to binary encoder in Problem 4.26, except write an architecture that uses a case statement that includes the assignment don't care output values for invalid input codes. Write an alternative architecture that uses an if statement and assigns don't care output values for invalid input codes.

5.19 A design entity named `nand_8` is an eight-input NAND that has the following entity declaration:

```
entity nand_8 is
    port(
        input : in std_logic_vector(7 downto 0);
        nand_out : out std_logic
         );
end nand_8;
```

Write a behavioral style architecture body that uses a loop to describe `nand_8`.

5.20 Write a behavioral style description of a 4-to-2 priority encoder. The input is the std_logic_vector `x` (3 down to 0) and the output is the std_logic_vector `a` (1 down to 0). Inputs are asserted low. The output is the binary equivalent of the index of the highest priority input that is asserted. Input `x(3)` has the highest priority. Use a for loop.

5.21 Write a description that has a behavioral style architecture for the 3-to-8 decoder described in Listing 4.8.1. The behavioral style architecture should be based on either a select, if, or loop statement. Before writing the actual code consider which construct is likely to provide the most efficient coding.

5.22 Type unsigned, defined in package NUMERIC_STD, is similar to type std_logic_vector in that its elements are type std_logic. It differs in that, in addition to the logical operators defined for std_logic_vector, it has arithmetic operators, including an addition operator. This addition operator can be used to add two unsigned vectors or to add an unsigned vector and an integer.

Consider a design entity having the following input and output:

```
inputs : in std_logic_vector(3 downto 0)
count : out std_logic_vector(1 downto 0)
```

Write an entity declaration named `ones_count` that has the input and output shown above. Write an architecture body named `looped` for entity `ones_count`. This architecture uses a loop inside a process labeled `p1` to determine the number of 1s in the input vector and outputs the binary representation of this number as the value `count`. The architecture must describe a combinational system.

5.23 Repeat the design of the thermometer to binary encoder in Problem 4.26, except write an architecture that primarily uses loop statements to implement the encoding. The approach can be based on the observation that when the number of consecutive 1s in the input (starting from `therm(0)`) is equal to the total number of 1s in the input, the input code is valid and the output is equal to that number. The description must assign the outputs don't care values for invalid input codes.

5.24 A combinational system takes as its inputs the votes from three judges and computes the winner of a competition. Each judge has two switches, one for contestant A and one for contestant B. Each judge must assign a 0 or a 1 to each contestant's performance. The score assigned by each judge can be any of the four possible switch combinations to indicate any of four possible evaluations by that judge.

B A	Outcome
0 0	no decision
0 1	A is the winner
1 0	B is the winner
1 1	tie

The switch outputs from each judge are combined to form two vectors, one for contestant A and one for contestant B. Each judge's evaluation contributes one element to each vector.

Design entity `tally` takes the two vectors as inputs and computes the two-bit output named `winner`.

Winner	Outcome
0 0	no decision
0 1	A is the winner—A has more 1 votes than B
1 0	B is the winner—B has more 1 votes than A
1 1	tie—A and B have an equal number of 1 votes

Given the following entity declaration for `tally`:

```
entity tally is
   port (scoresA, scoresB : in std_logic_vector (2 downto 0);
      winner : out std_logic_vector (1 downto 0));
end entity;
```

Write an architecture named `loopy` that computes the output `winner`. The computation to determine the winner must be performed using a loop statement. Use type integers in the intermediate computations for the final output value.

5.25 A system that is an eight-input AND is to be implemented by repeatedly iterating the two operand VHDL AND operator.

(a) Write a complete design description of this design entity using a behavioral style architecture with a for loop.

(b) If an eight-input NAND is required, can this be done by repeatedly iterating the two operand VHDL NAND operator? Give an explanation of the basis for your answer.

5.26 Write a complete behavioral style description that uses a loop instruction to implement an eight-input NAND gate. The input to the gate is an 8-bit vector name `inputs` and the output is named `nandout`. Name the entity `nand8` and the architecture `bhvloop`.

5.27 The 74AC280 is a 9-bit parity generator/checker IC that provides one output that indicates even parity and a second output that indicates odd parity. Given the entity declaration:

```
entity ic74ac280 is
   port (
      i: in std_logic_vector (8 downto 0);
      sum_even: out std_logic;
      sum_odd: out std_logic);
end ic74ac280;
```

(a) Write a behavioral style architecture that uses looping to describe a circuit that is functionally equivalent to a 74AC280.

(b) Describe how using a loop statement would compare with using either a case or if statement for the architecture.

5.28 Given the following truth table, write a complete VHDL design description of a design entity that accomplishes the function defined by the truth table below. Use a behavioral style architecture that consists of two processes, one to assign a value to x and the other to assign a value to y.

a	b	c	x	y
0	0	0	0	1
0	0	1	1	0
0	1	0	0	1
0	1	1	0	0
1	0	0	1	0
1	0	1	0	0
1	1	0	0	0
1	1	1	0	1

Chapter 6
Event-Driven Simulation

Simulation is the process of conducting experiments on a model of a system for the purpose of understanding or verifying the operation of the actual system.

Simulation is essential to the VHDL/PLD design methodology. Functional simulation verifies the functional operation of a description. Post-synthesis simulation verifies that the synthesizer accurately translated the description to logic. Timing simulation verifies that the synthesized logic, when mapped to the target PLD, will meet the system's timing requirements.

Understanding how an event-driven simulator executes a VHDL program helps clarify the semantics of many VHDL constructs. Construct semantics are defined in the IEEE Std 1076 Language Reference Manual (LRM) in terms of how an event-driven simulator that is compliant with the standard must execute the constructs. Indeed, the primary audiences for the LRM are simulation and synthesis tool developers.

Synthesis tool developers have as their primary objective creating a synthesizer that synthesizes logic that behaves the same way as does the functional simulation of a design description.

This chapter provides a conceptual description of the internal operation of an event-driven VHDL simulator. The simulator's data structures and operations are described. Simulations of some simple designs are discussed to illustrate each step in the simulator's operation.

6.1 SIMULATOR APPROACHES

VHDL is an event-driven language. A VHDL program represents the discrete event simulation of a physical system. Discrete event simulation allows deterministic modeling of concurrency.

Requirements for a VHDL Simulator

A VHDL simulator must provide data structures and algorithms that allow it to efficiently simulate the execution of concurrent statements in a VHDL program. These concurrent statements must be simulated by a simulator program that is being executed sequentially on a host computer. In other words, the simulator's sequential execution must be able to provide the illusion of concurrency.

A simulator must also efficiently produce an accurate representation of the output waveforms that result from the application of a set of input values to the design entity being simulated. Since the number of signals in a design can be very large, simulator memory efficiency and speed are important.

Simulation Time

During a simulation, changes in input and output signals occur in *simulation time*, the time value maintained by the simulator, as opposed to real time. Each change in the value of a signal is an *event*.

Kinds of Simulators

Based on their approach to performing simulation, there are three common kinds of simulators for digital systems: time driven, event driven, and cycle based.

For these different simulation approaches the arrows in Figure 6.1.1 indicate at which simulation times the simulator must execute simulation cycles to compute output values.

Time-Driven Simulation

In a *time-driven simulation*, simulation time is advanced in predetermined uniform increments. As time is advanced, any events in the next time step are simulated. The fixed size of the time step determines the resolution with which events can be observed. For example, simulation time might be advanced in 1 ns steps. During a time-driven simulation, there are typically a very large number of time steps during which there are no events. Nevertheless, a time-driven simulator must execute a simulation cycle for each time step.

Event-Driven Simulation

In an *event-driven simulation*, time is advanced in nonuniform steps whose sizes depend on when events occur (Figure 6.1.1). An event-driven simulator responds to each input event by executing a sequence of simulation cycles that determine when and to what values the simulated system's signals change.

Advantages of Event-Driven Simulation

Event-driven simulation eliminates the need for the simulator to evaluate the model at the empty time steps (those without events) that are evaluated in a time-driven simulation. Since the state of a system only changes in response to input events, these events are sufficient to determine when the model must be evaluated.

In event-driven simulation, simulation time is advanced in a single step from the current simulation time to the next scheduled event. Intervals of time in which there are no events are skipped, without affecting the validity of the simulation results. This approach results in faster and more precise simulations.

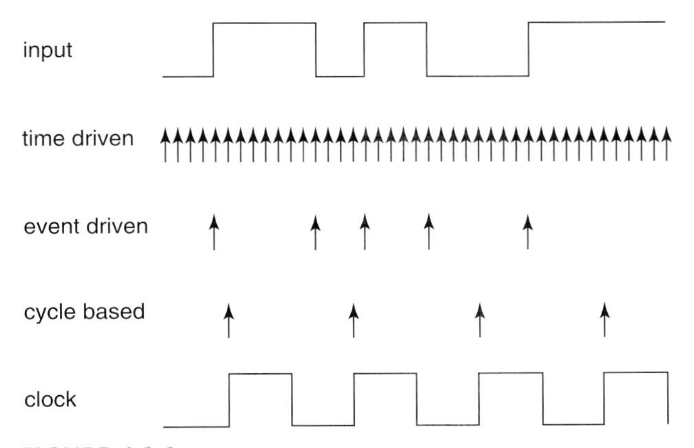

FIGURE 6.1.1
Time of execution of simulation cycles based on simulation approach.

Cycle-Based Simulation

Cycle-based simulation is even faster than event-driven simulation. However, cycle-based simulation is only applicable to the functional simulation of synchronous sequential systems that have a single clock. A *cycle-based simulator* collapses the logic that determines flip-flop input values in a sequential design into equations based on the present input values and present state of the system's flip-flops. The resulting model is evaluated only at the triggering edge of each clock cycle.

Execution of a model only at each clock triggering edge reduces simulation time substantially. However, all delay information is lost, and only signal values at clock triggering edges are available. Another tool, called a *static timing analyzer,* can be used to verify that flip-flop setup and hold times are met.

Because VHDL simulators are most commonly *event-driven simulators*, also known as *discrete event simulators*, this chapter focuses exclusively on them.

6.2 ELABORATION

A Conceptual Event-Driven Simulator

The LRM defines how an event-driven simulator that is compliant with the standard must execute the VHDL language. This chapter provides a conceptual discussion of the operation of such a simulator. Simulator vendors implement these concepts in vendor-specific ways. Simulators, from different vendors, that are compliant with IEEE Std 1076 (the LRM) should provide identical simulation results, independently of their specific implementations.

Steps in an Event-Driven Simulation

An event-driven simulator performs three steps to accomplish a simulation: elaboration, initialization, and repeated execution of simulation cycles. Each of these steps is described in a separate section of this chapter.

Simulation Processes

The process statement is the basic behavioral statement in VHDL. Every concurrent VHDL statement has an equivalent process representation. The term *simulation process* is used in this book to mean a process that is equivalent to some concurrent statement in a design entity's architecture and is used for the simulation of that statement.

Elaboration

Elaboration is the creation of a *simulation model* for a design entity from its VHDL description. This simulation model resides in the host computer's memory and consists of a net (network) of simulation processes.

Simulation Net

During elaboration, all concurrent statements are converted to equivalent simulation processes. Any VHDL program can ultimately be viewed as a collection of simulation processes communicating through signals, that is, a *simulation net*. The resulting simulation net is executed to simulate the design entity's behavior.

Flattening a Hierarchical Structure

During elaboration, a design that has a hierarchical structure is *flattened* until the entire design is described by a simulation net. In addition, all of the data objects declared in the description are created.

Elaboration starts with the top-level design entity. Each concurrent statement in the architecture of the top-level design entity is converted into corresponding simulation processes.

Creating Simulation Processes

The discussion that follows focuses on process, signal assignment, and component instantiation statements, but a transformation to a simulation process exists for all concurrent statements.

Converting a Process to a Simulation Process

A process statement is elaborated directly into a simulation process. The steps involved are:

- The process declarative part is elaborated; all the declared data objects are created and their initial values are determined.
- The drivers required by the process statement are created.
- The initial transaction, defined by the initial value specified in a signal's declaration or, if no initial value is specified, the default value, is placed in each signal's driver.

The creation of signal drivers is discussed in detail in the next section.

Converting a Concurrent Signal Assignment to a Simulation Process

A concurrent signal assignment statement is first converted to an equivalent process statement. A Boolean expression concurrent signal assignment statement has an implied sensitivity list that includes all the signals on the right-hand side of the assignment symbol. A selected signal assignment statement's implied sensitivity list includes all the signals in its select expression. A conditional signal assignment statement's implied sensitivity list includes all the signals in its conditions. In effect, a concurrent signal assignment statement is nothing more than a shorthand representation of a process statement. The equivalent process is then elaborated to a simulation process, as described above.

Converting a Component Instantiation to a Simulation Process

Only architectures containing components are not inherently behavioral. To convert component instances to simulation processes, each component instance in the top-level design entity is replaced by its associated design entity.

Primitive Components

Replacement of each component by its associated design entity is repeated at successively lower levels of the hierarchy until all components are ultimately replaced by design entities whose architectures contain only processes and/or concurrent signal assignment statements. These design entities are the *primitive components*. The concurrent signal assignment statements in each lower-level architecture are converted to equivalent simulation processes. These simulation processes are then elaborated.

Elaboration of an Inverter Design Description

To illustrate the elaboration of a hierarchical design, lets consider the functional simulation of a design entity having a trivial structural architecture. This design entity consists of the instantiation of a single inverter (NOT gate) component, Listing 6.2.1.

LISTING 6.2.1
A single inverter circuit to be simulated.

```
library ieee;
use ieee.std_logic_1164.all;

entity not_gate is  -- not gate entity
   port (x : in std_logic;
      o : out std_logic);
end;

architecture dataflow of not_gate is  -- not gate architecture
begin
   o <= not x;
end dataflow;

library ieee;
use ieee.std_logic_1164.all;

entity ckt is    -- top level entity
   port (a : in std_logic;
      f : out std_logic);
end;

architecture struct of ckt is  -- top level architecture
begin
   u0: entity not_gate port map (x => a, o => f);
end struct;
```

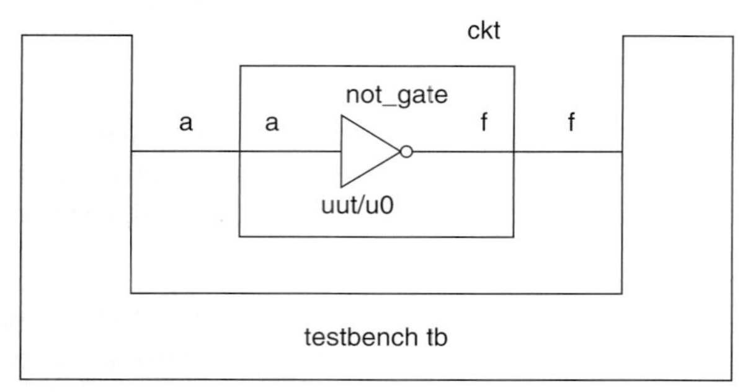

FIGURE 6.2.1
Relationship of the testbench to the circuit being tested.

This design file contains two design entities. The `not_gate` design entity appears first. The second design entity, `ckt`, is the circuit whose operation we wish to verify. Design entity `ckt` has an input `a` and an output `f`. The architecture for `ckt` is structural and consists of a single instantiation of a `not_gate`.

To simulate `ckt` we use a testbench named `testbench`. Design entity `ckt` is the UUT in design entity `testbench`. Since the testbench is now the top level of the hierarchy, what we will actually simulate is `testbench` (Figure 6.2.1). In this example, the same names, `a` and `f`, used for the UUT's ports are also used for the signals in the testbench that connect to these ports.

Listing 6.2.2, for `testbench`, has a mixed style architecture consisting of a single component instantiation and a single process.

The component instantiation, labeled `uut`, is an instance of the design entity `ckt`, whose operation we wish to verify. Since a component instantiation statement is not behavioral, the elaboration must go down a hierarchical level and elaborate the architecture for `ckt`.

LISTING 6.2.2
Testbench for inverter circuit of Listing 6.2.1.

```
library ieee;
use ieee.std_logic_1164.all;

entity testbench is
end testbench;

architecture behavior of testbench is
   signal a : std_logic;
   signal f : std_logic;

begin

   uut: entity ckt port map (a => a,f => f);
```

```
tb: process
   constant period: time := 20 ns;
   begin
      wait for period;  -- Wait 20 ns after initialization

      a <= '1'; -- Apply a 1 to ckt input

      wait for period; -- Wait another 20 ns
         assert (f = '0')
            report "test failed" severity error;

      a <= '0';-- Apply a 0 to ckt input

      wait for period; -- Wait another 20 ns
         assert (f = '1')
            report "test failed" severity error;

      wait;

   end process;
end behavior;
```

The architecture of ckt is structural and consists of a single component instantiation labeled u0. Component instance u0 is an instance of not_gate. Again, since a component instantiation statement is not behavioral, the elaboration must go down still another hierarchical level and elaborate the architecture of not_gate.

Design entity not_gate has a dataflow architecture which is, in effect, behavioral. Thus, not_gate is a primitive component. The elaboration can now create a simulation process uut/u0, which represents an instance of not_gate (u0) in an instance of ckt (uut). In the simulation process, the label uut/u0 represents the *hierarchical path* to u0.

A conceptual representation of the simulation process uut/u0 is:

```
uut_u0: process (a)
     begin
           f <= not a;
end process
```

The other concurrent statement in the architecture of testbench is a process statement labeled tb. This process assigns values to signal a, which connects to the input of ckt, and monitors signal f, which connects to ckt's output. Since a process statement is behavioral, the elaboration can directly create a simulation process called tb. The simulation process for tb is equivalent to the process statement tb in Listing 6.2.2.

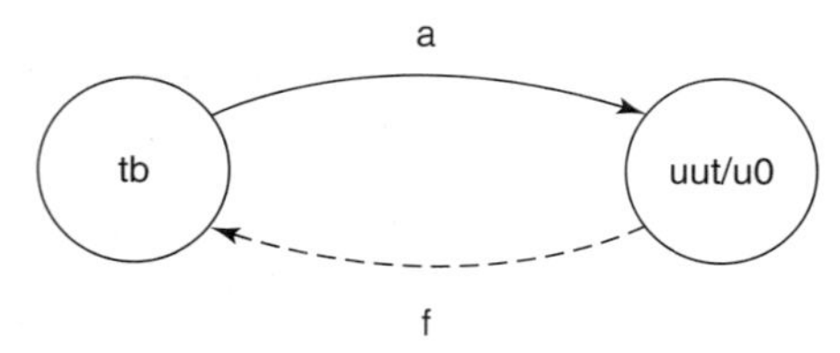

FIGURE 6.2.2
Simulation net from the elaboration of top-level design entity testbench in Listing 6.2.2.

Simulation Net for Inverter Testbench

The simulation net created by this elaboration is represented in Figure 6.2.2. Each circle represents a simulation process. The directed arrow from `tb` to `uut/u0` indicates that `tb` assigns a value to `a` and that `uut/u0` is sensitive to `a`. The dashed directed arrow from `uut/u0` to `tb` indicates that `uut/u0` assigns a value to `f` and that `tb` reads this value. A dashed directed arrow is used to distinguish that although `tb` reads `f`, it is not sensitive to `f` (does not have `f` in its sensitivity list).

6.3 SIGNAL DRIVERS

During elaboration, for each signal assigned a value in a simulation process a signal driver is created. This signal driver is associated with the simulation process containing the signal assignment statement. Only one signal driver is created for each signal assigned a value in a simulation process. This is true even if a simulation process contains multiple assignments to the same signal.

Multiply Driven Signals

If only one simulation process in a simulation model contains assignments to a particular signal, elaboration will produce only one driver for that signal. If two (or more) different simulation processes contain assignments to the same signal, elaboration will produce two (or more) different drivers for that signal, one associated with each simulation process. Such a signal is *multiply driven.*

Signal Drivers as Data Queues

In terms of a simulator's data structures, a signal driver is modeled as a queue. Each *signal driver queue* is defined by all of the signal assignment statements that assign values to a particular signal in a simulation process.

Transactions

Each time a signal assignment statement is executed, the simulator posts (schedules) a transaction in the signal's driver. Each *transaction* is a pair consisting of a value component and a time component.

Value and Time Transaction Components

The *value component* of a transaction represents the value that the signal driver is to assume at some future simulation time. The *time component* specifies the time at which the signal driver is to assume that value.

Projected Values of a Signal

Transactions are ordered in a signal's driver queue based on their time components. The ordered set of transactions specifies values of the signal at future simulation times, *projected values*. These values describe a *projected waveform*.

A driver always contains at least one transaction. This transaction, which is at the head of the queue, has a time component that is not greater than the current simulation time. This transaction's value is the current value of the driver. As simulation time is advanced, the value component of each transaction in the queue will, in turn, become the current value of the driver and transactions with earlier time components will be removed.

Active Driver

During a simulation cycle, if a signal driver queue has a transaction with a time component equal to the current simulation time, that driver is said to be *active* during that simulation cycle. This means that the driver will get a new current value during that simulation cycle. If the new value is different from the old value, then there is an event on the driver's signal.

Single and Multiple Signal Sources

A contributor to the value of a signal is called a *source*. A source can be either a driver or a port to which the signal is connected. If a signal has only a single source and that source is a driver, that driver's contents completely define the projected waveform for the signal. In this case, the current value of the driver is the current value of the signal.

Current Value of a Multiply Driven Signal

If a signal has multiple sources, the current values of all the signal's sources must be resolved to determine the current value of the signal. The simulator automatically performs this resolution operation. If one or more of the drivers of a signal are active, the signal is active.

Initial Values for Signals

If a signal's declaration specifies an initial value, the signal is initially set to this value. Otherwise, each signal is initially set to its default value. This value is the leftmost value for its type. For std_logic, this value is the uninitialized value `'U'`.

Signal Drivers for the Inverter Design Description

Returning to our inverter example, simulation process `tb` in Listing 6.2.2 contains two signal assignment statements that assign values to `a`. Since both assignments are in the same process, only a single driver for `a` is created.

Simulation process `uut/u0`, corresponding to the instantiation of `uut`, has a single signal assignment statement for `f`. Therefore, a single driver is also created for `f`. Thus, the elaboration of the testbench in Listing 6.2.2 creates two signal drivers,

Signal	Current Value	<	Signal Driver Transaction Queue		
After elaboration					
a		<	'U' @ 0		
f		<	'U' @ 0		

FIGURE 6.3.1
Signal drivers after elaboration of inverter.

one for `a` and one for `f`. A representation of the drivers for `a` and `f` immediately after elaboration is shown in Figure 6.3.1. The signal name is given in the first column. The rightmost three columns represent the signal driver queue. The head of each driver queue is at the left end of the queue (fourth column from the left). Values in this column represent the current values of the drivers.

Current Value Variable for a Signal

The simulator maintains a *current value variable* for each signal. This variable holds the current value of the signal. The current value variables are in the second column of Figure 6.3.1. When a signal's value must be accessed (read) during a simulation, its value is obtained from this variable.

Since each signal in this example has a single driver, the current value of each signal is the same as the current value of its driver.

Allocating Memory for Variables

In addition to allocating memory for the signal driver queues, the elaboration operation must also allocate memory for any variables declared in simulation processes. All variables declared in simulation processes that were assigned an initial value when they were declared are initialized to those values.

6.4 SIMULATOR KERNEL PROCESS

The *kernel process* is a conceptual representation of a simulator's controller. During a simulation, it coordinates the activities of the simulation processes. The kernel process is responsible for detecting events and causing the appropriate simulation processes to execute in response to those events.

The kernel process maintains the current value variable for each signal in the simulation model. It utilizes the resolution function to determine a multiply driven signal's current value from its drivers' current values.

Sequential Execution of Simulation Processes

As previously discussed, during elaboration the simulator creates at least one simulation process for each concurrent statement in the top-level architecture.

Concurrent statements, and their corresponding simulator processes, can execute in parallel in the system being described. For example, in a simple combinational circuit all the gates are simultaneously computing their outputs from their current inputs. In contrast, a simulator program is typically run on a host computer with a single CPU. Accordingly, during simulation only one simulation process can actually be executing at a time. So, the simulator must execute simulation processes sequentially, but in a way that models concurrent operation.

States of a Simulation Process

During a simulation, each simulator process can be in one of three states: *suspended, active,* or *running* (Figure 6.4.1):

- Suspended: simulation process is not running or active
- Active: simulation process is in the active processes queue waiting to be executed.
- Running: simulator is executing the simulation process.

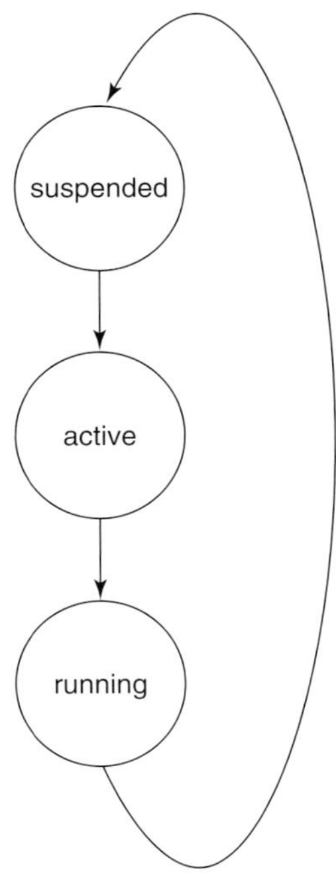

FIGURE 6.4.1
States of a simulation process.

Active Processes Queue

The simulator maintains an *active processes queue.* All simulation processes that need to be run during a particular simulation cycle are placed in the active processes queue by the kernel. The kernel then takes each simulation process from this queue and executes it until it suspends.

Waiting Processes Queue

If a simulation process is suspended because of the execution of a wait statement of the form `wait for time_expression`, that simulation process is put in the *waiting processes queue* in time order. The simulation process is, therefore, scheduled for activation after a period of time equal to `time_expression` from the current simulation time. A simulation process in the waiting processes queue is referred to as a *scheduled process* because it is scheduled to be resumed (be placed in the active processes queue) at a specific future time.

Nonsuspending Process

It is syntactically legal to have a process statement without a sensitivity list or wait statement(s). After elaboration, this results in a simulation process without a sensitivity list or wait statement(s). However, when simulated, such a process never suspends, simulation time does not advance, and the simulation hangs if a simulation cycle limit is not set.

6.5 SIMULATION INITIALIZATION

After elaboration is completed, simulation consists of an initialization phase followed by repetitive execution of simulation cycles. The time values referred to in the following discussions are simulation times, as maintained by the simulator's clock.

Initialization Phase

At the beginning of the *initialization phase,* the current time (Tc) is set to 0 (Figure 6.5.1).

Initial Execution of All Simulation Processes

The kernel places all of the simulation processes in the active processes queue. Each simulation process is then taken from this queue and executed until it suspends. The order of execution of simulation processes during initialization is not important. The initial execution of each simulation process ensures that all initial transactions are scheduled, so that the simulation may continue.

A simulation process is suspended either implicitly or explicitly. A process with a sensitivity list is suspended implicitly after its sequential statements have been executed to the end of the process. A process with one or more wait statements is suspended explicitly when its first wait statement is executed.

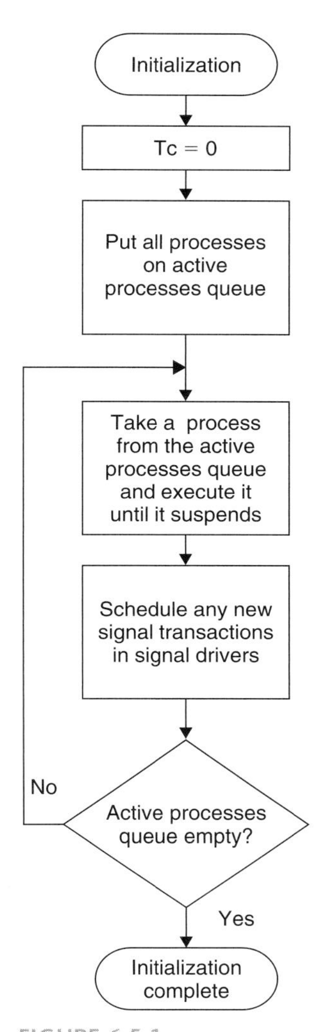

FIGURE 6.5.1
Initialization phase of an event-driven simulation.

Specifying Delays in Signal Assignments

In writing synthesizable descriptions, we use signal assignments with zero delay, for example:

```
x <= a and b;
```

In contrast, timing models of actual hardware and some testbenches use signal assignment statements that include an after clause. The after clause specifies the exact delay from the current time for the assignment to take effect. For example:

```
x <= a and b after 2 ns;
```

This statement might describe an AND gate in a PLD. The output changes 2 ns after a change in a or b. This delay is the result of the propagation delay of signals through the physical circuit.

It is not appropriate to use an assignment that specifies a delay in a description to be synthesized. At the time we write a description, we may not have chosen a target PLD. In addition, since the propagation delay in the physical implementation depends on many factors, including device technology, architecture, routing, and the operational enviornment, a synthesizer cannot guarantee that a specified delay can be achieved. Accordingly, a synthesizer simply ignores after clauses in signal assignment statements. The syntax of assignment statements with after clauses is discussed in Chapter 7 in the context of their use in testbenches.

Delta Delays

Execution of a simulation process may cause transactions to be added to the driver queues for some signals. If a signal assignment without an after clause is executed, the associated transaction is assigned a time component 1 delta (δ) delay greater than the current simulation time.

A *delta delay* is an infinitesimal period of time. It is used as a scheduling device to ensure a clear order of events in circuits with zero delay devices (devices with no delay specified). This is necessary so that a circuit's operation can be simulated in a correct and repeatable fashion.

Two-Dimensional View of Simulation Time

Since a delta delay is infinitesimal, when added to the current simulation time it does not advance the numerical time value. Conceptually, simulation time is viewed as two dimensional. If delta delays are added, simulation time increases on the Y axis, leaving its numerical time step value on the X axis unchanged.

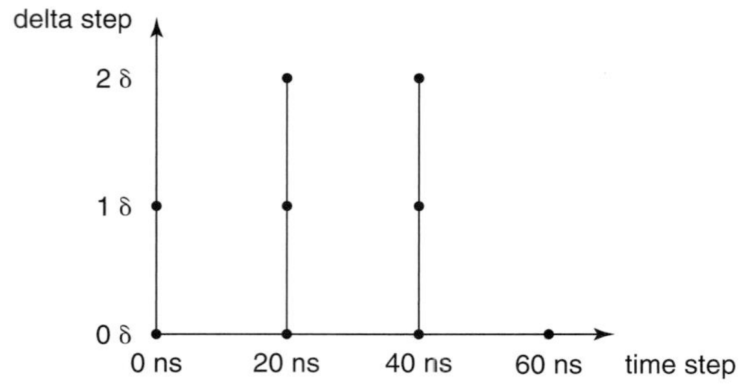

FIGURE 6.5.2
Deltas at each simulation time step for the inverter simulation.

Signal	Value	<	Signal Driver Transaction Queue		
After elaboration					
a	'U'	<	'U' @ 0		
f	'U'	<	'U' @ 0		
After initialization Tc = 0, Tn = 0 + 1δ					
a	'U'	<	'U' @ 0		
f	'U'	<	'U' @ 0	'U' @ 0 = 1 δ	

FIGURE 6.5.3
Signal drivers from elaboration through initialization of inverter.

This concept is illustrated in Figure 6.5.2, which represents one delta occuring at simulation time 0 and two deltas occurring at simulation time steps at 20 and 40 ns. The deltas are shown stacked on top of each other at each time step, producing a "two-dimensional" time representation. The stacking of deltas indicates that their occurrence does not advance the numerical value of simulation time. Delta delays are discussed further in Section 6.8.

Inverter Example Signal Drivers after Initialization

Returning again to our inverter example, the second entry (second group of two rows) in Figure 6.5.3 shows the drivers for the inverter example after initialization.

When simulation process `tb` is executed at initialization, the first statement is a wait statement that causes `tb` to suspend for 20 ns, without placing any transactions in any signal drivers. Accordingly, `tb` is placed in the waiting processes queue.

When simulation process `uut/u0` is executed at initialization, the statement:

```
f <= not a;
```

is executed and then the process suspends. This puts the transaction `'U' @ 0 + 1` δ in the driver for `f`. This transaction has the value component `'U'` because the **not** operator for std_logic operating on `'U'` returns `'U'`. Since the assignment statement does not have an after clause, the time component is `1` δ.

6.6 SIMULATION CYCLES

After the initialization phase, all simulation processes are in their suspended states. The first *simulation cycle* is then executed.

Simulation Cycle Phases

A simulation cycle consists of two phases: an update phase and an execution phase. These phases are described in detail in the following paragraphs.

Update Phase Overview

The update phase first determines the next value for the current simulation time and advances the simulator clock to this value. Based on this new simulation time, a determination is made as to whether the simulation is complete.

If the simulation is not complete, all signals scheduled to be updated at the current simulation time are updated.

A change in a signal's value can occur only during the update phase of a simulation cycle. If a signal's update results in an event, all simulation processes sensitive to this event are placed in the active processes queue. This determination is carried out for all signals that are updated. In addition, all simulation processes that are scheduled to resume at the current simulation time are placed in the active processes queue. The active processes queue now contains all of the simulation processes that need to be executed during the current simulation cycle.

Update Phase Details

Steps in the update phase are shown in detail in Figure 6.6.1. At the beginning of the update phase, the simulator clock is advanced to the next simulation time.

Next Simulation Time

The next simulation time (Tn) is determined as either:

1. One delta greater than the current simulation time (Tc + δ), if there is a transaction in any signal driver with a time component of Tc + δ. Since a delta is an infinitesimal increment in time, this does not advance the numerical value of the current simulation time.
2. The next time step. The next time step is the earliest of either:
   - TIME'HIGH,
   - The time component of the transaction whose time component is the smallest value greater than Tc + δ, or
   - The earliest time at which a process is scheduled to resume.

Simulation Complete?

The next simulation time is then checked against TIME'HIGH, which is the limit placed on how long the simulation is to run. If Tn equals TIME'HIGH, and there are no signal drivers with transactions scheduled for Tn or simulation processes scheduled to resume at Tn, the simulation is complete.

If the simulation is not complete, the current simulation time is set equal to the next simulation time (Tc = Tn). If Tn is equal to Tc + δ, the next simulation cycle is called a *delta simulation cycle*.

Update Signals

Each signal whose driver has a transaction with a time component equal to the current simulation time (is scheduled for the current simulation time) has its value updated to the value component of the transaction.

Determination of Processes to Resume

If none of these updates cause a signal's value to change, then there are no events. If there are no events and no processes are scheduled to resume, the simulation is complete.

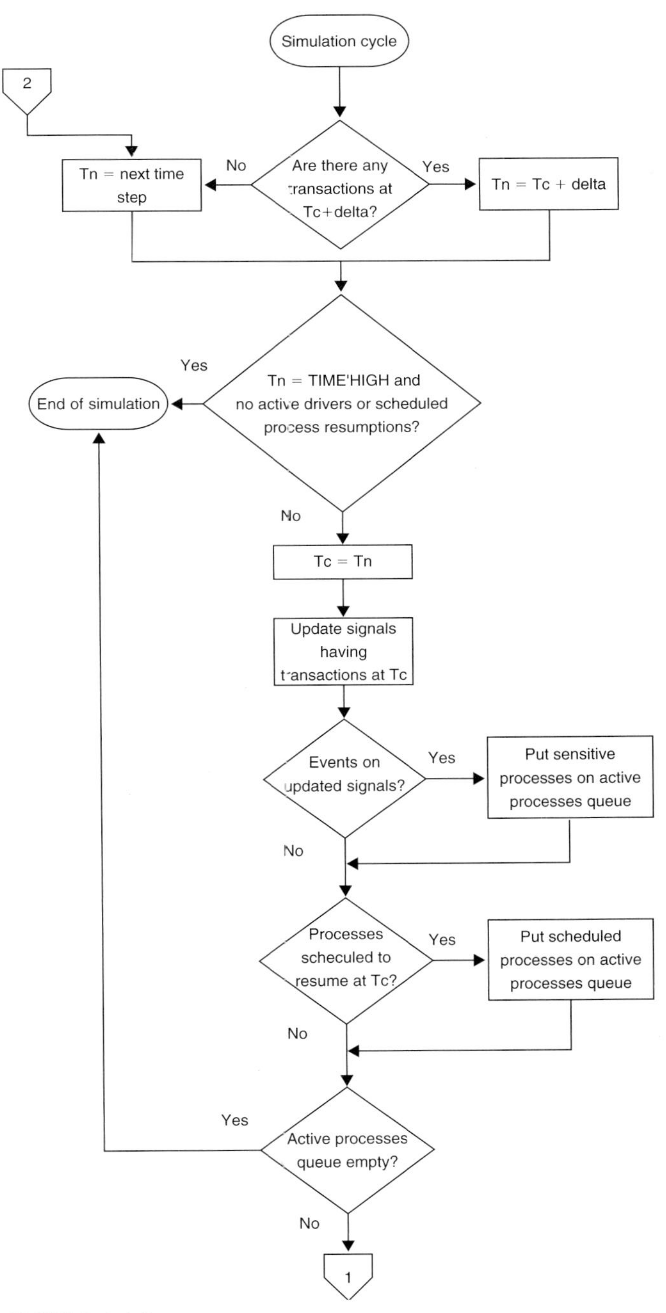

FIGURE 6.6.1
Simulation cycle update phase for an event-driven simulator.

If there are one or more events, each simulation process that is sensitive to those events (has a signal in its sensitivity list on which an event occurred) is placed in the active processes queue for execution.

If there are processes scheduled to resume at the current simulation time, they are also placed in the active processes queue. These processes become active independently of any events.

Execution Phase

During the *execution phase,* Figure 6.6.2, each simulation process in the active processes queue is taken from that queue and executed until it suspends. Execution of a simulation process may cause the execution of one or more signal assignment statements. Execution of a signal assignment statement causes one or more transactions to be scheduled for the specified signal's driver. After all simulation processes that were in the active processes queue have been executed, the simulator starts the next simulation cycle.

The order in which the kernel takes simulation processes from the active processes queue and executes them is not important, because these processes are concurrent.

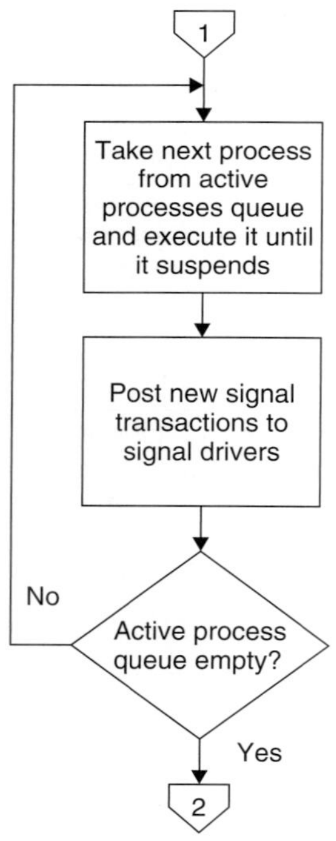

FIGURE 6.6.2
Simulation cycle execution phase.

When Do Signals Take New Values?

It is not until a simulation process suspends that the transactions resulting from signal assignments in that process are placed in the appropriate signal driver queues. Thus, in a single simulation process, if two or more signal assignment statements that assign values to the same signal are each executed, it is the last assignment executed before the simulation process suspends that determines the value component of the transaction placed in the driver.

It is not until the update phase of the next simulation cycle, after all the simulation processes that were in the active processes queue during the current simulation cycle have been executed and suspended, that any signals assigned values by these processes are updated. That is, values assigned by simulation processes in the execution phase of one simulation cycle don't take effect until the update phase of the next simulation cycle, at the earliest.

In our design descriptions, the time at which a value should be assigned to a signal is not specified in the signal assignment statement. Therefore, the time defaults to one delta delay from the current time. ***This means that the assignment of a new value to a signal cannot take effect until the update phase of the next simulation cycle. This is after the process containing the assignment has suspended. Therefore, the simulation process does not see the effect of any signal assignment until the next time it resumes.***

For a signal assignment with an after clause, the effect of the assignment does not take affect until simulation time has advanced by the amount equal to the time expression in the after clause.

From this discussion, we see that a signal's value is only indirectly modified by an assignment statement, with the effect occurring no sooner than 1 δ later.

Summary of Simulation Cycle Steps

A simulation cycle consists of the following steps:

1. The time of the next simulation cycle is determined. If Tn = Tc + δ, the next simulation cycle is a delta simulation cycle.
2. Simulation is complete if Tn = TIME'HIGH and there are no signal drivers with transactions scheduled for Tn (active drivers) or simulation processes scheduled to resume at Tn.
3. If the simulation is not complete, the current simulation time Tc is set to Tn.
4. The values of all signals having transactions scheduled for Tc are updated. Events may occur on some signals as a result of updating their values.
5. All processes that are sensitive to the events in (4) are put in the active processes queue.
6. All processes scheduled to resume at Tc are also put in the active processes queue.
7. Each simulation process is taken from the active processes queue and executed until it suspends. Execution of simulation processes may cause new transactions to be posted in signal drivers.
8. Return to step 1.

Simulation Cycles for Inverter Example

All the simulation cycles for simulation of the inverter, starting from elaboration, are given in Figure 6.6.3.

Signal	Value	<	Signal Driver Transaction Queue		
After elaboration					
a		<	'U' @ 0		
f		<	'U' @ 0		
After initialization Tc = 0, Tn = $0 + 1\delta$					
a	'U'	<	'U' @ 0		
f	'U'	<	'U' @ 0	'U' @ $0 + 1\delta$	
After simulation cycle Tc = $0 + 1\delta$, Tn = 20					
a	'U'	<	'U' @ 0		
f	'U'	<	'U' @ $0 + 1\delta$		
After simulation cycle Tc = 20, Tn = $20 + 1\delta$					
a	'U'	<	'U' @ 0	'1' @ $20 + 1\delta$	
f	'U'	<	'U' @ $0 + 1\delta$		
After simulation cycle Tc = $20 + 1\delta$, Tn = $20 + 2\delta$					
a	'1'	<	'1' @ $20 + 1\delta$		
f	'U'	<	'U' @ $0 + 1\delta$	'0' @ $20 + 2\delta$	
After simulation cycle Tc = $20 + 2\delta$, Tn = 40					
a	'1'	<	'1' @ $20 + 1\delta$		
f	'0'	<	'U' @ $20 + 2\delta$		
After simulation cycle Tc = 40, Tn = $40 + 1\delta$					
a	'1'	<	'1' @ $20 + 1\delta$	'0' @ $40 + 1\delta$	
f	'0'	<	'0' @ $20 + 2\delta$		
After simulation cycle Tc = $40 + 1\delta$, Tn = $40 + 2\delta$					
a	'0'	<	'0' @ $40 + 1\delta$		
f	'0'	<	'0' @ $20 + 2\delta$	'1' @ $40 + 2\delta$	
After simulation cycle Tc = $40 + 2\delta$, Tn = 60					
a	'0'	<	'0' @ $40 + 1\delta$		
f	'1'	<	'1' @ $40 + 2\delta$		
After simulation cycle Tc = 60, Tn = done					
a	'0'	<	'0' @ $40 + 1\delta$		
f	'1'	<	'1' @ $40 + 2\delta$		

FIGURE 6.6.3
Signal drivers during simulation of inverter.

During initialization, all processes are executed until they suspend. Simulation process `tb` suspends immediately, because its first statement is **`wait for`** `period` (where period equals 20 ns). This execution of `tb` causes no new transactions to be scheduled.

Simulation process `uut/u0` executes an assignment statement to assign a value to `f`. Since `f` is to be assigned the value **`not`** `a` and `a`'s value is `'U'`, `f` is assigned the value `'U'`. The value `'U'` is defined as the not of `'U'` in package STD_LOGIC_1164. Since there is no after clause in the assignment statement to specify when the assignment should take place, a transaction is scheduled to the driver for `f` with a value component of `'U'` and a time component 1 δ later than the current time ($0 + 1\ \delta$).

Accordingly, the next simulation cycle is at $0 + 1\ \delta$. During the update phase of this cycle, the value of `f` is updated to `'U'`. Since this value is the same as the previous value, no event occurs. So, no processes are placed in the active processes queue.

Because there were no events and process `tb` is waiting for 20 ns, the next simulation cycle time (Tn) is 20 ns (a new time step) and the next simulation cycle is started.

Simulation continues in this fashion until the current simulation time becomes equal to `40 + 2` δ. During the update phase at `40 + 2` δ, `f` changes from `'0'` to `'1'`, resulting in an event on `f`. None of the simulation processes is sensitive to `f`. However, simulation process `tb` is scheduled to resume at 60 ns.

When `tb` is executed at 60 ns it does not execute any signal assignment statements. In addition, it is suspended by a wait statement that does not schedule the process to resume. Since there are no future transactions on the drivers and no processes are scheduled to resume, Tn is advanced to TIME'HIGH and the simulation ends.

Waveforms for this simulation are given in Figure 6.6.4. Simulation waveforms only show signals changing at time steps, with no consideration of at which delta in a time step a signal changed. Since, practically speaking, a delta is infinitesimal and does not advance the step time, this is appropriate.

Simulation Cycles Listing

A simulator can produce a list of signals and the times at which each changed value in terms of both the time step and the particular delta in each time step. Table 6.6.1 is such a list for the inverter simulation example.

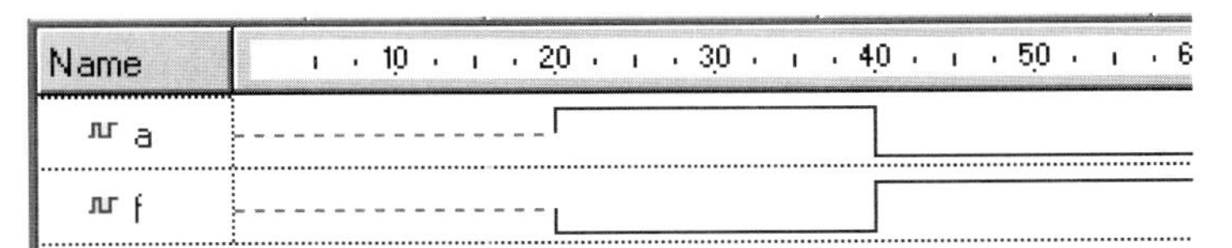

FIGURE 6.6.4
Waveform for functional simulation of inverter.

Table 6.6.1
Simulation cycles listing for inverter.

ns	δ	a	f
0	0	U	U
20	1	1	U
20	2	1	0
40	1	0	0
40	2	0	1

For conciseness, a simulation cycles list only shows deltas where an event occurred. Therefore, the transaction at 0 + 1 δ is not shown because that transaction assigned a value of `'U'` to `f`, but `f` previously had the value `'U'`.

Each time step after t = 0 results from the testbench simulation process `tb` becoming active at the end of a wait period (at 20, 40, and 60 ns). When `tb` executes at the 20 and 40 ns time steps, it schedules a transaction to the driver for `a` at the current simulation time + 1 δ. Since the value of `a` can't change until the update phase of the next simulation cycle, there are no events during these 0 δ cycles and, therefore, there are no 0 δ entries in the simulation cycles listing in Figure 6.6.1 at the 20 and 40 ns time steps.

Activation of `tb` at 60 ns does not appear in the list because that execution caused no signal assignments to be made, therefore no events could result.

Figure 6.6.5 represents the number of deltas (delta simulation cycles) that occur at each time step.

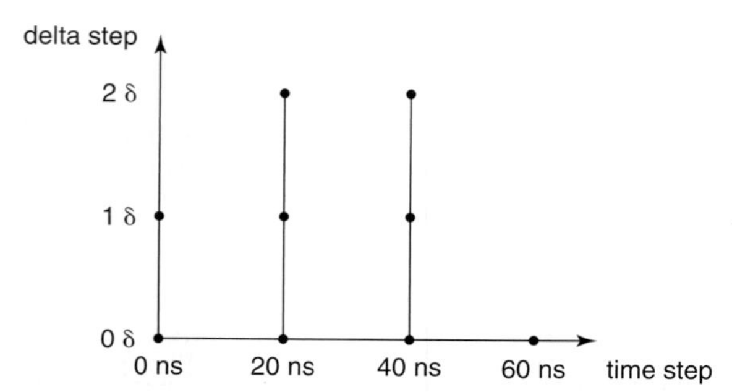

FIGURE 6.6.5
Delta cycles for each simulation time step for inverter simulation.

6.7 SIGNALS VERSUS VARIABLES

With a basic understanding of the operation of an event-driven simulator, we can more clearly understand the distinctions between signals and variables in a process.

When a signal assignment statement is executed, a transaction is posted in the signal's driver after the process suspends. If the assignment statement has no after clause, a time component 1 δ later than the current simulation time is assigned to the transaction. The signal's value is not changed during the current simulation cycle. Changes in signal values do not take effect until the update phase of the next simulation cycle, one delta later.

In contrast, when a variable assignment statement is executed, the variable takes its new value immediately, while the simulation process is being executed. Statements in the process that follow the variable assignment statement and access the variable see its new value.

To illustrate the impact of these differences, we consider several approaches to describing the combinational function specified by the truth table in Table 6.7.1.

Assigning an Input Aggregate to a Signal

Assume that we are given the entity declaration in Listing 6.7.1 and are asked to write an architecture body.

We decide to write the description using a case statement. Since the output is 0 in only two places in the truth table, we can write the case statement to assign the output a 0 for these two input combinations and default to 1 for all other combinations.

Table 6.7.1
Truth table for a function.

a	b	c	x
0	0	0	1
0	0	1	0
0	1	0	1
0	1	1	1
1	0	0	1
1	0	1	0
1	1	0	1
1	1	1	1

LISTING 6.7.1
Case description of truth table with an invalid sensitivity list.

```
library ieee;
use ieee.std_logic_1164.all;

entity truth_table is
   port ( a, b, c : in std_logic;
   x : out std_logic);
end truth_table;

architecture casey of truth_table is
signal abc :  std_logic_vector( 2 downto 0);
   begin

   tt: process (a, b, c)  -- invalid sensitivity list
   begin
      abc <= (a, b, c);
      case abc is
         when "001" | "101" => x <= '0';
         when others => x <= '1';
      end case;
   end process;

end casey;
```

The case statement is simpler to write if the inputs are treated as a vector inside the process. The architecture body in Listing 6.7.1 uses this approach. To accomplish this, signal abc is declared as a std_logic_vector in the architecture's declarative part and the first statement in the process assigns aggregate (a, b, c) to the std_logic_vector abc. Signal abc is then used as the expression in the case statement.

This description compiles without error. Simulation produces the waveforms in Figure 6.7.1. The testbench starts with all inputs as 0s, then counts through the input combinations in binary. Each input combination is applied for 50 ns.

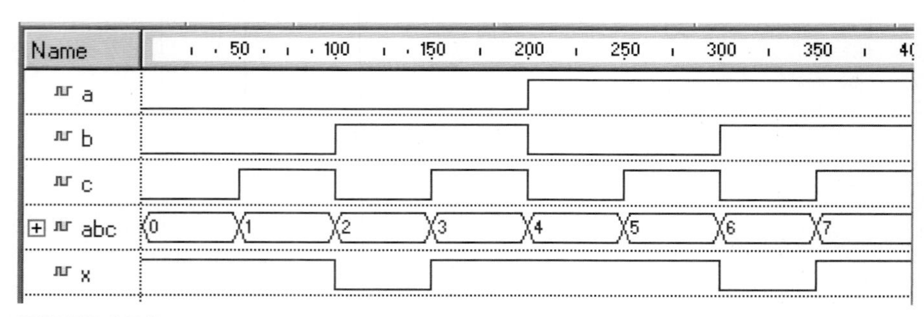

FIGURE 6.7.1
Timing waveform for description in Listing 6.7.1.

Examination of the simulation waveforms shows the output is 0 when the inputs are 010 (2) and 110 (6), rather than when they are 001 (1) and 101 (5), as desired. It appears that the correct output is delayed by one input count. In a sense, it is.

Further examination of the simulation cycles list reveals the source of the problem (Table 6.7.2). Assume that the inputs are 000 and that the testbench assigns input `c` a `'1'` at 50 ns (+ 0 δ). During the update phase of the next simulation cycle, one delta cycle later at 50 ns + 1 δ, the input is changed to 001. This event on `c` causes the process `tt` to be executed, since `c` is in the process's sensitivity list.

The first statement in the process assigns the new value 001 to signal `abc`. However, when this assignment statement is executed it only schedules a transaction in the driver for `abc` for a value of `"001"` at 50 ns + 2 δ. However, simulation time is still at 50 ns + 1 δ, so the value of `abc` is unchanged.

Next, the case statement is executed while `abc` is still `"000"`, so the others clause executes. The assignment statement in the others clause schedules a transaction to make `x` a `'1'` at 50 ns + 2 δ. The simulation cycle ends and simulation time changes to 50 ns + 2 δ.

Table 6.7.2
Simulation cycles list for description of Listing 6.7.1.

ns	delta	a	b	c	UUT/abc	x
0	0	U	U	U	UUU	U
0	1	0	0	0	UUU	1
0	2	0	0	0	000	1
50	1	0	0	1	000	1
50	2	0	0	1	001	1
100	1	0	1	0	001	1
100	2	0	1	0	010	0
150	1	0	1	1	010	0
150	2	0	1	1	011	1
200	1	1	0	0	011	1
200	2	1	0	0	100	1
250	1	1	0	1	100	1
250	2	1	0	1	101	1
300	1	1	1	0	101	1
300	2	1	1	0	110	0
350	1	1	1	1	110	0
350	2	1	1	1	111	1
400	1	0	0	0	111	1
400	2	0	0	0	000	1

During the simulation cycle at 50 ns + 2 δ, the value of `abc` is updated and changes from `"000"` to `"001"`. Thus, there is an event on `abc`. The signal `x` is also updated, but it remains `'1'` so there is no event on `x`. Since the only event is on `abc`, and `abc` is not in the `tt` simulation process's sensitivity list, the process is not executed during this (50 ns + 2 δ) simulation cycle. We are left with output `x` equal to `'1'` for an input of `"001"`, which is incorrect.

If we had included signal `abc` in our sensitivity list, simulation process `tt` would be executed during the execution phase of the simulation cycle at 50 + 2 δ. This missing execution of simulation process `tt` would have occurred with `abc` equal `"001"` and the case statement would schedule a transaction to assign the value `'0'` to `x` at 50 ns + 3 δ. The next simulation cycle would be at 50 ns + 3 δ. During the update phase of this simulation cycle, `x` would be changed to `'0'` and we would have the correct output.

The next simulation cycle is at 100 ns (+ 0 δ) when the testbench assigns input `b` a `'1'` and input `c` a `'0'`. One delta cycle later at 100 ns + 1 δ, the inputs are changed to `"010"`. Since both `b` and `c` change values, there are events on `b` and `c`. Either of these events is sufficient to cause simulation process `tt` to be executed during the execution phase of this simulation cycle. Execution of the assignment to `abc` schedules a transaction for `abc` to become `"010"` at 100 ns + 2 δ. The case statement is then executed. With `abc` still having the value of `"001"`, the first assignment statement in the case statement is executed. This assignment statement schedules a transaction to make `x` a `'0'` at 100 ns + 2 δ. The simulation cycle then terminates.

Since there are transactions in the drivers of `abc` and `x` with time components of 100 ns + 2 δ, the next simulation cycle is at 100 ns + 2 δ. During the update phase of this simulation cycle, `abc` is changed to `"010"` and `x` is changed to `'0'`. Now output `x` has the value it should have had for the previous input combination. Since neither `abc` nor `x` is in simulation process `tt's` sensitivity list, this process is not executed.

Signals Required in a Sensitivity List

This example emphasizes the fact that a process must include all relevant signals in its sensitivity list. For a combinational design, the sensitivity list must include all the signals read by the process. A process may read more signals than just those that are inputs to the design entity. In this example, the case statement reads `abc`, since `abc` is the case expression.

Including an Input Aggregate in Sensitivity List

We have already mentioned one approach to solving the problem. Simply, include signal `abc` in the sensitivity list. This may at first seem odd, since the signal `abc` is not an input to the circuit. However, the previous analysis shows that this is necessary for correct operation.

Placing the Assignment of the Aggregate Outside of the Process

Another approach would be to place the signal assignment statement that assigns the aggregate to `abc` outside of the process and make the process sensitive to only `abc`. This architecture body is shown in Listing 6.7.2.

LISTING 6.7.2
Case description of truth table with modified sensitivity list.

```
architecture casey2 of truth_table is
   signal abc :  std_logic_vector( 2 downto 0);
begin

   abc <= (a, b, c);     -- concurrent signal assignment

   process (abc)    -- valid sensitivity list
   begin
      case abc is
         when "001" | "101" => x <= '0';
         when others => x <= '1';
      end case;
   end process;

end casey2;
```

When simulated, this description gives the correct output waveform (Figure 6.7.2). The list file in Table 6.7.3 shows the additional delta cycles and correct changes in the output resulting from placing the signal assignment statement that assigns the input aggregate to abc outside of the process.

Type Qualifying Input Aggregate in Case Expression

Another approach is to remove the signal assignment statement that assigns the input aggregate to abc entirely and aggregate a, b, and c in the expression of the case statement and type qualify the aggregate:

```
case std_logic_vector'(a, b, c) is
```

Aggregate Inputs into a Variable Vector

Still another approach is to aggregate the scalar inputs into a variable vector instead of a signal vector. This approach is shown in Listing 6.7.3.

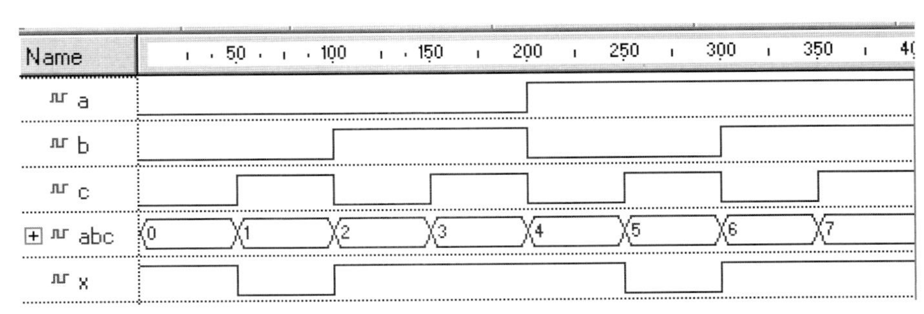

FIGURE 6.7.2
Timing waveform for description in Listing 6.7.2.

Table 6.7.3
Simulation cycles list from simulation of Listing 6.7.2.

ns	delta	a	b	c	abc	x
0	0	U	U	U	UUU	U
0	1	0	0	0	UUU	1
0	2	0	0	0	000	1
50	1	0	0	1	000	1
50	2	0	0	1	001	1
50	3	0	0	1	001	0
100	1	0	1	0	001	0
100	2	0	1	0	010	0
100	3	0	1	0	010	1
150	1	0	1	1	010	1
150	2	0	1	1	011	1
200	1	1	0	0	011	1
200	2	1	0	0	100	1
250	1	1	0	1	100	1
250	2	1	0	1	101	1
250	3	1	0	1	101	0
300	1	1	1	0	101	0
300	2	1	1	0	110	0
300	3	1	1	0	110	1
350	1	1	1	1	110	1
350	2	1	1	1	111	1
400	1	0	0	0	111	1
400	2	0	0	0	000	1

LISTING 6.7.3
Case description of truth table using a variable.

```
library ieee;
use ieee.std_logic_1164.all;

entity truth_table is
   port ( a, b, c : in std_logic;
   x : out std_logic);
end truth_table;
```

```
architecture casey of truth_table is
   begin

   process (a, b, c)

   variable abc : std_logic_vector(2 downto 0);

   begin
      abc := (a, b, c);
      case abc is
         when "001" | "101" => x <= '0';
         when others => x <= '1';
      end case;
   end process;

end casey;
```

Variable `abc` must be declared inside the process and is local to the process. When the variable assignment statement is executed, the variable's value is changed immediately. Next, when the case statement is executed, it uses the new value of `abc` that was assigned during the current simulation cycle.

Signal and Variable Distinctions

Some distinctions between signals and variables in an architecture are now summarized.

Signals in an architecture:

- Must be declared outside of any processes.
- Are visible to all processes (and other concurrent statements) in the architecture in which they are declared and are used to carry information between concurrent statements in the architecture.
- Have current and projected (future) values.
- Retain their values between executions of a process.
- A signal assignment schedules a transaction in the signal's driver queue. If the signal assignment has no after clause, the assigned value takes effect at the update phase of the next simulation cycle (one δ later). If there is an after clause in the assignment, the assigned value does not take effect until the update phase of the simulation cycle at the specified time.
- An initial value specified in a signal's declaration is not synthesizable.

Variables in an architecture:

- Can only be declared within processes or subprograms.
- Are not visible outside of the process or subprogram in which they are declared.

- Retain their values between executions of the process in which they are declared.
- Do ***not*** retain their values between executions of the subprogram in which they are declared.
- An initial value specified in a variable's declaration in a process is not synthesizable.

Another advantage of using variables rather than signals is that signals require large complicated data structures in the simulator to maintain their data queues. As a result, in complex designs using variables instead of signals results in reduced memory requirements and faster simulations.

6.8 DELTA DELAYS

We have seen how delta delays are used in event-driven simulation. At first, it may seem that they simply complicate matters. However, in this section we see why their use is necessary.

Delta Delay as a Scheduling Mechanism

A *delta delay* is a scheduling mechanism used to ensure a deterministic order of events in circuits having devices with no specified delays. This is necessary so that a circuit's operation can be simulated in a repeatable fashion.

Another, seemingly contradictory, way of stating this is that delta delays are necessary to allow order independence to be achieved in the simulation of concurrent statements. It has been stated that the order of execution of concurrent statements during a simulation does not affect simulation results. The use of delta delays allows a simulator to ensure that this is true when concurrent statements have signal assignments with no specified delays.

Simulation Process Execution Order Independence

It has also been stated that during a simulation cycle the order in which the simulation processes are executed does not affect simulation results. This follows from the requirement that the order in which concurrent statements in an architecture are executed does not affect simulation results.

Consider the two-level logic circuit in Figure 6.8.1(a), which is stimulated and monitored by a testbench. The corresponding simulation net is given in Figure 6.8.1(b).

If we were manually performing a functional evaluation of output `f` in terms of values applied to inputs `a`, `b`, and `c`, we would most likely first apply a new set of values to `a`, `b`, and `c`. Next we would compute the new value of `s1`. Then we would compute the new value of `f`. This would most likely be our order of evaluation, because we naturally think of information flowing from the inputs of the gates to their outputs. If we think of the simulation processes associated with the testbench and the gates in the simulation net, we would naturally think that we must execute these processes in this same order to determine the result.

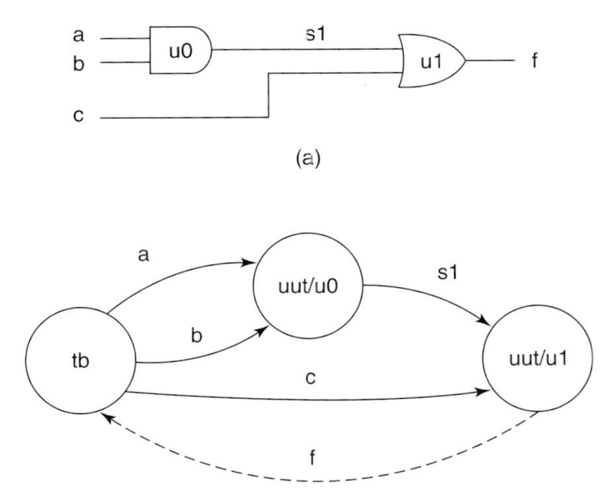

FIGURE 6.8.1
Simple AND/OR circuit and its simulation net.

However, consider performing the evaluation as a sequence of cycles where at the beginning of the sequence we apply a new set of values to `a`, `b`, and `c`. During the next cycle we compute the output of each gate, using the current value of the signals in the circuit. A new output computed by a gate cannot be used for computation of other gate outputs until the next cycle. The effect of this restriction is the same as associating a delay of 1 δ to each gate. We continue these computational cycles until we reach a cycle where all the signal values after the cycle are the same as they were at the start of the cycle. At that point, the circuit is quiescent and there will be no more changes in signal values in the circuit until the testbench applies a new value at `a`, `b`, or `c`.

In the computational algorithm just described the output of every gate is computed during every cycle. We can shorten this algorithm by realizing that, if the inputs to a gate at the beginning of a cycle are the same as they were at the beginning of previous cycle, we do not need to compute the gate's output because it will be the same. Since logic elements in the functional simulation of a synthesizable description have no delays specified, we can assign a default delay of 1 δ to each gate, and think of the time between the cycles that follow a change in `a`, `b`, or `c` as being 1 δ long. The result is the equivalent of an event-driven simulation.

As an example, consider the application of the first binary input values to `a`, `b`, and `c`, and the computation of `f` under the restriction that, during any simulation cycle where the outputs of both gates `u0` and `u1` must be computed, `u1`'s output is computed first and then `u0`'s output. This is the reverse of what we would consider the "natural" order of evaluation, and illustrates that the order of computation of gate outputs during a simulation cycle does not matter, as long as we use the current signal values for all computations during that cycle and continue the simulation cycles until there are no further changes in signal values (events).

In Figure 6.8.2, a copy of the circuit is given for each cycle resulting from the application of the first binary input values to a, b, and c. In Figure 6.8.2, new signal values are shown in parentheses to the right of current signal values only if the new value is different from the current value. The new values don't take effect (are not used) until the next cycle. The testbench and gates are shaded if they must execute during the current cycle.

As can be seen from Figure 6.8.2, the order of execution (computation of gate outputs) does not matter as long as each cycle computes using the current signal values and the cycles are continued until there are no more signal changes. For

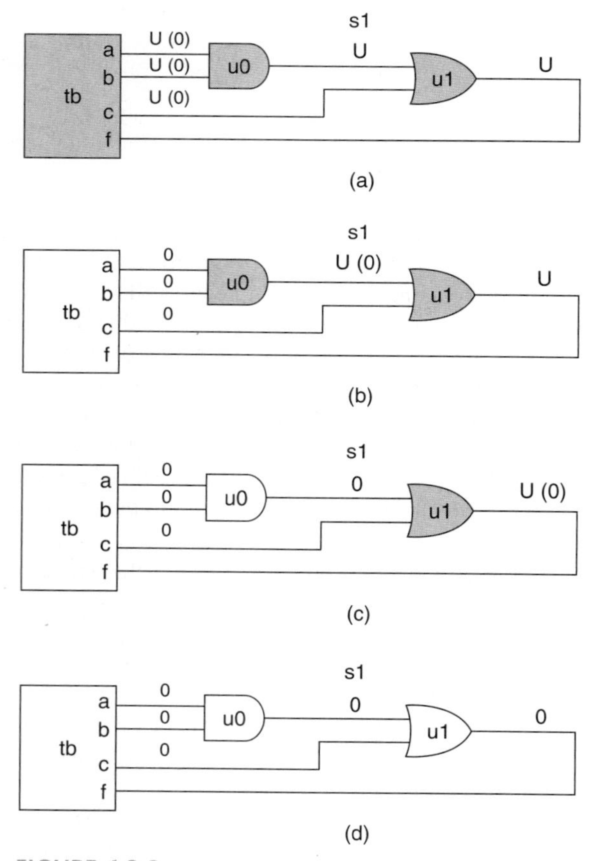

FIGURE 6.8.2
Cycles in the computation of AND/OR circuit output for application of one input combination.

example, in Figure 6.8.2(b) the output from u1 can be computed before the output from u0 and the result is the same.

Glitches in Zero Delay Circuits

As another example, consider a manual analysis of the operation of the circuit in Figure 6.8.3. Assume that we are interested in the functional operation of the circuit and do the analysis assuming the gate delays are zero.

The results of the manual analysis will depend on the order in which we evaluate the effect of signal changes as they propagate through the circuit. For example, assume input a is a 0, and has been for a substantial period of time. If a is changed from 0 to 1 at time t0, there is an event on a at t0. Does it matter whether we evaluate u0's or u1's output first? It certainly does! If the change at u0 is evaluated first, output f remains constant at 1.

In contrast, if the change at u1 is evaluated first, output f will go to 0. Then, evaluation of the change at the input to u0 causes the output of u1 to go to 0. This causes a change at input i1 of u1. Now u1's output must be reevaluated. This change at input i1 of u1 causes f to change back to 1. As a result, this order of analysis indicates that there is a narrow pulse (glitch) to 0 on output f when a changes from 0 to 1.

Does the 0 glitch really occur on f or not? One order of manual analysis says it does not and the other says it does. If it does, it could be very important because output f might be an input to a subsequent circuit that responds to a glitch.

If we reconsider this analysis in terms of a physical circuit, we know the gates would have some nonzero delay. If we associate any nonzero delay with u0, it is clear that input i2 of u1 changes to 1 before input i1 changes to 0, and that f has a glitch to 0 equal in length to the nonzero propagation delay through u1.

Use of delta delays solve this order of evaluation problem in zero delay components by associating a delta delay with each gate. So, the resulting simulation actually shows the glitch.

A structural description of the top-level entity of the circuit in Figure 6.8.3 is given in Listing 6.8.1.

An equivalent behavioral description is given in Listing 6.8.2. In this description, each process is equivalent to one of the gates in Figure 6.8.3. However, the processes are listed in a textual order the reverse of the order that we visualize the gates in Figure 6.8.3.

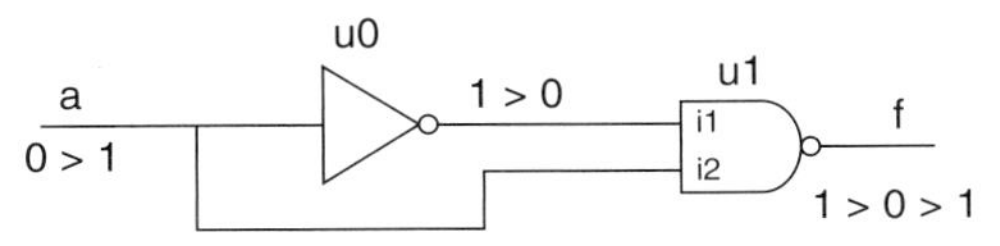

FIGURE 6.8.3
Gate circuit with two levels of delay.

LISTING 6.8.1
Structural description of two-level gate circuit from Figure 6.8.3.

```
library ieee;
use ieee.std_logic_1164.all;

entity delta_2gates is
   port (a : in std_logic;
      f : out std_logic);
end delta_2gates;

architecture struct of delta_2gates is
   signal s1  : std_logic;
begin

   u0: entity not_gate port map (a, s1);
   u1: entity nand_2 port map (s1, a, f);

end struct;
```

LISTING 6.8.2
Behavioral description of two-level gate circuit.

```
library ieee;
use ieee.std_logic_1164.all;

entity delta_2gates is
   port ( a : in std_logic;
      f : out std_logic);
end delta_2gates;

architecture behavioral of delta_2gates is
   signal s1 : std_logic;
begin

   u1: process ( a, s1 )
   begin
      f <= a nand s1;
   end process u1;

   u0: process ( a )
   begin
      s1 <= not a;
   end process u0;

end behavioral;
```

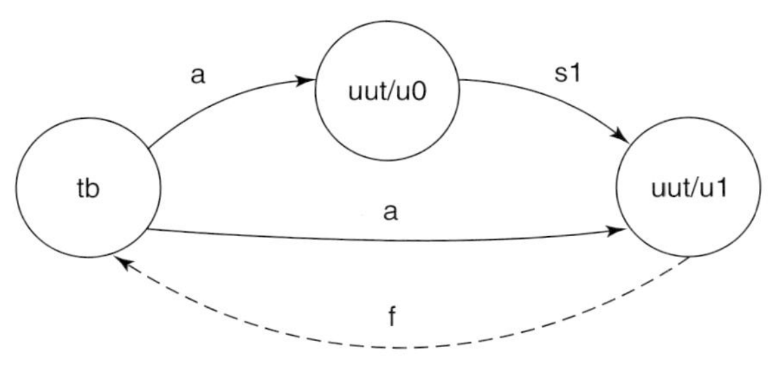

FIGURE 6.8.4
Simulation process net for two-level gate circuit.

The simulation net resulting from elaboration, including the testbench simulation process, is shown in Figure 6.8.4. This net is the same for either description,

With a previously '0', the testbench assigns a value of '1' to a at 20 ns. This assignment takes effect at 20 ns + 1 δ. The simulation cycles listing, Table 6.8.1, shows the glitch occurring at simulation time 20 ns + 2 δ.

This simulator waveform shows the glitch as a tick mark above the waveform for f at 20 ns (Figure 6.8.5).

6.9 DELTA DELAYS AND COMBINATIONAL FEEDBACK

As another example of the use of delta delays, consider the NAND gate circuit with combinational feedback in Figure 6.9.1. As previously discussed ("Avoiding Combinational Loops" on page 155), feedback is to be avoided in combinational design descriptions. Here we consider how the simulator handles combinational feedback when it is encountered.

Table 6.8.1
Simulation cycles listing for three gate circuit.

ns	δ	a	f
0	0	U	U
0	1	0	U
0	2	0	1
20	1	1	1
20	2	1	0
20	3	1	1

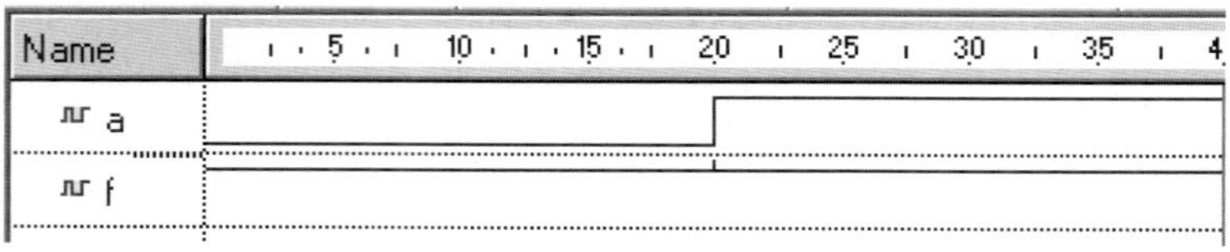

FIGURE 6.8.5
Waveform showing output glitch.

Assume input `enable` is `'0'` for 50 ns and then is changed to a `'1'` for 350 ns. Functional simulation produces the waveforms in Figure 6.9.2 and an error message: *"Delta count overflow - stopped. Try to increase the iterations limit in simulator preferences."* The simulation stopped after 50 ns, instead of continuing for 400 ns.

Closer examination reveals that when the simulation starts, `enable` is a `'0'` and the output of the NAND gate is a `'1'`. This `'1'` is fed back to the NAND gate's top input, but does not cause the gate's output to change. The circuit is stable.

However, at 50 ns, `enable` is changed to a `'1'` by the testbench. This input change takes effect at 50 ns + 1 δ. This causes the NAND gate's output (and its top input) to change to `'0'` at 50 ns +2 δ. This in turn causes the NAND gate's output (and its top input) to change to `'1'` at 50 ns +3 δ. In effect, when its bottom input is a `'1'`, the NAND simply inverts the value at its top input. The new output value is continually fed back and inverted. As a result, the NAND's output and its top input change state every delta. Thus, the circuit is unstable and oscillates.

The simulation process for this circuit has `enable` and `s1` in its sensitivity list. Since there is an event on `s1` during every delta cycle, the simulation process executes every delta cycle, scheduling a change in the value of `s1` for 1 δ later. This causes still another delta cycle.

Simulator Delta Cycle Limit

The simulator has a limit set on the maximum number of delta cycles in a time step. In this example, that limit is surpassed and simulation is never advanced past the 50 ns time step. Increasing the delta cycle limit will not solve the problem because the number of cycles required is infinite.

A timing simulation of the circuit targeted to a PLD with a 10 ns delay is given in Figure 6.9.3.

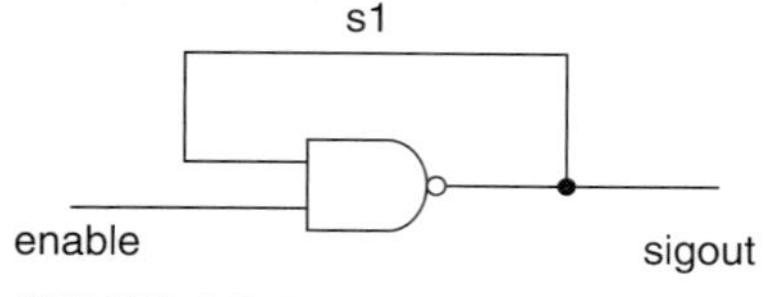

FIGURE 6.9.1
NAND circuit with combinational feedback.

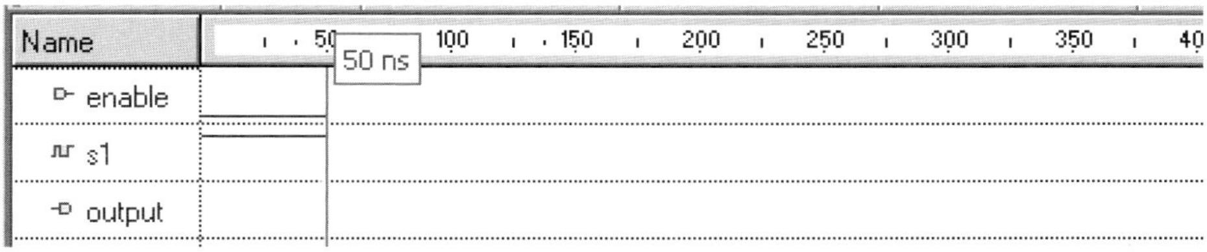

FIGURE 6.9.2
Output waveforms for functional simulation of NAND gate with feedback.

In contrast to the functional simulation, this timing simulation runs the desired 400 ns. In this simulation, when `enable` becomes a `'1'` at 50 ns + 1 δ, the simulation process executes and schedules a transaction to make the output `'0'` 10 ns later, at 60 ns. There are no more delta cycles at 50 ns and the simulation process suspends. The output remains unchanged until simulation time is advanced to 60 ns and the simulation process resumes. The output then changes to `'0'`. The resulting output is a square wave with a period of 20 ns.

NAND Latch Delta Delay Example

Another useful example to consider is a latch implemented using two cross-coupled NAND gates. The logic diagram is shown in Figure 6.9.4.

This circuit intentionally contains combinational loops and is an asynchronous sequential circuit. Its operation depends on both its present input values and on its past history of its input values. The circuit implements a simple memory element that remembers which of its inputs was last a 0. The operation of the circuit is given by the state table in Table 6.9.1.

Although it contains combinational loops, this circuit is stable. It has two stable states. The synthesizer will warn of the existence of combinational loops in the description, but here their existence is intentional.

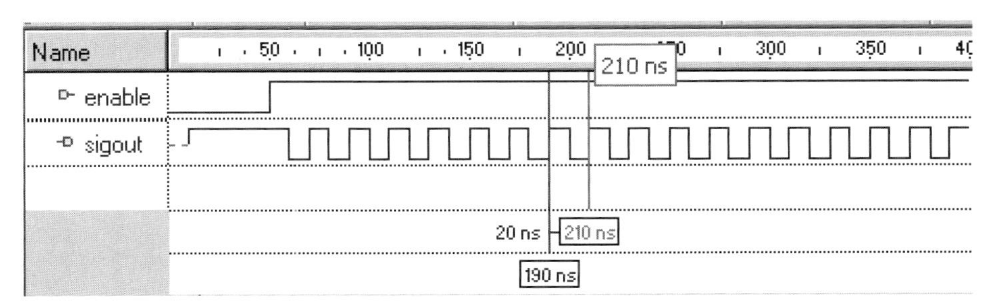

FIGURE 6.9.3
Output waveforms for timing simulation of NAND gate with feedback.

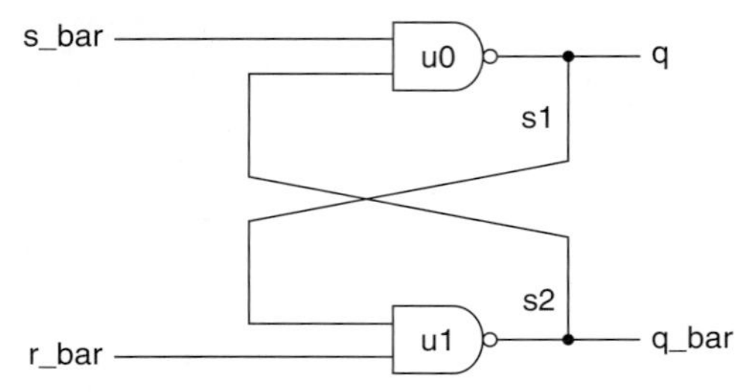

FIGURE 6.9.4
Set–reset latch from two cross-coupled NAND gates.

Table 6.9.1
State table for set–reset latch.

s_bar	r_bar	q	q_bar
0	0	?	?
0	1	1	0
1	0	0	1
1	1	q	q_bar

A behavioral description of the circuit is given in Listing 6.9.1.

An interesting aspect of this description is that its functional simulation depends on nonzero gate propagation delays to function properly. The simulator's use of delta delays meets this requirement.

LISTING 6.9.1
Behavioral description of NAND latch.

```
library ieee;
use ieee.std_logic_1164.all;

entity nand_latch is
   port (s_bar, r_bar : in std_logic;
         q, q_bar: inout std_logic);
end nand_latch;


architecture behavioral of nand_latch is
```

```
begin

u0: process (s_bar, q_bar)
begin
   q <= s_bar nand q_bar;
end process;

u1: process (q, r_bar)
begin
   q_bar <= q nand r_bar;
end process;

end behavioral;
```

6.10 MULTIPLE DRIVERS

In previous examples in this chapter, each signal had a single driver. A design description may have more than one simulation process that contains an assignment statement that assigns a value to the same signal. Each of these processes defines its own separate driver for that signal. For such a multiply driven signal, the simulator must create multiple driver queues for the same signal. The simulator uses a resolution function to determine, from the current value of each signal driver queue, what should be the current value of the signal.

Three-State Buffers with a Common Output

The description of two three-state buffers with their outputs connected, from Section 4.10, is an example of such a situation. A structural description of the circuit was given in Listing 4.10.2. An equivalent behavioral description is given in Listing 6.10.1.

LISTING 6.10.1
Behavioral description of two three-state buffers with their outputs connected.

```
library ieee;
use ieee.std_logic_1164.all;

entity three_state_bus is
   port (
      d_a: in std_logic;
      en_a_bar: in std_logic;
      d_b: in std_logic;
```

(Cont.)

LISTING 6.10.1 *(Cont.)*

```
        en_b_bar: in std_logic;
        d_bus: out std_logic
        );
end three_state_bus;

architecture three_state_bus of three_state_bus is
    begin

    u1: process (d_a, en_a_bar)
        begin
        if en_a_bar = '0' then
            d_bus <= d_a;
        else
            d_bus <= 'Z';
        end if;
    end process u1;

    u2: process (d_b, en_b_bar)
        begin
        if en_b_bar = '0' then
            d_bus <= d_b;
        else
            d_bus <= 'Z';
        end if;
    end process u2;

end three_state_bus;
```

Process `u1` has two assignment statements that assign values to `d_bus`. Since these assignment statements are in the same process, only one driver for `d_bus` is created by this process. Accordingly, elaboration creates a single driver for `d_bus` from `u1` `(u1/d_bus)`.

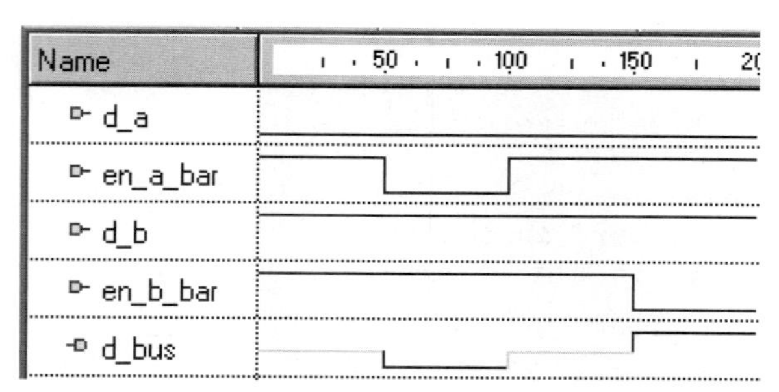

FIGURE 6.10.1
Waveform from simulation of two three-state buffers with a common output.

Table 6.10.1
Simulation cycles listing for simulation of two three-state buffers with common output.

ns	d	d_a	en_a_bar	d_b	en_b_bar	d_bus
0	0	U	U	U	U	U
0	1	0	1	1	1	Z
50	1	0	0	1	1	Z
50	2	0	0	1	1	0
100	1	0	1	1	1	0
100	2	0	1	1	1	Z
150	1	0	1	1	0	Z
150	2	0	1	1	0	1

Process `u2` also has two signal assignment statements for `d_bus`. Since these assignment statements are in a different process, a second driver for `d_bus` is created (`u2/d_bus`). The actual value of `d_bus` is the resolved value from the current values of drivers `u1/d_bus` and `u2/d_bus`.

The output waveform for a simulation of this circuit is shown in Figure 6.10.1. The input to three-state buffer `u1` is fixed at `'0'` and the input for three-state buffer `u2` is fixed at `'1'`. Both buffers are initially disabled. At 50 ns buffer `u1` is enabled. At 100 ns it is disabled. At 150 ns buffer `u2` is enabled.

The listing of delta cycles from the simulation is given in Table 6.10.1.

Signal drivers for the buffer outputs are represented in Figure 6.10.2. There are two separate drivers for `d_bus`, `u1/d_bus` and `u2/d_bus`. However, there is only one current value variable for `d_bus`. In Figure 6.10.2 the current value variable is shown to the left of the signal name.

After initialization, the testbench schedules transactions for the inputs at the 0 δ cycle of each time step. To reduce the amount of space required for the figure, the 0 δ cycles for each time step are not shown. Since transactions are only scheduled on the input drivers at those times, no events are possible during these delta cycles.

The value of the current value variable in column one is obtained by the std_logic resolution function operating on the current driver values for that simulation cycle.

6.11 SIGNAL ATTRIBUTES

VHDL's *attribute* feature allows us to extract information about an object (such as a signal, variable, or type) that may not be directly related to the object's value. This information can be used in a description or a testbench. Attributes also allow us to assign additional information to objects that can be useful for synthesis.

Value	Signal	<	Signal Driver Transaction Queue		
	After elaboration				
d_bus	u1/d_bus	<			
	u2/d_bus	<			
	After initialization Tc = 0, Tn = 0 + 1 δ				
'U'	u1/d_bus	<	'U' @ 0	'Z' @ 0 + 1 δ	
	u2/d_bus	<	'U' @ 0	'Z' @ 0 + 1 δ	
	After simulation cycle Tc = 0 + 1 δ, Tn = 50				
'Z'	u1/d_bus	<	'Z' @ 0 + 1 δ		
	u2/d_bus	<	'Z' @ 0 + 1 δ		
	After simulation cycle Tc = 50 + 1 δ, Tn = 50 + 2 δ				
'Z'	u1/d_bus	<	'Z' @ 0 + 1 δ	'0' @ 50 + 2 δ	
	u2/d_bus	<	'Z' @ 0 + 1 δ		
	After simulation cycle Tc = 50 + 2 δ, Tn = 100				
'0'	u1/d_bus	<	'0' @ 50 + 2 δ		
	u2/d_bus	<	'Z' @ 0 + 1 δ		
	After simulation cycle Tc = 100 + 1, Tn = 100 + 2 δ				
'0'	u1/d_bus	<	'0' @ 50 + 2 δ	'Z' @ 100 + 2 δ	
	u2/d_bus	<	'Z' @ 0 + 1 δ		
	After simulation cycle Tc = 100 + 2 δ, Tn = 150				
'Z'	u1/d_bus	<	'Z' @ 100 + 2 δ		
	u2/d_bus	<	'Z' @ 0 + 1 δ		
	After simulation cycle Tc = 150 + 1, Tn = 150 + 2 δ				
'Z'	u1/d_bus	<	'Z' @ 100 + 2 δ		
	u2/d_bus	<	'Z' @ 0 + 1 δ	'1' @ 150 + 2 δ	
	After simulation cycle Tc = 150 + 1 δ, Tn = done				
'1'	u1/d_bus	<	'Z' @ 100 + 2 δ		
	u2/d_bus	<	'1' @ 150 + 2 δ		

FIGURE 6.10.2
Signal drivers during simulation of three-state bus circuit.

Predefined and User-Defined Attributes

There are two classes of attributes: predefined and user defined. *Predefined attributes* are defined in IEEE Std 1076. *User-defined attributes* are defined outside of the standard, by tool vendors or by us. For example, we have seen attributes defined by a place and route tool vendor to assign signals to pin numbers in Listing 1.10.1.

One way of categorizing attributes is by the kind of results they return. When categorized this way there are five kinds of predefined attributes. *Value attributes* return a constant value. *Function attributes* call a function that returns a value. *Signal attributes* create a new implicit signal. *Type attributes* return a type. *Range attributes* return a range.

Predefined Signal Attributes

Predefined attributes are always applied to a prefix (such as a signal or variable name, or a type or subtype name). Predefined attributes can also be categorized by the kind of prefix involved. Given a signal S and a value t, VHDL defines the *signal attributes* listed in Table 6.11.1. Signal attributes provide information about the history of transactions and events on signals. A clear understanding of signal attributes is predicated on an understanding of simulation cycles in event-driven simulation. Accordingly, these attributes are introduced in this chapter and are used in later chapters.

Implicit Signals

The first four attributes create new signals from other signals. The signals created are called *implicit signals*; they have no explicit declarations.

The first attribute in Table 6.11.1, the transaction attribute, creates an implicit type bit signal that changes value (toggles) each time there is a transaction on its prefix signal.

Table 6.11.1
Predefined signal attributes.

Attribute	Result
S'Transaction	Implicit bit signal whose value is changed in each simulation cycle in which a transaction occurs on S (signal S becomes active).
S'Stable(t)	Implicit boolean signal. True when no event has occurred on S for t time units up to the current time, False otherwise.
S'Quiet(t)	Implicit boolean signal. True when no transaction has occurred on S for t time units up to the current time, False otherwise.
S'Delayed(t)	Implicit signal equivalent to S, but delayed t units of time.
S'Event	A boolean value. True if an event has occurred on S in the current simulation cycle, False otherwise.
S'Active	A boolean value. True if a transaction occurred on S in the current simulation cycle, False otherwise.
S'Last_event	Amount of elapsed time since last event on S, if no event has yet occurred it returns TIME'HIGH.
S'Last_active	Amount of time elapsed since last transaction on S, if no transaction has yet occurred it returns TIME'HIGH.
S'Last_value	Previous value of S immediately before last event on S.

(Cont.)

Table 6.11.1 *(Cont.)*

Attribute	Result
S'Driving	True if the process is driving S or every element of a composite S, or False if the current value of the driver for S or any element of S in the process is determined by the null transaction.
S'Driving_value	Current value of the driver for S in the process containing the assignment statement to S.

For example, assume that signal `sig` has been explicitly declared as a std_logic signal. Execution of the concurrent signal assignment statement

```
sig <= '0', '0' after 5 ns, '1' after 10 ns,
       '0' after 20 ns, '1' after 25 ns,
       '1' after 35 ns;
```

during the initialization phase of a simulation posts six transactions to the driver for `sig`. The transactions are for simulation times 0, 5, 10, 20, 25, and 35 ns.

The waveforms for signal `sig` and `sig'transaction` are given in Figure 6.11.1. For each transaction posted to `sig`, implicit signal `sig'transaction` toggles. Of the six transactions, only four cause events on `sig`. These events are at 0, 10, 20, and 25 ns. The event at simulation time 0 changes `sig`'s value from `'U'` to `'0'`.

The next three attributes in Table 6.11.1 take an optional time parameter. Attributes stable and quiet create implicit signals of type boolean. Attribute stable is true if there has been no event on its prefix signal for the specified time prior to the current time. In Figure 6.11.1 the time specified for the stable attribute is 5 ns. As a result, attribute `sig'stable(5ns)` is false until simulation time 5 ns, at which point it becomes true. However, since there is an event on `sig` at 10 ns, `sig'stable (5ns)` becomes false at that time.

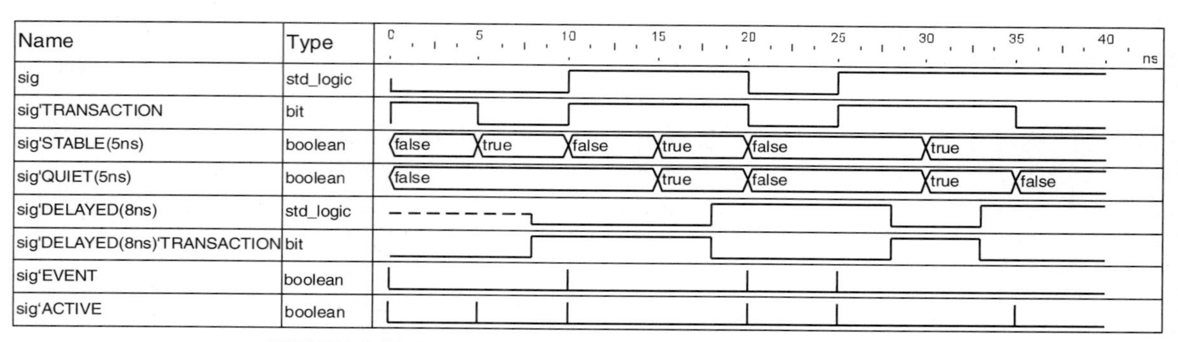

FIGURE 6.11.1
Some signal attributes for the signal `sig`.

Attribute quiet is true if there has been no transaction on its prefix signal for the specified time prior to the current time. In Figure 6.11.1 the time specified for the quiet attribute is 5 ns. As a result, attribute `sig'quiet(5ns)` is false until simulation time 15 ns, at which point it becomes true. However, since there is a transaction on `sig` at 20 ns, `sig'quiet(5ns)` becomes false at that time.

The delayed attribute creates an implicit signal that is the same type as its prefix, but is delayed by the time specified. In Figure 6.11.1, the signal corresponding to attribute `sig'delayed(8ns)` is shown. This is simply the signal `sig` shifted right 8 ns. For the first 8 ns this signal has the value `'U'`.

For the attribute delayed, if the time parameter is omitted, then $t = 0 + 1\ \delta$ is the default. That is, the delayed signal is delayed 1 δ. For the other two attributes, if this parameter is omitted, a default value of 0 fs is assumed.

Since an implicit signal is a signal, it also has attributes. Figure 6.11.1 shows the signal created by the attributed transaction applied to the implicit signal `sig'delayed(8ns)`. This creates the implicit signal `sig'delayed(8ns)'transaction`.

The event and active attributes return boolean values, not signals. The event attribute is true only during a simulation cycle where its prefix signal has an event. The active attribute is true only during a simulation cycle where there is a transaction on its prefix signal. Though they are not signals, the values of `sig'event` and `sig'active` are shown as waveforms on Figure 6.11.1.

Note that the implicit signals created are viewable on a waveform editor when any of the first four attributes in Table 6.11.1 are used in a description or testbench.

Attributes last_event and last_active return values of type time that correspond to the time elapsed since the last event or last transaction, respectively, on their prefix signals. This is not the time of the last event or transaction, but rather the amount of time that has passed since the last event or transaction. For example, if the current simulation time is 37 ns, `sig'last_event` is 12 ns and `sig'last_active` is 2 ns.

Attribute last_value gives the value its prefix signal before its last event. If there has been no event on its prefix signal, attribute last_value returns the current value of the prefix signal. For example, if the current simulation time is 37 ns, `sig'last_value` is `'0'`.

Signal attributes are often used in checking timing behavior within a model, as discussed in Chapters 7 and 14.

PROBLEMS

6.1 List the three common kinds of simulators for digital systems. What are the relative speeds of these simulators? Which one of these simulators cannot be used to simulate a combinational system and why?

6.2 What kind of simulator is specified in the VHDL LRM?

6.3 If the input to a system being functionally simulated changes twice in a 2 ms interval of simulation time and we wished to be able to simulate events with a resolution of 1 ns, how many times would a time-based simulator have to evaluate the model in the 2 ms time interval? For the same conditions, how many times would an event-driven simulator have to evaluate the model?

6.4 List the three steps an event driven-simulator performs to accomplish a simulation.

6.5 Given the following function:

$$y = a \cdot b + c$$

(a) Write an entity declaration and three alternative architectures for a system that computes the function. Each architecture must consist of a single concurrent signal assignment statement. The first architecture must use a Boolean expression. The second must use a selected signal assignment statement. The third must use a conditional signal assignment statement.

(b) For each of the concurrent signal assignment statements in part (a), write a corresponding simulation process.

6.6 Assume that the circuit described in Problem 5.7 is the UUT in a testbench. The testbench stimulus process named `tb` applies all possible input combinations to the UUT. Draw and label the simulation net that results from elaboration of the testbench.

6.7 Assume that a testbench applies every possible input combination to the half-adder design entity described in Listing 2.6.1. The testbench has a component instantiation with the label `uut` that instantiates the half adder and a process with the label `tb` that applies the stimulus.

(a) Write the code for the simulation processes that would result from elaboration of the testbench, except for process `tb`.

(b) Draw a completely labeled diagram of the simulation net (including `tb`) resulting from the elaboration.

(c) Draw the signal driver queues that would be produced by the elaboration.

(d) Assuming that when testbench process `tb` is first executed it assigns the value `'0'` to both `a` and `b`, list the values for all of the inputs and outputs for each simulation cycle from initialization through the end of the time step at 0 ns.

6.8 Assume that the design entity from Problem 2.29 is described by a structural style architecture. Also assume that a testbench has been written that uses a process named `tb` to apply inputs to the design entity and monitor its output.

(a) Write the code for all of the simulation processes created by elaboration, except for the process `tb`.

(b) Draw a diagram of the simulation net including `tb`.

(c) Assuming that when the testbench is simulated it first applies input combination "000" at time $0 + 0\ \delta$, write a list of all the signal values for all δs at time step 0.

6.9 A testbench is used to simulate the program in Problem 2.30. Design entity `gate_ckt` is instantiated as component `uut` in the testbench. Component `nand_2` is a two-input NAND function and component `invert` is an inverter (NOT) function. The first statement in stimulus process `tb` in the testbench applies inputs `a`, `b`, `c` = `"000"`. Every 50 ns `tb` applies a new binary test vector whose value is 1 greater than the previous test vector, until all input combinations have been applied.

(a) Draw a simulation net showing the simulation processes produced when the simulator elaborates the testbench.

(b) Write VHDL processes corresponding to the simulation processes that are produced during elaboration of `u0`, `u1`, and `u2`.

(c) Provide a listing of all the simulation deltas along with the signal values starting from initialization up to and including 50 ns.

6.10 Given the following architecture body:

```
architecture mixed of fcn is
   signal s1, s2 : std_logic ;
begin

   u0: s1 <= '1' when a = '0' and b = '1' else '0';

   u1: with std_logic_vector'(a, c) select
      s2 <= '1' when "11",
      '0' when others;

   u2: process (s1, s2)
   begin
      f <= s1 or s2;
   end process;

end mixed;
```

(a) Write the entity declaration for the design entity.

(b) Write each of the simulation processes created by elaboration.

(c) Assuming a testbench contains a process `tb`, which applies all possible input values and verifies the output, draw a diagram of the simulation net. Assume that the instantiation of the design entity in the testbench is labeled `uut`.

6.11 Under what conditions will a signal have multiple signal drivers?

6.12 List the possible states a simulation process can be in and what happens to the process in each state.

6.13 What happens during the initialization phase of a simulation and why?

6.14 Given the following architecture for a testbench for a half adder:

```
architecture waveform of testbench is

   signal a_tb, b_tb : std_logic;
   signal sum_tb, carry_out_tb : std_logic;

   begin

   UUT : entity half_adder
   port map (a => a_tb, b => b_tb, sum => sum_tb,
           carry_out => carry_out_tb );

   stim_a: a_tb <= '0', '1' after 40 ns;
   stim_b: b_tb <= '0', '1' after 20 ns, '0' after 40 ns, '1' after 60 ns;

end waveform;
```

(a) Draw the driver queues for a_tb and b_tb immediately after the initialization phase of the simulation.
(b) If the architecture for the half adder contains the following two statements:

```
u0:    sum <= a xor b;
u1:    carry_out <= a and b;
```

draw a diagram of the simulation net created after elaboration.
(c) Write the simulation process corresponding to the instruction labeled stim_b in the testbench.
(d) Provide the complete simulation cycle listing for time step 0 (include all the associated delta cycles).

6.15 Given the following design entity description:

```
library ieee;
use ieee.std_logic_1164.all;

entity and_or is
    port(
        a, b, c : in std_logic;
        f : out std_logic
        );
end and_or;

architecture csa of and_or is
    signal s1, s2 : std_logic ;
begin

    p0: s1 <= a and b;
    p1: s2 <= a and c;
    p2: f <= s1 or s2;

end csa;
```

and the testbench:

```
library ieee;
use ieee.std_logic_1164.all;

entity and_or_tb is
end and_or_tb;

architecture tb_architecture of and_or_tb is
    signal a, b, c, f : std_logic;
begin
    UUT : entity and_or port map (a => a, b => b, c => c, f => f);

    tb0: a <= '0', '1' after 20 ns;
```

```
    tb1: b <= '0', '1' after 40 ns, '0' after 60 ns;
    tb2: c <= '0', '1' after 80 ns;

end tb_architecture;
```

(a) Draw a logic diagram for a straightforward gate-level implementation of the design entity `and_or`.

(b) Write a simulation process for each of the concurrent statements `p0`, `p2`, and `tb1`.

(c) Draw a diagram of the simulation net for `and_or_tb`.

(d) Draw the stimulus waveforms for a simulation run for 100 ns.

(e) Create a table that lists the simulation time (including the delta number), values of inputs and outputs at the start of each simulation cycle, and the simulation processes executed during each simulation cycle, for **all** simulation cycles from 0 through 20ns. Follow the format of the table below.

time	delta	a	b	c	f	simulation processes

6.16 Write the simulation process for the NAND with a combinational loop in Figure 6.9.1 that is described by Listing 4.11.1.

6.17 The functional simulation in Figure 6.9.2 is stopped at 50 ns because of a delta count overflow. Can this problem be solved by increasing the simulator's delta cycle limit for a time step? Explain your answer.

6.18 Write a dataflow description of a simple NOR gate with feedback, similar to the NAND in Figure 6.9.1. Describe the output for a functional simulation where the input is `'1'` at 0 ns and is changed to a `'0'` at 40 ns and back to a `'1'` at 160 ns and remains at that value. Describe the output for a timing simulation with the same stimulus. Assume the target PLD has a 20 ns propagation delay.

6.19 Another way of categorizing predefined signal attributes is as transaction-related, event-related, or general. Specify into which category each of the attributes in Table 6.11.1 belongs.

6.20 Given the following concurrent signal assignment statement:

```
sigx <= '1', '1' after 5 ns, '0' after 8 ns,
          '0' after 12 ns, '1' after 20ns,
          '0' after 32 ns;
```

(a) Draw a figure similar to Figure 6.11.1 that shows `sigx` and the same signal attributes for `sigx` over the simulation time interval from 0 to 40 ns.

(b) Give the values for `sigx'last_event`, `sigx'last_active`, and `sigx'last_value` at a simulation time of 7 ns.

6.21 Given the following process:

```
p0: process
begin
   sigy <= '0'; wait for 4 ns;
   sigy <= '1'; wait for 6 ns;
   sigy <= '0'; wait for 15 ns;
   sigy <= '0'; wait for 5 ns;
   sigy <= '1'; wait for 5 ns;
   wait;
end process;
```

(a) Draw a figure similar to Figure 6.11.1 that shows `sigy` and the same signal attributes for `sigy` over the simulation time interval from 0 to 40 ns.

(b) Give the values for `sigy'last_event`, `sigy'last_active`, and `sigy'last_value` at a simulation time of 28 ns.

6.22 Restrictions on where a variable may be declared preclude declaring a variable in a manner that would make it accessible to more than one process. This is done to prevent a variable from being modified in an indeterminate order. Using an example of a variable declared in the declarative part of an architecture (a violation of the restrictions) containing two processes, explain, in the context of process simulation, how this could cause the value of the variable to be indeterminate.

Chapter 7

Testbenches for Combinational Designs

A design's functionality should be verified before its description is synthesized. After synthesis and place and route, the functionality and timing performance of the design mapped to the target PLD should be verified before a PLD is programmed.

A testbench is a program used to verify a design's functionality and/or its timing. An advantage of using VHDL is that the testbench is written in the same language as the design description. Since a testbench is not synthesized, it can be written using any of the constructs and features of VHDL. In terms of coding, the effort expended to write an appropriate testbench may exceed that required to write the design description.

This chapter focuses on writing testbenches for combinational designs. While the basic concepts introduced also apply to testbenches for sequential designs, that topic is covered separately in Chapter 14. Hierarchical testbenches are discussed in Chapter 15.

7.1 DESIGN VERIFICATION

Before getting into the details of writing combinational testbenches, some general design verification concepts are presented. These concepts pertain to both combinational and sequential designs.

We want to verify that our design is correct before the target PLD is programmed. The process performed to accomplish this is *design verification*.

There are two basic things that we need to verify. One is that our design achieves its intended functionality. The other is that our design, after being synthesized and mapped to the target PLD, will meet its timing requirements. Both of these aspects of a design can be verified through simulation using a testbench.

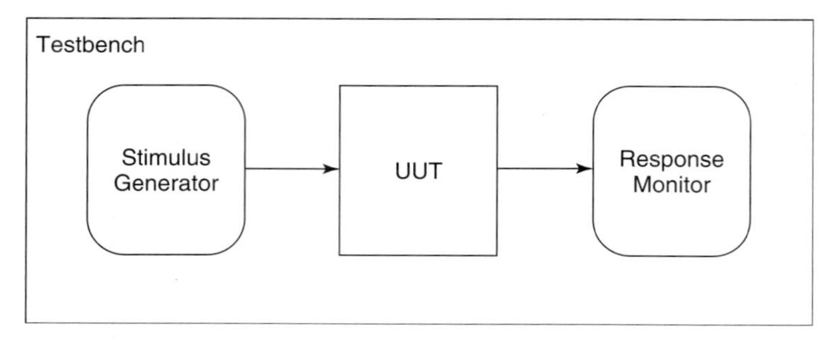

FIGURE 7.1.1
Typical constituents of a simple testbench.

Constituents of a Simple Testbench

The three basic constituents of a simple testbench—the UUT, stimulus generator, and response monitor—are shown in Figure 7.1.1.

UUT

A testbench is a top-level VHDL program that includes an instantiation of the design to be verified. This instance is labeled UUT (unit under test) in Figure 7.1.1. For a functional verification, the design description is the UUT. For a timing verification, the VHDL timing model generated by the place-and-route tool is the UUT.

Stimulus Generator

A testbench includes code that applies a sequence of predetermined stimulus values to the UUT's inputs. This code comprises a stimulus generator. In this book the term *stimulus* is used to refer to both a single combination of input values applied to the UUT and a sequence of such input values applied over time.

The stimulus must not contain any metalogical or high-impedance values. Logical values `'L'` and `'H'` are replaced by `'0'` and `'1'`, respectively. As a result, for combinational designs, only combinations of the values `'0'` and `'1'`, or sequences of such combinations, are applied as stimulus.

Response Monitor

In response to each stimulus, the UUT's output values must be checked to verify that they are identical to the expected output values.

Self Checking Testbench

In testbenches for very simple designs, verification can be accomplished by visual inspection of the UUT's outputs in the simulator's waveform editor window. However, a more efficient approach is to include code in the testbench to automatically check the actual UUT output values against the expected values. Such a testbench is referred to as being *self-checking*. Thus, the block labeled response monitor in Figure 7.1.1 could be either a person inspecting the simulator output waveforms or code that is part of the testbench.

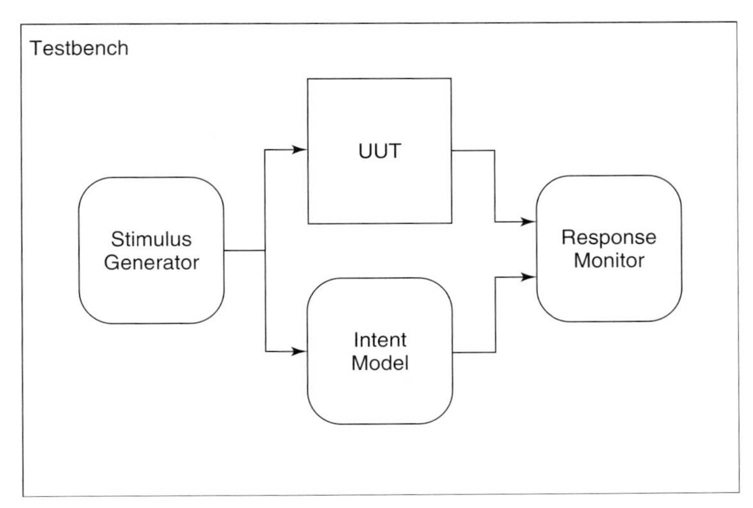

FIGURE 7.1.2
Self-checking testbench using an intent model.

Testbench Containing an Intent Model

The representation of a self-checking testbench in Figure 7.1.2 has an added block labeled "*intent model*." This block represents code used to model the intent of the design.

The intent model models the expected UUT response to the stimulus. It is also referred to as the *golden model*. Using an intent model, the response monitor simply compares the UUT's output with that of the intent model, for each stimulus applied.

Testbench Development

From the self-checking testbench representation in Figure 7.1.2, we can infer the primary tasks that we must accomplish in developing a testbench.

1. Determine the intent of the design from its specification.
2. Determine a strategy for verification of the design.
3. Determine the stimulus needed to achieve the desired verification.
4. Create a model that represents the intent of the design.
5. Create the stimulus generator.
6. Evaluate the adequacy of the verification.

Differences in Testbenches

The primary differences in testbenches are in how the stimulus is generated and how the expected output values are determined and compared to the actual output values.

What Is Actually Being Verified?

A natural language specification for a system usually contains ambiguities. That is, for certain scenarios the system's functionality may not be clearly stated in the specification.

In general, the more complex the system, the more ambiguities in the specification. In fact, complex systems are rarely completely specified. As a result, additional information must be developed during the design of the system to resolve ambiguities.

Our determination of the intent of a system from its natural language specification is our personal interpretation of that specification. If we write both the design description and its testbench and they are both based on the same incorrect interpretation of the system specification, then the verification may be successful, but the design is incorrect. We have verified only that our design meets our erroneous interpretation of the specification, not that it meets the specification's true intent.

Independent Verification

Ideally, the final testbench for a design is developed and written by persons other than those who wrote the design description. Then, if the two independent interpretations of design intent are not identical, the testbench should expose the differences, so that they can be properly resolved.

What Stimulus Is Required?

A testbench should, at a minimum, emulate the environment in which the design will ultimately operate. For a design that will be used as a general library component or sold as IP, the testbench must verify all the functionality claimed for the design.

Intent Models

The intent model provides the expected values in response to a stimulus. The intent model can be anything from a table to an executable specification. For example, design intent might be represented by a nonsynthesizable VHDL behavioral model.

When Is Verification Complete?

Since a combinational design's outputs at any time are a function only of its inputs at that time, the order in which input combinations are applied does not affect the verification results. For a combinational design, an *exhaustive verification* requires the application, in any order, of every possible input combination to the UUT and verification of its outputs for each input combination.

As previously stated, only the two values `'0'` and `'1'` from the nine std_logic values are applied to a UUT's inputs for verification. Thus, for a UUT with n inputs there are 2^n possible input combinations. Simulation time increases exponentially as the number of inputs to a combinational system is increased. However, since large systems are decomposed into less complex components, and components are individually designed and verified before verification of the entire system, exhaustive verification is practical for most combinational components.

In contrast, exhaustive verification of a sequential design requires that every possible sequence of input combinations be applied to the UUT and its outputs verified. For an even moderately complex sequential design, it is impractical to perform an exhaustive verification because of the enormous number of different stimuli that would have to be applied. Therefore, it is usually impossible to prove that a complex sequential design is error free. Consequently, a determination must be made as to what constitutes an adequate verification for a particular design.

7.2 FUNCTIONAL VERIFICATION OF COMBINATIONAL DESIGNS

Functional verification is performed to determine if a design meets a system's functional specifications. It is not concerned with any timing delays that result from mapping synthesized logic to a target PLD. Accordingly, functional verification should be performed immediately after a design description is written, as was shown in Figure 1.6.1. This allows us to find logical errors early in the design flow and prevents our wasting time performing synthesis, place and route, and timing simulation for a description that is not functionally correct.

In addition to detecting logical errors resulting from our misinterpreting a system's specification, functional verification allows us to detect errors that result from our misunderstanding the semantics of VHDL statements that we have used. When first learning VHDL, semantic errors can be a common occurrence in the descriptions that we write. It is a good rule to never assume that a design description is too simple to justify functional verification.

VHDL Constructs Used in Testbenches

Since a testbench is only simulated, not synthesized, the full range of VHDL language constructs and features can be used. Some additional nonsynthesizable VHDL features useful in testbenches include multidimensional arrays and text I/O.

Counting Approach

A simple approach to performing exhaustive verification is for the testbench to treat all the UUT's inputs as a single vector. Stimulus vectors are applied to the UUT starting with the combination where all elements are 0s. The stimulus vector is subsequently incremented through all of its possible binary combinations. For each vector applied, the UUT's outputs are verified.

Functionality Approach

While counting is straightforward, it is not always the best approach. Treating all of a UUT's inputs as a single vector may obscure the functional relationship between individual inputs and the output values, making it harder for others to understand the design's logic. In some situations it is better to take into account the functionality of the design being tested when determining the order for applying input combinations.

Both the counting and functionality approaches are used in testbench examples in this chapter.

7.3 A SIMPLE TESTBENCH

The simplest testbenches don't include code for a response monitor or an intent model. They just apply stimulus to the UUT. We must then visually verify UUT output values by inspecting waveforms using the simulator's waveform editor.

LISTING 7.3.1
Simple testbench for a half adder using projected waveforms.

```
library ieee;
use ieee.std_logic_1164.all;

entity testbench is -- testbench entity has no ports
end testbench;

architecture waveform of testbench is

   -- Stimulus signals - to connect testbench to UUT input ports
   signal a_tb, b_tb : std_logic;

   -- Observed signals - to connect testbench to UUT output ports
   signal sum_tb, carry_out_tb : std_logic;

   begin

   -- Unit Under Test port map
   UUT : entity half_adder port map (a => a_tb, b => b_tb,
            sum => sum_tb, carry_out => carry_out_tb );

   -- Signal assignment statements generating stimulus values
   a_tb <= '0', '1' after 40 ns;
   b_tb <= '0', '1' after 20 ns, '0' after 40 ns, '1' after 60 ns;

end waveform;
```

In order to simulate the operation of a UUT using a testbench, we must actually simulate the testbench, since the UUT is simply a component in the testbench.

A simple testbench that generates stimulus to exhaustively verify a half-adder design is given in Listing 7.3.1.

In this example, the testbench entity is named `testbench`. Often a convention is followed where the testbench's name is the same as the UUT's name with the addition of the suffix `_tb`. The entity declaration for `testbench` has no ports.

In the declarative part of the architecture, signals `a_tb`, `b_tb`, `sum_tb`, and `carry_out_tb`, used to connect the UUT to the testbench, are declared. The testbench assigns values to `a_tb` and `b_tb` and we must observe the values of `sum_tb` and `carry_out_tb`.

For clarity in our early testbench examples, the suffix `_tb` is used so that local testbench signals can be distinguished from UUT ports in component instantiation port maps. In later testbench examples, these local signals are usually declared with names identical to the UUT's port names.

The statement part of the architecture contains three concurrent statements. The first is the component instantiation statement for the UUT. The label `UUT` used on this statement is not significant; any meaningful name could be used.

Half-Adder Instantiation in a Testbench

Named association is used in the UUT's instantiation to map the UUT ports to local testbench signals:

```
port map ( a => a_tb, b => b_tb, sum => sum_tb,
       carry_out => carry_out_tb );
```

Here the signals' `_tb` suffixes make the syntax of the association clear. That is, the item to the left of the association symbol is a component port and the item to the right is the signal to which it is connected.

Named association is used instead of positional association so that the post place-and-route model used later in timing simulations, which will not likely have its ports listed in the same order, can be instantiated as the UUT in the same testbench.

Stimulus Generated by Projected Signal Waveforms

As discussed in Chapter 6, VHDL signals are *projected waveforms.* A signal has a current value, and future scheduled values. Two simple ways of generating a stimulus are:

- Using a projected signal assignment
- Using a process with wait statements

In Listing 7.3.1, two projected signal assignments are used for stimulus generation. These statements correspond to signal assignment statements having multiple waveform elements with after clauses. This syntax is given in Figure 7.3.1.

After Clause

The optional after clause specifies when its associated value should become the new value of a signal. The execution of an assignment statement with multiple waveform elements results in a sequence of transactions being posted to the target signal's driver. The time component of each transaction is the sum of the current simulation time (Tc) and the value of the corresponding after clause's time expression. Transactions in the signal assignment statement must be in ascending order with respect to time. After clauses are not synthesizable.

The time expression in an after clause must evaluate to a value of type time. Type time is a predefined physical type that is described in more detail in the next section.

The second and third concurrent statements in the architecture of Listing 7.3.1 are signal assignment statements that use after clauses to assign waveforms to signals `a_tb` and `b_tb`:

```
a_tb <= '0', '1' after 40 ns;
b_tb <= '0', '1' after 20 ns, '0' after 40 ns,
             '1' after 60 ns;
```

signal_assignment_statement ::= target <= *value*_expression [**after** time_expression] ~~{, *value*_expression [**after** time_expression]}~~;

FIGURE 7.3.1
Syntax for a signal assignment statement with multiple waveform elements.

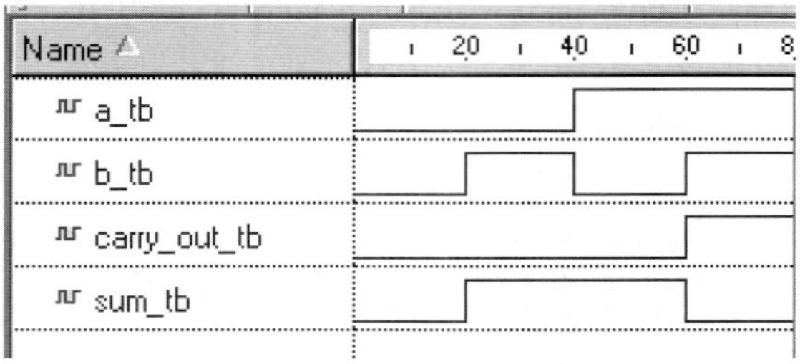

FIGURE 7.3.2
Timing waveform from simulation of half-adder testbench.

These two statements are executed only once, at system initialization, during a simulation.

The first statement describes the waveform, assigned to `a_tb`. This waveform becomes `'0'` at the current simulation time and remains `'0'` for 40 ns. Then it becomes and remains a `'1'`.

The second statement applies a `'0'` to `b_tb` at the current simulation time, then applies a `'1'` after 20 ns, a `'0'` after 40 ns, and, finally, a `'1'` after 60 ns. This assignment statement describes a waveform that alternates between 0 and 1 every 20 ns until 60 ns after the current simulation time, at which time it remains constant as a `'1'`.

When viewed together, these two signals are seen to count the two inputs through all four possible input combinations. When simulated for 80 ns, each input combination is applied for 20 ns.

If the entity `testbench` is simulated for 80 ns, the waveforms shown in Figure 7.3.2 are produced.

There are no statements in this testbench to stop the simulation; we can either set the maximum simulator time limit prior to starting the simulation, or let the simulator run until there are no further stimulus changes to process.

While using projected waveforms to generate stimulus is fine for combinational designs with very few inputs, for designs with numerous inputs it becomes tedious and error prone. Another drawback to the testbench in Listing 7.3.1 is that we must verify the outputs visually.

7.4 PHYSICAL TYPES

Let us now consider physical types in greater detail. Physical types are numeric types used to represent real-world physical quantities, such as time, frequency, voltage, and current. Physical types are like integer types with associated physical units. The syntax for a physical type declaration is given in Figure 7.4.1.

The strike-through on the left-hand side of the production in Figure 7.4.1 indicates that physical type declarations are not supported for synthesis.

```
physical_type_declaration ::=
type identifier is range simple_expression (to | downto) simple_expression
        units
                identifier ;   -- primary unit declaration
                { identifier = physical_literal ; }   -- secondary_unit_declarations
        end units [ physical_type_simple_name ]]

physical_literal ::= [ abstract_literal ] unit_name
```

FIGURE 7.4.1
Physical type declaration syntax.

Primary and Secondary Units

The *primary unit* of measure, listed immediately after keyword **units**, is the smallest unit represented by the physical type. The range specifies the multiples of the primary unit that are included in the type. Any value of a physical type is an integral multiple of its primary unit. Any *secondary units* are listed after the primary unit and specified as integer multiples of the primary unit.

Using a physical type declaration, we can create our own physical types appropriate for particular designs.

Type Time

The physical type that we are currently interested in is type time. Type time is the only predefined physical type. It is used extensively to specify delays. Type time is declared in package STANDARD as:

```
type time is range implementation_defined
     units
          fs;-- femtosecond
          ps = 1000 fs;-- picosecond
          ns = 1000 ps;-- nanosecond
          us = 1000 ns; -- microsecond
          ms = 1000 us; -- millisecond
          sec = 1000 ms;-- second
          min = 60 sec; -- minute
          hr = 60 min; -- hour
     end units;
```

The range for type time is defined by the particular VHDL compiler (implementation). The primary unit is `fs` for femtoseconds (10^{-15} seconds). All secondary time units are integer multiples of `fs`. All of the secondary units in this declaration, except one, are specified in terms of other secondary units.

The primary unit femtosecond is the resolution limit for type time. Any time value whose absolute value is smaller than 1 femtosecond is truncated to 0 time units. A simulator may allow a secondary time unit to be selected as the resolution limit.

Subtype Delay_length

Package STANDARD also defines a subtype of time named delay_length:

```
subtype delay_length is time range
                            0 fs to time'high;
```

The range for the subtype `delay_length` is from 0 `fs` to the highest value of type time.

Physical Literals

Physical literals can be written as integral or nonintegral multiples of primary or secondary units. If the value written is not an exact multiple of the primary unit, it is rounded down to the nearest multiple. Note that there must be a space before the unit name when physical values are written as literals.

Arithmetic Operations on Physical Types

Arithmetic operators—addition, subtraction, identity, and negation—can be applied to physical types. A value of a physical type can also be multiplied or divided by a number of type integer or type real. All of the preceding operations yield a result of the same physical type. For example, assuming that constant `prop_delay` is type time, we could write any of the following assignments:

```
s1 <= '0' after prop_delay * 2;
s2 <= '1' after prop_delay / 4;
```

If two values of the same physical type are divided, the result is type integer. Direct multiplication of two physical types is not permitted, because the result would not be dimensionally correct. If two physical types must be multiplied, their values must first be converted to abstract integers, these integers multiplied, and the result then converted to the final physical type. Examples of these types of operations are given in Chapter 14.

7.5 SINGLE PROCESS TESTBENCH

An easy approach to generating stimulus and checking responses uses a *single process testbench*. Such a testbench includes a process that applies a stimulus to the UUT, waits an appropriate length of time, and then checks the UUT's outputs. These actions are repeated until all stimulus values have been applied and all UUT outputs verified. The wait between applying each stimulus results in waveforms being generated. Thus, the functionality of the stimulus generator and response monitor in Figure 7.1.1 are provided by the single process.

A single process testbench has two concurrent statements in its architecture body. One instantiates the UUT. The other is the process that applies the stimulus and verifies the UUT output values. Since the UUT instantiation and the process are both concurrent statements, either one can appear first in the architecture body. However, instantiation of the UUT first is the more common practice.

Single process testbenches differ in how the process generates or obtains each stimulus value and determines the expected response. Some read test vectors from a table of constants in the testbench. Others execute algorithms that compute stimulus and expected output values during the simulation. Still others read test vectors from a file.

The testbench in Listing 1.5.1 used a process to apply each possible input combination to a half adder over time and to check the half-adder's outputs. This testbench is repeated in Listing 7.5.1. Stimulus values and expected response values are simply written into the code as literals. The process is written without a sensitivity list. Wait statements are used to realize a specific delay between the application of each stimulus value. After the delay, an assertion statement checks the outputs.

LISTING 7.5.1
Testbench for a half adder using a process to apply stimulus and assertion statements to check outputs.

```
libraryieee;
use ieee.std_logic_1164.all;

entity testbench is
end testbench;

architecture behavior of testbench is

   -- Declare local signals to assign values to and to observe
   signal a_tb, b_tb, sum_tb, carry_out_tb :  std_logic;

   begin
   -- Create an instance of the circuit to be tested
   uut: entity half_adder port map(a => a_tb, b => b_tb,
      sum => sum_tb, carry_out => carry_out_tb);

   -- Define a process to apply input stimulus and verify outputs
   tb : process
      constant period: time := 20 ns;

      begin -- Apply every possible input combination

      a_tb <= '0'; --apply input combination 00 and check outputs
      b_tb <= '0';
      wait for period;
      assert ((sum_tb = '0') and (carry_out_tb = '0'))
      report "test failed for input combination 00" severity error;

      a_tb <= '0'; --apply input combination 01 and check outputs
      b_tb <= '1';
```

(Cont.)

LISTING 7.5.1 *(Cont.)*

```
        wait for period;
        assert ((sum_tb = '1') and (carry_out_tb = '0'))
        report "test failed for input combination 01" severity error;

        a_tb <= '1'; --apply input combination 10 and check outputs
        b_tb <= '0';
        wait for period;
        assert ((sum_tb = '1') and (carry_out_tb = '0'))
        report "test failed for input combination 10" severity error;

        a_tb <= '1'; --apply input combination 11 and check outputs
        b_tb <= '1';
        wait for period;
        assert ((sum_tb = '0') and (carry_out_tb = '1'))
        report "test failed for input combination 11" severity error;

        wait; -- indefinitely suspend process

    end process;
end;
```

For example, the following statements assign a stimulus of all `'0'`s to the UUT and check its outputs:

```
a_tb <= '0';
b_tb <= '0';
wait for period;
assert ((sum_tb = '0') and (carry_out_tb = '0'))
report "test failed for input combination 00"
       severity error;
```

Wait Statements in a Testbench

After a stimulus value is assigned to the half-adder's inputs, a wait for statement is used to cause the simulator to wait for a time equal to the constant `period`. There are three reasons why this statement is required.

First, the signal assignments can't take effect until after the process suspends. Without the wait for statement, the response would be checked during the current simulation cycle and would not be the response to the newly assigned input values.

Second, waiting for a time interval allows the new input values to appear in the simulator waveforms for this time period. This makes it possible to view the waveform. Otherwise, all the actions in the process would take place in zero simulation time. As a result, the waveform's time axis would be compressed to zero length and the waveforms would not be viewable.

Third, the wait for statements allow this same testbench to be used later, without modification, for a timing simulation. For a timing simulation, the instantiated UUT is the timing model. During the simulation, we want the simulator to wait a period of time, determined from the design's timing specification, before the UUT's outputs are checked. The maximum propagation delay of the design mapped to the target PLD must be less than or equal to this period of time, or the timing simulation fails.

The final wait statement, without a for clause, suspends the process indefinitely. This causes the simulation to stop. Without this final wait statement, the process would repeat forever, or until the simulator time limit is reached.

Assertion Statements in a Testbench

Use of assertion statements eliminate the need to visually inspect timing waveforms. The condition in the first assertion statement requires that:

```
(sum_tb = '0') and (carry_out_tb = '0')
```

If this condition is true, the next statement in the process is executed. If it is false, the message

```
"test failed for input combination 00"
```

is generated and severity level `error` is assigned. This message aids debugging by making it clear for which input combination the failure occurred.

The use of a wait statement to produce a delay between the application of each stimulus, and the use of an assertion statement to automatically verify the UUT's response, is common practice. Wait statements and assertion statements are examined in detail in the next two sections, prior to our examining other testbench techniques.

7.6 WAIT STATEMENTS

A *wait statement* suspends and resumes execution of the process containing the statement. A wait statement can also appear in a procedure. A wait statement in a procedure suspends and resumes the process that contains the call to the procedure.

Wait statements find widespread use in testbenches, but limited use in synthesizable code. The syntax for a wait statement is given in Figure 7.6.1.

As shown by its syntax, any combination of the three optional clauses (or no optional clause) can be included in a wait statement. These clauses specify under what conditions the process resumes execution.

If only one, or none, of the optional clauses is included, four forms result:

```
wait on sensitivity_list;
wait until boolean_expression;
wait for time_expression;
wait;
```

```
wait_statement ::=
wait [sensitivity_clause] [condition_clause] [timeout_clause] ;

sensitivity_clause ::= on signal_name {, signal_name}
condition_clause ::= until boolean_expression
timeout_clause ::= for time_expression.
```

FIGURE 7.6.1
Syntax for a wait statement.

Wait On

A wait on statement has a sensitivity list. While a process is suspended due to the execution of a wait on statement, all signals in the statement's sensitivity list are monitored by the simulator. When an event occurs on any one of these signals, the process resumes execution at the statement following the wait on statement. Note that a wait on statement's sensitivity list is not enclosed in parentheses.

Wait Until

A wait until statement causes a process to suspend until its associated boolean expression is true. A wait until statement has an *implied sensitivity list* involving all the signals in the Boolean expression. These signals are monitored by the simulator while the process is suspended, due to execution of the wait until statement.

When an event occurs on any signal in the implied sensitivity list, the boolean expression is evaluated. If it evaluates true, the suspended process resumes execution at the statement following the wait until statement.

If its boolean expression is already true, when a wait until statement is executed, the process will still suspend. There must be a subsequent event on one of the signals in the implied sensitivity list that causes the boolean expression to again be true for the process to resume execution. For example, execution of the statement

```
wait until interrupt = '1';
```

requires that the signal `interrupt` has changed value (during the current simulation cycle) and its new value is `'1'` for the process to resume execution.

If a wait statement contains both a sensitivity clause and a condition clause, the boolean expression is evaluated only when an event occurs on a signal in the sensitivity clause. For example, the statement

```
wait on ie until interrupt = '1';
```

will resume the process when signal `ie` changes value and signal `interrupt = '1'`. In this case, `interrupt` does not have to change value for the process to resume.

Wait For

A wait for statement causes a process to suspend for a *time-out interval* equal to its associated `time_expression`. After the time-out interval elapses, the process automatically resumes. We have previously used this form of wait statement in testbenches to delay for a time interval after a stimulus has been applied, before checking the UUT's outputs. For example, the testbench of Listing 7.5.1 contains the following wait for statement:

```
wait for period;
```

where `period` was a constant equal to 20 ns.

Multiple Condition Wait

A wait statement can contain all three clauses. For example, the statement

```
wait on ie until interrupt = '1' for period;
```

causes its process to resume when either `ie` changes value and `interrupt = '1'`, or a time-out interval equal to `period` has expired, whichever occurs first.

Wait (Forever)

The keyword **`wait`** by itself causes a process to suspend forever. We used this form of wait statement at the end of Listing 7.5.1 to stop the testbench's execution.

A Process as an Infinite Loop

A process with no wait statement or sensitivity list executes its statements in sequence, starting at the first statement. After the last statement in the process is executed, the process immediately continues execution at its first statement, forming an infinite loop. As a result, the process never suspends.

Wait Statements in Synthesizable Code

Std 1076.6 2004 supports synthesis of all wait statements, except wait for statements. The earlier Std 1076.6 1999 only supported the wait until form for synthesis. Furthermore, Std 1076.6 1999 only allowed one wait until statement per process and it had to be the first statement in the process. An example of the use of a wait statement for synthesis is given in Chapter 8 in the section "Using a Wait Statement to Infer a Flip-flop" on page 319.

7.7 ASSERTION AND REPORT STATEMENTS

The true power of a testbench is realized when it not only generates stimulus, but automatically verifies output values. A process can check UUT output values against expected values. This is accomplished by placing assertion statements at appropriate points in the process. An *assertion statement* checks whether a specified condition (the assertion) is true. If it is not true, a message is displayed. This approach can eliminate the need to visually inspect simulator waveforms. The syntax rule for an assertion statement is given in Figure 7.7.1.

An assertion statement always has the required assert clause. The report and severity clauses are optional. Note that an assertion statement is a single statement and only one semicolon is used, and must be at the very end of the statement.

```
assertion_statement ::=
assert condition [ report expression ] [ severity expression ] ;
```

FIGURE 7.7.1
Syntax for an assertion statement.

We use an assertion statement to assert a condition that must be true for correct operation of the design entity. When the condition is written in terms of correct behavior, it is simply written as the desired condition. When the condition is written in terms of incorrect behavior, it is written as the `not` of the undesired condition.

When verifying output values in a testbench, the condition we put in an assertion statement equates the actual output value to its expected value. When an assertion statement is executed, its condition is evaluated. If the condition evaluates true, execution continues with the statement following the assertion statement.

Assertion Violation

If the condition in the assert statement evaluates false, an *assertion violation* occurs. If the assertion statement has no report clause, the simulator displays the default error message "Assertion violation." If the assertion statement includes a report clause, the string that comprises the expression in the report clause is displayed. This allows us to provide more specific information in the message about what caused the assertion violation.

Severity Levels

The severity clause allows a *severity level* to be assigned. This level indicates the degree to which an assertion violation affects operation of the system and what actions the simulator must take. The levels are enumeration literals of the type *severity_level*, which is predefined in package STANDARD as:

```
type severity_level is ( note, warning, error,
    failure);
```

Note is simply used to display informative messages during a simulation. *Warning* is used to indicate an unusual situation where the simulation can continue but may produce unusual results. *Error* is used to indicate a situation where corrective action should be taken. *Failure* is used to indicate a situation that should never arise. If the severity clause is omitted, the default level is error.

VHDL Std 1076 defines the severity levels, but not the action to be taken in response to a particular level. Simulators allow a severity level threshold to be set where any assertion violation with a severity equal to or above the threshold causes the simulation to stop execution. In Listing 7.5.1, if the simulator threshold is set at error and an assertion violation occurs, the message in the report clause is displayed and the simulator halts.

If the simulation completes and no reports are displayed, it is assumed that the UUT performed as expected.

In using assertion statements to check outputs, we should place an assertion statement in the process at each point where a new UUT output value is expected. The new expected value of the output is specified in the assertion condition. A unique message should be provided in each assertion statement.

Report Statement

In addition to the report clause in an assertion statement, there is a separate statement called a report statement. A *report statement* displays a message every time it is executed. If its severity clause is omitted, its severity level defaults to note (instead of error). Whenever executed, a report statement always generates the message associated with the report expression.

A report statement has the syntax given in Figure 7.7.2.

Report Clause Message Text

Report statements and report clauses in assertion statements only accept a single string. However, this string can be produced by concatenating substrings. The string is displayed as a single line, unless the special character constants CR (carriage return) and/or LF (line feed) are used in the concatenation of substrings. For example, the report statement

```
report "test failed" & CR LF & "for input combination 01"
             severity error;
```

will print the two lines:

```
test failed
for input combination 01
```

To aid in debugging, it is helpful if the string displayed by an assertion statement indicates the value of the input stimulus or other conditions that caused the assertion violation. In Listing 7.5.1 it was not difficult to include the stimulus values as constants in the report string. In general, we want to be able to display the actual value of the signal or variable associated with each assertion violation.

Image Attribute

Values that appear in a report clause are not limited to literal values, but can be an expression that is dynamically computed during simulation. Unfortunately, VHDL has limited provisions for formatting types for display. For any scalar type, the *image attribute* can be used to produce a string representation of the scalar type. T'image(x) returns the string representation of the expression x of type T.

report_statement ::=
report expression [**severity** expression] ;

FIGURE 7.7.2
Syntax for a report statement.

For example, the report statement

```
report "test failed for a_tb = " &
       std_logic'image(a_tb) & " and b_tb = " &
       std_logic'image(b_tb);
```

will display the values of `a_tb` and `b_tb` at the time that the statement is executed. For example, if the assertion violation occurs while `a_tb` equals `'1'` and `b_tb` equals `'0'`, the message displayed is:

```
test failed for a_tb = '1' and b_tb = '0'
```

To display values of composite data types, we must provide type conversion functions to convert the data types to formatted strings. This approach is discussed further in Chapter 12.

Concurrent Assertion Statements

The assertion statements that we have seen in examples have been in processes and are, therefore, sequential statements. Assertion statements can also appear as concurrent statements in an architecture body. A concurrent assertion statement is executed when one of the signals in its condition clause has an event. If the assertion condition evaluates false, the report expression and severity expressions are displayed.

7.8 RECORDS AND TABLE LOOKUP TESTBENCHES

A simple way to specify both stimulus and expected response values is to use a table of constants. This approach is similar to the table lookup technique used in Section 4.9 to describe combinational designs.

When used in a testbench, table entries usually include both the input and output values of the combinational function's truth table. Initial elements in a row are input values to be applied to the UUT. The remaining elements in that same row are the corresponding expected output values. Each combination of input values and corresponding expected output values (each row) in the table comprises a *test vector.*

The testbench process reads each row in turn. Input values from a row are applied to the UUT. Expected output values from the same row are compared to the UUT's outputs.

Records

A convenient way to represent a row of the table is as a record. A *record* is a composite type that consists of named elements that may be of different types. Elements in a record are selected by name, not by an index number. The syntax for a record declaration is given in Figure 7.8.1.

```
record_declaration ::=
type identifier is record
    identifier_list : element_subtype_definition ;
    { identifier_list : element_subtype_definition ; }
end record [ record_type_simple_name ]
```

FIGURE 7.8.1
Syntax for a record declaration.

The identifiers of all the elements of a record type must be distinct. The value of a record object is the composite value of its elements.

Selected Names

The entire value of a signal, variable, or constant record can be assigned to a signal or variable of the same record type using an assignment statement. When a single element of a record is to be selected for assignment to an object of the same element type or a value is to be assigned to a single element of a record, that element's selected name is used.

A *selected name* consists of a prefix and a suffix separated by a period, as shown in the syntax in Figure 7.8.2.

For an element of a record, the prefix is the record name and the suffix is the element name.

Test Vector Records

Each test vector is defined as a record. This approach requires us to:

1. Declare a record type
2. Declare an array of records of the record type
3. Declare a constant of the array of records type

For example, to declare a record for a test vector for the half adder we can write:

```
type test_vector is record
        a : std_logic;
        b : std_logic;
        sum : std_logic;
        carry_out : std_logic;
end record;
```

This record has four elements, all of which are type std_logic. The first two elements are values for inputs `a` and `b`. The second two elements are the corresponding values for outputs `sum` and `carry_out`.

```
selected_name ::=
prefix . suffix
```

FIGURE 7.8.2
Syntax for a selected name.

The table of constants that comprises all of the test vectors is an array of records of type `test_vector`. This unconstrained array is declared as follows:

```
type test_vector_array is array
      (natural range <>) of test_vector;
```

The actual table of constants, declared using named association, is

```
constant test_vectors : test_vector_array := (
-- a, b, sum, carry_out
(a => '0', b => '0', sum => '0', carry_out =>'0'),
(a => '0', b => '1', sum => '1', carry_out =>'0'),
(a => '1', b => '0', sum => '1', carry_out =>'0'),
(a => '1', b => '1', sum => '0', carry_out =>'1'));
```

or more concisely, using positional association, is

```
constant test_vectors : test_vector_array := (
-- a, b, sum, carry_out
('0', '0',  '0', '0'),
('0', '1',  '1', '0'),
('1', '0',  '1', '0'),
('1', '1',  '0', '1'));
```

The first two elements in each vector correspond to sequencing the inputs through all the possible combinations required for an exhaustive verification. The second two elements are the expected output values for those input values.

The complete testbench is given in Listing 7.8.1.

LISTING 7.8.1
Table lookup testbench for a half adder.

```
library ieee;
use ieee.std_logic_1164.all;

entity testbench is
end testbench;

architecture table of testbench is

   -- Stimulus signals
   signal a : std_logic;
   signal b : std_logic;
   -- Observed signals
   signal sum : std_logic;
   signal carry_out : std_logic;
```

```
   -- Declare record type
   type test_vector is record
      a : std_logic;
      b : std_logic;
      sum : std_logic;
      carry_out : std_logic;
   end record;

   type test_vector_array is array (natural range <>) of test_vector;

   constant test_vectors : test_vector_array := (
   -- a, b, sum, carry_out
   ('0', '0', '0', '0'),
   ('0', '1', '1', '0'),
   ('1', '0', '1', '0'),
   ('1', '1', '0', '1'));

begin

   UUT : entity half_adder
port map (a => a, b => b, sum => sum, carry_out => carry_out );

   verify : process
   begin
      for i in test_vectors'range loop

         a <= test_vectors(i).a;
         b <= test_vectors(i).b;

         wait for 20 ns;

         assert (( sum = test_vectors(i).sum )
                      and (carry_out = test_vectors(i).carry_out))
         report "test vector " & integer'image(i) & " failed"&
                      " for input a = " & std_logic'image(a)
                      & " and b = " & std_logic'image(b)
         severity error;
      end loop;
      wait;
   end process;
end table;
```

In the declarative part of the architecture body, types `test_vector` and `test_vector_array` are declared along with the constant `test_vectors`, as described at the beginning of this section.

Following the UUT's instantiation is the process `verify` that applies values to the UUT's inputs and verifies the UUT's outputs.

Looping Through Input Combinations

The statement part of the process consists primarily of a single for loop. The loop's index range corresponds to the range of the constant `test_vectors`. In this example, that range is from 0 through 3, since there are four rows in table `test_vectors`.

Each time through the loop, the input values from a test vector determined by the loop index are assigned to the inputs of the UUT, using the selected names of the elements in the record:

```
a <= test_vectors(i).a;
b <= test_vectors(i).b;
```

After a wait of 20 ns, the assertion statement checks whether the UUT's outputs have the expected values, as specified by the test vector. If either UUT output is not equal to its expected value, then the message "test vector *n* failed" is displayed by the simulator.

The report clause uses the predefined image attribute, `integer'image(i)`, to convert the loop index to a string equal to the index value, so that the index of the specific test vector that failed can be displayed. It also uses the image attribute to print the values of `a` and `b` for test vectors that failed. For example, if the fourth test vector (with index 3) fails, the message output is:

```
test vector 3 failed for input a = '1' and b = '1'
```

In this example, the vector index and values for `a` and `b` in the report clause provide redundant information; only one is really needed.

The loop repeats until all test vectors from the table have been processed.

7.9 TESTBENCHES THAT COMPUTE STIMULUS AND EXPECTED RESULTS

In previous testbenches, expected output values were predetermined for each input combination and included in the testbench as literals. Alternatively, a testbench can be written so that the expected output values are computed during the simulation. In a single process testbench, the same process is used to apply stimulus and compute and verify expected results.

Listing 7.9.1 is a testbench for the half adder that computes the expected results.

This testbench differs from the one in Listing 7.5.1 only in its process. In the beginning of the process, a constant n is defined that is equal to the number of UUT inputs. A for loop, whose index value ranges from 0 to $2^n - 1$, is then used. The exponentiation operator symbol is the double asterisk (**). Inside the for loop, the function `to_unsigned` is used to convert the integer index value to an unsigned

LISTING 7.9.1
Half-adder testbench that computes expected results.

```
library ieee;
use ieee.std_logic_1164.all;
use ieee.numeric_std.all;

entity testbench is
end testbench;

architecture behavior of testbench is
   -- Declare signals to assign values to and to observe
   signal a_tb, b_tb, sum_tb, carry_out_tb :  std_logic;

begin
   -- Create an instance of the circuit to be tested
   uut: entity half_adder port map(a => a_tb, b => b_tb,
      sum => sum_tb, carry_out => carry_out_tb);

   -- Define a process to apply input stimulus and test outputs
   tb : process
      constant period: time := 20 ns;
      constant n: integer := 2;
   begin -- Apply every possible input combination
      for i in 0 to 2**n - 1 loop
         (a_tb, b_tb) <= to_unsigned(i,n);
         wait for period;
         assert ((sum_tb = (a_tb xor b_tb)) and (carry_out_tb = (a_tb and b_tb))
)
         report "test failed" severity error;
      end loop;
      wait; -- indefinitely suspend process
   end process;
end;
```

vector of length *n*. The value of this vector is then assigned to the aggregate formed from the UUT's inputs `(a_tb, b_tb)`:

```
(a_tb, b_tb) <= to_unsigned(i,n);
```

Using Function `to_unsigned` to Generate Stimulus

Function `to_unsigned` was previously introduced in the subsection "Functions for Conversion to and from Type Integer" on page 109. Recall that this function has two parameters. The first parameter is the integer to be converted. The second is the length of the returned unsigned vector. This vector's element values are the binary equivalent of the integer value passed to the function.

In the previous assignment statement, the unsigned vector value returned by the `to_unsigned` function is assigned to an aggregate made up of the scalar input signals. Since each of these scalar inputs is type std_logic and the elements of an unsigned vector are also type std_logic, the assignment is valid.

We will often use the function `to_unsigned` inside a loop in a testbench to apply all possible input combinations to a UUT. The package NUMERIC_STD and type unsigned are discussed in more detail in Chapter 13.

After waiting for an interval of time, the UUT's outputs are checked. The expected values are computed from the input values using Boolean expressions. The condition in the assertion requires that the expressions

```
sum_tb = (a_tb xor b_tb)
```

and

```
carry_out_tb = (a_tb and b_tb)
```

evaluate true for the assertion to be valid.

Using a loop, a conversion function, and Boolean expressions to compute the output leads to a more concise testbench.

When the preceding testbench is used for functional verification of a half adder whose architecture is behavioral style or structural style, it is very effective. But, if this testbench is used for functional verification of a dataflow style half adder whose architecture uses the same equations, it is not at all effective, because we are simply verifying the Boolean expressions against themselves.

However, if the UUT is the synthesized logic for the dataflow style half adder, the specific logic primitives of the hardware implementation are not usually identical to those of the equations and the testbench again provides an effective verification.

The computation of expected results can also be done using a process separate from the one that generates stimulus. This separate process may simply be an intent model. We will see an example of this in Section 7.12, after shift operators are introduced in the next section.

7.10 PREDEFINED SHIFT OPERATORS

In addition to its logical operators, VHDL provides a number of shift and rotate operators. The shift and rotate operators are collectively called shifts, and are defined in package STANDARD for one-dimensional arrays whose elements are either of type bit or type boolean, Table 7.10.1.

The operators in Table 7.10.1 are not overloaded for std_ulogic operands in package STD_LOGIC_1164, as VHDL's logical operators are. So, we cannot use them on type std_logic_vector. Package NUMERIC_STD does overload these operators for types unsigned and signed, and these types can be used when we want to synthesize these operatons.

Table 7.10.1
VHDL shift operators.

Operator	Description
`sll`	shift left logical
`srl`	shift right logical
`sla`	shift left arithmetic
`sra`	shift right arithmetic
`rol`	rotate left logical
`ror`	rotate right logical

However, as we shall see, the predefined shift operators using type bit operands can be very useful in testbenches. Accordingly, they are discussed here.

Logical Shifts

The left operand for each operator must be a one-dimensional array, either a bit vector or boolean vector. The right operand must be an integer. For bit vectors, logical shifts shift the bit vector *n* positions in the direction specified by the right operand. *n* `'0'`s are shifted in, where *n* is the value of the integer. The bits shifted out are discarded. If *n* is a negative integer, the direction of the shift is opposite of that specified by the operator. For example:

```
b"10101101" sll 3 = b"01101000"

b"10101101" sll -3 = b"00010101"
```

For boolean vectors, `false` is shifted in instead of `'0'`.

Arithmetic Shifts

For arithmetic shifts, a copy of the bit in the original operand at the end opposite the direction of shift is shifted in. For example. for shift right arithmetic, a copy of the leftmost bit is shifted in:

```
b"10101101" sra 2 = b"11101011"
```

For a shift right arithmetic operation this preserves the sign bit, assuming the bit vector is interpreted as a two's complement number. Arithmetic shifts are common in the assembly language instruction sets of computers.

In assembly language, there is no distinction between a shift left arithmetic operation and a shift left logical operation. In contrast, in VHDL a shift left arithmetic operation shifts in a copy of the rightmost bit of the original operand. For example:

```
b"10101101" sla 2 = b"10110111"
```

Rotates

The **ror** and **rol** rotate operations are similar to the **srl** and **sll** shift operations, except the bit rotated out of one end of the vector is rotated into the other end.

The shift operators defined in package STANDARD cannot be applied directly to std_logic_vectors. However, by using type conversion functions we can use shift operators with std_logic_vector signals. An example of this is given in the next section.

7.11 STIMULUS ORDER BASED ON UUT FUNCTIONALITY

In previous testbenches the stimulus values were applied by aggregating all inputs into a single vector and counting this vector through all of its possible values starting from 0. It is often preferable, particularly for more complex combinational systems, to apply stimulus values in an order that is related to the functionality of the UUT. That is, values are applied to certain UUT control inputs to put the UUT in a particular mode of operation. Other inputs are then sequenced through all of their values. This verifies each different mode of operation of the UUT in a logical fashion. While this may lead to a textually longer stimulus process, it is often more readable and understandable to others.

For example, consider the 74F539 design "Example of a 74F539 Decoder Equivalent" on page 175. Using the truth table in Table 5.4.1 as the specification of the design's functionality, we might first make `oe_bar` = 1, which should cause all UUT's outputs to be high impedance (`Z`), independent of any other input's value. Then we could sequence all the other inputs through every combination, verifying that each output is `Z` for all combinations of these inputs.

Next, the stimulus process could then make `oe_bar` = 0 and `e_bar` = 1 and verify that the decoder functionality is not enabled and that all outputs should be equal to input `p`.

Finally, the stimulus process could make `e_bar` = 0, enabling the decoder and causing the output selected by `a` to be the complement of `p` and all other outputs equal to `p`. Input `a` would then be sequenced through all of its possible values.

Each part of the stimulus process would be documented to indicate exactly what aspect of the decoder's functionality was being verified by that part. A testbench that follows this approach is given in Listing 7.11.1.

LISTING 7.11.1
Testbench for ic74f539 with stimulus application order based on UUT functionality.

```
library ieee;
use ieee.numeric_std.all;
use ieee.std_logic_1164.all;

entity ic74f539_tb is
end ic74f539_tb;
```

```
architecture tb_architecture of ic74f539_tb is
   -- Stimulus signals
   signal a : std_logic_vector(1 downto 0);
   signal p : std_logic;
   signal e_bar : std_logic;
   signal oe_bar : std_logic;
   -- Observed signals
   signal o : std_logic_vector(3 downto 0);
    -- stimulus application interval
   constant period : time := 20 ns;

begin

   -- Unit Under Test port map
   UUT : entity ic74f539
      port map (a => a, p => p, e_bar => e_bar, oe_bar => oe_bar,
         o => o);

    -- Stimulus and monitor process
   stim_mon: process
    variable o_exp :  std_logic_vector(3 downto 0);
    begin

        -- verify output enable functionality
        oe_bar <= '1';
        for i in 0 to 15 loop
            (p, e_bar, a(1), a(0)) <= to_unsigned(i, 4);
            wait for period;
            assert (o = "ZZZZ") report "output enable failed"
            severity error;
        end loop;

        -- verify enable unasserted functionalify
        oe_bar <= '0'; e_bar <= '1';
        for i in 0 to 7 loop
            (p, a(1), a(0)) <= to_unsigned(i, 3);
            wait for period;
            o_exp := p & p & p & p;
            assert (o = o_exp)  report "enable unasserted failed"
            severity error;
        end loop;

        -- verify enable asserted decoder functionality
        e_bar <= '0';
        for i in 0 to 7 loop
            (p, a(1), a(0)) <= to_unsigned(i, 3);
            wait for period;
```

(Cont.)

LISTING 7.11.1 *(Cont.)*

```
            o_exp := (p & p & p & p) xor
            to_stdlogicvector("0001" rol to_integer(unsigned(a)));
            assert (o = o_exp)
            report "enable asserted failed" severity error;
        end loop;
        wait;
    end process;

end tb_architecture;
```

In the stimulus process, a variable `o_exp` is used to hold the computed expected output value that is compared to the actual output value `o`.

There are two type conflicts that must be resolved to use the **rol** operator. First, the right operand of a **rol** operator must be an integer, but `a` is a std_logic_vector. To eliminate this type conflict, `a` is first converted to unsigned. Function `to_integer` is then used to convert the unsigned value to an integer.

Converting a Bit_vector to a Std_logic_vector

The result of a **rol** operation is a bit_vector. In this example, the result must be assigned to a std_logic_vector. To eliminate this second type conflict, the function `to_stdlogicvector`, from package STD_LOGIC_1164, is used. Function `to_stdlogicvector` converts a bit_vector to a std_logic_vector.

Verifying Don't Care Output Values

Stimulus application ordered by UUT functionality is also useful when a combinational system has some outputs that are defined as don't cares for certain input (stimulus) values. For example, when verifying the outputs of a BCD to seven-segment decoder, we would separate the stimulus into two sets: one for valid inputs and the other for invalid inputs (those assigned don't care output values). We would need two different testbenches, one for functional simulation and another for post-sythesis and timing simulation.

The testbench for functional simulation would use two loops, one to apply the valid inputs and the other to apply the invalid inputs. The first loop would check for the expected decoder output values. The second loop would check for all outputs being `'-'`. It is important that we verify the `'-'` outputs for inputs corresponding to the second loop, because we want the synthesizer to be able to take advantage of these don't cares to simplify the logic.

The testbench for post-sythesis and timing simulation would be the same as the first testbench, except that it would not include the second loop. We don't want to verify the outputs from the synthesized logic corresponding to invalid inputs, because we don't know or care what those values will be.

7.12 COMPARING A UUT TO A BEHAVIORAL INTENT MODEL

A behavioral style model, that may not be synthesizable, can be used as an intent model to verify a synthesizable design description. The output of the intent model is compared to the output of the synthesizable design. This approach is advantageous when it is possible to write a concise nonsynthesizable behavioral style model. The intent model and the UUT receive the same stimulus and their corresponding outputs are simply compared for equality (Figure 7.1.2).

For example, consider the 3-to-8 decoder in Listing 4.8.1. A functionally equivalent model of the 3-to-8 decoder can be written using the **ror** shift operator. This operator is not defined for std_logic_vector type. A description of the 3-to-8 decoder using this operator is given in Listing 7.12.1.

The approach here is to rotate a bit_vector literal that has a single `0 (b"11111 110")` a number of places to the left, equal to the binary value of the select inputs `c`, `b`, `a`. This operation is conditioned on the decoder being enabled. This puts the single `'0'` in the appropriate position of the output vector. If the decoder is not enabled, the output is assigned the value `"11111111"`.

If the binary equivalent of the inputs `c`, `b`, `a` were 4, then the `'0'` would be rotated four positions to the left and the output would be `"11101111"` as desired. The operation here has to be a rotate and not a logical shift, because shift operations shift in `'0'`s.

LISTING 7.12.1
3-to-8 decoder nonsynthesizable intent model.

```
library ieee;
use ieee.std_logic_1164.all;
use ieee.numeric_std.all;

entity decoder_3to8_mod is
   port (
      c, b, a : in std_logic;              -- select inputs
      g1, g2a_bar, g2b_bar: in std_logic; -- enable inputs
      y_mod : out std_logic_vector(7 downto 0)
      );
end decoder_3to8_mod;

architecture rotate of decoder_3to8_mod is
begin

   y_mod <= to_stdlogicvector (B"1111_1110" rol  to_integer(unsigned'(c, b, a)))
            when unsigned'(g1, g2a_bar, g2b_bar) = "100" else "11111111";

end  rotate;
```

There are two type conflicts that must be resolved to use the **rol** operator. First, the right operand of a **rol** operator must be an integer, but c, b, a is an aggregate. To eliminate this type conflict, the aggregate is first type qualified as unsigned. Function to_integer is then used to convert the unsigned value to an integer.

The result of a **rol** operation is a bit_vector. In this example, the result must be assigned to a std_logic_vector. To eliminate this second type conflict, the function to_stdlogicvector, from package STD_LOGIC_1164, is used.

The testbench is given in Listing 7.12.2.

LISTING 7.12.2
A testbench that compares the outputs of two models.

```
library ieee;
use ieee.std_logic_1164.all;
use ieee.numeric_std.all;

entity decoder_3to8_tb is
end decoder_3to8_tb;

architecture tb_architecture of decoder_3to8_tb is

   -- Stimulus signals
   signal c : std_logic;
   signal b : std_logic;
   signal a : std_logic;
   signal g1 : std_logic;
   signal g2a_bar : std_logic;
   signal g2b_bar : std_logic;
   -- Observed signals
   signal y : std_logic_vector(7 downto 0);
   signal y_mod : std_logic_vector(7 downto 0);

begin
   -- Unit Under Test port map
   UUT : entity decoder_3to8 port map (c => c, b => b, a => a,
      g1 => g1, g2a_bar => g2a_bar, g2b_bar => g2b_bar, y => y );

   -- Behavioral Model port map
   UMOD : entity decoder_3to8_mod port map (c => c, b => b, a => a,
      g1 => g1, g2a_bar => g2a_bar,g2b_bar => g2b_bar, y_mod => y_mod );

   -- Stimulus and verification process
   tb: process
      constant period : time := 100 ns;
   begin
      for i in 0 to 63 loop
         (c, b, a, g1, g2a_bar, g2b_bar) <= std_logic_vector(to_unsigned(i, 6));
```

```
         wait for period;
         assert y = y_mod
         report "error for input vector " & integer'image(i)
         severity error;
      end loop;
      wait;
   end process;
end tb_architecture;
```

In the statement part of the architecture body there are three concurrent statements. Two are the instantiations of the components UUT and UMOD, where UMOD is the intent model. The third is the process that provides the stimulus and verifies the outputs. The loop inside this process applies each possible input combination simultaneously to both UUT and UMOD. An assert statement compares output vectors y and y_mod from UUT and UMOD, respectively, for each input combination. If—for any combination—they are not equal, an error is reported.

7.13 CODE COVERAGE AND BRANCH COVERAGE

When developing a testbench to functionally verify a design, we would like to know the effectiveness of the verification. We have seen that, for combinational designs, it is not difficult to create a testbench that applies all possible input combinations. We will see later, in Chapters 14 and 15, that for complex sequential designs, exhaustive verification is usually impractical.

Code Coverage Tool

A software tool that is very helpful in evaluating testbench effectiveness is a code coverage tool. A *code coverage tool* is a program that allows us to determine what source code is executed during a simulation. While the simulation is running, the code coverage tool gathers information about how many times each statement is executed. When the simulation is complete, this information can be displayed. We can obtain code coverage information on the UUT, all of its subcomponents, and the testbench itself.

We can use code coverage to determine whether our testbench causes all of the statements in a design description (UUT) to be executed. If it does not, we can determine how the testbench should be modified to achieve 100% code coverage for the UUT.

While our half adder and its testbench are so simple that they do not require the use of code coverage, they can be used to illustrate some of the basic capabilities provided by such a tool.

After completion of the execution of the testbench in Listing 7.5.1 we can open the code coverage viewer to see the code coverage results. The Hierarchy pane displays the hierarchical structure of the testbench (Figure 7.13.1). At the root of the

Hierarchy	CC [%]	CC with ...
Root : testbench (behavior)	94.44	95.00
uut : half_adder (dataflow)	100.00	100.00
line__12	100.00	100.00
line__13	100.00	100.00
tb	94.44	94.44

FIGURE 7.13.1
Hierarchy window of code coverage viewer.

hierarchy tree is the testbench itself and its process `tb`. Beneath the root we have the instantiation of the UUT. Within the UUT we have two simulation processes associated with the two concurrent signal assignment statements in the architecture body.

Code Coverage as a Metric

Code coverage is the ratio of the number of statements executed to the total number of executable statements, expressed as a percentage. As seen from Figure 7.13.1 (CC[%] column), the code coverage for each simulation process in the UUT is 100% and, therefore, the code coverage for the UUT is 100%.

When an instance name in the hierarchy pane is selected, its source code is displayed. This provides further details of the code coverage. With the UUT selected, the source code in Figure 7.13.2 is displayed.

We can see from this figure that each concurrent signal assignment statement is executed five times. We might have expected that each statement would be executed only four times, since the testbench only applies the four possible binary combinations of two inputs. However, the first execution of these statements is not due to the application of a binary input combination by the testbench, but rather is the execution caused by the initialization phase of the simulation.

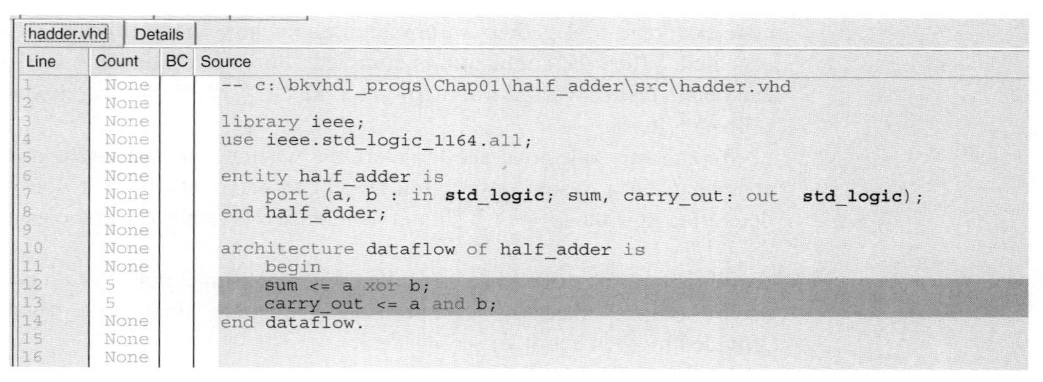

FIGURE 7.13.2
Code coverage view of source code for half adder.

The testbench itself is indicated in Figure 7.13.1 as having 94.44% coverage (17 of its 18 instructions were executed). If the testbench is selected in the Hierarchy pane, its source code is displayed (Figure 7.13.3). From this figure it can be seen that the end process statement is not executed because of the last wait statement.

Though not readily discernible in Figures 7.13.2 and 7.13.3, this code coverage tool uses color coding to indicate which statements were executed. Executed statements have a green background. Statements that were not executed have a red background. Nonexecutable statements have a gray background.

half_adder_tb.vhd | Details

Line	Count	BC	Source
1	None		`-- c:\kls\Bkvhdl_progs\Chap01\half_adder\src\half_adder_tb.vhd`
2	None		
3	None		`library ieee;`
4	None		`use ieee.std_logic_1164.all;`
5	None		
6	None		`entity testbench is`
7	None		`end testbench;`
8	None		
9	None		`architecture behavior of testbench is`
10	None		
11	None		`-- Declare signals to assign values to and observe`
12	None		`signal a_tb, b_tb, sum_tb, carry_out_tb : std_logic;`
13	None		
14	None		`begin`
15	None		`-- Create an instance of the circuit to be tested`
16	None		`uut: entity half_adder port map(a => a_tb, b => b_tb,`
17	None		`sum => sum_tb, carry_out => carry_out_tb);`
18	None		
19	None		`-- Define a process to apply input stimulus and test outputs`
20	None		`tb : process`
21	None		`constant period: time := 20 ns;`
22	None		
23	None		`begin -- Apply every possible input combination`
24	None		
25	1		`a_tb <= '0'; --apply input combination 00 and check outputs`
26	1		`b_tb <= '0';`
27	1		`wait for period;`
28	1		`assert ((sum_tb = '0') and (carry_out_tb = '0'))`
29	None		`report "test failed for input combination 00" severity error;`
30	None		
31	1		`a_tb <= '0'; --apply input combination 01 and check outputs`
32	1		`b_tb <= '1';`
33	1		`wait for period;`
34	1		`assert ((sum_tb = '1') and (carry_out_tb = '0'))`
35	None		`report "test failed for input combination 01" severity error;`
36	None		
37	1		`a_tb <= '1'; --apply input combination 10 and check outputs`
38	1		`b_tb <= '0';`
39	1		`wait for period;`
40	1		`assert ((sum_tb = '1') and (carry_out_tb = '0'))`
41	None		`report "test failed for input combination 10" severity error;`
42	None		
43	1		`a_tb <= '1'; --apply input combination 11 and check outputs`
44	1		`b_tb <= '1';`
45	1		`wait for period;`
46	1		`assert ((sum_tb = '0') and (carry_out_tb = '1'))`
47	None		`report "test failed for input combination 11" severity error;`
48	None		
49	1		`wait; -- indefinitely suspend process`
50	0		`end process;`
51	None		`end;`

FIGURE 7.13.3
Code coverage view of half-adder testbench.

habhv.vhd | Details

Line	Count	BC	Source
1	None		library ieee;
2	None		use ieee.std_logic_1164.all;
3	None		
4	None		entity half_adder is
5	None		Port (a, b : in std_logic;
6	None		sum, carry_out: out std_logic);
7	None		end entity half_adder;
8	None		
9	None		architecture behavior of half_adder is
10	None		begin
11	None		ha: process (a,b)
12	None		begin
13	5	2t 3f	if a = '1' then
14	2		sum <= not b;
15	2		carry_out <= b;
16	None		else
17	3		sum <= b;
18	3		carry_out <= '0';
19	5		end if;
20	5		end process ha;
21	None		
22	None		end architecture behavior;

FIGURE 7.13.4
Branch coverage view of behavioral half adder.

Branch Coverage

If any branch of an if statement or case statement is executed, the entire statement is counted as having been executed in terms of code coverage. This provides insufficient information, since we would like to know whether the testbench causes every branch of these statements to be executed. This problem is solved by *branch coverage*. For an if statement, branch coverage displays the number of times the condition in each branch of the if statement evaluates true and the number of times it evaluates false.

For example, Figure 7.13.4 is the code coverage view of the simulation of the behavioral version of the half adder from Listing 2.6.2 that used a single if statement.

The testbench causes the if statement to be executed five times. The if condition evaluates true two times and the assignment statements in the if branch are executed two times. The if condition evaluates false three times and the assignment statements in the else branch are executed three times.

For case statements, branch coverage displays the number of when branches covered divided by the total number of when branches in the case statement. For each branch, branch coverage also displays the total number of times the branch is taken.

7.14 POST-SYNTHESIS AND TIMING VERIFICATIONS FOR COMBINATIONAL DESIGNS

Previous discussions have focused on functional verification of combinational design descriptions. Once a testbench has been developed for this purpose, it can also be used for verification of the post-synthesis model generated by the synthesizer. We simply replace the design description as the UUT with the post-synthesis VHDL model.

Since the post-synthesis model does not include delays and since the purpose of post-synthesis verification is to verify that the synthesizer has successfully translated a design description to gate-level logic, use of the same testbench used for functional verification is appropriate.

If functional verification of the post-synthesis model is skipped, verification of the synthesized logic can be accomplished as part of the timing verification. If the delay between application of each stimulus and verification of the corresponding UUT outputs was appropriately chosen when the functional verification testbench was originally written, this testbench may be directly usable for the timing verification. However, it may be preferable to modify this testbench by adding assertion statements to automatically verify the UUT's timing.

Hazards in Combinational Circuits

Before considering use of assertion statements to verify timing performance, it is important to consider hazards in combinational circuits. A *hazard* is an output glitch caused by the gate-level structure of a circuit and the propagation delays of its individual gates.

Static Hazard

A *static hazard* occurs when a change in the input values to a combinational circuit causes an output to briefly change value when functionally it should have remained the same. This brief change (glitch) is caused by differences in propagation delays through different signal paths in the circuit. Static hazards are illustrated in Figure 7.14.1(a) and (b).

A static 0 hazard occurs when an output that should remain a 0 briefly transitions to a 1 in response to an input change. A static 1 hazard occurs when an output that should remain a 1 briefly transitions to a 0.

Dynamic Hazard

A *dynamic hazard* occurs when a change in the input values to a combinational circuit causes an output to briefly change value multiple times when it should have

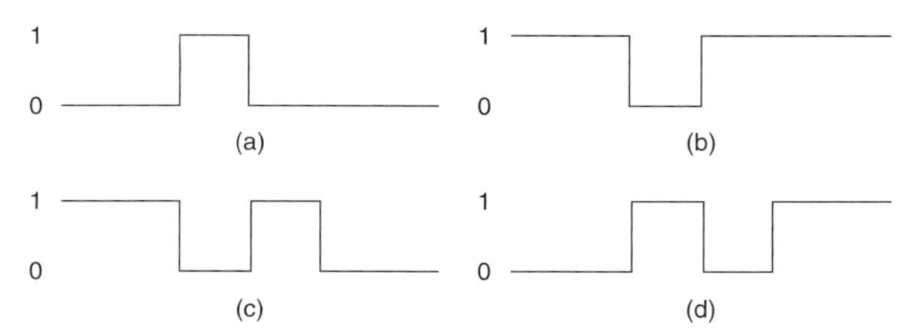

FIGURE 7.14.1
Static and dynamic hazards: (a) static 0 hazard, (b) static 1 hazard, (c) dynamic 0 hazard, and (d) dynamic 1 hazard.

changed value only once (Figure 7.14.1(c) and (d)). A dynamic 0 hazard results when, instead of a single transition from 1 to 0, there are multiple transitions. A dynamic 1 hazard results when, instead of a single transition from 0 to 1, there are multiple transitions.

Hazards in Functional and Post-synthesis Simulations

We have seen that functional simulation of the circuit in Figure 6.8.3 shows a static 1 hazard when the input is changed from 0 to 1. This hazard can be seen during functional simulation because the simulator associates a δ delay with each gate in the circuit and the propagation delays of the two signal paths from the circuit's input to the inputs of the NAND gate differ by 1 δ. Because of this difference, a hazard of 1 δ duration occurs.

The duration of a hazard in a functional or post-synthesis simulation is one or more δs and does not extend beyond the time step. Since we verify the UUT's outputs after a delay that is much greater than the duration of the hazard, the hazard never causes an incorrect value to be seen by an assertion statement.

Hazards in Timing Simulations

In a timing model, the gate delays and propagation delays of signal paths are those of the logic mapped to the target PLD. These delays are relatively significant and the duration of any hazards will be for corresponding magnitudes of time.

Two Pass Logic

In SPLDs of limited logic capacity, the place-and-route tool may feed an output back to one of the SPLD's inputs to provide the additional logic required to realize a particular Boolean expression. This second pass through a SPLD can result in static hazards having durations comparable to the propagation delay of the SPLD.

When visually verifying the output of the timing model of a combinational system, we may see output values that are "incorrect" for significant periods of time due to hazards.

Hazard Free Design

In traditional digital design, when a hazard free circuit is desired, we can modify the design to achieve this objective. This is typically done using a Karnaugh map and adding implicants so that any two adjacent 1s in the map are covered by the same implicant. This corresponds to adding terms to the sum-of-products expression for the function. The result is to add gates (hazard-suppressing logic) to the circuit. This produces a circuit that is not minimal cost.

Synthesis tools do not produce hazard free designs. If we start with a hazard free description, the synthesis tool will likely remove the redundant terms in the sum-of-products expression during optimization. This removes the hazard-suppressing logic and the remaining logic is not hazard free.

Creating circuits with hazards is usually a problem only if we are implementing asynchronous sequential systems. However, in this book, when designing sequential systems our interests are limited to synchronous systems.

Using Assertions to Verify Timing in Combinational Systems

When using assertions statements to verify timing, we must consider the possibility of hazards affecting the verification. For a timing simulation, we must verify the UUT's outputs at the time that they are required by the system's specification to be stable with their new values. After we do this, we need to verify that the output values do not subsequently change as the result of a hazard.

The testbench process in Listing 7.9.1 is modified to verify a half-adder's timing in Listing 7.14.1.

In this process, the time between each application of stimulus is given by the constant `period`. The maximum allowed propagation delay, taken from the system's specification, is given by the constant `tpd_spec`.

After a new stimulus is applied, the process suspends for `tpd_spec`. When the process resumes, it uses an assertion statement to verify the UUT's output values. The process then suspends for a time equal to `period - tpd_spec`.

LISTING 7.14.1
Process to verify logic and timing of half-adder timing model.

```
  tb : process
     constant tpd_spec : time := 11 ns;
     constant period: time := 20 ns;
     constant n: integer := 2;

  begin
     -- Apply every possible input combination
     for i in 0 to 2**n - 1 loop
        (a_tb,b_tb) <= to_unsigned(i,n);

        -- Verify output values at specified time
        wait for tpd_spec;
        assert ((sum_tb = (a_tb xor b_tb))
           and (carry_out_tb = (a_tb and b_tb)))
        report "test failed for a_tb = " & std_logic'image(a_tb)
        & " and b_tb = " & std_logic'image(b_tb)
        severity error;

        -- Verify that outputs do not subsequently chnage
        Wait for period - tpd_spec;
        assert sum_tb'quiet(period - tpd_spec)
           and carry_out_tb'quiet(period - tpd_spec)
        report "propagation delay specification exceded"
        severity error;

     end loop;
     wait;
   end process;
end;
```

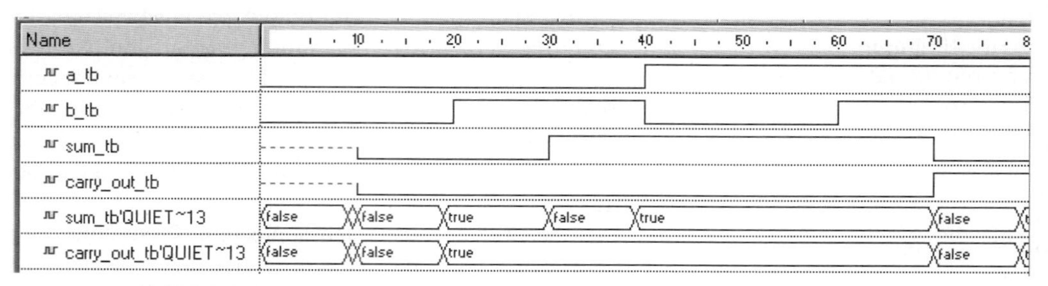

FIGURE 7.14.2
Timing waveforms showing implicit signals created by quiet attributes.

Signal Attribute Quiet

When the process again resumes, it verifies that the outputs have not changed since they were last verified. This is accomplished using the signal attribute `quiet`. Recall that attribute `quiet` creates an implicit signal that has a boolean value of `true` only if no transactions have occurred on its associated signal for the specified time up to the current time.

Implicit signals created by signal attributes can be displayed in waveforms. The implicit signals created by the process in Listing 7.14.1 can be seen in Figure 7.14.2.

7.15 TIMING MODELS USING VITAL AND SDF

Timing Model

To perform a timing simulation, the UUT must be the *timing model* automatically generated by the place-and-route tool. This is a structural VHDL netlist that describes how components, which are primitives of the target PLD architecture, are interconnected to implement the design.

A SPLD timing model may include specification of timing delays within the VHDL netlist description. More commonly for CPLDs and FPGAs, the timing model consists of two files. One is a delayless structural VHDL netlist and the other is a non-VHDL text file that contains timing delay information for each component instance in the netlist.

Timing Model for a SPLD

The architecture of a SPLD is relatively simple. As a result, it's fairly easy to create its timing model. Interconnect paths in a SPLD are relatively short and of similar lengths. The associated interconnect path delays are insignificant relative to the delays of the component primitives.

Generics

Generics allow a design entity to be described so that, for each use of the design entity, its structure and/or behavior can be altered by the choice of generic values. Generics can be any type, but most often are types integer, boolean, or time.

entity_declaration ::=
  entity identifier **is**
    [**generic** (*generic*_interface_list) ;]
    [**port** (*port*_interface_list) ;]
  end [**entity**] [*entity*_simple_name] ;

*generic*_interface_list ::=
[**constant**] identifier_list : [**in**] subtype_indication [:= *static*_expression]
  {[**constant**] identifier_list : [**in**] subtype_indication [:= *static*_expression]}

FIGURE 7.15.1
Expanded syntax for an entity declaration that includes a formal generic clause.

Generics are similar to constants, except their default values can be overridden from outside of the design entity in which they are declared. Default generic values for a design entity are specified inside its entity declaration. If not overridden, these values apply. Default values can be overridden from outside of the design entity in either:

- The component instantiation statement that instantiates the design entity
- A configuration declaration.

Generic Clause

Generics are declared in a formal generic clause in an entity declaration. An expanded syntax for an entity declaration is given in Figure 7.15.1. When generics are used, the generic clause precedes the port clause.

Once declared, a generic can be used in the entity declaration itself and in all architectures associated with the entity.

Generics in a Timing Model

A timing model may make use of generics to specify delay times. As a very simple example, a description of a trivial timing model for a half-adder design is given in Listing 7.15.1. For simplicity, this model uses concurrent signal assignment statements rather than component instantiation statements.

LISTING 7.15.1
Trivial timing model for a half adder.

```
library ieee;
use ieee.std_logic_1164.all;

entity half_adder is
   generic ( tpd : time := 10 ns);
   port (a, b : in std_logic; sum, carry_out: out std_logic);
end half_adder;
```

(Cont.)

LISTING 7.15.1 *(Cont.)*

```
architecture dataflow of half_adder is
   begin
   sum <= a xor b after tpd;
   carry_out <= a and b after tpd;
end dataflow;
```

This model has a single generic named `tpd`. It is used to specify the input to output delay of both the **xor** operation and the **and** operation. Assume that a SPLD actually implements the **xor** operation with an XOR gate and the **and** operation with an AND gate. In a physical implementation, these gates would likely have different propagation delay values. Therefore, for a more accurate model, a separate generic delay value would be specified for each gate. In addition, the propagation delay value for a gate output changing from 0 to 1 may be different than for the same output changing from 1 to 0. This would require separate timing generics for each kind of transition.

The advantage of using generics is that we can use the same timing model for different speed grades of the same PLD by simply using the appropriate generic values.

Effect of Placement and Routing on Timing in CPLDs and FPGAs

With CPLDs and FPGAs the programmable interconnect allows a greater choice of possible pathways between two primitive components. The actual pathways used are determined by the place-and-route tool. The lengths of pathways can be substantially different, leading to significant differences in interconnect path delays between one component and another. In addition, as PLD technology has improved, PLDs have became larger and their primitive components faster. Thus, interconnect path delays have become significant relative to primitive component delays. As a result, for FPGAs in particular, the routing for a design is a significant factor in its timing performance. The effect of a particular routing has to be taken into account to produce an accurate timing model. The actual routing delays can only be computed after the primitive components in the EDIF netlist have been placed and routed.

Back Annotation

In general, *back annotation* is the process of extracting specific data from a design representation and incorporating it in other representations of the design.

Our interest is in having the path delays that are computed by the place-and-route tool after it does the placement and routing of the EDIF netlist back annotated to the timing model. This timing delay information is usually placed in a separate file that the simulator reads and uses to modify the value of the timing model's generics. Back annotation occurs immediately after elaboration and the updated generic values then remain constant during the simulation.

VITAL

The LRM does not specify any standard method for describing timing behavior. As a result, there was also no standard method of describing back annotation. The *VITAL* standard was created to solve this problem. VITAL is the acronym for VHDL initiative toward ASIC libraries. It is an IEEE standard that allows modeling accurate timing at the gate level. VITAL specifies a standard method for writing ASIC and PLD libraries so that timing information can be back annotated. The timing information is contained in a separate file in a format called Standard Delay Format (SDF). This is the same file format Verilog uses for storing and annotating timing values.

VITAL Libraries

VITAL libraries, used with a VITAL-compliant VHDL simulator, allow accurate gate-level timing simulations to be performed. CPLD and FPGA manufacturers provide VITAL-compliant libraries with models of their PLDs. VITAL descriptions are nonsynthesizable descriptions written in a standard style. They make use of functions and procedures from two VITAL packages. The *VITAL Timing Package* contains procedures and functions for accurate delay modeling, timing checks, and timing error reporting. The *VITAL Primitive Package* contains built-in primitives that are optimized for simulator performance. This package makes available all the gate-level primitives found in Verilog. Because the primitives are in a standard package, they can be optimized by a VHDL compiler for faster simulation.

VITAL allows two styles of modeling that can be back annotated with SDF timing data for timing-accurate simulation. The first style, VITAL level 1, uses only VITAL primitives for modeling the behavior of the design. The second, VITAL level 0, uses behavioral statements to describe the functionality of the design. VITAL level 1 descriptions can be accelerated by VITAL-compliant simulators that have the primitives package built into the simulator.

Timing Simulations Using VITAL and SDF

From the EDIF netlist input the PLD vendor's place-and-route tool outputs two files. One is the delayless VHDL netlist that describes the interconnection of the PLD's primitive components to implement the design. Each component instance contains a number of generics that can receive timing information.

The second output is the timing accurate SDF back annotation file. This file contains the delay information for all the generics for all component instances in the netlist that need data passed to them. This delay information includes not only the delay of each component, but also the wiring delays for all the paths that bring signals to the component's inputs. These wiring delays are a result of the routing carried out by the place-and-route tool and are unique for each component instance.

Full Adder Timing Model

If the full adder design is targeted to a SPLD, the timing model will likely include the timing information as generics within the VHDL netlist. If it is targeted to a CPLD or FPGA, the timing model will likely consist of a VHDL netlist and a separate SDF file. A block diagram of the half-adder netlist generated by a Xilinx place-and-route tool for a SPARTAN II 2s15vq100 FPGA is shown in Figure 7.15.2.

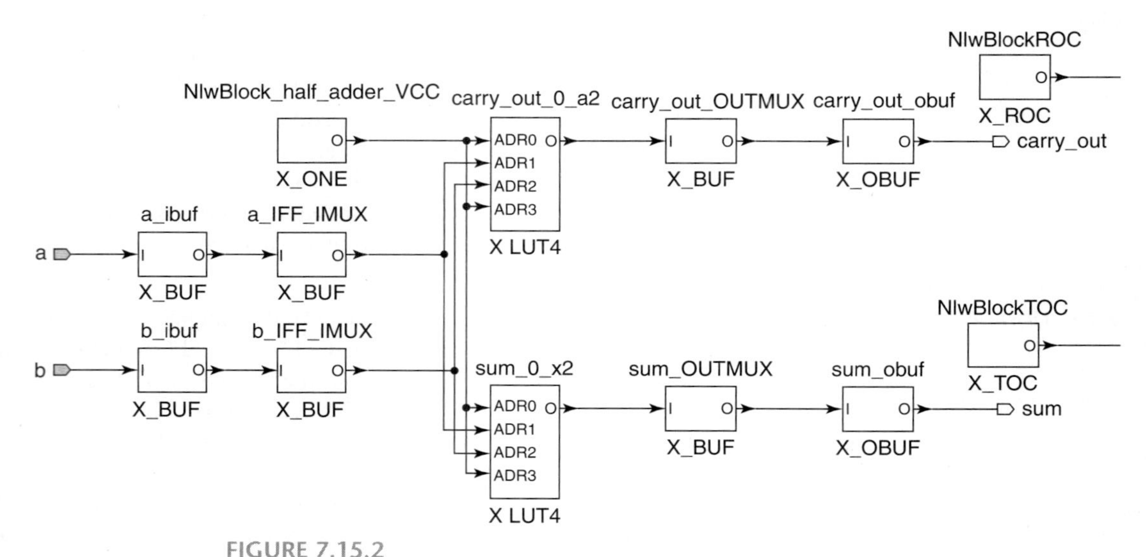

FIGURE 7.15.2
Block diagram of interconnection of SPARTAN II primitives to implement a half adder.

There are multiple instances of three different kinds of primitives. One is a four-input lookup table (LUT), which is the primary kind of logic element for this FPGA. The other two kinds are buffers; one kind is an internal buffer and the other is an output buffer.

For brevity, only a small portion of the VHDL netlist generated by the Xilinx place-and-route tool is given in Listing 7.15.2. The entity declaration and declarative part of the architecture body are complete. The statement part of the architecture body has been stripped of all but three of its component instances. These are the three components on the lower right-hand side of Figure 7.15.2.

LISTING 7.15.2
Partial VHDL netlist generated by a place-and-route tool.

```
library IEEE;
use IEEE.STD_LOGIC_1164.ALL;
library SIMPRIM;
use SIMPRIM.VCOMPONENTS.ALL;
use SIMPRIM.VPACKAGE.ALL;

entity half_adder is
  port (
    sum : out STD_LOGIC;
    a : in STD_LOGIC := 'X';
```

```
    b : in STD_LOGIC := 'X';
    carry_out : out STD_LOGIC
  );
end half_adder;

architecture Structure of half_adder is
  signal a_c_0 : STD_LOGIC;
  signal b_c_0 : STD_LOGIC;
  signal carry_out_OUTMUX_0 : STD_LOGIC;
  signal sum_OUTMUX_1 : STD_LOGIC;
  signal a_c : STD_LOGIC;
  signal b_c : STD_LOGIC;
  signal sum_c : STD_LOGIC;
  signal carry_out_c : STD_LOGIC;
  signal VCC : STD_LOGIC;
begin
-- several component instances have been removed

sum_0_x2 : X_LUT4
    generic map(INIT => X"5A5A", LOC => "CLB_R4C1.S0" )
    port map (ADR0 => b_c_0, ADR1 => VCC, ADR2 => a_c_0,
      ADR3 => VCC, O => sum_c);

sum_OUTMUX : X_BUF
    generic map(LOC => "PAD85")
    port map (I => sum_c, O => sum_OUTMUX_1 );

sum_obuf : X_OBUF
    generic map(LOC => "PAD85")
    port map (I => sum_OUTMUX_1, O => sum );

end Structure;
```

The components instantiated in the architecture are declared in package `VCOMPONENTS` in the library `SIMPRIM` provided by Xilinx.

The component instantiations' generic maps in this netlist assign values to generics related to the placement of each instance on the FPGA. These are not timing generics. The timing generics are declared in the models of the instantiated components.

The first instantiation shown is the instance `sum_0_x2` of the `X_LUT4` primitive component. This is the four-input LUT that computes the half-adder's sum.

The second instantiation shown is the instance `sum_OUTMUX` of the `X_BUF` primitive component. This represents multiplexing in the output block that routes the sum to the output buffer that drives the output pin.

The third instantiation is the instance `sum_obuf` of an output buffer primitive `X_OBUF`. This buffer drives the output pin.

Each of these primitives has a VITAL compliant model in the library `SIMPRIM`. The VITAL model of the X_OBUF is given in Listing 7.15.3.

LISTING 7.15.3
VITAL model of X_OBUF primitive component.

```
library IEEE;
use IEEE.std_logic_1164.all;


library IEEE;
use IEEE.Vital_Primitives.all;
use IEEE.Vital_Timing.all;

library simprim;
use simprim.Vcomponents.all;
use simprim.VPACKAGE.all;

entity X_OBUF is
  generic(
      Xon    : boolean := true;
      MsgOn : boolean := true;
      LOC    : string  := "UNPLACED";

      tipd_I : VitalDelayType01 := (0.000 ns, 0.000 ns);
      tpd_GTS_O : VitalDelayType01z := (0.000 ns, 0.000 ns, 0.000 ns, 0.000 ns,
0.000 ns, 0.000 ns);
      tpd_I_O : VitalDelayType01 := (0.000 ns, 0.000 ns);

      PATHPULSE : time := 0 ps
    );

  port(
    O    : out std_ulogic;
    I    : in  std_ulogic
    );

  attribute VITAL_LEVEL0 of
    X_OBUF : entity is true;
end X_OBUF;

architecture X_OBUF_V of X_OBUF is
--  attribute VITAL_LEVEL1 of
  attribute VITAL_LEVEL0 of
    X_OBUF_V : architecture is true;

  signal GTS_resolved : std_ulogic := 'X';
  signal I_ipd    : std_ulogic := 'X';
begin

  GTS_resolved <= TO_X01(GTS);
```

```
WireDelay       : block
begin
  VitalWireDelay (I_ipd, I, tipd_I);
end block;

VITALBehavior              : process (GTS_resolved, I_ipd)
  variable O_zd            : std_ulogic;
  variable O_GlitchData    : VitalGlitchDataType;
  variable I_GlitchData    : SimprimGlitchDataType;
  variable O_prev          : std_ulogic;
  variable InputGlitch     : boolean := false;
  variable I_ipd_reg       : std_ulogic;
  variable FIRST_TRANSITION_AFTER_GTS_ACTIVE : boolean := false;
begin
  I_ipd_reg     := TO_X01(I_ipd);
  if (falling_edge(GTS_resolved)) then
    FIRST_TRANSITION_AFTER_GTS_ACTIVE                          := true;
  end if;

  if (GTS_resolved = '0') then
    if (FIRST_TRANSITION_AFTER_GTS_ACTIVE = true) then
      FIRST_TRANSITION_AFTER_GTS_ACTIVE := false;
    else
      if ((tpd_I_O(tr01) < PATHPULSE) or (tpd_I_O(tr10) < PATHPULSE)) then
      else
        SimprimGlitch
          (
          GlitchOccured => InputGlitch,
          OutSignal     => O,
          GlitchData    => I_GlitchData,
          InSignalName  => "I",
          NewValue      => I_ipd_reg,
          PrevValue     => O_Prev,
          PathpulseTime => PATHPULSE,
          MsgOn         => false,
          MsgSeverity   => warning
          );
      end if;
    end if;
  end if;
  if (InputGlitch = false) then
    O_prev := TO_X01(O_zd);
  end if;

  O_zd := VitalBUFIF0 (data => I_ipd, enable => GTS_resolved);
```

(Cont.)

LISTING 7.15.3 *(Cont.)*

```
    VitalPathDelay01Z (
      OutSignal     => O,
      GlitchData    => O_GlitchData,
      OutSignalName => "O",
      OutTemp       => O_zd,
      Paths         => (0 => (GTS_resolved'last_event, tpd_GTS_O, true),
                         1   => (I_ipd'last_event, VitalExtendToFillDelay(tpd_I_O
), (GTS_resolved = '0'))),
      Mode          => VitalTransport,
      Xon           => Xon,
      MsgOn         => MsgOn,
      MsgSeverity   => warning,
      OutputMap     => "UX01ZWLH-");

    if (InputGlitch = true) then
      InputGlitch := false;
    end if;
  end process;
end X_OBUF_V
```

As is clear from Listing 7.15.3, the VITAL model for even a simple primitive component can be complex. In general, the architecture for a VITAL model is separated into four sections: wire delay, timing violation, function description, and path delay. Not all models contain all four sections. A combinational model, such as the half adder, does not require a timing section.

Listing 7.15.3 is provided just to give an overview of how generics are used to provide timing information in a VITAL model. In the entity declaration three timing generics are declared: `tipd_I`, `tpd_GTS_O`, and `tpd_I_O`. For all of these generics the default values given in the model are 0 ns.

VITAL timing generic names follow a defined naming convention. Each name starts with a standard prefix that defines what kind of timing parameter it is. There are approximately 17 different prefixes defined by the standard. Prefix `tipd` indicates an interconnect delay, and prefix `tpd` indicates a pin-to-pin delay within a component.

The prefix is followed by the name(s) of the signals to which the delay applies. In Listing 7.15.3 generic `tipd_I` is the interconnect delay modeled as a delay at input port `I`. Generic `tpd_I_O` is pin-to-pin delay from the input of the buffer `I` to its output `O`. The types of these delays are `VitalDelayType01`, which is an array of elements of type time. The first elements in the delay time for an output transition from 0 to 1. The second element is the delay time for an output transition from 1 to 0.

The timing generic `tpd_GTS_O` is related to a global control signal (GTS) that puts all output pins in their high impedance state. GTS is a pin on the FPGA that can be asserted during board-level testing. It is of no further interest in this example.

SDF

Standard Delay Format (SDF) was originally developed by Cadence Design Systems as a file format for conveying timing and delay information between EDA tools for simulation of designs written in the Verilog language. It is now an IEEE standard titled *IEEE Std 1497 IEEE Standard for Standard Delay Format (SDF) for the Electronic Process*. The data in the ASCII file is represented in a tool and language independent way and includes path delays, timing constraint values, and interconnect delays. SDF is now used to convey timing and delay information into both VHDL and Verilog simulations. SDF contains constructs for the description of computed timing data for back annotation and the specification of timing constraints for forward annotation.

An individual component in a technology library is called a *cell*. For each component instance in the VHDL structural netlist the SDF file contains cell data. Each cell contains information specifying the type of cell, the instance name in the netlist, and timing values to be back annotated to the design.

A partial listing of the SDF file for the half adder is given in Listing 7.15.4.

LISTING 7.15.4
Partial SDF file for half-adder simulation.

```
(DELAYFILE
  (SDFVERSION "3.0")
  (DESIGN "half_adder")
  (DATE "Tue May 29 15:43:27 2007")
  (VENDOR "Xilinx")
  (PROGRAM "Xilinx SDF Writer")
  (VERSION "I.24")
  (DIVIDER /)
  (VOLTAGE 2.375)
  (TEMPERATURE 85)
  (TIMESCALE 1 ps)
)
(CELL (CELLTYPE "X_LUT4")
  (INSTANCE sum_0_x2)
   (DELAY
    (ABSOLUTE
     (PORT ADR0 (568))
     (PORT ADR2 (1022))
     (IOPATH ADR0 O (284))
     (IOPATH ADR1 O (284))
     (IOPATH ADR2 O (284))
     (IOPATH ADR3 O (284))
     )

(CELL (CELLTYPE "X_BUF")
 (INSTANCE carry_out_OUTMUX)
   (DELAY
```

(Cont.)

LISTING 7.15.4 *(Cont.)*

```
  (DELAY
   (ABSOLUTE
    (IOPATH I O (2603))

 (CELL (CELLTYPE "X_OBUF")
  (INSTANCE sum_obuf)
   (DELAY
    (ABSOLUTE
     (PORT I (1042))
      (IOPATH I O (3428))
```

The file starts with a header that provides information about the design and the target FPGA to which the file applies. The header is followed by a list of cells that provide values for the generics. Each cell applies to a component instance in the netlist. Only the three cells that apply to the component instances we have previously discussed are shown in the listing. The others have been removed.

The last cell is for the instance `sum_obuf` of the primitive `X_OBUF`. The SDF keyword ABSOLUTE precedes the delay values that are to replace the default values during back annotation. The keyword PORT followed by the port name I gives the value for generic `tipd_I`. The header indicates that the TIMESCALE is picoseconds (ps). Therefore, this delay is 1042 ps or 1.042 ns. Keyword IOPATH specifies an input to output propagation delay. In this case the delay is from I to O of the instance `sum_obuf` (IOPATH I O). Therefore, this delay value of 3.428 ns is for generic `tpd_I_O`.

For each SDF keyword there is a corresponding VITAL prefix (for example, IOPATH and `tpd`).

When the simulation is performed, the VITAL-compliant VHDL simulator reads the SDF back-annotation file and matches the timing information to the generics in the VHDL netlist to peform a timing-accurate simulation.

Running a VITAL Simulation

To run a VITAL simulation, the VITAL library must have been previously compiled into a simulator library. The VHDL timing model is then compiled. When you have more than one model of the UUT design entity, you must make sure that the desired one is used in the simulation. At this point, you would have both a functional model and a timing model of the UUT. There are several approaches to make sure that you simulate the desired model. One is to take advantage of default binding. If the timing model is the last of the two models compiled, it will be the one used by default. Another approach is to use a configuration declaration. Configuration declarations are discussed in Chapter 15.

Next the testbench is compiled. Finally, the SDF file must be specified. This might be done in the command line that invokes the simulator or, in a simulator with an IDE, a menu is used to specify which SDF file is to be used. Once this is done, the

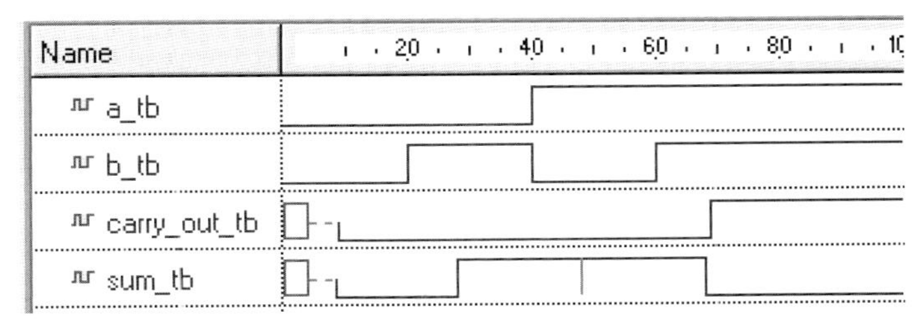

FIGURE 7.15.3
Waveforms from timing simulation of half adder implemented in a SPARTAN II FPGA.

simulation can be performed. The remainder of the simulation process is the same as for a functional simulation. However, the resultant simulation waveforms show the actual delays.

Waveforms from the timing simulation of the half adder implemented in a Xilinx XC2S15 SPARTAN II FPGA with a nominal 6 ns delay (6 ns speed grade) are shown in Figure 7.15.3. The actual input to output delay as read from the waveforms is about 8.157 ns. Also, notice the glitch in the `sum` output when `a` is changed from 0 to 1 simultaneously with `b` being changed from 1 to 0. This glitch does not appear in the functional simulation, as it is a result of primitive component propagation delays in the target PLD.

Writing VITAL timing models is a specialized area of VHDL coding and is not one of the objectives of this text. We simply use the model's created by the PLD vendor's place-and-route tool in our timing simulations.

PROBLEMS

7.1 What are the two basic things that need to be verified about a design?

7.2 What are the three primary constituents of a testbench and what is the purpose of each? What fourth constituent must be added to make a testbench self-checking?

7.3 Which constituent of a testbench is changed depending on the kind of verification being performed and what is it changed to for each kind of verification?

7.4 What is exhaustive verification and why is it significantly more difficult to achieve for a sequential design than for a combinational design?

7.5 Why would it be preferable to have the testbench for a design written by someone other than the person who wrote the design description?

7.6 Given a UUT with std_logic inputs, which of the nine std_logic values are used by a testbench to form stimulus combinations?

7.7 A combinational design has n inputs. As a function of n, how many different input combinations must be applied by a testbench to achieve exhaustive verification? In which order must these input combinations be applied? What would be the number of input combinations to be applied for the following values of n: 4, 9, 14, and 32?

7.8 Modify the statements in the testbench in Listing 7.3.1 to provide stimulus for the exhaustive verification of a UUT named `three_inputs` that has three inputs `a`, `b`, and `c`.

7.9 Write a testbench for the simulation described in Figure 6.9.3. Use a single concurrent signal assignment statement to provide the required stimulus.

7.10 For the following signal assignment statement:

```
c_tb <= '0', '1' after 10 ns, '0' after 25 ns, '1' after 45 ns;
```

assuming the simulation time is 150 ns when this statement is executed, at what time will each new signal value become effective?

7.11 Write a set of concurrent signal assignment statements that could be used in a testbench to generate a stimulus for UUTs with up to six inputs. The stimulus starts from all 0s and increments in binary. Use constants `n` and `delay` to specify the number of inputs and the delay between application of each new input, respectively. For a particular use, the values of `n` and `delay` would be appropriately specified and all assignment statements beyond the *n*th statement would be commented out.

7.12 Modify the testbench in Listing 7.5.1 to be a testbench for the comparator in Listing 3.2.1. Make the time between application of each stimulus value 15 ns.

7.13 In terms of the application of each stimulus value in a testbench, what is the purpose of a wait statement? In terms of checking the actual output value against the expected output value, what is the purpose of an assertion statement?

7.14 Write an architecture that uses a process containing a wait statement so that the architecture is exactly equivalent to that in Listing 2.6.1. Discuss why you chose the form of wait statement used and any restrictions that must be placed on the wait statement's location.

7.15 Write a process that replaces the two concurrent signal assignment statements that generate `a_tb` and `b_tb` in the testbench architecture in Listing 7.3.1. The architecture must use only sequential signal assignment statements without **`after`** clauses and wait statements.

7.16 What is the difference between an assertion statement and a report statement?

7.17 If an assertion statement has a report clause and a severity clause together, all three clauses comprise one statement and there must be only one semicolon used and it must follow the severity clause. What is the effect in a testbench if a semicolon is inadvertently placed after the assert clause in addition to after the severity clause?

7.18 Rewrite the assertion statements in Listing 7.5.1 so that the conditions are written in terms of incorrect behavior. In general, how might the approach of writing conditions in terms of the incorrect behavior lead to a situation where the assertions don't indicate an error when the circuit is actually incorrect?

7.19 Assume that a behavioral style design description named `combfcn` has been written for the function defined by the table that follows. Write a single process testbench named `combfcn_tb` that uses a loop to apply all possible input combinations to the UUT and verifies the outputs. If the UUT fails for a particular stimulus value, an integer number must be displayed that represents the decimal equivalent of the binary stimulus. The expected UUT outputs used in the verification must be determined by a table lookup. Allow 20 ns between the application of each stimulus vector.

a	b	c	w	y
0	0	0	1	0
0	0	1	0	1
0	1	0	0	1
0	1	1	1	0
1	0	0	1	1
1	0	1	0	0
1	1	0	0	1
1	1	1	1	0

7.20 Write a table lookup testbench for the fuel encoder described in Problem 4.12.

7.21 Write a table lookup testbench to exhaustively verify the binary-to-bargraph converter in Problem 4.27.

A demultiplexer has the following entity declaration:

```
entity demux_1to4 is
   port (
      data_in: in std_logic;
      sel: in std_logic_vector (1 downto 0);
      demuxout: out std_logic_vector (3 downto 0)
   );
end demux_1to4;
```

The input `sel` specifies, in binary, the index of the output at which a value equal to `data_in` appears. All other outputs are 0s.

Write a testbench that uses a constant array of records and a table lookup to generate the stimulus and expected outputs for the testbench. Assume the record is defined as:

```
type test_vector is record
   data_in : std_logic;
   sel : std_logic_vector(1 downto 0);
   demuxout : std_logic_vector(3 downto 0);
end record;
```

7.22 Write a testbench that provides exhaustive verification of the `ones_count` design entity in Problem 4.19. The testbench must use a loop to generate all possible input combinations and use another loop to compute the expected outputs.

7.23 Write a testbench that computes all of the valid inputs to the `therm2bin` design entity in Problem 4.26 and verifies the output for all valid inputs.

7.24 Write a testbench that provides exhaustive verification of the 9-bit parity generator/checker in Problem 5.27. The testbench must use a loop to generate all possible input combinations and use equations to compute the expected outputs.

7.25 Write a testbench that provides exhaustive verification of the `tally` scoring system in Problem 5.24. The testbench must use a loop to generate all possible input combinations and use equations to compute the expected outputs.

7.26 Two different testbenches are to be written for the BCD-to-bargraph decoder `bcd_2_bar` in Problem 4.27. Each uses one of the design descriptions for the decoder that has don't care outputs. The approaches used in the two testbenches to compute the expected result are table lookup and computed.

Name the two testbenches based on the technique used: `table` and `computed`. The stimulus values applied to the UUT by the testbench must be generated by one or more loops. Each testbench must provide an exhaustive verification of the decoder's functionality, except that the outputs for address input values of 1010 to 1111 are generated but not verified against any predefined values. Each testbench must use assertion statements so that any failures of the UUT are reported to the console along with the input combination causing the failure.

7.27 Write a testbench to exhaustively verify the binary-to-bargraph converter in Problem 4.27 by computing its expected outputs using a "shift" operation.

7.28 Write a testbench that provides an exhaustive functional verification of the BCD to seven-segment decoder in Listing 4.8.2. This testbench must use two loops. The first loop generates all valid input combinations and uses a table lookup to verify the output. The second loop generates all invalid inputs and verfies that these outputs are all don't cares. How must this testbench be modified for post-synthesis and timing simulations?

7.29 A design description and a testbench for a 4-to-1 multiplexer must be written. The entity declaration for the multiplexer is:

```
entity mux_4_to_1 is
   port(
      i : in std_logic_vector(3 downto 0);     -- input data
      s : in std_logic_vector(1 downto 0);     -- input channel selection
      y : out std_logic   -- output data
      );
end mux_4_to_1;
```

(a) Write a dataflow architecture for `mux_4_to_1` that consists of a single conditional signal assignment statement.

(b) Write a self-checking testbench that provides a comprehensive and functionally meaningful verification of the design entity and is not simply an exhaustive verification. The testbench must first make one data input a 0 (and all other data inputs 1s) and verify the output for all possible values of the selection input. This verification must be repeated with the single 0 applied to each data input.

The testbench must then make one data input a 1 (and all other data inputs 0s) and check the output for all possible values of the selection input. This verification must be repeated with the single 1 applied to each data input.

7.30 Assume that you have written a design description of a combinational circuit that has the following entity declaration:

```
entity synthfcn is
port (inputs : in std_logic_vector (2 downto 0);
   outputs : out std_logic_vector (2 downto 0));
end entity;
```

You have also been given a nonsynthesizable VHDL model for the system that is known to be correct. It has the following entity declaration:

```
entity golden is
port (goldnin : in std_logic_vector (2 downto 0);
   goldnout : out std_logic_vector (2 downto 0));
end entity;
```

Write a test bench that exhaustively verifies `synthfcn`. If the verification fails the testbench should display an indication for which input stimulus values it fails.

7.31 Write a testbench to exhaustively verify the binary-to-bargraph converter in Problem 4.27 by comparing its outputs with those of a behavioral intent model. The behavioral intent model need not be synthesizable.

Chapter 8
Latches and Flip-Flops

Previous chapters focused on combinational systems. However, most digital systems are sequential. A sequential system's outputs are a function of both its present input value and past history of its input values. Memory elements are required in a sequential system to store information that characterizes the system's past history of input values.

In this chapter, basic latch and flip-flop memory elements are discussed. Using VHDL, a memory element can be described either explicitly or implicitly. An explicit description corresponds to the instantiation of a memory element component. In contrast, an implicit description uses a programming style that describes behavior corresponding to that of a memory element. In this later case, a synthesizer infers that a memory element is needed to achieve the described behavior. This chapter focuses on descriptions that infer either latch or flip-flop memory elements.

There are several important practical considerations when describing memory elements. One consideration is whether a particular memory element can actually be synthesized for the target PLD. This depends on the characteristics of the primitive memory elements available in the PLD. Another consideration is that the timing parameters of a PLD memory element are critical to its proper operation. These practical issues are also discussed.

8.1 SEQUENTIAL SYSTEMS AND THEIR MEMORY ELEMENTS

Digital systems can be classified as either combinational or sequential. As we have seen, a combinational system's outputs are completely determined by its present input values.

State of a Sequential System

In contrast, a *sequential system's* outputs depend not only on its present input values, but also on its past history of input values. This history is represented by the binary *present state* value stored in memory elements in the sequential system. Accordingly, the output of a sequential system is a function of both its present input values and present state.

This chapter focuses on memory elements and their VHDL descriptions. How these memory elements are used to create more complex sequential systems is discussed in Chapters 9, 10, and 11.

Memory Element

A single memory element is, by itself, a simple sequential system that stores 1 bit of information. A *memory element* has two stable *states*, corresponding to its storing either a 0 or a 1. When a memory element stores a 0, it is said to be *clear*. When it stores a 1, it is said to be *set*. A memory element's output indicates its present state.

Synchronous Inputs to a Memory Element

A memory element has inputs that allow it to be placed into either of its two states. *Synchronous inputs* are inputs whose values can change the memory element's state only in response to a clock pulse at its clock input.

After a memory element has been placed in a particular state, a change in the values of its synchronous inputs cannot change its state until the next clock pulse occurs. In other words, the memory element stores (remembers) its state value until the next clock pulse.

Pulse

A *pulse* is a short duration change in a signal's value. For example, the signal in Figure 8.1.1(a) is normally logic 0. It transitions (changes) to 1 for a short time and then returns to 0. Such a pulse is called a *positive pulse*. The duration of the pulse is its *pulse width*. In Figure 8.1.1(b) the signal is normally logic 1. It transitions to 0 for a short time and then returns to 1. Such a pulse is called a *negative pulse*.

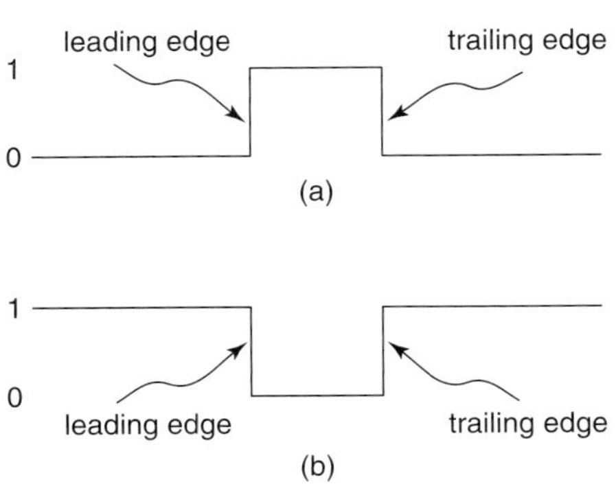

FIGURE 8.1.1
A signal pulse: (a) positive pulse, (b) negative pulse.

A signal transition is also called an edge. A transition from 0 to 1 is a *rising edge* or *positive edge*. A transition from 1 to 0 is a *falling edge* or *negative edge*. The first edge of a pulse to occur in time is its *leading edge* and the last edge to occur is its *trailing edge*. For a positive pulse, the leading edge is a 0 to 1 transition and the trailing edge is a 1 to 0 transition.

Clock Signal

A *clock signal* is a train (sequence) of pulses used as a timing signal. In a fully synchronous sequential system, a single clock signal controls when all the memory elements can change state. Therefore, all the memory elements that change state do so at the same time. Since a synchronous system can advance from one state to the next only in response to a clock pulse, the clock imposes a synchronous timing behavior on the system.

Periodic and Nonperiodic Signals

Examples of *periodic* and *nonperiodic* signals are shown in Figure 8.1.2. A typical clock signal is periodic. The time between corresponding edges of a periodic signal is constant (Figure 8.1.2(a)). Therefore, changes of state occur at regular intervals. This time interval is the *clock's period* (T), which is measured in seconds. Each period of a clock is also referred to as a *clock cycle*.

The reciprocal of the clock's period ($1/T$) is its *frequency*, which is measured in hertz (Hz). The amount of time a periodic signal is 1 during its period (t_H) is also constant. This time is the *clock's width*. The ratio of the clock's width to its period (t_H/T), expressed as a percent, is its *duty cycle*.

The time between corresponding edges of a nonperiodic clock signal is not constant (Figure 8.1.2(b)). A nonperiodic signal's clock width may also vary.

IEEE Std 1076.6 Compliant Clock Signals

IEEE Std 1076.6 limits the types allowed for a clock signal to bit, std_ulogic, and their subtypes (including std_logic) that have a minimum subset of values consisting of `'0'` and `'1'`. Only the values `'0'` and `'1'` from these types are allowed in expressions representing clock levels or clock edges.

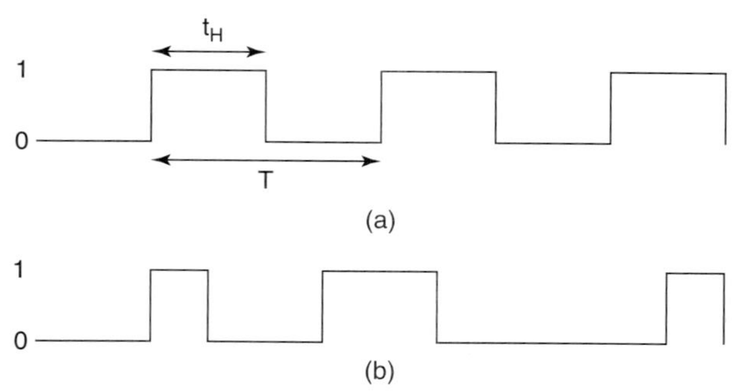

FIGURE 8.1.2
Clock signal characteristics: (a) periodic clock, (b) nonperiodic clock.

Latch versus Flip-flop

Latches and flip-flops are the basic kinds of memory elements. A memory element is categorized as either a latch or a flip-flop, based on to which characteristic of its clock signal it is sensitive. The primary difference between the two is that a latch's state can be changed by its synchronous inputs during the entire time its clock is asserted. In contrast, a flip-flop's state can be changed by its synchronous inputs only at the edge of a clock pulse. Flip-flops are used much more extensively as memory elements in PLD-based designs than are latches.

Clocked (Gated) Latch

A clocked (gated) *latch* is sensitive to its clock's level. Accordingly, it displays a *level-sensitive* synchronous behavior. That is, the state of the latch can be changed by the values of its synchronous inputs during the entire time the clock is at its asserted level. The state of the latch at the time its clock changes to its unasserted level is stored in the latch.

If a latch is sensitive to a high or *positive clock level*, its clock signal is considered asserted when it is 1. If a latch is sensitive to the low or *negative clock level*, its clock signal is considered asserted when it is 0.

The function table for a positive-level D latch is given in Table 8.1.1. The first two columns are the values of the latch's D and CLK inputs. The third column (Q_{t+1}) is the value of the latch's Q output in response to the input values. The value Q_t is the previously stored value of Q. That is, the value of Q when CLK was previously changed to its unasserted level.

When a latch's clock is at its unasserted logic level, its synchronous inputs have no effect on its state.

Flip-flop

A *flip-flop* is a memory element whose output can change only at the time of its clock's transition (edge). Accordingly, it has a *transition-sensitive* synchronous behavior. If a flip-flop is sensitive to a 0-to-1 transition of its clock (*rising edge*), the flip-flop is said to be *positive edge triggered*. If a flip-flop is sensitive to a 1-to-0 transition of its clock (falling edge), the flip-flop is said to be *negative edge triggered*.

The function table for a positive-edge-triggered D flip-flop is given in Table 8.1.2. An up arrow symbol ↑ in the table indicates a positive-edge transition of a signal. The third column (Q_{t+1}) is the value of the flip-flop's Q output in response to the input values D and CLK. The value Q_t is the value of Q prior to the positive edge of the clock.

Table 8.1.1
Function table for a positive-level D latch.

D	CLK	Q_{t+1}
0	0	Q_t
0	1	0
1	0	Q_t
1	1	1

Table 8.1.2
Function table for a positive-edge-triggered D flip-flop.

D	CLK	Q_{t+1}
0	↑	0
1	↑	1
X	0	Q_t
X	1	Q_t

Triggering Clock Edge

Whichever edge of its clock a flip-flop is sensitive to is called its *triggering edge* or *active edge.* Therefore, a flip-flop's output can change only once during a clock cycle, at the triggering edge of its clock.

Asynchronous Inputs

A memory element may also have asynchronous inputs. *Asynchronous inputs* affect a memory element's state independently of its clock or its synchronous inputs. When power is first applied to a memory element, its initial state is unpredictable. Asynchronous inputs are used to force a memory element into a desired initial state at power on. It is considered bad design practice to use asynchronous inputs to change the state of a memory element subsequent to power on. That is, during normal system operation.

8.2 D LATCH

Logic symbols for a positive-level D latch and a negative-level D latch are shown in Figure 8.2.1(a) and (b), respectively. A latch's clock input is often called an *enable input,* and labeled EN, G, or LE (latch enable).

When a positive-level D latch's clock input (CLK) is a 1, its Q output value follows its D input value (Table 8.1.1). Under this clock condition, any change at D appears at Q (after a short delay in a physical device). The input data is said to "flow through" to the output, and the latch is termed *transparent.* That is, the output tracks the input (Figure 8.2.2). Such a latch may also be referred to as a *transparent high latch.*

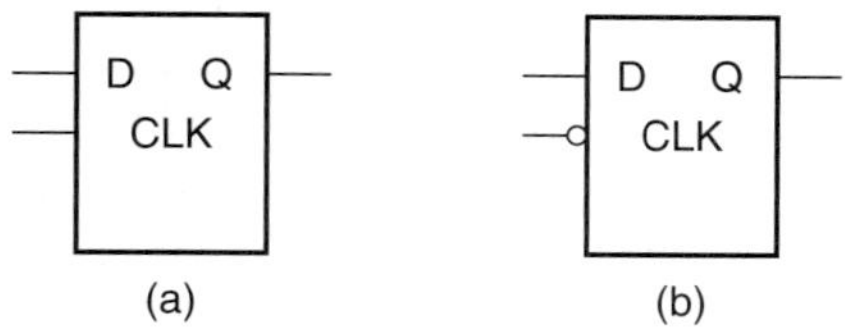

FIGURE 8.2.1
Logic symbols for: (a) positive-level D latch, (b) negative-level D latch.

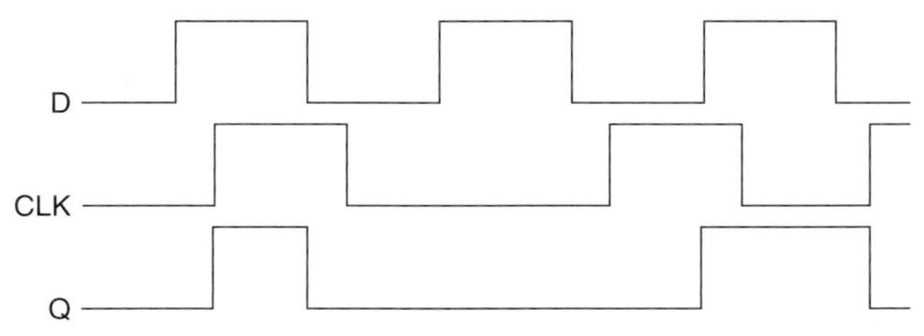

FIGURE 8.2.2
Waveforms representing the functional behavior of a positive-level D latch.

When CLK is changed from 1 to 0, Q keeps the value it had at the time CLK was last equal to 1. Under this clock condition, the value of D when CLK was last a 1 is stored. As long as CLK remains 0, any subsequent changes in D have no effect on Q.

Since changes in Q, corresponding to changes in D, are conditioned on the clock level, D is a synchronous input. Since the value stored in a positive-level D latch is always equal to the value of D when CLK was last equal to 1, the D input is thought of as the data input. A D latch may also have a $\overline{Q}$ output, which is, by definition, the complement of Q.

Latch Inference

We can cause a memory element to be synthesized by writing code that is recognized as implying the need for a memory element. The technique used by a synthesizer to determine, from its interpretation of the VHDL code, when a latch must be synthesized is called *latch inference.*

When Is a Latch Inferred

In general, a latch is inferred when a statement conditionally makes an assignment to a signal or variable and the condition does not involve a clock edge. Such an assignment is called an *asynchronous assignment*.

For example, an if statement whose if clause contains an asynchronous assignment to a signal, but which has no terminating else clause, causes a synthesizer to infer that a latch is required. A latch is required because, when the condition is not true, the signal is not assigned a new value. Therefore, the signal must retain its previous value, necessitating a latch. When such an if statement is synthesized, the synthesizer may give a warning such as *"Latch generated from process for signal q, probably caused by a missing assignment in an if or case statement."* The signal (or variable) that is conditionally assigned is the output of the inferred latch.

Latch Template Using an If Statement

An if statement template for synthesizing a latch is given in Figure 8.2.3.

The clock-level condition following keyword `if` is written to evaluate true when the clock is asserted. Included in the then clause are one or more assignment statements. Any signals or variables assigned values by these statements are implemented as outputs of latches.

```
[process_label:] process (<clock_signal>, <input_signals>)
    <declarations>
begin
    if <clock_level> then
      <sequence_of_statements>
    end if;
end process [process_label];
```

FIGURE 8.2.3
Template for latch inference using an if statement.

D Latch Description

A D latch description using the template in Figure 8.2.3 is given in Listing 8.2.1. The order of the clock and the other inputs in the sensitivity list is not important. ***However, the sensitivity list must contain all of the signals read within the process.***

This architecture has a single process. Both `d` and `clk` must be included in the process' sensitivity list, since, if either changes value, it is possible for the output of the D latch to change value.

If there is an event on either input, the process is executed. If `clk` = `'1'`, then `q` gets assigned the value of `d`. Since there is no else clause, if `clk` is not `'1'`, there is no assignment to `q`. This later condition implies that the old value of `q` must be remembered, and hence the need for a latch.

LISTING 8.2.1
D latch description.

```
library ieee;
use ieee.std_logic_1164.all;

entity d_latch is
   port (
      d: in std_logic;       -- data input
      clk: in std_logic;     -- clock input
      q: out std_logic       -- output
      );
end d_latch;

architecture behavioral of d_latch is
   begin
   process (d, clk)
      begin
      if clk = '1' then
         q <= d;
      end if;
   end process;
end behavioral;
```

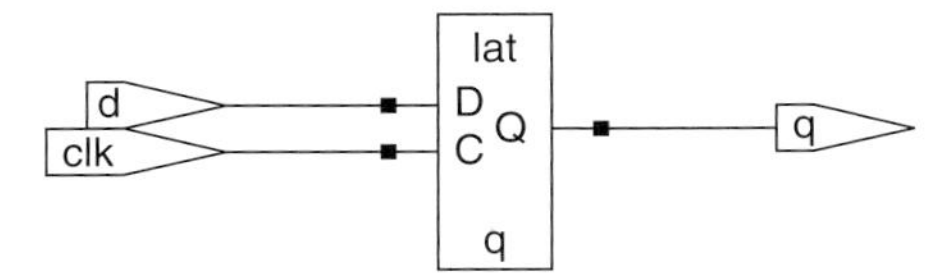

FIGURE 8.2.4
Latch inferred by synthesizer from description in Listing 8.2.1.

Since we intended to describe a D latch, the else clause was intentionally left out so that the synthesizer infers the need for a latch. From Listing 8.2.1 the synthesizer synthesizes the latch in Figure 8.2.4.

Equivalent D Latch Descriptions

The incomplete if statement in Listing 8.2.1 is essentially equivalent to the following complete if statement

```
if clk = '1' then
      q <= d;
else
      q <= q;
end if;
```

that assigns the signal `q` to itself when the clock condition is not true. However, this if statement can be used only when the mode of `q` is either buffer or inout. Some compilers may treat the

```
q <= q;
```

self assignment as redundant and simply remove it.

An if statement that is exactly equivalent to the one in Listing 8.2.1 is:

```
if clk = '1' then
      q <= d;
else
      null;
end if;
```

This if statement can be used with `q` being mode out. Recall that the **`null`** keyword means nothing is to be done.

All three if statements infer the same D latch. However, technically, the if statement that assigns `q <= q` when the clock condition is false causes all pending transactions on the driver for `q` to be replaced with the transaction assigning the current value of `q` to `q`, 1 δ later. In contrast, use of the null statement leaves any pending transactions for `q` on its driver.

Negative-Level Latch Description

To describe a negative-level latch, the if statement's clock-level condition is simply changed to:

```
clk = '0'
```

LISTING 8.2.2
Latched multiplexer.

```
library ieee;
use ieee.std_logic_1164.all;

entity latched_mux is
   port (clk, a, b, sel : in std_logic;
      q: out std_logic);
end latched_mux;

architecture behavioral of latched_mux is
begin
   process (clk, a, b, sel)
   begin
      if clk = '1' then
         if sel = '0' then
            q <= a;
         else
            q <= b;
         end if;
      end if;
   end process;
end behavioral;
```

Combinational Input Logic for a Latch

A D latch description can be written so that combinational logic is synthesized prior to the latch's D input. For example, a description of a latched multiplexer is given in Listing 8.2.2.

The inner if statement is complete and is synthesized to a multiplexer. The outer if statement is incomplete, causing the multiplexer's output to be latched. The logic synthesized from Listing 8.2.2 is shown in Figure 8.2.5.

Asynchronous Set and Clear Inputs

A latch may also have an asynchronous set input, an asynchronous clear input, or both. These inputs are used, independently of the clock level, to force Q to be a 1 or a 0. The term *preset* is synonymous with set and the term *reset* is synonymous with

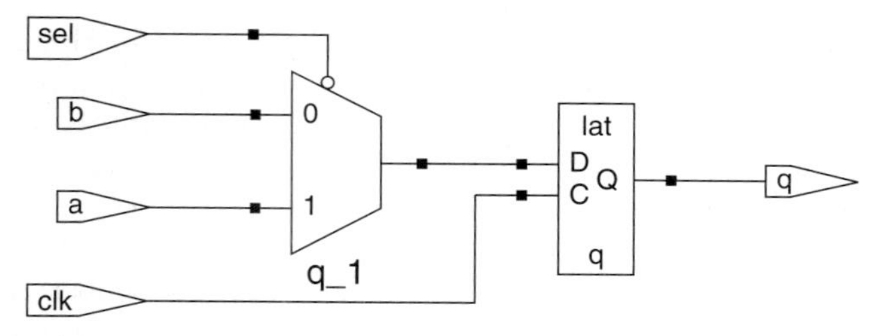

FIGURE 8.2.5
Latched multiplexer logic synthesized from Listing 8.2.2.

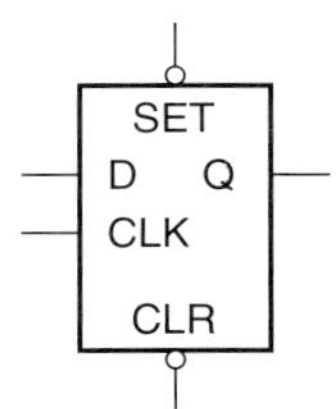

FIGURE 8.2.6
D latch with asserted low-asynchronous set and clear inputs.

clear. Any asynchronous inputs must be kept unasserted for normal synchronous operation. We can add an asynchronous set or clear input, or both, to a D latch symbol as symbolized in Figure 8.2.6.

The D latch description in Listing 8.2.3 includes asserted-low asynchronous set and clear inputs.

Since all of the latch's inputs are in the process's sensitivity list, an event on any input causes the process to execute. The if statement first checks whether `set_bar` is asserted. If it is, `q` is assigned a `'1'`. If not, the first elsif clause checks whether `clear_bar` is asserted. If `clear_bar` is asserted, then `q` is assigned a `'0'`.

LISTING 8.2.3
D latch with asynchronous set and clear inputs.

```
library ieee;
use ieee.std_logic_1164.all;

entity d_latch is
   port (
      d, clk, set_bar, clear_bar : in std_logic;
      q: out std_logic);
end d_latch;

architecture behavioral of d_latch is
   begin
   process (d, clk, set_bar, clear_bar)
      begin
      if set_bar = '0' then        -- asynchronous set
         q <= '1';
      elsif clear_bar = '0' then -- asynchronous clear
         q <= '0';
      elsif clk = '1' then       -- d latch
         q <= d;
      end if;
   end process;
end behavioral;
```

Set Dominant Latch

Because of the precedence associated with an if statement, when both `set_bar` and `clear_bar` are simultaneously asserted, the latch is set. This happens because the set condition is tested first. Such a latch is *set dominant.*

Because this D latch is set dominant, it does not behave exactly like a conventional D latch with asynchronous set and clear inputs. For a conventional D latch, having both its `set_bar` and `clear_bar` inputs simultaneously asserted is an input condition that is not allowed. That is, any circuit that uses the latch must be designed so that it does not simultaneously assert `set_bar` and `clear_bar`. This is because when they are both asserted, Q equals $\overline{Q}$. In addition, if they are simultaneously unasserted, the resulting state of the latch is indeterminate (unknown).

In Listing 8.2.3, if neither asynchronous input is asserted, the second elsif clause checks the clock. If the condition `clk = '1'` is true, then `d` is assigned to `q`. Since the if statement has no terminating else clause, a latch is inferred. The branches of the if preceding the clock branch represent asynchronous set and clear logic. The branch with the clock condition must be the last branch in the if statement.

A diagram of the logic synthesized from Listing 8.2.3 is shown in Figure 8.2.7.

The synthesizer created a D latch primitive with an active low set and an active high reset (clear). The synthesizer included a gate to qualify the signal to the reset input so that reset is asserted if `set_bar` is not asserted and `clear_bar` is asserted. This logic implements set dominance.

Using a Case Statement to Infer a Latch

A case statement can also be used to infer a latch, but it becomes unwieldy when the latch must also have asynchronous inputs. This leaves if statements as the preferred method for inferring a latch.

Unwanted Latches

If a process contains code where a variable is read before it is assigned, it implies that the old value must be available for reading and a synthesizer infers a latch. The latch is necessary because it is the value from the previous execution of the process that is being read and, therefore, this value must be saved between process executions. In contrast, if a variable is assigned a value before it is read, a latch should not be inferred, because the value from the previous process execution is not needed. It is recommended that this style of latch inference be avoided. In addition, when this is done inadvertently while trying to describe combinational logic, a latch is unintentionally inferred.

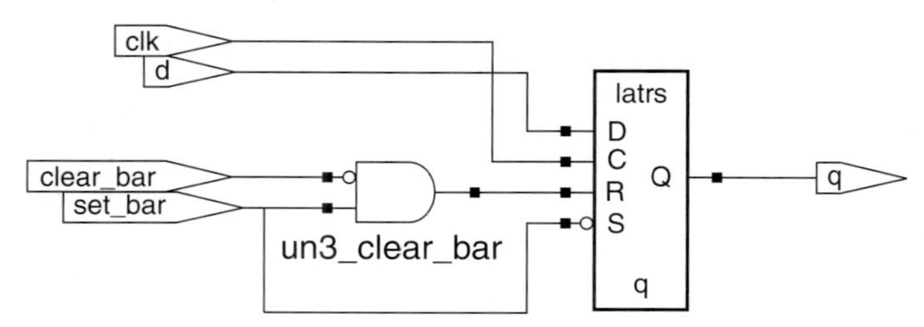

FIGURE 8.2.7
Logic synthesized from Listing 8.2.3 for D latch with asynchronous set and clear inputs.

Using a Concurrent Signal Assignment to Infer a Latch

A concurrent signal assignment statement can also be used to infer a latch, although this is not a preferred approach. For example, the process statement in Listing 8.2.3 can be replaced by the single conditional signal assignment statement

```
q <= '1' when set_bar = '0' else
     '0' when clear_bar = '0' else
     d when clk = '1';
```

to cause the latch in Figure 8.2.7 to be synthesized.

Unaffected Keyword

An equivalent, but less concise, way to write the previous concurrent signal assignment is:

```
q <= '1' when set_bar = '0' else
     '0' when clear_bar = '0' else
     d when clk = '1' else
     unaffected;
```

This statement uses the keyword `unaffected`. Keyword `unaffected` is used in concurrent statements to indicate an assignment that does not affect the driver of the target. It has the same effect that `null` has in sequential assignments.

8.3 DETECTING CLOCK EDGES

Since a flip-flop is only sensitive to its synchronous inputs at the occurrence of a triggering clock edge, its description requires detection of each triggering clock edge.

Positive Clock Edge Conditions

There are several boolean type expressions that can be used to describe a condition that corresponds to the occurrence of a clock edge. These expressions are also called synchronous conditions. Expressions for a positive clock-edge condition are summarized in Table 8.3.1.

Assuming a clock signal named `clk`, the first expression in Table 8.3.1 is written:

```
clk'event and clk = '1'
```

Table 8.3.1
Expressions for a positive clock-edge condition.

```
clock_signal_name'event and clock_signal_name = '1'
not clock_signal_name'stable and clock_signal_name = '1'
rising_edge(clock_signal_name)
```

Event Attribute

This expression is the **and** of two subexpressions. The first is an event attribute. Recall that an *event attribute* is an expression that is true when there is a change in the associated signal's value. An event attribute for a signal S (S'event) is true only if S changes value during the current simulation cycle; see "Signal Attributes" on page 241.

The other subexpression, `clk = '1'`, evaluates true when `clk` is a `'1'`. These two conditions can be true simultaneously only if `clk` has just changed (an event) to a `'1'`. Assuming that during a simulation `clk` is assigned only the values `'0'` or `'1'`, these conditions being true mean that clock had a positive-edge transition during the current simulation cycle.

The order of the subexpressions is of no consequence. So,

```
clk = '1' and clk'event
```

is equivalent to the previous expression.

Stable Attribute

Another boolean expression for a positive edge uses the signal attribute stable. The *stable attribute* forms an expression that is true only when there is no change in the associated signal's value during the current simulation cycle. Accordingly, the expression

```
not clk'stable
```

is true when `clk` is not stable during a simulation cycle, which is equivalent to an event on `clk`. The **and** of this expression with the expression `clk = '1'` is true when there is a positive edge on `clk`.

rising_edge Function

The first two expressions in Table 8.3.1 are fine for synthesis, but have a drawback when simulating std_logic signals. Both expressions detect a transition to a `'1'` state from any other state. Therefore, they are true for a change from a metalogic value, such as `'U'` to `'1'`. This could give a misleading result during a simulation.

This problem is eliminated by using the function `rising_edge`, defined in package STD_LOGIC_1164. This function is true only if `clk` makes a transition from `'0'` to `'1'`. It is not true for a transition from any other value to a `'1'`.

Negative Clock-Edge Conditions

Expressions for a negative clock-edge condition are given in Table 8.3.2.

Table 8.3.2
Expressions for a negative clock-edge condition.

```
clock_signal_name'event and clock_signal_name = '0'
not clock_signal_name'stable and clock_signal_name = '0'
falling_edge(clock_signal_name)
```

falling_edge Function

The function `falling_edge` serves the same purpose in detecting a negative clock edge as the function `rising_edge` does for a positive clock edge.

8.4 D FLIP-FLOPS

A D flip-flop is an edge-triggered memory element that transfers the value on its D input to its Q output when a triggering edge occurs at its clock input. This output value is stored until the next triggering clock edge. Thus, a D flip-flop can change its state only once during a clock cycle.

Logic symbols for positive- and negative-edge-triggered D flip-flops are given in Figure 8.4.1. The *dynamic input indicator*, a triangle, indicates that the clock input is edge sensitive (edge triggered).

Synchronous Flip-flop Inputs

Synchronous inputs to a flip-flop are sampled only at a triggering clock edge. Thus, the flip-flop's output can change value in response to synchronous inputs only at a triggering clock edge. Figure 8.4.2 illustrates this property for a positive-edge-triggered D flip-flop. Output Q assumes the value of the D input at each triggering clock edge.

Synchronous Assignment

A VHDL statement causes a *synchronous assignment* when a signal or variable is updated as a direct result of a clock edge condition evaluating true. During synthesis, a signal or variable updated by a synchronous assignment results in a flip-flop being inferred.

Using an If Statement to Infer a Flip-flop

A positive-edge-triggered flip-flop can be described using the if statement template in Figure 8.4.3. A D flip-flop description using this template is given in Listing 8.4.1.

Since the output of a D flip-flop that has no asynchronous inputs can only change at a triggering clock edge, `d` is not included in the process's sensitivity list. The process needs to be executed only when `clk` changes. In order to store the input only

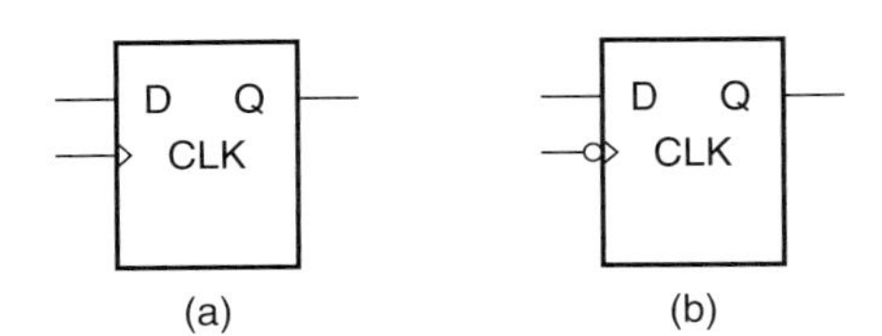

FIGURE 8.4.1
Logic symbols for (a) positive-edge-triggered D flip-flop, (b) negative-edge-triggered D flip-flop.

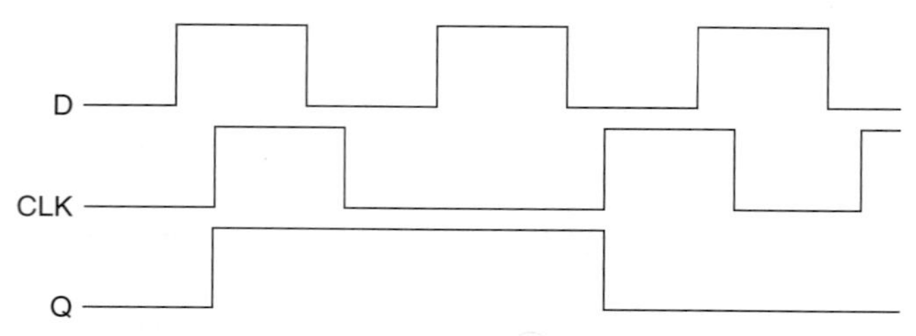

FIGURE 8.4.2
Waveforms representing the functional behavior of a positive-edge-triggered D flip-flop.

at positive clock edges, the if statement condition is changed from that of the D latch in Listing 8.2.1 to:

```
clk'event and clk = '1'
```

Synthesizers That Ignore Sensitivity Lists

In this example, the process has only `clk` in its sensitivity list; therefore, the process will not execute unless there is an event on `clk`. As a result, it would seem that `clk'event` would not be needed as part of the if statement's condition, and that `clk = '1'` would be sufficient. However, some synthesizers may ignore sensitivity lists, thus the requirement that `clk'event` explicitly be part of the condition.

From the description in Listing 8.4.1, the logic in Figure 8.4.4. is synthesized.

The triangle on the D flip-flop symbol's `clk` input in Figure 8.4.4 indicates that the storage element is edge triggered (a flip-flop). Absence of an inversion circle at the `clk` input further indicates that it is the positive edge of the clock that triggers this flip-flop.

The condition `clk'event and clk = '1'` in Listing 8.4.1 can be replaced with `rising_edge(clk)` as shown in Listing 8.4.2.

As previously stated, use of the function `rising_edge` is often preferred, because it returns true only for an event corresponding to a change in a signal's value from `'0'` to `'1'`.

```
[process_label:] process (<clock_signal>)
   <declarations>
begin
   if <clock_edge> then
      <sequence_of_statements>
   end if;
end process [process_label];
```

FIGURE 8.4.3
Template for an edge-triggered flip-flop using an if statement.

LISTING 8.4.1
Positive-edge-triggered D flip-flop.

```
library ieee;
use ieee.std_logic_1164.all;

entity d_ff_pe is
   port (
      d, clk: in std_logic;
      q: out std_logic;
end d_ff_pe;

architecture behavioral of d_ff_pe is
   begin
   process (clk)
      begin
      if clk'event and clk = '1' then
         q <= d;
      end if;
   end process;
end behavioral;
```

Negative edge triggering is easily accomplished by changing the if condition to

```
clk'event and clk = '0'
```

or to:

```
falling_edge(clk)
```

Using a Wait Statement to Infer a Flip-flop

Edge triggering can also be accomplished using a wait statement. A template for inferring a flip-flop using a wait until statement is given in Figure 8.4.5.

When a single wait statement is used in a process to infer a flip-flop, it must be either the first or last statement in the process.

For example, the architecture for a positive-edge-triggered D flip-flop is described in Listing 8.4.3 using a wait until statement. The logic synthesized is the same as for the description in Listing 8.4.1.

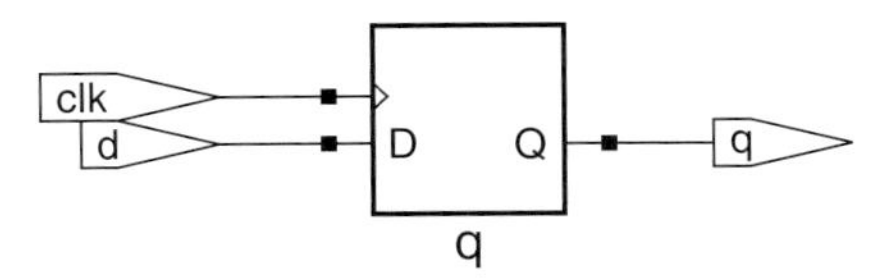

FIGURE 8.4.4
Logic synthesized from Listing 8.4.1 for a positive-edge-triggered D flip-flop.

LISTING 8.4.2
Positive-edge-triggered D flip-flop architecture using rising_edge function.

```
architecture behavioral2 of d_ff_pe is
   begin
   process (clk)
      begin
      if rising_edge(clk) then
         q <= d;
      end if;
   end process;
end behavioral2;
```

Clock Edge Conditions in a Wait Until Statement

Any of the conditions in Tables 8.3.1 or 8.3.2 can be used in a wait until statement to specify a clock edge. In addition, the condition `clk = '1'` or `clk = '0'` is sufficient to represent a positive or negative clock edge, respectively. This is true because the wait until statement suspends execution of the process until there is an event on a signal in its condition and the condition evaluates true. (See "Wait Until" on page 264.)

Each signal assigned a value by an assignment statement following the wait statement infers a separate flip-flop.

Using a Variable Assignment to Infer a Flip-flop

The interpretation of whether a flip-flop should be inferred from an assignment to a variable following a wait statement is more complex. If the value of the variable is read before the variable is assigned a value, a flip-flop should be inferred.

Asynchronous Set and Clear Inputs

Asynchronous set and clear inputs can be added to D flip-flop descriptions, just as they were added to D latch descriptions.

A signal (or variable) that is synchronously assigned may also be asynchronously assigned to realize asynchronous set and clear operations. A template for this is given in Figure 8.4.6.

```
[process_label:] process
   <declarations>
begin
   wait until <clock_edge>
     <sequence_of_statements>
end process [process_label];
```

FIGURE 8.4.5
Template for inferring a flip-flop using a wait until statement.

LISTING 8.4.3
Positive-edge-triggered D flip-flop architecture using a wait until statement.

```
architecture behavioral of d_ff_pe_wait is
   begin
   process
      begin
      wait until clk = '1';
         q <= d;
   end process;
end behavioral;
```

The conditions associated with the branches prior to the last elsif branch don't involve a clock edge. Assignment statements in these branches make asynchronous assignments. The last elsif condition is a clock-edge condition and assignment statements in this branch are synchronous assignments.

The sensitivity list must include the clock signal, all signals read in the conditions of the asynchronous elsif branches, and all signals read in the sequential statements controlled by those conditions. No other signals should be included in the sensitivity list.

A D flip-flop with asynchronous set and clear is described in Listing 8.4.4.

Limitations When Using Wait Statements to Infer Flip-flops

A drawback in using a wait statement to infer a flip-flop is that it is not possible to add asynchronous set and/or clear inputs, because all assignments executed after the wait are synchronous assignments. However, synchronous set and/or clear inputs can be described using a wait statement.

```
[process_label:] process (<clock_signal, asynch_signals>)
   <declarations>
begin
   if <condition1> then
     <sequence_of_statements>
   elsif <condition2> then
     <sequence_of_statements>
     . . .
   elsif <conditionn> then
     <sequence_of_statements>

   elsif <clock_edge> then
     <sequence_of_statements>
   end if;
end process [process_label];
```

FIGURE 8.4.6
Template for an edge-triggered flip-flop with asynchronous set or clear.

LISTING 8.4.4
D flip-flop with asynchronous set and clear.

```
library ieee;
use ieee.std_logic_1164.all;

entity d_ff_pe_asc is
   port (
      d, clk: in std_logic;
      set_bar, clear_bar: in std_logic;    -- asynchronous set and clear inputs
      q: out std_logic);
end d_ff_pe_asc;

architecture behavioral of d_ff_pe_asc is
begin
   process (clk, set_bar, clear_bar)
   begin
      if set_bar = '0' then
         q <= '1';
      elsif clear_bar = '0' then
         q <= '0';
      elsif rising_edge(clk) then
         q <= d;
      end if;
   end process;
end behavioral
```

Synchronous Set and Clear Inputs

Synchronous set and clear inputs can be used with D flip-flops in place of, or in addition to, asynchronous set and clear inputs. A synchronous set or clear input has an effect only at a triggering clock edge. Accordingly, synchronous set and clear inputs are not included in the process's sensitivity list. Listing 8.4.5 is a description of a D flip-flop with synchronous set and clear.

LISTING 8.4.5
D flip-flop with synchronous set and clear.

```
library ieee;
use ieee.std_logic_1164.all;

entity d_ff_pe_ssc is
   port (
      d, clk: in std_logic;
      set_bar, clear_bar: in std_logic;    -- synchronous set and clear inputs
      q: out std_logic);
end d_ff_pe_ssc;
```

```
architecture behavioral of d_ff_pe_ssc is
   begin
   process (clk)
      begin
      if rising_edge(clk) then
         if set_bar = '0' then
            q <= '1';
         elsif clear_bar = '0' then
            q <= '0';
         else
            q <= d;
         end if;
      end if;
   end process;
end behavioral
```

The synthesized logic for the D flip-flop with synchronous set and clear in Listing 8.4.5 is shown in Figure 8.4.7.

Using a Concurrent Signal Assignment Statement to Infer a Flip-flop

A conditional signal assignment statement can also be used to infer a flip-flop: For example, the process statement in Listing 8.4.4 can be replaced by the single conditional signal assignment statement:

```
q <= '1' when set_bar = '0' else
      '0' when clear_bar = '0' else
      d when rising_edge(clk);
```

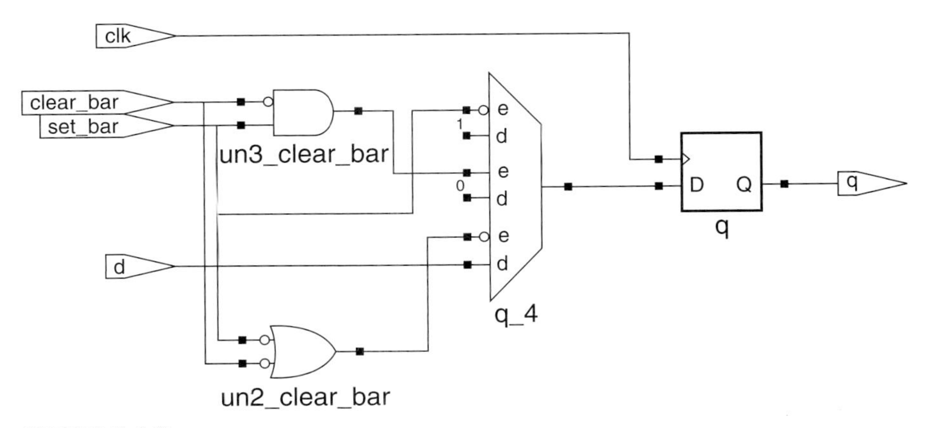

FIGURE 8.4.7
Synthesized logic for D flip-flop with synchronous set and clear.

8.5 ENABLED (GATED) FLIP-FLOP

All previous D flip-flop examples stored the data at their D inputs at each triggering clock edge. Thus, data was stored for no longer than one clock cycle. Often there is a need to have a flip-flop store its data for more than one clock cycle. This is achieved by having the flip-flop store its input data only at selected triggering clock edges and not at others. At the other triggering clock edges, we want the flip-flop to remain in its previous state. To accomplish this, we add an enable (gate) input to the flip-flop. A D flip-flop with an *enable input* stores its input data at a triggering clock edge only if its enable input is asserted.

Approaches to Enabling a Flip-flop

There are two common approaches to creating an enabled flip-flop: gated clock or gated data. The gated clock approach, which we discuss first, is generally to be avoided. Instead, the gated data approach is preferred.

Gated Clock Approach

An obvious, but risky, approach to creating an enabled flip-flop is to gate the clock signal using the enable signal, so that the clock signal reaches the flip-flop's clock input only when the enable input is asserted. A description using this approach is given in Listing 8.5.1.

LISTING 8.5.1
Gating a flip-flop's clock—considered a poor design practice.

```
library ieee;
use ieee.std_logic_1164.all;

entity d_ff_gated is
   port (
      d, clk: in std_logic;
      en: in std_logic;    -- enable input
      q: out std_logic;
end d_ff_gated;


architecture behavioral of d_ff_gated is
   signal gated_clk : std_logic;
begin
   gated_clk <= clk when en = '1' else '1';
   process (gated_clk)
   begin
      if gated_clk'event and gated_clk = '1' then
         q <= d;
      end if;
   end process;
end behavioral;
```

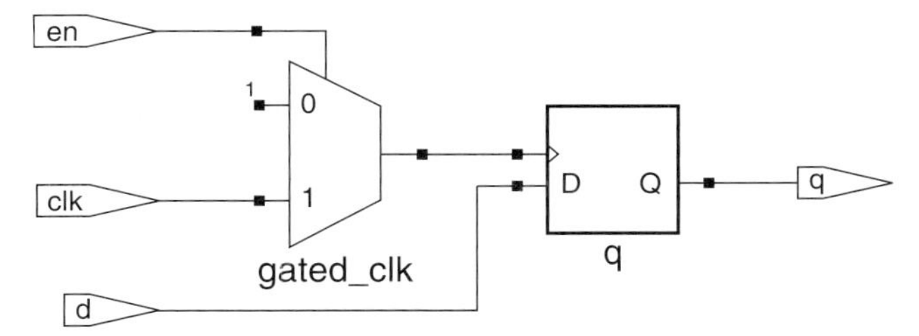

FIGURE 8.5.1
Hierarchical view of logic synthesized from Listing 8.5.1.

The architecture contains two concurrent statements. The conditional signal assignment statement assigns `gated_clk` the value of `clk` when the enable input `en` is `'1'`. Otherwise, `gated_clk` is held at `'1'`. The process statement infers an edge-triggered flip-flop with `gated_clk` as its clock signal. The logic synthesized from this description is shown in Figure 8.5.1.

The synthesized logic appears to do what is required. When `en` is a `'0'`, the clock input of the flip-flop is held at `'1'`. When `en` is a `'1'`, the clock signal passes through the multiplexer to the flip-flop's clock input.

If the signal at the `en` input has glitches, there is a problem with this approach. Often, this signal is the output of a combinational system, which might produce glitches as its inputs change. As discussed in "Post-Synthesis and Timing Verifications for Combinational Designs" on page 284, synthesizers do not create hazard-free combinational logic.

For example, a 0-1-0 glitch (a static 0 hazard) at the `en` input while the clock is 0 produces a 1-0-1 glitch at the flip-flop's clock input, causing it to store the data at its D input. Because the flip-flop can be incorrectly triggered by glitches from the clock gating circuit, using this approach is a poor design practice. Note that any glitches on `en` while the clock is high will not clock the flip-flop. Thus, a period of time equal to the clock width is available for any glitches on the `en` input to have passed without adversely affecting the circuit's operation.

There are situations in the design of low-power circuits where a clock is intentionally gated. In such cases, it is the designer's responsibility to ensure that the circuit producing the enable signal is glitch free or that the clock is high for a sufficient time for any glitches to have ended.

Gated Data Approach

The preferred approach to enabling a flip-flop is to gate the data to the flip-flop instead of gating the clock. A description of a gated data flip-flop is given in Listing 8.5.2.

This description consists of a single process. Signal `q_sig` is declared so that output `q` can be read. The outer if statement infers a flip-flop. The inner if statement says that the flip-flop should store `d` only if `en = '1'`. Otherwise, it should store

LISTING 8.5.2
Gating a flip-flop's data using an enable input.

```
library ieee;
use ieee.std_logic_1164.all;

entity d_ff_enabled is
   port (d, clk, en: in std_logic;
      q: out std_logic);
end d_ff_enabled;

architecture behavioral of d_ff_enabled is
   signal q_sig : std_logic;
begin
   process (clk)
   begin
      if clk'event and clk = '1' then
         if en = '1' then
            q_sig <= d;
         else
            q_sig <= q_sig;
         end if;
      end if;
   end process;
   q <= q_sig;
end behavioral;
```

q_sig. This latter action has been explicitly specified by the inclusion of an else clause containing a statement assigning the current value of q_sig to q_sig.

The synthesis of Listing 8.5.2 produces the logic in Figure 8.5.2.

The flip-flop's output is fed back to one of the multiplexer's inputs. The other multiplexer input is d. The value of en determines whether d or q_sig (the value

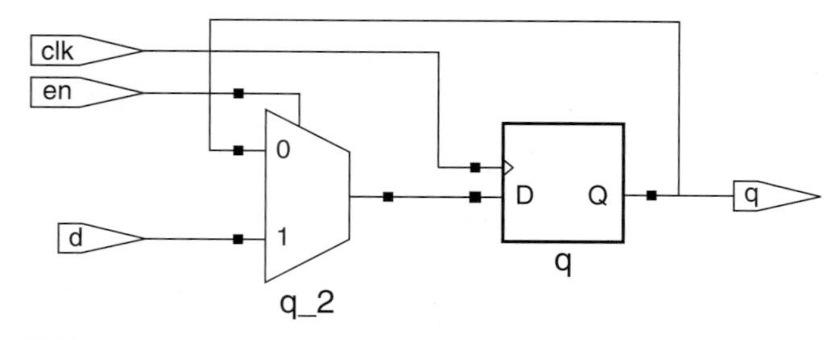

FIGURE 8.5.2
Synthesized logic for a D flip-flop with an enable (gated data).

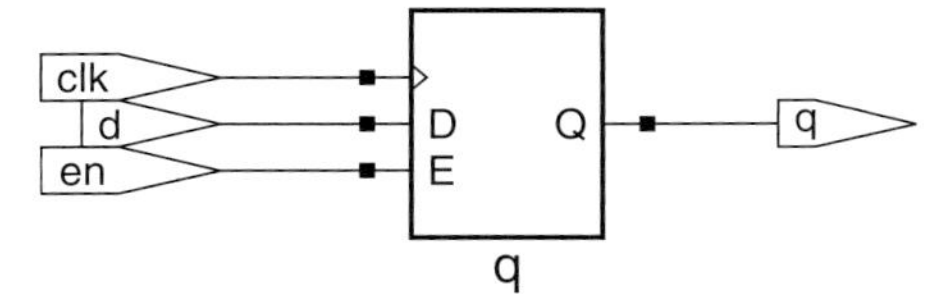

FIGURE 8.5.3
Logic symbol for a positive-edge-triggered flip-flop with an enable.

currently stored in the flip-flop) appears at the flip-flop primitive's D input. When the enable condition is false, the data currently stored in the flip-flop is recirculated through the multiplexer at each triggering clock edge. Note that the flip-flop primitive in Figure 8.5.2 actually stores the value at its D input at every triggering clock edge.

Glitches on the `en` input do not affect the state of the flip-flop, as long as `en` is stable for a time equal to the flip-flop's setup time prior to the positive edge of the clock. If the enable signal is derived from logic clocked by the same clock, any changes (including glitches) in its value should occur immediately after the positive clock edge. Accordingly, any glitches in the enable signal will have ended long before the setup time of the next triggering clock edge.

An equivalent representation of the logic in Figure 8.5.2 is given in Figure 8.5.3, where the symbol for the memory element is that of a positive-edge-triggered flip-flop with a built-in enable.

Functionally equivalent, but simpler coding for the architecture in Listing 8.5.2, is given in Listing 8.5.3.

LISTING 8.5.3
Simpler architecture for positive-edge-triggered flip-flop with enable.

```
architecture behavioral2 of d_ff_enabled is
begin
   process (clk)
   begin
      if clk'event and clk = '1' then
         if en = '1' then
            q <= d;
         end if;
      end if;
   end process;
end behavioral2;
```

8.6 OTHER FLIP-FLOP TYPES

There are four basic types of flip-flops that have traditionally been used in digital systems: D, S-R, J-K, and T. These flip-flops differ primarily in the number and function of their synchronous inputs.

In the traditional design of finite state machines (FSMs), where typically only a few flip-flops were provided in each IC package (using small-scale integration), the use of S-R, J-K, or T flip-flops instead of D flip-flops often allowed simplification of the combinational logic required to drive the flip-flops' inputs. This would typically result in the reduction of the number of IC packages required to implement the system.

Characteristic Equations

Its *characteristic equation* specifies a flip-flop's next state (output) as a function of its synchronous inputs. The characteristic equations for the four different flip-flops are given in Table 8.6.1.

Today, most systems are designed using D flip-flops because of their simplicity and their prominence as the primitive memory elements in PLDs.

Some PLDs contain primitive memory elements that can be directly configured to implement any of the four types of flip-flops. If the target PLD does not contain primitive memory elements that are directly configurable for the type of flip-flop described, the synthesizer can synthesize combinational logic to drive the primitive memory element's synchronous inputs to realize the desired flip-flop behavior.

S-R Flip-flop

A *S-R flip-flop* has two synchronous inputs, S and R. It is set when S = 1 and R = 0 and cleared when S = 0 and R = 1. Having S and R both 1 simultaneously is an unallowed condition that must be avoided. When S and R are both 0s, the flip-flop remains in its previous state. Thus, unlike the basic D flip-flop, there is an input combination that allows the state of the flip-flop to remain unchanged for more than one clock cycle. Listing 8.6.1 is a description of a S-R flip-flop.

Table 8.6.1
Flip-flop characteristic equations.

Flip-flop Type	Characteristic Equation
D	$Q = D$
S-R	$Q = S + \overline{R}Q$
J-K	$Q = J\overline{Q} + \overline{K}Q$
T	$Q = T\overline{Q} + \overline{T}Q$

LISTING 8.6.1
S-R flip-flop description using characteristic equation.

```
library ieee;
use ieee.std_logic_1164.all;

entity srff is
   port(
      s, r, pre_bar, clr_bar, clk : in std_logic;
      q : out std_logic
      );
end srff;

architecture behavioral of srff is
   signal q_sig : std_logic ; -- local signal to assign and read
begin
   process (clk, pre_bar, clr_bar)
   begin
      if pre_bar = '0' then  -- asserted low asynchronous preset
         q_sig <= '1';
      elsif clr_bar = '0' then  -- asserted low asynchronous clear
         q_sig <= '0';
      elsif rising_edge(clk) then
         q_sig<= s or (not r and q_sig);  -- characteristic equation
      end if;
   end process;
   q <= q_sig;      -- assign local signal to output
end behavioral;
```

The S-R flip-flop described has an asserted-low preset and an asserted-low clear; both are asynchronous.

Figure 8.6.1 shows the logic generated when the S-R flip-flop is synthesized to a 22V10 SPLD. The primitive memory element in a 22V10 is a positive-edge-triggered D flip-flop with an asserted-low asynchronous set (labeled S on the symbol for the flip-flop primitive in Figure 8.6.1) and an asserted-high asynchronous reset (labeled R on the symbol for the flip-flop primitive). The AND gate and OR gate that feed the D input of the flip-flop primitive implement the characteristic equation for an S-R flip-flop.

J-K Flip-flop

A *J-K flip-flop* has two synchronous inputs, J and K. It is set when J = 1 and K = 0 and cleared when J = 0 and K = 1. When J and K are both 1s, the flip-flop toggles (changes to the opposite state). When J and K are both 0s, the flip-flop remains in its previous state. A description of a J-K flip-flop can be created from Listing 8.6.1 by simply changing the synchronous input labels and replacing the characteristic equation with that of a J-K flip-flop.

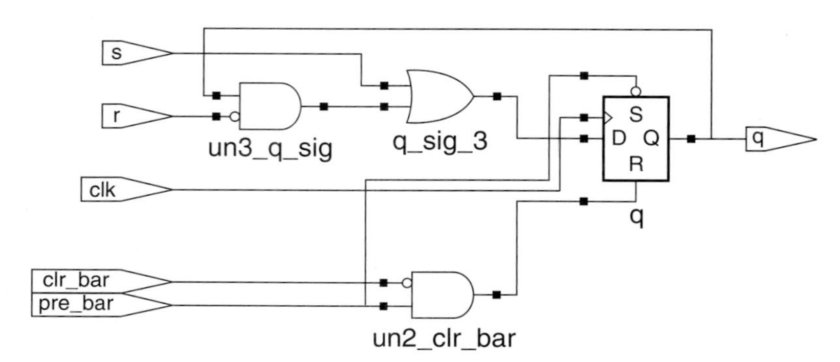

FIGURE 8.6.1
Logic synthesized to implement a S-R flip-flop in a 22V10 SPLD.

Logic synthesized from a description of a J-K flip-flop obtained by modifying Listing 8.6.1, as described in the previous paragraph and targeted to a 22V10, is shown in Figure 8.6.2.

Toggle Flip-flop

A *toggle flip-flop* (T flip-flop) changes state at each triggering clock edge if its toggle input is asserted. Otherwise, it remains in its previous state. A description of a T flip-flop can be created from Listing 8.6.1 by simply changing the synchronous inputs to a single T input and replacing the characteristic equation with that of a T flip-flop.

Logic synthesized for a T flip-flop is shown in Figure 8.6.3. The input logic consists of a multiplexer that feeds the flip-flop primitive's output back to its D input, if its T input is unasserted. If the T input is asserted, the complement of the flip-flop's output is fed back.

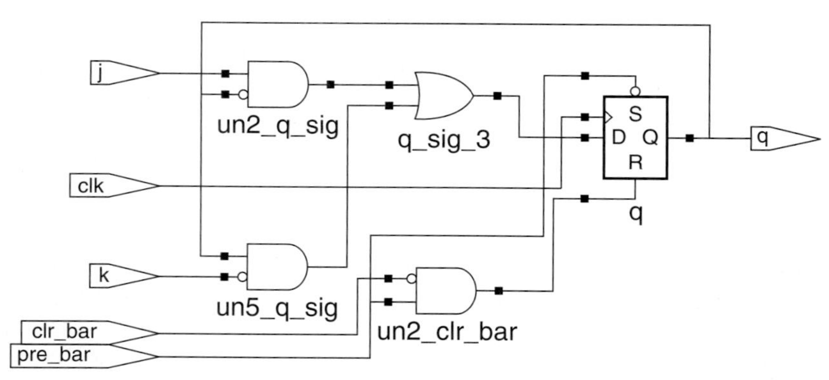

FIGURE 8.6.2
Logic synthesized to implement a J-K flip-flop in a 22V10 SPLD.

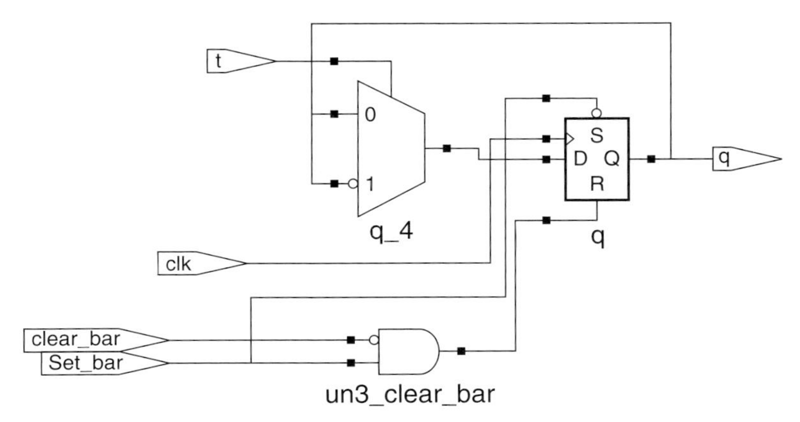

FIGURE 8.6.3
Logic synthesized to implement a T flip-flop in a 22V10 SPLD.

8.7 PLD PRIMITIVE MEMORY ELEMENTS

We have seen that we can write VHDL descriptions for any kind of latch or flip-flop. From these descriptions logic in a technology independent form can be synthesized. However, whether the synthesized logic can be mapped to a particular PLD is dependent on the PLD's primitive memory elements. Generally, the more complex the PLD, the more versatile are its primitive memory elements in terms of configurability. The configurability of primitive memory elements in CPLDs and FPGAs is much greater than it is in SPLDs.

For example, if we write a description that infers a D latch and try to map it to a 22V10 SPLD, the attempt fails. The 22V10's primitive memory element is positive-edge triggered and its clock input must come directly from a specific pin of the IC; therefore, it cannot implement level-triggered logic.

However, various types of flip-flops can be synthesized and mapped to a 22V10. For example, a J-K flip-flop can be synthesized and mapped, as illustrated by the logic in Figure 8.6.2. The synthesizer automatically synthesizes combinational logic prior to the D input of the 22V10 flip-flop primitive to create the appropriate value at its D input, as a function of the J and K inputs.

Similar limitations exist for asynchronous inputs. For example, the primitive memory element in a 22V10 has an asynchronous reset and a synchronous preset. As a result, it is impossible to map a flip-flop design with an asynchronous preset to a 22V10.

Other architectural limitations of a PLD might limit whether its individual memory elements can have independent clocks and/or independent asynchronous inputs. For example, in a 22V10 all the flip-flops share a single clock input.

Use of Flip-flops versus Latches

Generally, the use of latches as simple storage elements is not recommended, especially S-R latches. Flip-flops are used instead. Traditionally, latches had the advantage in custom IC design of requiring fewer transistors, thus requiring half as much IC area.

However, latches in CPLDs and FPGAs are typically implemented using memory elements that can be configured as either a latch or flip-flop. The amount of chip area taken by a configurable memory element in a PLD is fixed, regardless of whether it is configured as a latch or flip-flop. More importantly, use of latches increases the timing complexity of a design. This necessitates much more careful design techniques and timing analysis to ensure correct operation. Finally, few SPLDs have memory elements that can be configured as latches.

If the target PLD for a design is predetermined, we must determine what kinds of memory elements it can support and write our description to use only those kinds.

8.8 TIMING REQUIREMENTS AND SYNCHRONOUS INPUT DATA

Ideal Memory Elements

Previous latch and flip-flop descriptions represented ideal devices, ones with no timing requirements or propagation delays. For example, during functional simulation a flip-flop's D input can be changed to a new value simultaneously with (during the same simulation cycle as) the occurrence of a triggering clock edge, and the new data value will be stored in the flip-flop.

Physical Memory Elements

In contrast, physical memory elements have timing requirements that must be met to ensure their proper operation. For example, in an actual PLD, input data to a D flip-flop must be stable for a period of time (*setup time*, t_{su}) before a triggering clock edge and must remain stable for a period of time (*hold time*, t_h) after a triggering clock edge. There is also a minimum clock width requirement (*clock width*, t_{wh}) that must be met. These three timing requirements are illustrated in Figure 8.8.1 for a positive-edge-triggered D flip-flop.

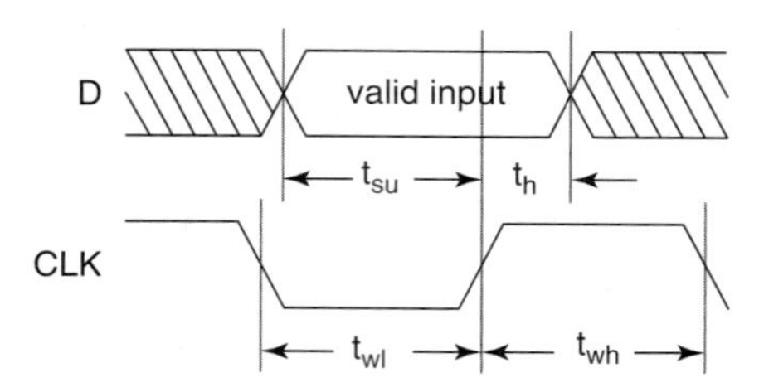

FIGURE 8.8.1
Basic timing requirements for a positive-edge-triggered flip-flop.

Positive values for setup time and hold time define a timing "window" with respect to the triggering clock edge. Data must be stable throughout this entire timing window to guarantee predictable operation of the flip-flop.

Violating Setup Time or Hold Time Requirements

One result of not meeting setup time or hold time requirements is the possibility of a metastable event. The term *metastable event* denotes the situation where a flip-flop's output either remains between valid states for an unusually long period of time or fails to complete the transition to the desired state after a triggering clock edge. Metastable events can result in incorrect system operation.

Synchronous Output Data

The Q output of a D flip-flop can change value only at a triggering clock edge. Accordingly, a change in Q's value is said to be *synchronized* to the clock. In an ideal flip-flop, the output changes simultaneously with the triggering clock edge; there is no propagation delay. In contrast, in a physical D flip-flop, there is always a propagation delay from the triggering clock edge until the output data is stable (*clock to output delay,* t_{co}).

Thus, synchronous output data in a physical system does not become valid until some delay after the triggering clock edge. However, this delay is relatively small compared to the clock's period. When the timing model of a D flip-flop mapped to a target PLD is simulated, this delay is observable on the timing waveforms.

Synchronous Input Data

If the Q output of one flip-flop provides input data to a second flip-flop that uses the same clock signal, the input data to the second flip-flop is *synchronous input data* (Figure 8.8.2). This is true even if the input data to the second flip-flop is generated as a combinational function of the outputs of multiple flip-flops all using the same clock signal. However, in this case, the delay from the triggering clock edge until the signal at the D input of the second flip-flop becomes stable is increased by the added propagation delay of the combinational logic.

When multibit input data to a sequential system comes from flip-flop outputs of another sequential system using the same clock, this input data is also synchronous. With synchronous input data and a sufficiently long clock period it is easy to ensure that a flip-flop's setup time and hold time requirements are met.

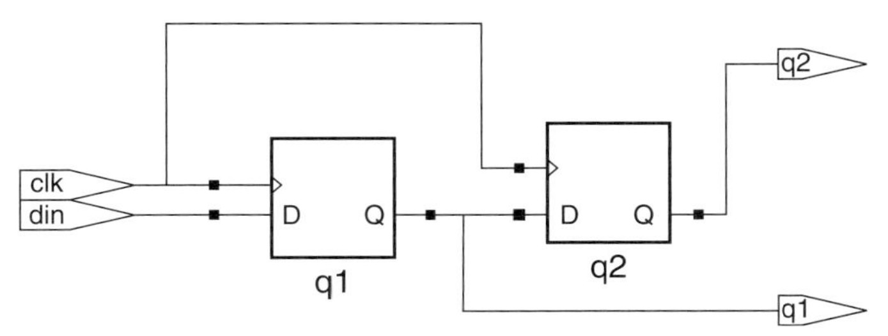

FIGURE 8.8.2
Synchronous input data.

Asynchronous Input Data

In a practical system it is common that some input data that comes directly from the outside world changes asynchronously to the system clock (*asynchronous input data*). If this is the case, it is possible that sometimes the setup or hold time requirements for a flip-flop will not be met. When a system has asynchronous input data, additional logic is used to synchronize the asynchronous input data to the system clock to reduce the probability of metastable operation. This logic, called a *synchronizer*, is identical to the logic in Figure 8.8.2, with the asynchronous input signal connected to `din` and `q2` being the synchronized version of the asynchronous input.

To simplify matters, in our subsequent discussions of sequential systems, unless otherwise stated, it is assumed that all data input to the synchronous inputs of a flip-flop are synchronous and that setup and hold timing requirements for all flip-flops are met.

PROBLEMS

8.1 Briefly state, in terms of their functional operation, the differences between a combinational system and a sequential system.

8.2 How is the past sequence of input values to a sequential system represented inside the system?

8.3 How many stable states does a single memory element have? What are the terms used to describe those states?

8.4 A clock signal has a 100 ns period and is high for 25 ns during each clock cycle. What is the clock's frequency and what is its duty cycle?

8.5 Under IEEE Std 1076.6, what VHDL types can be used to represent a clock signal? What values from these types are allowed in expressions representing clock levels or clock edges?

8.6 Explain the similarities and differences between a D latch and a D flip-flop.

8.7 Write the function table for a negative-level latch.

8.8 What determines which edge of a clock pulse is the triggering edge?

8.9 Write the function table for a negative-edge-triggered flip-flop.

8.10 What are the differences between synchronous and asynchronous inputs to a memory element?

8.11 Draw the waveform for output Q of a negative-level latch that has input waveforms D and CLK corresponding to those in Figure 8.2.2.

8.12 What does transparency mean with regard to a D latch? Under what conditions is a D latch transparent?

8.13 When does a D latch with a negative-level clock input "latch" (store) its D input value?

8.14 What is latch inference? What are the general criteria for a synthesizer to infer a latch when there is an assignment to a signal or a variable in a process?

8.15 What signals should be in the sensitivity list of a process describing a latch and why? Is the order of appearance of signals in the sensitivity list significant?

8.16 Using an if statement, write a description of a negative-level D latch with no asynchronous set or clear inputs and outputs Q and $\overline{Q}$.

8.17 Is it possible to modify the D latch description in Listing 8.2.1 so that it has both Q and $\overline{Q}$ outputs and these outputs function like those of a traditional D latch with asynchronous active-low set and clear inputs? That is, when both asynchronous inputs are asserted, Q equals $\overline{Q}$ and neither asynchronous input has priority over the other. If so, write the modified code. If not, explain why.

8.18 Write a description of a D latch using a case statement. Can such a D latch be provided with asynchronous set and clear inputs?

8.19 Leaving `d` out of the sensitivity list for the latch in Listing 8.2.1, simulate the description and the post-synthesis netlist (or, alternatively, the post place-and-route netlist). Determine from your simulations whether the two models differ functionally. Pay particular attention to their transparency.

8.20 Write a description of a D latch using a selected signal assignment statement, including asserted-low set and clear inputs. Is either the set or clear input dominant?

8.21 Write a behavioral description of a negative-level-triggered clocked S-R latch that is true to the traditional function of a S-R latch. That is, when both its S and R inputs are 1s and the clock is asserted, both the Q and $\overline{Q}$ outputs are 1s.

8.22 Write two different boolean expressions that can be used as conditions to detect a negative clock edge.

8.23 What is the advantage, if any, of using the rising_edge function to detect a positive clock edge rather than using the boolean condition `clk'event and clk = '1'`?

8.24 How many times can a D flip-flop change its state during one clock cycle? How many times can a D latch change its state during one clock cycle? Explain your answer.

8.25 Explain why the `d` input is not included in the sensitivity list for the positive-edge-triggered flip-flop in Listing 8.4.1, but is included in the sensitivity list of the positive-level D latch in Listing 8.2.1.

8.26 Would including the `d` input in the positive-edge-triggered flip-flop description in Listing 8.4.1 change the functional operation of the flip-flop? Would inclusion of `d` affect the simulation time when simulating this description?

8.27 Using the signal attribute stable, write a description of a negative-edge-triggered D flip-flop that has no preset or clear input.

8.28 Write a description of a negative-edge-triggered D flip-flop that has an asserted-low asynchronous reset and an asserted-high synchronous preset. Use a process containing an if statement.

8.29 Explain why the statement:

```
wait until clk = '0'
```

is sufficient to detect a negative clock edge.

8.30 Can a wait statement be used to implement a positive-edge-triggered D flip-flop with an asynchronous set and clear? Explain your answer.

8.31 Write a description that uses a wait statement to describe a positive-edge-triggered D flip-flop with a synchronous set and clear.

8.32 Can a selected signal assignment statement be used to describe a D flip-flop without an asynchronous set or clear input? If not, explain why. If so, write the description.

8.33 Can a selected signal assignment statement be used to describe a D flip-flop with an asynchronous set and clear input? If not, explain why. If so, write the description.

8.34 What is the advantage of a flip-flop with an enable input?

8.35 What are the two approaches to implementing a flip-flop with an enable input? Which approach is preferred and why?

8.36 Write a description of a positive-edge-triggered D flip-flop that has two asserted-high enables `en1` and `en2`. Both enables must be simultaneously asserted for the flip-flop to store the value at its `d` input at its clock's triggering edge. The flip-flop also has an asserted-low asynchronous reset (`clr_bar`).

8.37 Simulate the description in Listing 8.6.1 and verify that it describes an S-R flip-flop.

8.38 For the diagram of the synthesized logic in Figure 8.6.1:
(a) Trace through the logic that generates the signal at the D input of the flip-flop primitive and verify that it implements the characteristic equation for a S-R flip-flop.
(b) Trace through the logic that generates the signal at the asserted-high clear input R of the flip-flop primitive and determine its function and how it corresponds to the description.

8.39 Modify Listing 8.6.1 to describe a J-K flip-flop. Verify the resulting J-K flip-flop's functionality through simulation.

8.40 For the diagram of the synthesized logic in Figure 8.6.2:
(a) Trace through the logic that generates the signal at the D input of the flip-flop primitive and verify that it implements the characteristic equation for a J-K flip-flop.
(b) Trace through the logic that generates the signal at the asserted-high clear input R of the flip-flop primitive and determine its function and how it corresponds to the description.

8.41 Modify Listing 8.6.1 to describe a T flip-flop. Verify the resulting T flip-flop's functionality through simulation.

8.42 For the diagram of the synthesized logic in Figure 8.6.3:
(a) Trace through the logic that generates the signal at the D input of the flip-flop primitive and verify that it implements the characteristic equation for a T flip-flop.
(b) Trace through the logic that generates the signal at the asserted high clear input (R) of the flip-flop primitive and determine its function and how it corresponds to the description.

8.43 Using data sheets, determine the setup time and hold time for a 74HC74 IC and for a 22V10 SPLD. For the 22V10 specify the complete part number of the device you use, particularly its speed grade.

Chapter 9

Multibit Latches, Registers, Counters, and Memory

Either latches or flip-flops can be interconnected to form memory structures that store multiple bits of information as a unit. Flip-flops can also be combined to form shift registers, shift register counters, and counters. These simple sequential systems find widespread use as components in more complex systems. Their descriptions are discussed in this chapter.

In some sequential systems, it is desirable to initiate an action in response to an edge of a signal other than the clock. This requires detection of a non–clock signal edge. Techniques for accomplishing this are presented.

With the preceding concepts as a foundation, descriptions of a digital retriggerable single shot and of a microprocessor compatible programmable pulse width signal generator are developed.

Finally, descriptions of large bit capacity RAM and ROM memory structures are discussed.

9.1 MULTIBIT LATCHES AND REGISTERS

Two or more D latches with their clock inputs connected together form a *multibit D latch* or *latch array.* A multibit latch stores multiple bits of data simultaneously, in response to its clock signal (Figure 9.1.1).

Another structure that stores multiple data bits is a register. A D flip-flop is a 1-bit register. Two or more D flip-flops with their clock inputs connected together form a *multibit register* (or simply a register).

The number of memory elements, or bits, in a multibit latch or register determines its *width.* While latches and registers can be of any width, 4, 8, 16, 32, and 64 bit widths are common in computing systems.

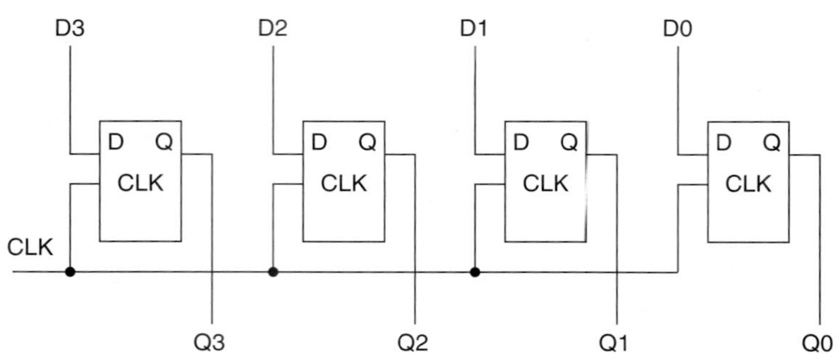

FIGURE 9.1.1
A multibit D latch.

Multibit Latches

We can easily convert the D latch in Listing 8.2.3 into a multibit latch by simply changing the types of signals `d` and `q` from std_logic to std_logic_vector. Listing 9.1.1 describes an octal (8-bit) D latch with asynchronous set and clear.

The `d` input and `q` output are simply expressed as vectors. A description of an equivalent latch of any width is obtained from Listing 9.1.1 by changing the size of vectors `d` and `q` and the constant values assigned to `q`.

LISTING 9.1.1
An octal D latch with asynchronous set and clear.

```
library ieee;
use ieee.std_logic_1164.all;

entity octal_d_latch is
   port (
      d: in std_logic_vector(7 downto 0);
      clk, set_bar, clear_bar: in std_logic;
      q: out std_logic_vector(7 downto 0));
end octal_d_latch;

architecture behavioral of octal_d_latch is
   begin
   process (d, clk, set_bar, clear_bar)
      begin
      if set_bar = '0' then
         q <= (others => '1');
      elsif clear_bar = '0' then
         q <= (others => '0');
      elsif clk = '1' then
         q <= d;
      end if;
   end process;
end behavioral;
```

LISTING 9.1.2
An octal register functionally equivalent to a 74HC574.

```
library ieee;
use ieee.std_logic_1164.all;

entity ic74hc574 is
   port (
      d: in std_logic_vector(7 downto 0);
      clk: in std_logic;
      oe_bar: in std_logic;    -- output enable control
      q: out std_logic_vector(7 downto 0));
end ic74hc574;

architecture behavioral of ic74hc574 is
signal q_int: std_logic_vector(7 downto 0);
begin
   process (clk)
      begin
      if rising_edge(clk) then
         q_int <= d;
      end if;
   end process;

   q <= q_int when oe_bar = '0' else (others => 'Z');

end behavioral;
```

Registers

A description of a register that is functionally equivalent to a 74HC574 octal positive-edge-triggered D flip-flop with three-state outputs is described in Listing 9.1.2.

In Listing 9.1.2, the process describes an octal D flip-flop whose output is the local signal `q_int`. Outside of the process, the conditional signal assignment statement describes the three-state output `q` by assigning `q` the value of `q_int` when `oe_bar` is asserted; otherwise each bit of `q` is assigned the high impedance (`'Z'`) value.

Registers with Enable Inputs

In Section 8.5, an enable input was added to a flip-flop, so that at a triggering clock edge the flip-flop stored the value at its D input only if its enable input was asserted. An enable input can be added to a register in a similar fashion. The flip-flop with an enable input described in Listing 8.5.2 can be easily converted to a register with an enable input. Scalar signals `d`, `q`, and `q_sig` are simply converted to vectors of the appropriate width. Registers with enable inputs find widespread use in the RTL designs described in later chapters.

9.2 SHIFT REGISTERS

Bits stored in a simple *shift register* are shifted one bit position (right or left) at each triggering clock edge. If the shift is to the right, all bits are shifted one position to the right, the input value is shifted into the leftmost bit position, and the original value in the rightmost bit position is shifted out of the register. The primary uses for shift registers are serial-to-parallel conversion, parallel-to-serial conversion, and synchronous delay.

A shift register that shifts in only one direction is easily implemented by connecting the Q output of one flip-flop to the D input of the next flip-flop in the direction of the shift (Figure 9.2.1). Each flip-flop in a shift register (and any logic associated with the flip-flop) is referred to as a *stage*.

In Figure 9.2.1, the input data is shifted into the leftmost stage at each triggering clock edge. Data in the rightmost stage is shifted out. Data at input SI appears at output SOUT four clock cycles later. In this sense, an *n*-stage shift register can be thought of as implementing a *n* clock period *synchronous delay.*

Shift registers are not constructed from D latches, because doing so would require that the clock pulse width be very precisely controlled. If the clock pulse were too wide, data would shift through multiple stages during one clock cycle.

Shift Register Using a Buffer Mode Port

A description of a 4-bit right shift register with a synchronous clear input and buffer mode outputs is given in Listing 9.2.1.

This shift register's serial data input is `si`. Its serial output is taken from `qout(0)`. The parallel outputs of the shift register are `qout(3)` down to `qout(0)`. To describe the shift, we need to be able to read the present value in the shift register. If its mode is out, port `qout` cannot be directly read. Instead, `qout` is declared as mode buffer.

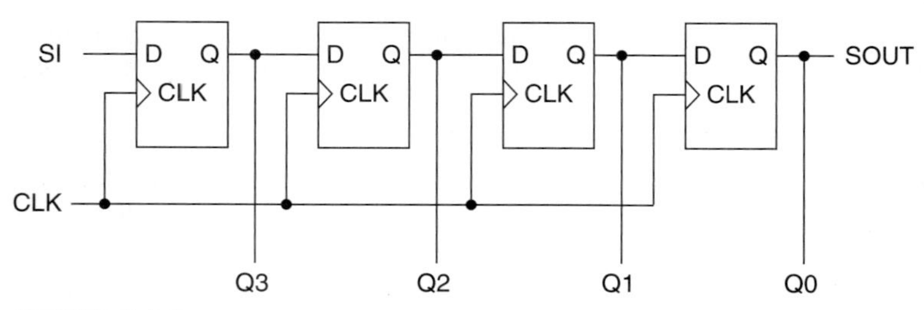

FIGURE 9.2.1
Shift register that shifts to the right.

LISTING 9.2.1

A 4-bit right shift register with synchronous clear using a buffer mode port.

```
library ieee;
use ieee.std_logic_1164.all;

entity shiftreg_rb is
   port (si, clr_bar, clk : in std_logic;
      qout : buffer std_logic_vector(3 downto 0));
end shiftreg_rb;

architecture behavior of shiftreg_rb is
begin
   process (clk)
   begin
      if rising_edge(clk) then
         if clr_bar = '0' then
            qout <= "0000";        -- clear shift register
         else
            qout(0) <= qout(1);    -- shift right
            qout(2) <= qout(3);
            qout(1) <= qout(2);
            qout(3) <= si;
         end if;
      end if;
   end process;
end behavior;
```

Prior Restrictions on Mode Buffer

Prior to IEEE Std 1076-2002 there was a restriction that, in a port map, a formal port of mode buffer could be associated with an actual that was itself a port only if the actual port was also of mode buffer. In practical terms, this meant that when an entity with a port of mode buffer was used as a component, it could only be connected to a signal or to a port of mode buffer of the enclosing entity. This limited the use of entities with buffer mode ports as components. IEEE Std 1076-2002 removed this restriction. The actual port associated with a formal port of mode buffer in a port map association is now allowed to be mode out, inout, or buffer. However, some versions of simulators and synthesizers may still enforce the prior restrictions on ports of mode buffer.

Using a Signal to Read a Mode Out Port

As a consequence of the prior restriction on mode buffer port associations, a local signal was often used to allow the value assigned to a mode out port to be read (Figure 9.2.2). The signal is declared in the declarative part of the architecture and is written and read in the statement part of the architecture. At the end of the process, the value of this signal is assigned to the mode out port.

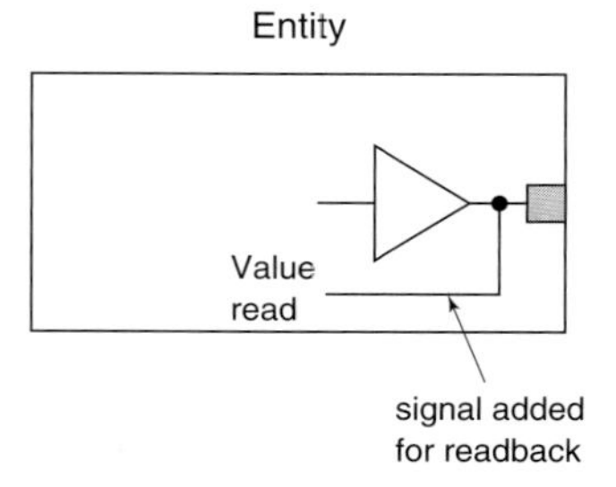

FIGURE 9.2.2
Using a signal to read the value assigned to an output port within an architecture.

Shift Registers Using Readback Signals

Accordingly, in Listing 9.2.2, signal `q` is declared and used as the shift register output. Outside of the process, a concurrent signal assignment statement assigns the value of `q` to port `qout`, which is mode out.

LISTING 9.2.2
A 4-bit right shift register using a signal to "read" a mode out port.

```
library ieee;
use ieee.std_logic_1164.all;

entity shiftreg_rs is
   port (si, clr_bar, clk : in std_logic;
      qout : out std_logic_vector(3 downto 0));
end shiftreg_rs;

architecture behavior of shiftreg_rs is
   signal q : std_logic_vector(3 downto 0);
begin
   process (clk)
   begin
      if rising_edge(clk) then
         if clr_bar = '0' then
            q <= "0000"; -- clear shift register
         else
            q(0) <= q(1); -- shift right
            q(2) <= q(3);
            q(1) <= q(2);
            q(3) <= si;
         end if;
      end if;
   end process;
   qout <= q;
end behavior;
```

Order of "Sequential" Signal Assignment Statements

In Listings 9.2.1 and 9.2.2, the sequences of assignment statements in the else clauses are intentionally written in an order other than from the rightmost flip-flop to the leftmost. For clarity, it would have been preferable that the assignments be listed in that order. Instead, they were listed in a random order. This was done to reinforce the point that, since these are signal assignments, their values don't change during the current execution of the process (simulation cycle). Therefore, the order in which the flip-flop outputs are read and their values assigned does not matter.

Shift Registers Using Variables

Another technique to allow reading the value assigned to a port of mode out is to use a variable, rather than a signal. If a variable is used, the order of assignment to elements of the variable vector is critical, because each assignment takes effect immediately. Assignments for a right shift must first assign variable `q(1)` to `q(0)`, then `q(2)` to `q(1)`, `q(3)` to `q(2)`, and finally `si` to `q(3)`, as shown in Listing 9.2.3.

LISTING 9.2.3
A 4-bit right shift register using variables.

```
library ieee;
use ieee.std_logic_1164.all;

entity shiftreg_rv is
   port (si, clr_bar, clk : in std_logic;
      qout : out std_logic_vector(3 downto 0));
end shiftreg_rv;

architecture behavior of shiftreg_rv is
begin
   clock: process (clk)
      variable q : std_logic_vector(3 downto 0);
   begin
      if rising_edge(clk) then
         if clr_bar = '0' then    -- clear shift register
            q := "0000";
         else
            q(0) := q(1);    -- shift right
            q(1) := q(2);
            q(2) := q(3);
            q(3) := si;
         end if;
         qout <= q;
      end if;
   end process;
end behavior;
```

Any change in the order of the assignments to variables in the else clause results in a design that doesn't function correctly, because the contents of one or more stages of the shift register are overwritten before they are read. The shift register descriptions in Listings 9.2.1, 9.2.2, and 9.2.3 all synthesize to the same logic.

Use of the Concatenation Operator to Shift

A shift operation can be written more concisely using the concatenation operator and slices. Recall that the concatenation operator is predefined for all one-dimensional array types. The sequence of instructions

```
q(0)  := q(1);  -- shift right
q(1)  := q(2);
q(2)  := q(3);
q(3)  := si;
```

can be replaced by a single assignment statement:

```
q := si & q(3 downto 1); -- shift right
```

This assignment takes the values in the leftmost three bit positions, prior to the triggering clock edge, and appends them to the right of the serial input value to create a right shifted result that is assigned to `q`. That is, `si` is assigned to `q(3)` and slice `q(3)` down to `q(1)` is assigned to `q(2)` down to `q(0)`.

Shift Register Digital Noise Filter

Figure 9.2.3 shows a logic representation of a digital noise filter. The purpose of this filter is to reject short-duration noise spikes at its data input `cx`. The filter's output `y` can change only after its input has had the same value at three consecutive triggering clock edges. Input changes that have a duration of two or fewer clock cycles are rejected (don't cause a change in the output). Examination of the logic diagram indicates that a valid input change is delayed by five clock periods, one for each flip-flop, before it is seen at the output.

A description of a functionally equivalent system is given in Listing 9.2.4.

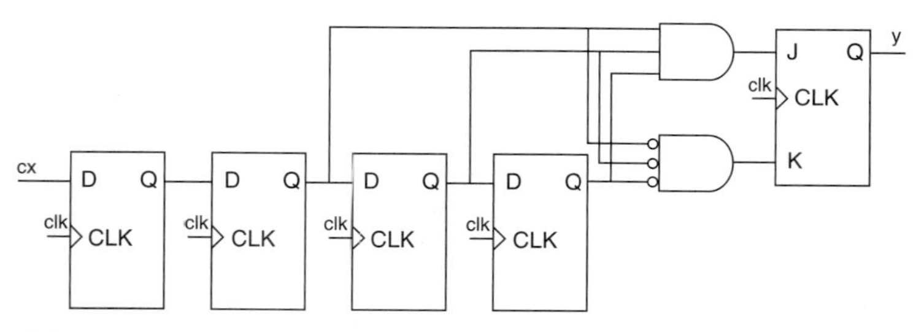

FIGURE 9.2.3
Digital noise filter logic diagram.

LISTING 9.2.4
Shift register digital filter.

```
library ieee;
use ieee.std_logic_1164.all;

entity filter is
   port (cx, clk, rst_bar : in std_logic;
      y : out std_logic);
end;

architecture behavior of filter is
begin
   synch: process (clk)
      variable q : std_logic_vector(3 downto 0);
   begin
      if rising_edge(clk) then
         if rst_bar = '0' then
            q := "0000";
            y <= '0';
         else
            if q(2 downto 0) = "111" then      -- filter
               y <= '1';
            elsif  q(2 downto 0) = "000" then
               y <= '0';
            else
               null;
            end if;
            q := cx & q(3 downto 1);       -- right shift
         end if;
      end if;
   end process;
end behavior;
```

This code describes the functionality of the 4-bit shift register and J-K flip-flop with gated inputs in Figure 9.2.3. The first flip-flop of the shift register is used to reduce the chances of a metastable event if the input data is asynchronous.

The filtered output is computed from the rightmost 3 bits of the shift register. If all three of these bits are zeros `"000"`, then output `y` is assigned a `'0'`. If all three of these bits are ones `"111"`, then `y` is assigned a `'1'`. For any other combination of these 3 bits, `y` remains unchanged.

The filter operation is accomplished by the innermost if statement. The shift operation is accomplished by the assignment statement for `q` that uses a concatenation operation.

9.3 SHIFT REGISTER COUNTERS

Using feedback, a shift register can be configured to produce a characteristic sequence of states. Such a shift register functions like a counter with a characteristic count sequence and is often called a *shift register counter.* Two examples of shift register counters implemented by connecting their serial outputs back to their serial inputs are given in this section.

Johnson Counter

A *Johnson counter* simply connects the complement of its serial output back to its serial input. If a Johnson counter is cleared when reset, its serial output is 0. Since the complement of its serial output appears at its serial input, at each subsequent triggering clock edge a 1 is shifted in, until the counter is filled with 1s. With its serial output now a 1, a 0 is shifted in at each subsequent triggering clock edge, until the counter is filled with 0s, making its serial output a 0. From this point, the count sequence repeats.

Counter Modulus

The number of unique states a counter has is its *modulus*. The count (or state sequence) for a Johnson counter is $2 \times n$ counts long (modulus $2 \times n$), where n is the number of stages in the shift register. A description of a 4-bit Johnson counter is given in Listing 9.3.1.

LISTING 9.3.1
Johnson counter.

```
library ieee;
use ieee.std_logic_1164.all;

entity jhnsncntr is
   port (clr_bar, clk : in std_logic;
      qout : out std_logic_vector(3 downto 0));
end;

architecture behavior of jhnsncntr is
begin
   process (clk)
      variable q : std_logic_vector(3 downto 0);
   begin
      if rising_edge(clk) then
         if clr_bar = '0' then
            q := "0000";
         else
            q := not q(0) & q(3 downto 1);   -- right shift
         end if;
      end if;
      qout <= q;
   end process;
end behavior;
```

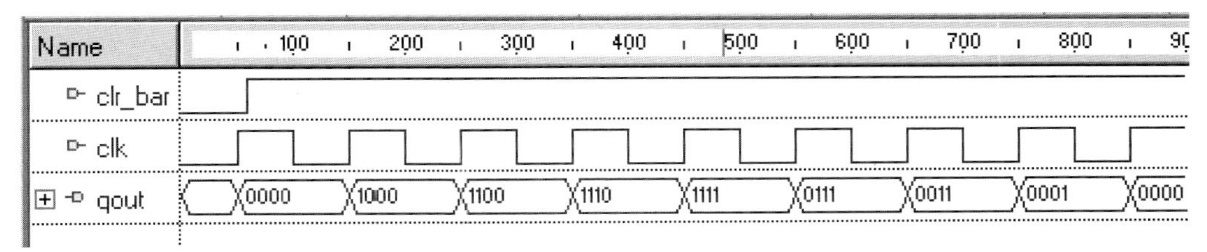

FIGURE 9.3.1
Output sequence from a 4-bit Johnson counter.

In Listing 9.3.1, the assignment statement that accomplishes the right shift assigns the complement of the serial output `q(0)` to the serial input:

```
q := not q(0) & q(3 downto 1);   -- right shift
```

The resulting state sequence is shown in the waveforms in Figure 9.3.1.

Ring Counter

A *ring counter* feeds its serial output directly back to its serial input. If at reset the shift register is loaded with a single 1, this 1 is shifted around the register over and over. The length of the count sequence is simply n. A description of a 4-bit ring counter is given in Listing 9.3.2.

LISTING 9.3.2
Ring counter.

```
library ieee;
use ieee.std_logic_1164.all;

entity ringcntr is
   port (rst_bar, clk : in std_logic;
      qout : out std_logic_vector(3 downto 0));
end;

architecture behavior of ringcntr is
begin
   process (clk)
      variable q : std_logic_vector(3 downto 0);
   begin
      if rising_edge(clk) then
         if rst_bar = '0' then
               q := "1000";
         else
               q := q(0) & q(3 downto 1);   -- right shift
```

(Cont.)

LISTING 9.3.2 *(Cont.)*

```
            end if;
        end if;
        qout <= q;
    end process;
end behavior;
```

This code for the ring counter is essentially the same as that for the Johnson counter in Listing 9.3.1, except that this shift register is reset to "1000" and the serial output is fed back to the serial input without being complemented. The count sequence for this ring counter is illustrated in the waveforms in Figure 9.3.2. Each pattern or codeword in the sequence has a single bit that is a 1. Such a set of code words is referred to as a *one-hot code*.

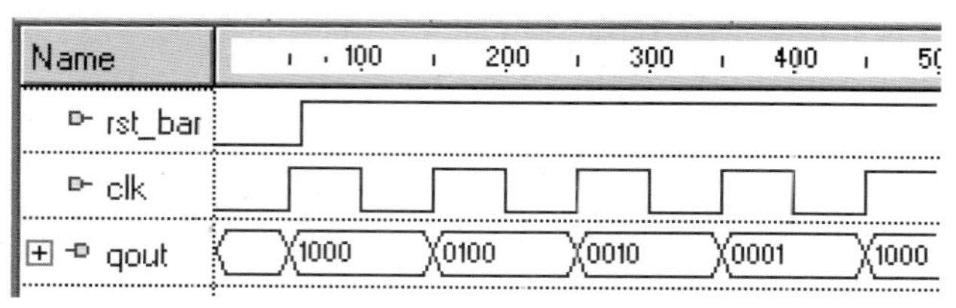

FIGURE 9.3.2
Simulation waveforms for ring counter.

9.4 COUNTERS

Counters are simple examples of finite state machines (FSMs). FSMs are discussed at length in Chapter 10. Counters are described more simply in this chapter.

Among other things, counters are used to count events, generate time intervals, generate events, and divide down an input signal to create a lower frequency output signal.

A counter's *next state* is the state it goes to from its *present state* at a triggering clock edge. A simple counter transitions from one unique state to another, until it eventually returns to its initial state, after which its state transitions repeat.

Simple Counters

The simplest counter has no inputs other than a clock and a reset. The counter's outputs are taken directly from its flip-flops. After reset, the counter changes state at each triggering clock edge.

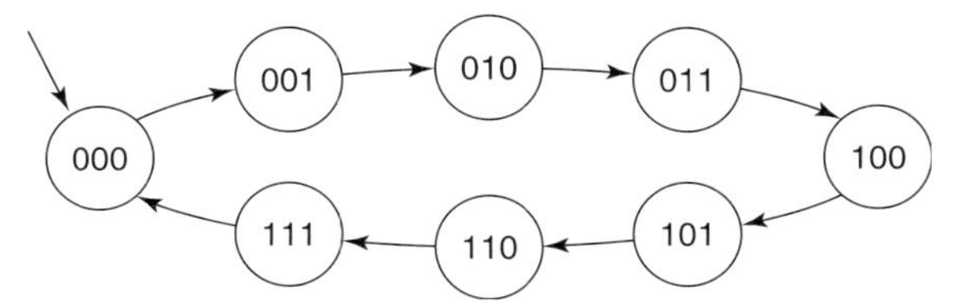

FIGURE 9.4.1
State diagram for a 3-bit counter.

Counter State Diagram

The state transitions of a counter are clearly depicted by a state diagram. A state diagram for a 3-bit binary up counter is given in Figure 9.4.1. The counter is initialized to 000 at reset and counts up to 111 before recycling to 000. A 3-bit binary counter has a modulus of 8, or is, equivalently, a modulo-8 counter.

A 3-bit binary counter requires three flip-flops. The counter's output consists of the Q outputs of its flip-flops. With n flip-flops we can create a counter with a modulus of up to 2^n. Counters can also be designed to have a *truncated sequence* or *truncated modulus*, so that the counter's modulus is less than 2^n.

Synchronous Counters

A *synchronous counter* is one in which all the flip-flops are clocked simultaneously by a common clock signal. All counters and other FSMs considered in this book are synchronous. Synchronous designs are easier to verify than asynchronous designs, particularly with respect to timing. In addition, most PLD architectures are not well suited to implementing asynchronous sequential systems.

Counting by Addition or Subtraction

A straightforward approach to the description of a counter involves use of *arithmetic operators.* An up counter is simply a register whose value is incremented. Incrementing a register is accomplished by adding 1 to its present value. A down counter is a register whose value is decremented. Decrementing a register is accomplished by subtracting 1 from its present value.

Adding or Subtracting Std_logic_ vectors

In VHDL, addition and subtraction operations are predefined for only the types integer, real, and time. Since we are using std_logic or std_logic_vector types for inputs and outputs, we cannot use the predefined + and – operators directly. Two ways to solve this problem are:

1. Use type integer signals or variables along with the predefined + and – operators, then use functions to convert the integer results to std_logic.
2. Use type unsigned from the package NUMERIC_STD, which also overloads the + and – operators for this type, then convert the unsigned results to std_logic_vector.

LISTING 9.4.1
A 4-bit binary up counter using an integer signal.

```
library ieee;
use ieee.std_logic_1164.all;
use ieee.numeric_std.all;

entity counter_4bit is
   port (clk, reset_bar: in std_logic;
      count: out std_logic_vector (3 downto 0));
end counter_4bit;

architecture behav_int of counter_4bit is
   signal count_int : integer range 0 to 15;
begin

   cnt_int: process (clk, reset_bar)
   begin
      if reset_bar = '0' then
         count_int <= 0;
      elsif rising_edge(clk) then
         if count_int = 15 then
            count_int <= 0;
         else
            count_int <= count_int + 1;   --read and increment count_int
         end if;
      end if;
   end process;

   count <= std_logic_vector(to_unsigned(count_int,4));

end behav_int;
```

Counter Using an Integer Signal

The first approach uses an integer signal or a variable to represent the count and the predefined + and – operators. An example is shown in Listing 9.4.1 for a 4-bit binary up counter.

Signal `count_int` is declared as type integer with a range of 0 to 15. This range is consistent with that of a 4-bit binary counter. Signal `count_int` maintains the counter's value and is incremented by the process at each triggering clock edge. Use of signal `count_int` is another example of the previously introduced technique used to read the value assigned to a mode out port.

Counter Initialization

A counter, like any other FSM, needs to be initialized to operate properly. However, an IEEE Std 1076.6 compliant synthesizer ignores initial values in declarations. It does so because it cannot know whether the initialization specified can be accomplished at power on reset in the target PLD. Accordingly, proper initialization requires the inclusion of either a reset or a load input (or both). A reset input can be asynchronous or synchronous, as desired. The VHDL code for a counter reset is similar to that used for a single flip-flop or for a multibit register.

The process in Listing 9.4.1 is sensitive to `clk` and `reset_bar`. If an event occurs on either `clk` or `reset_bar`, the process is executed. If `reset_bar` is `'0'`, then `count_int` is assigned the value `0`, asynchronously resetting the counter.

If the counter is not reset and there is a rising edge on `clk`, the present value of `count_int` is checked to see if it is equal to `15`. If so, then `count_int` is already at its maximum value and incrementing it must cause it to roll over to `0`. If this check is not made, `count_int` will increment from `15` to `16` and be out of its declared range (and the range of a 4-bit counter). This would cause a dynamic semantic error, which would be detected during simulation. If `count_int` is not equal to `15`, then it is incremented by `1`.

The outer if statement has no else clause and, therefore, infers a register. The inferred register must have the number of bits necessary to represent an integer as large as `15` encoded in binary. Accordingly, the synthesizer infers a 4-bit register.

Counter Represented as a Register and Adder

A technology independent representation of the inferred logic is shown in Figure 9.4.2.

Part of the counter is a 4-bit register. The input to this register is from a 4-bit adder. The adder's inputs are the signal `count_int` and a constant of value 1. The adder symbol is a circle with a + sign inside. At each triggering clock edge, the sum of the present value in the register and the constant 1 is loaded into the register, producing the effect of a counter.

Signal `count_int` is needed for two reasons. First, we want an integer signal so that we can use the predefined + operator. This integer signal is then converted to a std_logic_vector for output. Second, in order to increment this signal, we need to be able to read the present value of the counter. We can't read that value directly from `count`, because `count` is a mode out port.

The concurrent assignment statement

```
count <= std_logic_vector(to_unsigned(count_int,4));
```

first uses the function `to_unsigned` to convert integer signal `count_int` into a 4-bit unsigned vector. The value returned by this function is then converted to type

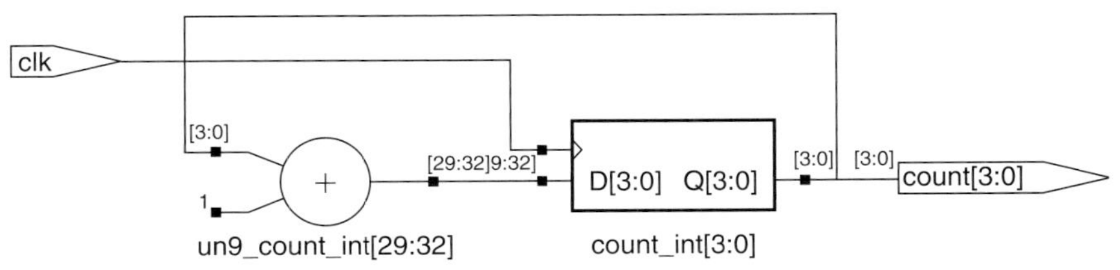

FIGURE 9.4.2
Hierarchical representation of synthesized 4-bit counter.

std_logic_vector and assigned to output count. Since this assignment statement is a concurrent function call, it executes anytime there is an event on the parameter count_int of the function to_unsigned.

Counter Using Unsigned Signal

The second approach to describing a counter is simpler and more direct. A signal is again used, but its type is unsigned.

A description of a 4-bit counter using an unsigned signal is given in Listing 9.4.2. The count is incremented by the statement:

```
count_us <= count_us + 1;
```

In this statement, the integer 1 is added to unsigned signal count_us. Package NUMERIC_STD overloads the + operator so that it can be used to add an unsigned type to a natural type. This package also overloads the + operator so that it can add two unsigned types. So this statement could also have been written as:

```
count_us <= count_us + "0001";
```

LISTING 9.4.2
A 4-bit binary counter using an unsigned signal.

```
library ieee;
use ieee.std_logic_1164.all;
use ieee.numeric_std.all;

entity counter_4bit is
   port (clk, reset_bar: in std_logic;
      count: out std_logic_vector (3 downto 0));
end counter_4bit;

architecture behav_us of counter_4bit is

   signal count_us : unsigned(3 downto 0);

begin

   cnt_us: process (clk, reset_bar)
   begin
      if reset_bar = '0' then
         count_us <= "0000";
      elsif rising_edge(clk) then
         count_us <= count_us + 1;
      end if;
   end process;
```

```
   count <= std_logic_vector(count_us);

end behav_us;
```

Advantage of Using Type Unsigned for Counting

An advantage of using an unsigned type in a binary counter is that we do not have to check whether the count is at its maximum value to force it to 0 on the next count. An unsigned vector naturally rolls over to 0 on the next count after it has reached all 1s ($2^n - 1$).

Counter Using a Variable

Instead of using a signal to hold the count, we can use a variable. Listing 9.4.3 describes a 4-bit binary counter using an unsigned variable.

Listing 9.4.3 reinforces an important concept regarding variables. ***A variable maintains its value between executions of a process (except for a variable declared in a subprogram).***

All three of the previous descriptions of a 4-bit counter create the same logic when synthesized.

LISTING 9.4.3
A 4-bit binary counter using an unsigned variable.

```
library ieee;
use ieee.std_logic_1164.all;
use ieee.numeric_std.all;

entity counter_4bit is
   port (clk, reset_bar: in std_logic;
      count: out std_logic_vector (3 downto 0));
end counter_4bit;

architecture behav_var of counter_4bit is
begin
   cnt_var: process (clk, reset_bar)
      variable count_v : unsigned(3 downto 0);
   begin
      if reset_bar = '0' then
         count_v := "0000";
      elsif rising_edge(clk) then
         count_v := count_v + 1;
      end if;
      count <= std_logic_vector(count_v);
   end process;
end behav_var;
```

Down Counters

Any of the previous binary counters can be changed to a down counter by replacing the + operator with the – operator. If the counter uses an integer to hold the count, then a check must be made to determine if the present count is 0. If so, then the next count must be equal to $2^n - 1$ (all 1s in binary).

Up/Down Counters

A control input can be added to make an up/down counter, which can count in either direction. For example, an `up` input could be added. When a triggering clock edge is detected, the inner if would determine if `up = '1'`. If so, the counter is incremented. If not, the counter is decremented.

Of course, we can easily create counters that count by values other than 1. Also, the amount an up/down counter increments can be different from the amount it decrements.

Up/Down Counter with an Enable Input

A count enable input can also be added to a counter, so the counter only counts at a triggering clock edge when it is enabled. If the counter is not enabled, its count remains the same at the triggering clock edge.

A 12-bit binary up/down counter with a count enable input and a synchronous reset is described in Listing 9.4.4. This counter's count direction can be reversed at any point in its count sequence.

LISTING 9.4.4
A 12-bit binary counter with count enable and synchronous reset.

```
library ieee;
use ieee.std_logic_1164.all;
use ieee.numeric_std.all;

entity binary_cntr is
   port (clk, cnten, up, rst_bar: in std_logic;
      q: out std_logic_vector (11 downto 0));
end binary_cntr;

architecture behavioral of binary_cntr is
begin
   cntr: process (clk)
      variable count_v : unsigned(11 downto 0);
   begin
      if rising_edge(clk) then
         if rst_bar = '0' then
            count_v := (others => '0');
         elsif cnten = '1'  then
            case up is
               when '1' => count_v := count_v + 1;
               when others => count_v := count_v - 1;
            end case;
```

```
         end if;
      end if;
      q <= std_logic_vector(count_v);
   end process;
end behavioral;
```

If an event on `clk` is a positive edge, the inner if statement determines if the counter should be reset. If not, the count enable signal `cnten` is checked to determine whether the counter is enabled to count. If so, a case statement determines whether to count up or down. If the counter is not enabled to count, its value remains unchanged.

Loading an Initial Value

Sometimes the initial state of a counter needs to be a value other than 0. Another approach to providing an initial state, one that solves this problem, is to provide a load input. If such an input is synchronous, then, when it is asserted and a triggering clock edge occurs, instead of incrementing (or decrementing), a predefined value is loaded into the counter.

Truncated Sequence Counters

A *truncated sequence counter* counts through a sequence that is not a power of 2. If at a triggering clock edge the present count is the last count desired in the sequence, the counter is loaded with the starting (initial) count value, instead of being incremented to the next binary count.

Frequency Divider

A truncated sequence down counter is often used as a programmable *frequency divider* (Listing 9.4.5).

LISTING 9.4.5
Modulo m counter used for frequency division.

```
library ieee;
use ieee.std_logic_1164.all;
use ieee.numeric_std.all;

entity freq_div is
   port (clk, reset_bar: in std_logic;
      divisor: in std_logic_vector(3 downto 0);
      q: out std_logic);
end freq_div;

architecture behavioral of freq_div is
begin
   div: process (clk)
      variable count_v : unsigned(3 downto 0);
   begin
```

(Cont.)

LISTING 9.4.5 *(Cont.)*

```
        if rising_edge(clk) then
            if reset_bar = '0' then
                count_v := unsigned(divisor);
                q <= '0';
            else
                case count_v is
                    when "0010" =>
                    count_v := count_v - 1;
                    q <= '1';
                    when "0001" =>
                    count_v := unsigned(divisor);
                    q <= '0';
                    when others =>
                    count_v := count_v - 1;
                    q <= '0';
                end case;
            end if;
        end if;
    end process;
end behavioral;
```

The counter in Listing 9.4.5 counts the input load value (`divisor`) down by one at each triggering clock edge. When the present count is `"0010"` at a triggering clock edge, the count goes to `"0001"` and simultaneously output `q` becomes `'1'`. When the present state is `"0001"`, the counter loads the value of `divisor` at the next triggering clock edge and simultaneously `q` becomes `'0'`.

In effect, `q` functions as a *terminal count* signal. The `'1'` output at `q` lasts until the next triggering clock edge, at which time `q` becomes `'0'`. Thus, the duration of the pulse at `q` is equal to one clock period. It takes a number of clock cycles equal to the value of `divisor` to produce one pulse at `q`. The overall effect is that the frequency of output `q` is `1/divisor` times the frequency of the clock. Figure 9.4.3 shows the waveforms for a divisor of value 3.

Variable `count_v`, which maintains the count, never actually has the value `"0000"`. Instead, the case statement checks whether `count_v` is `"0001"` at a triggering clock edge occurs. If this is true, `count_v` is assigned the value of `divisor`. This is equivalent to counting the value of `count_v` down, determining that its new value is `"0000"`, and reloading `count_v` with the value of `divisor`, all in one clock cycle.

BCD Counters

A two-digit (two-decade) *BCD counter* is described in Listing 9.4.6. Since it counts in decimal, a two-digit BCD counter counts from 00 decimal to 99 decimal. On the next count, it rolls over to 00 decimal.

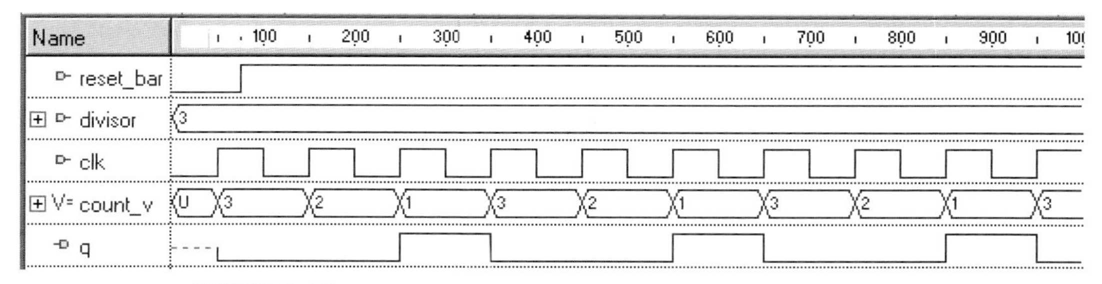

FIGURE 9.4.3
Waveforms for a divide-by-3 frequency divider.

The least significant four bits of the counter represent the least significant decimal digit (0 to 9) encoded in BCD. The most significant four bits represent the most significant decimal digit, encoded in BCD. Since a BCD digit ranges from 0000 to 1001 in binary, when the counter's value is 00001001 (09 decimal), its value on the next count must change to 00010000 (10 decimal).

LISTING 9.4.6
Two-digit BCD counter.

```
library ieee;
use ieee.std_logic_1164.all;
use ieee.numeric_std.all;

entity counter_bcd is
   port (
      clk, reset: in std_logic;
      count: out std_logic_vector (7 downto 0)
      );
end counter_bcd;

architecture behavioral of counter_bcd is
begin
   cnt: process (clk, reset)
      variable low_digit_v, high_digit_v : unsigned(3 downto 0);
   begin
      if reset = '1' then
         low_digit_v := b"0000";
         high_digit_v := b"0000";
```

(Cont.)

LISTING 9.4.6 *(Cont.)*

```
      elsif rising_edge(clk) then
         low_digit_v := low_digit_v + 1;     -- increment low digit
         if low_digit_v = b"1010" then       -- low digit > 9?
            low_digit_v := b"0000";          -- yes, make low digit = 0
            high_digit_v := high_digit_v + 1;    -- increase high digit by 1
         else
            null;
         end if;
         if high_digit_v = b"1010" then      -- high digit > 9?
            high_digit_v := b"0000";         -- yes, make high digit 0
         else
            null;
         end if;
      end if;
      count(3 downto 0) <= std_logic_vector(low_digit_v);
      count(7 downto 4) <= std_logic_vector(high_digit_v);
   end process;
end behavioral;
```

In Listing 9.4.6, two unsigned variables are declared, one for the low digit and one for the high digit. The outer if statement first checks to see if `reset` is asserted. If so, the low- and high-digit variables are set to 0.

If `reset` is not asserted, the elsif branch of the outer if statement checks for a rising clock edge. At a rising clock edge, the low-digit variable is incremented. The first inner if statement then checks to see if the low digit was incremented to 1010 (10 decimal). If so, the low-digit variable is assigned the value 0000 (0 decimal) and the high-digit variable is incremented.

The second inner if statement checks to see if the high digit was incremented to 1010 (10 decimal). If so, the high digit is assigned the value 0000 (0 decimal).

At the end of the process, the values of these variables are assigned to the output port signal in two slices:

```
count(3 downto 0) <= std_logic_vector(low_digit_v);
count(7 downto 4) <= std_logic_vector(high_digit_v);
```

Modulo-32 BCD Counter

As a last counter example in this section, we consider a modulo-32 BCD counter. This two-digit BCD counter counts from 00 to 31 and then rolls over to 00. The count direction can be up or down. The counter has two count enable inputs; both must be asserted for the counter to count. Integer variables are used to store the count.

The counter description is given in Listing 9.4.7.

LISTING 9.4.7
Modulo-32 two-digit BCD counter.

```
library ieee;
use ieee.std_logic_1164.all;
use ieee.numeric_std.all;

entity bcd_2dec is
   port(
      clk : in std_logic;
      rst_bar : in std_logic;      -- reset input
      up_bar : in std_logic;        -- direction control
      cnt_en1_bar : in std_logic;      -- count enable
      cnt_en2_bar : in std_logic;      -- count enable
      qbcd0 : out std_logic_vector(3 downto 0);    -- ls decade output
      qbcd1 : out std_logic_vector(3 downto 0)     -- ms decade output
      );
end bcd_2dec;

architecture behavioral of bcd_2dec is
begin
   process (clk)
      variable bcd0_v : integer range 0 to 9 ;
      variable bcd1_v : integer range 0 to 3;
   begin
      if rising_edge(clk) then
         if rst_bar = '0' then
            bcd0_v := 0;
            bcd1_v := 0;
         elsif
            cnt_en1_bar = '0' and cnt_en2_bar = '0' then
            if up_bar = '0' then
               if bcd1_v = 3 and bcd0_v = 1 then
                  bcd1_v := 0;
                  bcd0_v := 0;
               elsif bcd0_v = 9 then
                  bcd0_v := 0;
                  bcd1_v := bcd1_v + 1;
               else
                  bcd0_v := bcd0_v + 1;
               end if;
            elsif up_bar = '1' then
               if bcd1_v = 0 and bcd0_v = 0 then
                  bcd1_v := 3;
                  bcd0_v := 1;
               elsif bcd0_v = 0 then
                  bcd0_v := 9;
                  bcd1_v := bcd1_v - 1;
```

(Cont.)

LISTING 9.4.7 *(Cont.)*

```
               else
                  bcd0_v := bcd0_v - 1;
               end if;
            end if;
         end if;
      end if;
      qbcd0 <= std_logic_vector(to_unsigned(bcd0_v, 4));
      qbcd1 <= std_logic_vector(to_unsigned(bcd1_v, 4));
   end process;
end behavioral;
```

9.5 DETECTING NON-CLOCK SIGNAL EDGES

Detecting clock signal edges was discussed in Section 8.3. Often a design requires that some action be carried out when an event occurs on a signal other than the clock signal. Prior to IEEE Std 1076.6-2004, only one signal was allowed to be used as the clock in a process. As a result, a non-clock signal could not simply be treated like a second clock signal so that its edge could be detected. IEEE Std 1076.6-2004 now allows multiple clock signals in a process, as long as the occurrence of the clock edges is mutually exclusive.

Since some tools and target PLDs don't support multiple clocks in a process, we will look at an approach that uses a synchronous edge detector that requires only a single clock to detect the occurrence of an edge on some other signal. A *synchronous edge detector* generates a synchronous pulse of one clock duration whenever it detects the required edge of a specified signal.

The logic of a synchronous edge detector is shown in Figure 9.5.1. The objective is to detect positive edges of signal a. To do this, a version of signal a delayed by one

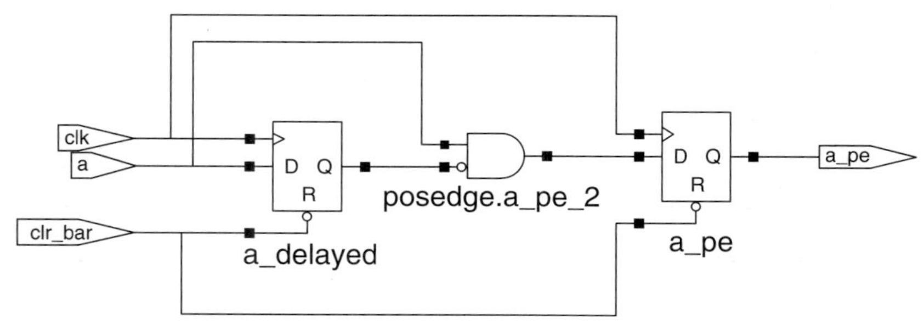

FIGURE 9.5.1
Synchronous positive edge detector.

LISTING 9.5.1
Positive edge detector system.

```
library ieee;
use ieee.std_logic_1164.all;

entity posedge is
   port(
      a : in std_logic;
      clr_bar : in std_logic;
      clk : in std_logic;
      a_pe : out std_logic);
end posedge;

architecture behavioral of posedge is
   signal a_delayed : std_logic;
begin
   posedge: process (clr_bar, clk)
   begin
      if clr_bar = '0' then
         a_pe <= '0';
         a_delayed <= '0';
      elsif rising_edge (clk) then
         a_delayed <= a;
         a_pe <= not a_delayed and a;
      end if;
   end process;
end behavioral;
```

clock cycle, called `a_delayed`, is created. This is done by storing `a` in a flip-flop at each triggering clock edge. The complement of the delayed version of signal `a` is ANDed with signal `a` itself. This AND function is a 1 only if `a` was 0 at the previous triggering clock edge and is currently a 1. If the output of the AND is a 1, the output of the edge detector (`a_pe`) becomes a 1 at the next triggering clock edge. Thus, `a_pe` is a `'1'` only if a positive edge occurred on `a` between the last two triggering clock edges.

A description of a positive edge detector is given in Listing 9.5.1.

This edge detector system detects each positive edge of signal `a`, as long as the positive edge is not part of a narrow pulse that occurs between two adjacent triggering clock edges. The operation of this system is illustrated by the waveforms in Figure 9.5.2. The narrow pulse that occurs just before 800 ns in Figure 9.5.2 is missed because this pulse occurs between two triggering clock edges (one at 750 ns and the other at 850 ns).

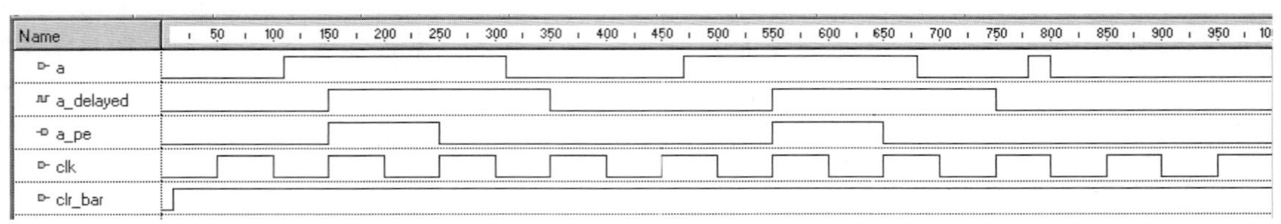

FIGURE 9.5.2
Waveforms of synchronized edge detector.

Synchronous Edge Detector Used to Enable a Counter

Output `a_pe` of the synchronous edge detector in Listing 9.5.1 can be connected to enable input `cnten` of the `binary_counter` in Listing 9.4.4 to create a system that counts the 0-to-1 transitions of a non–clock signal.

Synchronous Digital Single Shot

A *single shot* (*one shot* or *monostable multivibrator*) is a system that, when triggered (fired), generates an output pulse of a predetermined duration. SSI single shots use an external resistor and capacitor combination to set the duration of the output pulse. At the termination of its output pulse, a single shot is said to have *timed out*.

Retriggerable Single Shot

A *retriggerable single shot* is one whose output pulse duration is extended if the single shot is triggered again (retriggered) before its current output pulse times out. The duration of the current output pulse is extended at the occurrence of an additional trigger by a length of time equal to the predetermined pulse duration.

Nonretriggerable Single Shot

Once it has been triggered, a *nonretriggerable single shot* does not respond to additional triggers until after its current pulse has timed out.

74LS221 Single Shot

A 74LS221 IC contains two independent retriggerable single shots. Each single shot has two triggering inputs `a` and `b`. A single shot is triggered if its `a` input is low and a positive edge occurs on its `b` input, or if its `b` input is high and a negative edge occurs on its `a` input. Each single shot also has an overriding active-low clear input that, when asserted, terminates the output pulse.

The pulse duration for a 74LS221 single shot is an analog function set by the values of an external resistor R and capacitor C. The width of its output pulse is approximately $0.7 \times R \times C$ seconds.

74LS221 Digital Equivalent

A block diagram for an all-digital single shot that is functionally equivalent to one single shot in a 74LS221 IC is given in Figure 9.5.3.

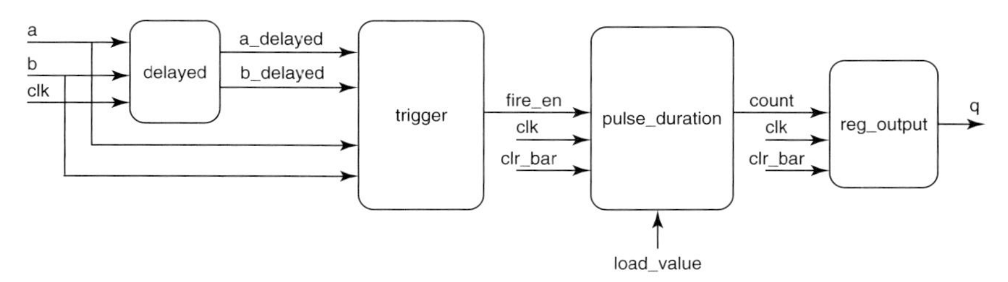

FIGURE 9.5.3
Block diagram of processes for an all-digital retriggerable single shot.

Each of the blocks with rounded corners in Figure 9.5.3 is described by a separate process. The VHDL code is given in Listing 9.5.2 and combines a number of concepts presented in this chapter. The triggering conditions are the same as for the 74LS221. However, instead of using an external resistor and capacitor to set the pulse duration, it is set by a constant. The single shot's output pulse duration is the product of the constant and the clock period.

LISTING 9.5.2
Retriggerable single shot functionally similar to a 74LS221.

```
library ieee;
use ieee.std_logic_1164.all;
use ieee.numeric_std.all;

entity single_shot is
   port(
      a : in std_logic;
      b : in std_logic;
      clr_bar : in std_logic;
      clk : in std_logic;
      q : out std_logic);
end single_shot;

architecture behavioral of single_shot is
   signal a_delayed, b_delayed, fire_en : std_logic;
   signal count : unsigned (7 downto 0);
begin
   delayed: process (clk)
   begin
      if rising_edge (clk) then
         a_delayed <= a;
         b_delayed <= b;
```

(Cont.)

LISTING 9.5.2 *(Cont.)*

```
      end if;
   end process;

   trigger: process (a, a_delayed, b, b_delayed)
      variable a_tran, b_tran : std_logic_vector(1 downto 0);
      variable a_nedg, b_pedg : std_logic;
   begin
      a_tran := a_delayed & a;  -- concatenate signals to form case expression
      b_tran := b_delayed & b;

      case a_tran is
         when "10" => a_nedg := '1';
         when others => a_nedg := '0';
      end case;

      case b_tran is
         when "01" => b_pedg := '1';
         when others => b_pedg := '0';
      end case;

      fire_en <= (a_nedg and b) or (b_pedg and not a);
   end process;

   pulse_duration: process(clk, clr_bar)
      constant load_value : unsigned := b"0000_0101";
   begin
      if clr_bar = '0' then
         count <= (others => '0');
      elsif rising_edge (clk) then
         if fire_en = '1' then
            count <= load_value;
         else
            if count /= b"0000_0000" then
               count <= count - 1;
            end if;
         end if;
      end if;
   end process;

   reg_output: process(clk, clr_bar)
   begin
      if clr_bar = '0' then
         q <= '0';
      elsif rising_edge (clk) then
         if count /= b"0000_0000" then
            q <= '1';
```

```
            else
                q <= '0';
            end if;
        end if;
    end process;

end behavioral;
```

The description consists of four processes that communicate using the signals a_delayed, b_delayed, fire_en, and count.

Process delayed generates the signals a_delayed and b_delayed, which are versions of signals a and b, respectively, delayed by one clock period. These signals are used by process trigger.

Process trigger detects the required transitions in a and b and generates the signal fire_en. Signal fire_en is asserted for one clock period whenever either of the conditions to fire the single shot are met. Note the creation of two vectors, one of a_delayed concatenated with a and the other of b_delayed concatenated with b. These vectors are used in two case statements to determine the occurrence of the required conditions to trigger the single shot.

Process pulse_duration counts out the predetermined duration of the output pulse. If clr_bar is asserted, the counter is cleared. If not, clk is checked for a rising edge. If there is a rising edge on clk and fire_en is asserted, the counter is loaded with constant load_value, which determines the output pulse's duration. If fire_en is not asserted and the counter's value is not zero, then the counter is decremented. If the single shot has been fired but has not timed out and fire_en is again asserted (because the single shot has been retriggered), the counter is reloaded with the constant value. This accomplishes the desired retriggering feature.

The process reg_out sets a flip-flop if the counter output is not 0 and clears the flip-flop if it is 0. This flip-flop's output is the output of the single shot. A functional simulation of Listing 9.5.2, with constant load_value equal to five, is given in Figure 9.5.4.

The logic synthesized for Listing 9.5.2 is given in Figure 9.5.5.

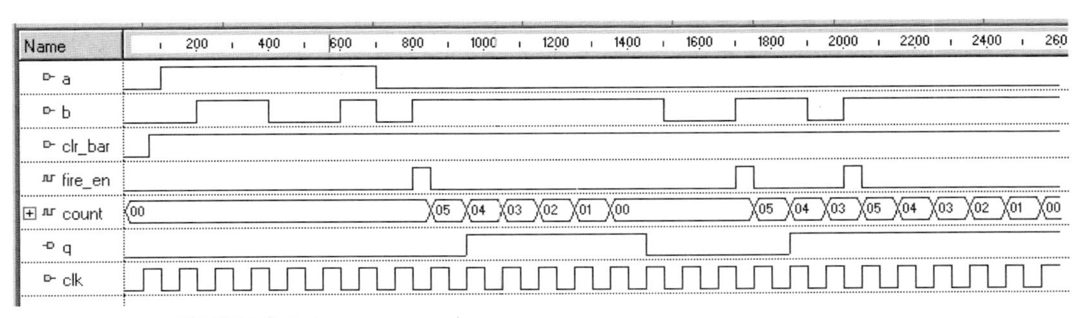

FIGURE 9.5.4
Simulation waveforms for single shot.

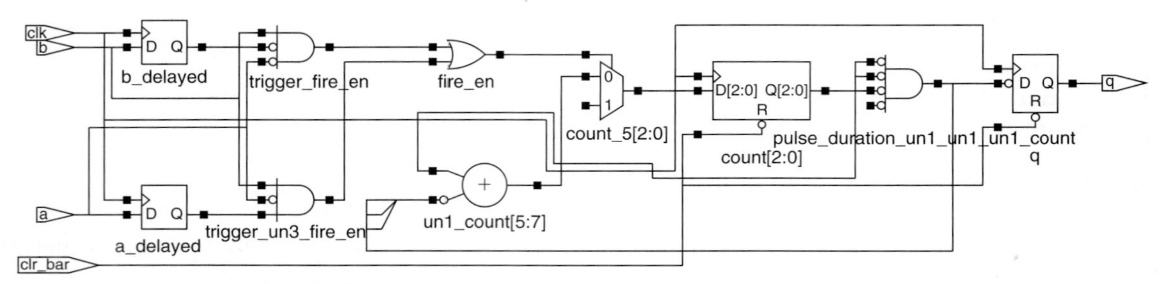

FIGURE 9.5.5
Logic synthesized for single shot.

9.6 MICROPROCESSOR COMPATIBLE PULSE WIDTH MODULATED SIGNAL GENERATOR

A pulse width modulated (PWM) signal has a fixed period, but its duty cycle can be changed (Figure 9.6.1).

There are many applications for PWM signals in electronic systems. A common one is to control the speed of a DC motor. The PWM signal may be used to drive the gate of a HEXFET transistor that turns the motor ON when the signal is 1 and OFF when it is 0. If the period of the PWM signal is sufficiently short, the motor's speed is proportional to the signal's duty cycle. If the period of the PWM signal is too large, the motor's speed is no longer controlled by the average value of the PWM signal. Instead, the motor runs full speed when the PWM signal is a 1 and will, eventually, stop when it is a 0.

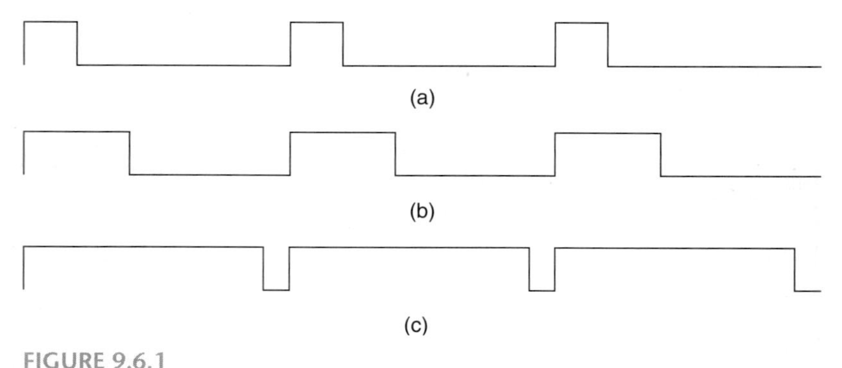

FIGURE 9.6.1
Pulse width modulated signal waveforms: (a) 20% duty cycle, (b) 40% duty cycle, (c) 90% duty cycle.

In this section, a microprocessor compatible PWM signal generator (PWMSG) design based on counter and register functions is presented. The PWMSG is designed to be interfaced to the bus of a microprocessor. The microprocessor writes a data byte to the PWMSG to specify the duty cycle. The PWMSG then generates an output signal with the specified duty cycle. The output signal's duty cycle remains constant until the microprocessor writes a new data byte to the PWMSG.

A block diagram of the system is given in Figure 9.6.2. The microprocessor's clock, `clk`, is used to drive the clock input of the PWMSG. A prescalar divides down the microprocessor clock to create a lower frequency signal, `ps`. This lower frequency signal is used as the enable to the period counter, which, when enabled, counts at the triggering edge of the microprocessor clock.

The 8-bit period counter sets the period of the PWM signal to be equal to 256 times the period of the prescalar output `ps`. The period counter's output `period_cnt` is an 8-bit value that continuously cycles from 0 to 255 in binary. The time it takes `period_cnt` to cycle from 0 through 255 and then roll over to 0 determines one period of the PWM output signal `pwm_out`.

The duty cycle subsystem contains an 8-bit register, `duty_cycle`. This register is written by the microprocessor to control the duty cycle. The microprocessor's data bus provides the data to be written at the `data` input of the duty cycle subsystem and controls the write operation using the `cs_bar` and `we_bar` inputs. The duty cycle of `pwm_out` is equal to the value written to the `duty_cycle` subsystem divided by 256. For example, to obtain a 25% duty cycle the microprocessor must write the value 64 to the `duty_cycle` register.

The comparator compares the 8-bit value `period_cnt` from the period counter with the 8-bit contents of register `duty_cycle`. If `period_cnt` is less than `duty_cycle`, then `pwm_out` is a `'1'`; otherwise, it is a `'0'`.

Listing 9.6.1 is the description corresponding to Figure 9.6.2.

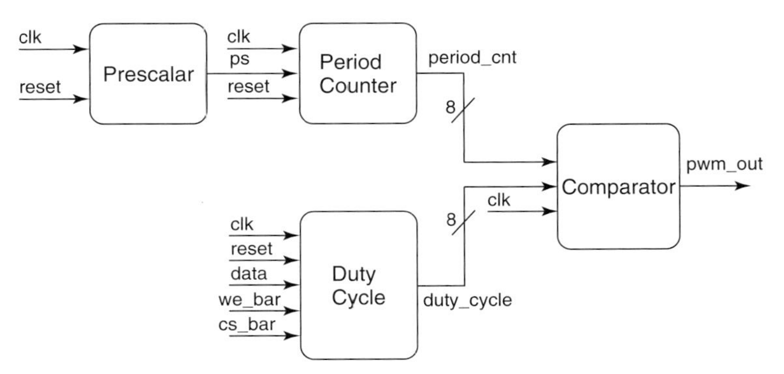

FIGURE 9.6.2
Microprocessor compatible pulse width modulated signal generator.

LISTING 9.6.1
Pulse width modulated signal generator.

```
library ieee;
use ieee.std_logic_1164.all;
use ieee.numeric_std.all;
entity pwm is
   port (
      data: in std_logic_vector (7 downto 0);
      cs_bar: in std_logic;
      we_bar: in std_logic;
      clk: in std_logic;
      reset: in std_logic;
      pwm_out: out std_logic);
end pwm;

architecture behavioral of pwm is
   signal ps : std_logic;
   signal period_cnt : std_logic_vector(7 downto 0);
   signal duty_cycle : std_logic_vector(7 downto 0);
begin

   prescalar : process (clk)
      variable pscount_v : unsigned(7 downto 0);
      variable ps_v : std_logic;
   begin
      if rising_edge(clk) then
         if reset = '1' then
            pscount_v := "10000100";
            ps_v := '0';
         else
            case pscount_v is
               when "00000001" =>
               pscount_v := "10000100";
               ps_v := '1';
               when others =>
               pscount_v := pscount_v - 1;
               ps_v := '0';
            end case;
         end if;
         ps <= ps_v;
      end if;
   end process;

   period_cntr: process (clk)
      variable count_v : unsigned(7 downto 0);
   begin
      if rising_edge(clk) then
```

```
            if reset = '1' then
               count_v := (others => '0');
            elsif ps = '1'  then
               count_v := count_v + 1;
            end if;
            period_cnt <= std_logic_vector(count_v);
         end if;
      end process;

      duty : process (clk)
      begin
         if rising_edge(clk) then
            if reset = '1' then
               duty_cycle <= (others => '0');
            elsif we_bar = '0' and cs_bar = '0'  then
               duty_cycle <= data;
            end if;
         end if;
      end process;

      comparator : process (clk)
      begin
         if rising_edge(clk) then
            if reset = '1' then
               pwm_out <= '0';
            elsif period_cnt < duty_cycle  then
               pwm_out <= '1';
            else
               pwm_out <= '0';
            end if;
         end if;
      end process;

   end behavioral;
```

Each subsystem in Figure 9.6.2 is described by a separate process. The four processes communicate via signals. Each process describes a sequential subsystem that is clocked by the microprocessor clock.

These four processes are modifications of previous designs. The prescalar process is a variation of the frequency divider in Listing 9.4.5. The primary difference between them is that the divisor value was an input in the frequency divider design entity and is a constant for this prescalar. The value of the prescalar's divisor constant is determined based on the microprocessor's clock frequency and the desired period for `pwm_out`.

The period counter process is an 8-bit up counter with an enable. It is a simplification of the 12-bit up/down counter with an enable described in Listing 9.4.4.

The duty cycle process describes a simple 8-bit register with a synchronous load capability. If `we_bar` and `cs_bar` are both asserted at a triggering clock edge, the register is loaded with the data from the microprocessor.

The comparator process describes a comparator that has a registered output. The comparator compares two 8-bit vectors and produces a `'1'` output as long as `period_cnt` is less than `duty_cycle`; otherwise, the output is `'0'`.

Code Reuse

The four processes in Listing 9.6.1, being variations of previous processes we have written, are examples of code reuse. The more code that we write, the more we find that the functionality needed for new subsystems is similar to that of some subsystem we designed in the past. The problem is keeping track of and maintaining the processes that we have previously written so they can be reused.

Instead of using separate processes for each subsystem, we can use separate design entities and connect them in a structural architecture. The advantage of separate design entities is that they can be compiled to libraries, simplifying management of the code for reuse. This approach is used in later chapters.

9.7 MEMORIES

Earlier in this chapter, we considered memory structures that were individual multibit latches or registers. These structures are one-dimensional arrays of memory elements (bits). In this section, we consider *memories* that consist of multiple multibit latches or multiple registers. Such a memory is a single storage unit that consists of a two-dimensional array of memory elements (Figure 9.7.1). This two-dimensional array of bits is equivalent to a one-dimensional array of multibit latches or registers. To be uniquely identified, each multibit latch or register has an address.

Memory Organization

Bits in a single row (multibit latch or register) of the array comprise a *word*. All the bits in a word are written or read simultaneously.

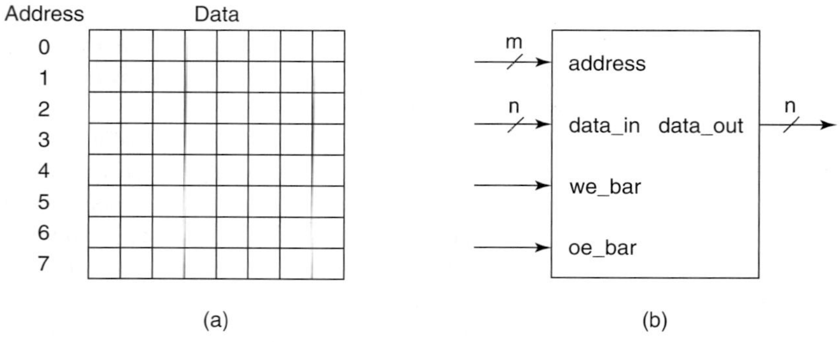

FIGURE 9.7.1
Memory: (a) conceptual representation, (b) logic representation.

The *external organization* of a memory is expressed as the number of words × the number of bits per word in the memory. This is also referred to as the length × width, or depth × width, or *aspect ratio* of the memory. It is typically written as $2^m \times n$, where m is the number of address inputs used to select a single word from the array of words and n is the number of data bits in each word. For example, a 16 × 8 memory contains 16 words each 8 bits wide. It has 4 address bits (inputs).

RAM and ROM

Technically, RAM stands for *random access memory*. A random access memory is one in which the amount of time it takes to read (access) a word from any one address (location in the memory) is essentially the same as for reading from any other address in the memory. This is in contrast to what is the case for a serial memory. However, the term RAM is commonly used to mean a memory that can be written and read during its normal operation. This is in contrast to read-only memory (ROM), which although random access, can only be read during its normal operation. A ROM's contents are defined only once; therefore, a ROM stores constant values.

Asynchronous RAM Using a Signal

A description of an 8 × 8 asynchronous RAM with separate data inputs and outputs is given in Listing 9.7.1.

LISTING 9.7.1
Asynchronous RAM using a signal.

```
library ieee;
use ieee.std_logic_1164.all;
use ieee.numeric_std.all;

entity memory is
   port(
      we_bar : in std_logic; -- write enable
      oe_bar : in std_logic; -- output enable
      data_in : in std_logic_vector(7 downto 0);
      address : in std_logic_vector(2 downto 0);
      data_out : out std_logic_vector(7 downto 0));
end memory;

architecture behavioral of memory is
type mem_array is array (0 to 2**(address'length) - 1)
      of std_logic_vector (7 downto 0);
   signal mem_s : mem_array;
begin

   write: process(we_bar, data_in, address)
   begin
      if we_bar = '0' then
         mem_s(to_integer(unsigned(address))) <= data_in;
```

(Cont.)

LISTING 9.7.1 *(Cont.)*

```
        end if;
    end process ;

    read: process (oe_bar, address, mem_s)
    begin
        if oe_bar = '0' then
            data_out <= mem_s(to_integer(unsigned(address)));
            else
            data_out <= (others => 'Z');
        end if;
    end process;

end behavioral;
```

Each word in the RAM is realized as a negative-level multibit latch. When the write enable input (`we_bar`) is asserted, the data at the data input (`data_in`) is written to the multibit latch specified by `address`. When the output enable input (`oe_bar`) is asserted, the data at the addressed location is output to the data output (`data_out`).

The key to describing a RAM is that it can be described as an array of vectors. In Listing 9.7.1 a type named `mem_array` is declared

```
type mem_array is array (0 to 2**(address'length) - 1)
   of std_logic_vector (7 downto 0);
```

which is an array of vectors. Each vector represents one word in the RAM. The array of vectors represents the entire memory. A signal `mem_s` is then declared that is type `mem_array`.

The architecture of the memory consists of two processes, one to write the memory and the other to read the memory. Since the memory elements are level sensitive, the sensitivity list of each process contains all the signals that are read in that process.

The `write` process is the code that actually causes the memory elements to be inferred, since it makes a conditional assignment to the signal `mem_s`. The element (location) in the array of vectors `mem_s` that is written is indexed (selected) by `address`.

The `read` process describes purely combinational logic, since the if statement in that process is complete. The output is driven by the contents of a memory location when the memory is being read. When the memory is not being read, the output is in its high-impedance state.

Asynchronous RAM Using a Variable

The RAM in Listing 9.7.1 is described using a signal. The data structure required for simulating a signal takes significantly more CPU memory than does the data structure for a variable. As a result, it is advantageous to use a variable to describe memory. This reduces the amount of CPU memory required for simulation and the

LISTING 9.7.2
Asynchronous memory using a variable.

```
library ieee;
use ieee.std_logic_1164.all;
use ieee.numeric_std.all;

entity memory is
   port(
      we_bar : in std_logic;-- write enable
      oe_bar : in std_logic;-- output enable
      data_in : in std_logic_vector(7 downto 0);
      address : in std_logic_vector(2 downto 0);
      data_out : out std_logic_vector(7 downto 0));
end memory;

architecture behavioral of memory is
   type mem_array is array (0 to 2**(address'length) - 1)
   of std_logic_vector (7 downto 0);
begin
   write_read: process(we_bar, oe_bar, data_in, address)
      variable mem_v : mem_array;
   begin
      if we_bar = '0' then
         mem_v(to_integer(unsigned(address))) := data_in;
      end if;

      if oe_bar = '0' then
         data_out <= mem_v(to_integer(unsigned(address)));
      else
         data_out <= (others => 'Z');
      end if;
   end process;
end behavioral;
```

simulation time, particularly for large memories. Listing 9.7.2 describes the same memory as Listing 9.7.1, except a variable is used.

Synchronous RAM

Synchronous RAM can be created by revising the write process so that it is conditioned on the occurrence of a triggering clock edge.

ROM

ROM can easily be described using either a one-dimensional array of constant vectors or using a one-dimensional array with data in a single case statement.

An example of using a constant array to define ROM was given in Listing 4.9.2, which is a table lookup description of a code converter. The hierarchical view of the logic synthesized from that description showed a ROM symbol with its address inputs

LISTING 9.7.3
ROM description using a one-dimensional array with data in a case statement.

```
library ieee;
use ieee.std_logic_1164.all;
use ieee.numeric_std.all;

entity rom is
   port(
      oe_bar : in std_logic;-- output enable
      address : in std_logic_vector(2 downto 0);
      data_out : out std_logic_vector(7 downto 0));
end rom;

architecture behavioral of rom is
begin
   read: process (oe_bar, address)
   begin
      if oe_bar = '0' then
         case address is
            when "000" => data_out <= x"83";
            when "001" => data_out <= x"a5";
            when "010" => data_out <= x"c2";
            when "011" => data_out <= x"49";
            when others => data_out <= x"00";
            end case;
         else data_out <= (others => 'Z');
      end if;
   end process;
end behavioral;
```

driven by the code to be converted and its data output providing the conversion result (Figure 4.9.1).

An example of using a one-dimensional array with data in a case statement to describe an 8×8 ROM is given in Listing 9.7.3.

The entity declaration has no `data_in` input or `we_bar` input since ROM cannot be written. If `oe_bar` is asserted, `data_out` is assigned a constant value based on the value of `address`. In this example, hexadecimal bit string literals are used to specify the constant values. The first four address locations have been assigned nonzero values and the last four have been assigned values of 0. If `oe_bar` is not asserted, `data_out` is in its high impedance state.

Block RAM in PLDs

If a design requires one or more large memories, it is usually preferable to choose a target PLD that provides on-chip *block RAM*. Some CPLDs and many FPGAs provide blocks of on-chip RAM that can be configured to provide the desired memory aspect ratios. Memory elements in these block RAMs have been optimized for their intended purpose and are much more area efficient than RAM created by a synthesizer logically combining a PLD's distributed memory elements (individual latches or flip-flops).

While some synthesizers can infer block RAM from RAM and ROM descriptions like those in this section, it is more common to instantiate each block RAM as a component. For a PLD with block RAM, the manufacturer provides parameterized RAM and ROM components in a library. When a block RAM is instantiated, generics are used to specify its aspect ratio. Alternatively, the PLD manufacturer may provide a program that uses a wizard to generate the VHDL code for a block ram component from selections made in a menu.

Block RAM on a PLD can be used to implement RAM or ROM. When implementing ROM, values for the contents of the block RAM must be specified. When implementing RAM, if values are specified for its contents, they are initial values. Block RAM contents are typically specified in a separate file in a format created for specifying memory contents, such as Intel's HEX format.

The drawback in instantiating block RAM from a PLD manufacturer's library is that the resulting design has limited portability to other target PLDs.

PROBLEMS

9.1 What is the characteristic of an interconnection of latches or flip-flops that make them form a multibit latch or register, respectively?

9.2 Rewrite Listing 8.5.2 to create a description of a positive-edge-triggered 8-bit register named `reg8_pe`. This register has an asserted high enable input (`en`) and an asserted-low clear input (`clear_bar`).

9.3 Consult the data sheet for a 74HC573 octal transparent D latch with three-state outputs. Write a description of a functionally equivalent device. Use the same names for ports in your description as the pin names used on the data sheet. If a pin name uses an overbar, use the same name with a `_bar` suffix. Specify the D inputs and Q outputs as vectors.

9.4 Consult the data sheet for a 100353 8-bit register. Write a description of a functionally equivalent device. Use the same names for ports in your description as the pin names used on the data sheet. If a pin name uses an overbar, use the same name with a `_bar` suffix. Specify the D inputs and Q outputs as vectors.

9.5 A port of an entity is declared as mode out. The value output by the port must be read in the entity's architecture to compute a new output value. How can this be accomplished without changing the port's mode?

9.6 In terms of when the output actually shifts, does the operation of the shift register in Listing 9.2.2 change if the assignment statement `qout <= q` is placed immediately after:

(a) the inner if statement?

(b) the outer if statement?

Base your analysis on how a simulator would simulate the description. Also, would a synthesizer generate different logic for the three descriptions?

9.7 Examine Listing 9.2.4. If variable q were changed to a signal would the behavior of the system remain the same? Provide an explanation to justify your answer.

9.8 Examine Listing 9.2.4. How is the operation of the system changed if the innermost if statement that describes the filter operation is interchanged with the assignment statement that describes the right shift? Explain the differences in behavior in terms of VHDL's semantics.

9.9 Examine Listing 9.2.4. How is the operation of the system changed if:

(a) the assignment statement that describes the right shift is moved up one line so that it precedes the end if statement that closes the outermost if statement?

(b) the assignment statement that describes the right shift is moved up two lines so that it follows the end if statement that closes the innermost if statement?

Explain any differences in behavior in terms of VHDL's semantics.

9.10 Write a description for an 8-bit shift register. The shift register has a serial input `ser_in` and a serial output `ser_out`. The shift register also has eight parallel inputs `din7` down to `din0`. The shift register is synchronously loaded from these inputs when its `load` input is asserted. The shift register also has a synchronous clear input `clear`. When its `shift_r` input is asserted the shift register shifts its contents one position to the right at the next positive edge at its `clk` input.

9.11 Consult the data sheet for a 74AC299 Universal Shift/Storage Register. Write a complete VHDL description of a functionally equivalent device. Use the same names for ports in your description as the pin names used on the data sheet. If a pin name uses an overbar, use the same name with a `_bar` suffix attached. Specify IO0 to IO7 as a vector.

9.12 A fully synchronous doubler system is to be designed. This system can load an initial 10-bit value, and then double the value on each positive edge of the input clock, if the system's enable input is asserted. The doubled value is available as an output from the system. If the result after doubling exceeds 10 bits, then the overflow flag is set and the result modulo 2**10 (2^{10}) is output. If an attempt is made to double the result after overflow, the output is made all 0s and the overflow flag is cleared.

The entity declaration for the doubler system is:

```
entity doubler is
   port(
      clk : in std_logic;-- clock input
      enable : in std_logic;-- enable doubling input
      load : in std_logic;-- enable loading of value to be doubled
      load_val : in std_logic_vector(9 downto 0); -- initial value to be doubled
      overflow : out std_logic;-- overflow flag
      q : out std_logic_vector(9 downto 0)-- doubled value
      );
end doubler;
```

Write a behavioral architecture for the `doubler` design entity.

9.13 Write a description of a 4-bit ring counter named `ring_cntr`. `ring_cntr` circulates a single 0 from right to left when it is enabled. Its asserted-low reset is synchronous. The architecture that you create for this design entity must be behavioral. The entity declaration is:

```
entity ring_cntr is
   port (
      rst_bar : in std_logic;     -- reset active low
      clk : in std_logic;         -- clk
      enable : in std_logic;      -- enable active high
      qout : out std_logic_vector(3 downto 0)    -- output
      );
end;
```

9.14 Write a description of a 5-bit ring counter named `ring_cntr5`. When reset, design entity `ring_cntr5` contains the pattern 00011. The contents of this ring counter are circulated one bit position to the left when it is enabled. The asserted-low reset is synchronous. You must create a behavioral style architecture. This design entity has the following entity declaration:

```
entity ring_cntr5 is
   port (
      rst_bar : in std_logic;     -- reset active low
      clk : in std_logic;         -- clk
      enable : in std_logic;      -- enable active high
      qout : out std_logic_vector(4 downto 0)    -- output
      );
end;
```

9.15 Write a description of a 4-bit up/down counter that does not overflow or underflow. When the count reaches its maximum value it remains at that value if the count mode is up. When the count reaches zero, it remains zero if the count mode is down. Use a control input named up to control the direction of the count.

9.16 Write a description of a 4-bit up counter named `by1or2`. The counter increments on a positive clock edge only if its active low enable (`cnten_bar`) is asserted. If its `by2` input is asserted, the counter increments by 2; otherwise, it increments by 1. If the counter overflows it continues to count from 00000. The counter has an active-low asynchronous reset (`rst_bar`) and an active-low synchronous clear (`clr_bar`).

9.17 A counter is to be designed that can count up or down when enabled. When counting up, the counter counts in binary from 3 to 6 and repeats. When counting down, it counts from 6 down to 3 and repeats. When reset, the counter's value is 3.

The entity declaration for the counter is:

```
entity cntr_trunc_seq is
   port(
       clk : in std_logic;           -- system clock
       cten_bar : in std_logic;      -- enable counting
       up_bar : in std_logic;        -- count direction
       rst_bar : in std_logic;       -- synchronous reset
       count : out std_logic_vector (2 downto 0)      -- output count
       );
end cntr_trunc_seq;
```

Write a context clause and an architecture body for the counter.

9.18 Write a description of a frequency divider that produces an output `fdiv` that is equal to its input clock frequency divided by a factor of either 3, 8, 12, or 16. The divisor value is selected by a 2-bit input vector named `divby`. The division factor determined by `divby` is given in the table.

divby	divisor
00	3
01	8
10	12
11	16

If `divby` is changed, the current output period is completed before the new output period takes effect.

9.19 Consult the data sheet for a 74HC191 counter. Write a design description of a functionally equivalent device.

9.20 Modify the counter in Problem 9.19 to add an input `bin` that controls whether the counting is one in binary or BCD.

9.21 You must design a BCD synchronous up/down counter having the following entity declaration:

```
entity bcd_cntr is
   port(
      clk : in std_logic;     -- clock
      clear_bar : in std_logic;  -- synchronous clear counter
      up_bar : in std_logic; -- count direction control
      cten_bar : in std_logic;  -- count enable
      max_min_bar : out std_logic;   -- 0 for maximum/minimum count
      qdcba : out std_logic_vector(3 downto 0) -- counter output
      );
end bcd_cntr;
```

Write a complete description of the counter. Use a process containing an if statement and use an integer signal to represent the count. The `max_min_bar` output is asserted when the output reaches the maximum count when counting up or reaches the minimum count when counting down.

9.22 An event detector system indicates whether an event (positive or negative edge) has occurred on an input signal. If an event has occurred just prior to the last triggering clock edge, the output is a 1 for one clock period; otherwise, it remains a 0. The entity declaration for the system is:

```
entity event_detector is
    port(
        input : in std_logic;     -- input
        clk : in std_logic;       -- system clock
        rst_bar : in std_logic; -- asynchronous reset
        event : out std_logic     -- output
        );
end event_detector;
```

Write a behavioral design description of this entity.

9.23 A system needs to carry out an action for 0.5 seconds. To do this it will use a component called `delay` to measure the 0.5 s. When a positive edge occurs at its `fire` input, component `delay` generates a positive pulse of one clock cycle duration at its `delayed_out` output 0.5 s later. Assume the system clock is 1 MHz and the system is to be fully synchronous. Write a complete description of `delay` using a behavioral architecture body. This architecture body must be as simple as possible. Use an integer signal to count out the delay.

9.24 Assuming that the microprocessor provides an 8 MHz square wave for `clk`, what is the period, frequency, and duty cycle of the local signal `ps` and what is the period of the output signal `pwm_out` for the PWM signal generator of Listing 9.6.1?

9.25 The comparator process for the PWM signal generator in Listing 9.6.1 was written so that its output is latched. Write a process description that provides an alternative purely combinational comparator subsystem. This combinational approach has simpler hardware since it does not include the output latch. Discuss the advantages and disadvantages of these two approaches in terms of various applications (the subsystem driven by `pwm_out`).

9.26 What would be the effect on the simulation of the program in Listing 9.7.1 of leaving `data_in` out of the sensitivity list of the `write` process? What would be the effect on the synthesized logic of this omission?

9.27 Modify the program in Listing 9.7.1 to include a chip select input `cs_bar`. When `cs_bar` is asserted, the memory can be either written or read. When `cs_bar` is not asserted, no operation can be carried out on the memory and its output is in its high-impedance state.

9.28 Modify the program in Listing 9.7.2 so that the memory has common data I/O; in other words, a single bidirectional data bus. In normal operation it would be assumed that both `oe_bar` and `we_bar` are not simultaneously asserted. However, write your description so that if this were the case, the write operation would predominate and RAM would not drive the data bus during this time.

9.29 Modify the program in Listing 9.7.2 to produce RAM that is written synchronously on a positive clock edge, but is read asynchronously. This requires the addition of a clock signal, `clk`.

9.30 RAM can be both written and read synchronously. Modify the program in Listing 9.7.2 to produce RAM that is both written and read synchronously on a positive clock edge.

9.31 Rewrite the ROM description in Listing 9.7.3 to use a variable to hold the ROM data.

Chapter 10
Finite State Machines

A sequential system is also known as a finite state machine (FSM). The FSM approach to sequential system design is a straightforward generic approach that allows any sequential system to be designed. In theory, a sequential system of any complexity can be designed as a single FSM. In practice, complex systems are decomposed into components, where one or more are FSMs.

A FSM's operation may be graphically represented by a state diagram. Development of a state diagram from a FSM's specification is the first step in the FSM's design.

After a brief review of FSM concepts and state diagrams, the VHDL description of both Mealy and Moore type FSMs using a three-process approach is presented. One and two process approaches are also discussed.

During the development of a state diagram, states are usually represented by descriptive names. If so, these same names are used in the FSM's VHDL description. During synthesis, each state name must be replaced by a binary code. The choice of binary values to encode the states can be specified by the designer or left to the synthesizer. The effect of this choice on the complexity of the synthesized logic is also considered.

10.1 FINITE STATE MACHINES

A *sequential system* is a *finite state machine (FSM)*. It is a digital system whose outputs are a function of both its present input values and its past history of input values.

State of a FSM

Information characterizing the effect of a FSM's past history of input values is provided by its *state*. Physically, a FSM's state is represented by the value stored in a register called the *state register*. This value is referred to as the *present state* of the system or simply the *state* of the system.

Synchronous versus Asynchronous FSM

A sequential system may be either asynchronous or synchronous. An *asynchronous sequential system* has no clock and its state can change value as soon as any of its inputs changes value. In contrast, a *synchronous sequential system* is controlled by a system clock and its state can change value only at a triggering clock edge. In a synchronous system, asynchronous actions are limited to the external reset used to force the synchronous system into an initial state at power on.

In theory asynchronous systems can operate faster and use less power. However, they are significantly more difficult to design and their successful operation is very dependent on the specific delays of their components. Furthermore, PLD architectures are not appropriate for implementing asynchronous systems.

Fully Synchronous FSM

The majority of sequential systems are synchronous and our interest is further limited to fully synchronous systems. In a *fully synchronous FSM* all registers are clocked by a single clock signal. Synthesis tools perform well in synthesizing fully synchronous designs. Thus, in this book, when the term FSM is used without elaboration, a fully synchronous FSM is assumed.

Inputs

A FSM's inputs are sampled at each triggering clock edge. Accordingly, a FSM's input sequence is ordered in time by its clock. Figure 10.1.1 is a block diagram of a

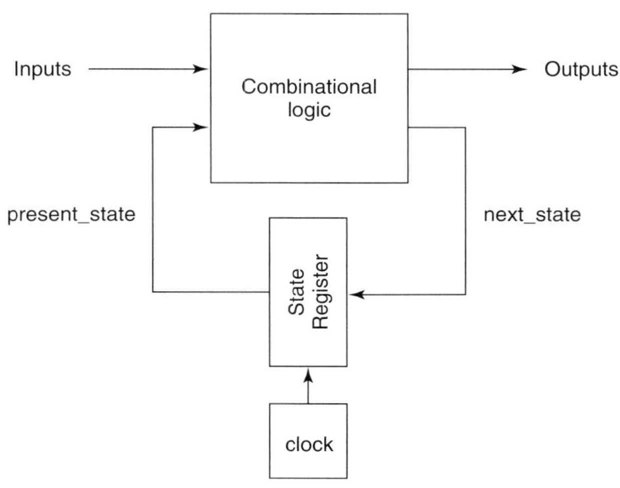

FIGURE 10.1.1
Fully synchronous finite state machine (FSM) block diagram.

fully synchronous FSM. The block diagram separates the FSM into a state register and a combinational logic block.

Present State

A FSM's *present state* provides all the information needed about its past input history to determine its current output and next state. The present state is stored as a binary value in the *state register*.

The number of memory elements in the state register is finite. If the state register has n memory elements, the maximum number of unique states the FSM can have is 2^n. This number is also finite, hence the name finite state machine.

In theory, the state register could be comprised of any type of clocked memory element. The simplest type to use is a D flip-flop. If the state register is comprised of D flip-flops, the input to the state register is the next-state value and its output is the present-state value.

For ease of design, a register is used, rather than a multibit latch, to hold the present state. A multibit latch's transparency would require precise control of the clock pulse width to prevent a change in state, when the clock is asserted, from feeding through the combinational logic and causing one or more additional changes in state before the clock is unasserted.

Outputs

A FSM's outputs are computed by combinational logic that receives two sets of inputs. One set is the external inputs to the FSM. The other set is the FSM's present state. The state register's outputs provide the present state value to the combinational logic. It is clear from Figure 10.1.1 that the outputs of the FSM depicted are a function of both its inputs and present state.

Next State

The combinational logic also computes the *next state* value from the FSM's inputs and present state. This value is provided to the inputs of the state register. At each triggering clock edge, the next state value is clocked into the state register and becomes the new present state value. While the next state signal is a feedback signal, it is not combinational feedback, because the feedback loop is broken by the state register.

Output Logic and Next State Logic

Figure 10.1.2 separates the combinational logic of the FSM in Figure 10.1.1 into two independent subsystems. One computes the FSM's outputs from the inputs and present state. The other computes the FSM's next state from the inputs and present state.

From Figure 10.1.2 it is clear that, when considered separately, we already know how to design the components of an FSM using VHDL, since we know how to design combinational systems and we know how to design registers, but, of course, there are additional considerations that are covered in this chapter.

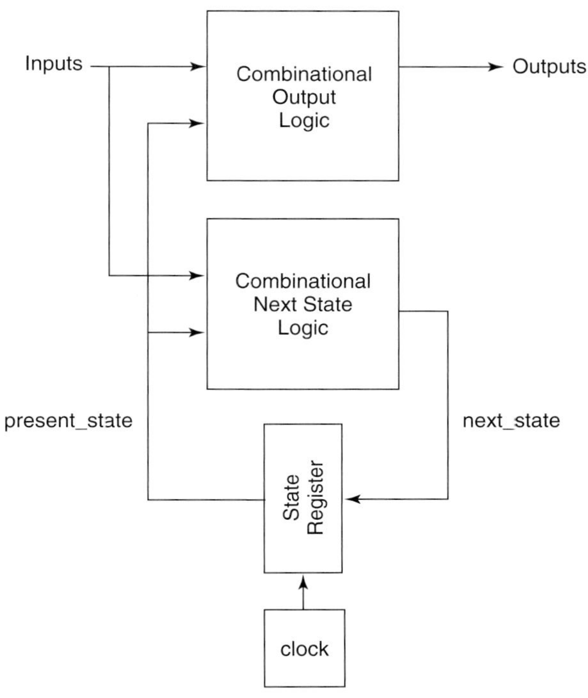

FIGURE 10.1.2
Separation of output combinational logic from next state combinational logic.

FSM Clock

A FSM's clock is periodic. It provides a synchronizing train of pulses to the state register's clock input. Accordingly, the state value can change only at a triggering clock edge.

Synchronous Inputs Assumption

For proper operation, a FSM's inputs must be stable for a sufficient time before and after each triggering clock edge to meet the setup and hold timing requirements for the state register's flip-flops. For now, we assume that input changes are synchronized to the system clock, so these timing requirements are always met. The effect of removal of this input synchronization assumption is discussed later.

State Time

Figure 10.1.3 shows a single cycle of a FSM's clock. The clock's cycle defines a *state time* of one clock period duration. At the triggering edge of the clock the next state value is stored in the state register, becoming the new present state. In a physical circuit there is a delay after the triggering clock edge before the new present state value appears at the output of the state register.

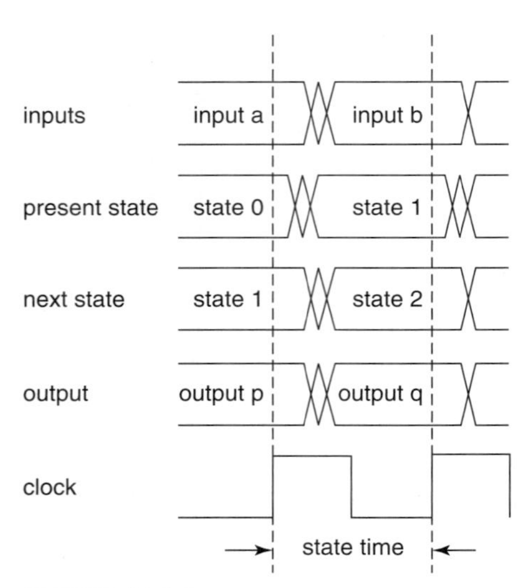

FIGURE 10.1.3
State time as defined by a FSM's clock.

Since a FSM's outputs and next state are a function of its present state, they will be unstable for a short time after each triggering clock edge. This is caused by propagation delays through the flip-flops and the combinational logic that computes these values.

The present state, outputs, and next state are defined only during that portion of the clock cycle where these signals are stable.

FSM Operation

In operation, a FSM transitions from one state to another at each triggering clock edge, as determined by its present state and input values just prior to the triggering clock edge (Figure 10.1.4). For example, assume that letters a, b, and c represent different FSM binary input values and letters p, q, and r represent different binary output values.

When the FSM is in state 0 and its input is a, its output, computed by the output logic, is p. In state 0, the contents of the state register is the bit pattern representing state 0. The next state logic computes a next state value of state 1, and the bit pattern for that state appears at the inputs to the state register.

The triggering clock edge in this example is the positive edge. At the next triggering clock edge, the state register stores state 1 and a new input value of b appears at the input.

Because of signal propagation delays in a physical circuit, immediately after a triggering clock edge there is a period of time when the new present state is unstable (clock to output time of the flip-flops in the state register). During this time and the propagation delay time of the combinational logic, the new output and new next state values are also unstable. After these signal transitions have subsided, the system has the new output value q and new next state value, state 2.

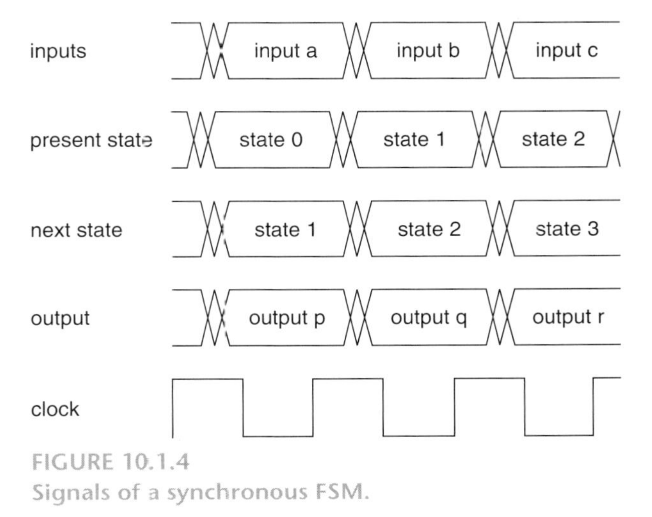

FIGURE 10.1.4
Signals of a synchronous FSM.

Combinational logic propagation delays do not affect the proper operation of a synchronous FSM, as long as the setup time and hold time requirements of the state register are met.

Mealy FSMs

FSMs can be classified as either Mealy or Moore. These two kinds of FSMs differ in how their outputs are computed. The FSM in Figure 10.1.2 is a Mealy FSM. A *Mealy FSM*'s outputs are a function of both its inputs and its present state. Because of this, the outputs of a Mealy FSM can change as soon as any of its inputs change. However, its state still cannot change until a triggering clock edge.

Moore FSMs

A *Moore FSM*'s outputs are a function of only its present state. The block diagram of the Mealy FSM in Figure 10.1.2 is modified to represent a Moore FSM by removing all external inputs to the output logic, Figure 10.1.5. It is clear from this diagram that since a Moore FSM's outputs are a function of only its present state, they can change only at a triggering clock edge.

Mealy and Moore Functional Equivalence

In general, any sequential function can be implemented by either a Mealy FSM or a Moore FSM. However, while a Moore FSM is often conceptually simpler, it usually requires more states than a Mealy FSM to implement the same function. Because a Mealy FSM's outputs are a function of its inputs and state, an output can have different values in a single state.

A FSM does not have to be exclusively Mealy or Moore. It can be a combination of both, having some outputs that depend on the present state and present inputs (*Mealy outputs*) and other outputs that depend on only the present state (*Moore outputs*).

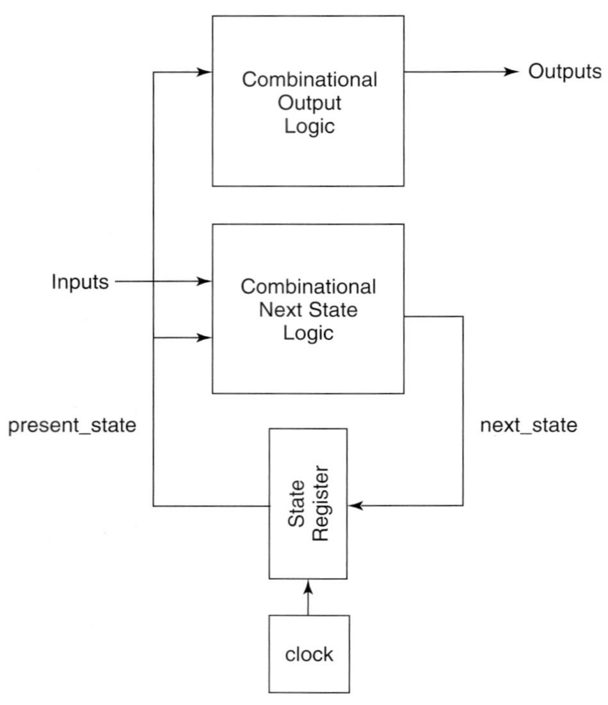

FIGURE 10.1.5
A Moore FSM's output is a function of only its present state.

10.2 FSM STATE DIAGRAMS

The design of a FSM might be the ultimate objective of a particular effort. More often, a FSM is designed as a component used to perform control functions in a larger system. In either case, the FSM design usually starts with the development of a state diagram or, alternatively, an ASM chart. ASM charts are discussed in Chapter 11.

State Diagram

A *state diagram* provides an abstract graphical representation of the operation of a FSM. It allows the conceptualization of the FSM's operation to be separated from its implementation. Each individual state of the FSM is represented by a *state circle*, with the state's name or its encoding located inside.

Example state diagrams for a Mealy and a Moore FSM are given in Figure 10.2.1. The Mealy FSM has two states, A and B, and the Moore FSM has three states, A, B, and C.

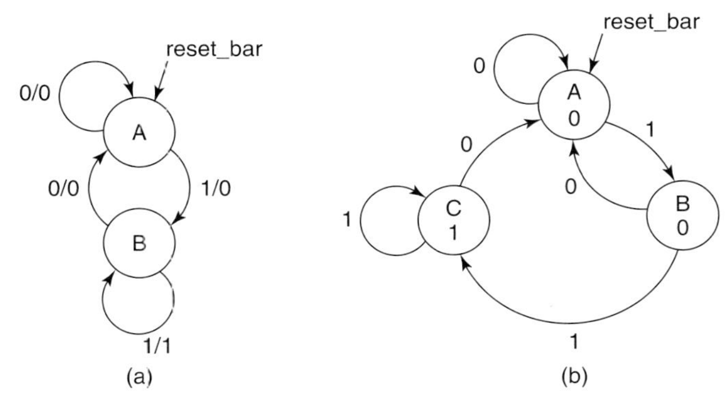

FIGURE 10.2.1
State diagrams of a (a) Mealy machine and (b) Moore machine.

Directed arcs show the possible transitions from one state to another and under what input conditions each transition will occur. Since a FSM can be in only one state at a time, from a given present state only one of the transitions depicted occurs at the next triggering clock edge. The FSM's clock is not explicitly shown on a state diagram.

A Completely Specified State Diagram

For a state diagram to be *completely specified*, for every state the output and next state transition must be shown for all possible input combinations.

The initial or reset state has a directed arc showing a transition to the state from nowhere; for example, state A in Figure 10.2.1.

Unconditional State Transitions

In some FSMs a transition from one particular state to another may be unconditional. The directed arc for such a transition will not be labeled with an input value.

Mealy State Diagram

A state diagram for a Mealy FSM has each directed arc labeled with an input/output value pair (Figure 10.2.1(a)). This value pair indicates the FSM's output when it is in the state from which the arc emanates and has the specified input value. This clearly indicates that the FSM's outputs are a function of its inputs and present state.

Moore State Diagram

Since a Moore FSM's outputs are a function of only its present state, its output values are associated with the state circles, rather than the directed arcs. These output values are either written inside the state circle below the state name (Figure 10.2.1(b)), or in a box attached to the state circle. Each directed arc is labeled with the input values that cause the transition to the next state.

Table 10.2.1
State/output table for FSM in Figure 10.2.1(a).

Present State	Input	Next State	Output
A	0	A	0
A	1	B	0
B	0	A	0
B	1	B	1

Reset State

A state diagram must have a reset state. This is the state in which the FSM is placed in response to the assertion of its reset signal. The reset signal is asserted at power on and places the FSM in a known initial state. If the FSM is reset at a later time in its operation, it will transition from whatever state it is in to its reset state. For simplicity, directed arcs representing transitions from every state to the reset state, in response to assertion of the reset input, are omitted from state diagrams.

Creating a State Diagram

Creating a state diagram is sometimes a mostly heuristic process. In such cases, there is no simple algorithm to follow. Detailed examples of this process are given later in this chapter. The efficiency of a FSM is a function of the efficiency of the state diagram used to represent it.

State Diagram Editor

A *state diagram editor* is an EDA software tool designed for graphically creating and editing state diagrams. A state diagram editor with VHDL output can automatically generate VHDL code corresponding to a state diagram. The form of the automatically created code may not necessarily follow that of the three-process FSM template discussed in the next section.

The primary drawback to the use of a state diagram editor is that the diagram itself is not portable. However, the VHDL code that it generates is portable.

State/Output Table

The same information presented in a state diagram can be presented in a table. A *state/output table* for the Mealy FSM in Figure 10.2.1(a) is given in Table 10.2.1.

10.3 THREE-PROCESS FSM VHDL TEMPLATE

Once a FSM's state diagram has been created, its VHDL code can be written. There are many ways a FSM can be described in VHDL. One straightforward way is to code each block in Figure 10.1.2 (excluding the clock) as a separate process. The result is a description consisting of three processes. One describes the state register. The other two describe the output combinational logic and the next state combinational logic.

A template for a three-process FSM is given in Listing 10.3.1.

LISTING 10.3.1
Three-process template for a general finite state machine.

```
library ieee;
use ieee.std_logic_1164.all;

entity fsm_template is
   <port_declarations>
end fsm_template;


architecture three_processes of fsm_template is
type state is (enumeration_of_state_names);
signal present_state, next_state : state;
begin

state_reg: process (clk)
begin
   if rising_edge(clk) then
      if reset_bar = '0' then
         present_state <= <initial_state>;
      else
         present_state <= next_state;
      end if;
   end if;
end process;

outputs: process (present_state, <inputs>);
begin
   case present_state is
      -- one case branch required for each state
      when <state_value_i> =>
         if <input_condition_1> then
            -- assignments to outputs
         elsif <input_condition_2> then
            -- assignments to outputs
         else
            -- assignments to outputs;
         end if;

         ...

      -- default branch
      when others =>
            -- assignments to outputs

   end case;
end process;
```

(Cont.)

LISTING 10.3.1 *(Cont.)*

```
nxt_state: process (present_state, <inputs>);
begin
   case present_state is

      -- one case branch required for each state
      when <state_value_i> =>
         if <input_condition_1> then
            -- assignment to next_state
         elsif <input_condition_2> then
            -- assignment to next_state
         else
            -- assignment to next_state
         end if;

         ...

      -- default branch
      when others =>
         -- assign initial_state to next_state

   end case;
end process;

end three_processes;
```

Enumerated States

In this template, states of the FSM are enumerated. In the declarative part of the architecture, type `state` is declared as an enumeration of state names. Two signals, `present_state` and `next_state`, are then declared of type `state`. When the FSM is synthesized, each enumerated state is converted to a unique binary code.

State Register Process

In the statement part of the architecture are the three processes. The first, labeled `state_reg`, describes the state register. The sensitivity list for this process includes only the clock signal. All FSMs must have a reset signal. The reset may be either asynchronous or synchronous. If it is asynchronous, the reset signal must also be included in the sensitivity list. In this template, the reset is synchronous and, therefore, is not included in the sensitivity list.

In response to any event on input `clk` the process `state_reg` is executed. If the event on `clk` is anything other than a rising edge, the execution of the process has no effect on the `present_state`.

If the event on `clk` is a rising edge and if `reset_bar` is `'0'`, state register `present_state` is assigned the value of `initial_state`. If the event on `clk` is a rising edge and `reset_bar` is not `'0'`, `present_state` is assigned the value of `next_state`. This later action corresponds to the FSM transitioning from one state to the next at a triggering clock edge.

Output Process

The second process, labeled `outputs`, describes the combinational subsystem that generates the FSM's outputs from its inputs and present state. Accordingly, for a Mealy FSM, the sensitivity list for this process includes `present_state` and all of the FSM's inputs. For a Moore FSM, the sensitivity list includes only `present_state`.

Process `outputs` consists of a case statement. The case expression is `present_state`. The case statement has a branch for each state in the FSM's state diagram (each value of `present_state`).

In each branch of the case statement there is an if statement. Each branch within a particular if statement has a Boolean condition involving the FSM's inputs and assigns a particular set of values to the outputs. The last branch of each if statement is an else branch that assigns default output values for all input conditions not previously specified.

The output process template in Listing 10.3.1 is appropriate for a Mealy FSM. For a Moore FSM, this template is simplified by removing the if statements from the branches of the case statement. This simplification is warranted because, for a Moore FSM, the outputs are not a function of the inputs, only of the state. Accordingly, each branch of the case statement simply contains assignments to the outputs appropriate for the present state represented by that branch.

The When Others Branch

The final branch of the case statement is a when others branch. In this template, the case expression `present_state` is an enumerated type. If all of the members of this type are not explicitly specified as choices in the previous branches, then the when others branch is required for syntactical correctness. If all of the members of this type are specified as choices in the previous branches, then the when others branch is not required for syntactical correctness, but its inclusion will not cause an error.

Later we will use variations of this template where `present_state` is a std_logic_vector and the when others branch is required for the case statement to be syntactically correct.

The effect of the when others branch on synthesis, when the state encoding has binary combinations that correspond to unused encoding values, is discussed later in Section 10.7.

Next State Process

The third process is the next state process, labeled `nxt_state` in Listing 10.3.1. This process is similar to the output process in Listing 10.3.1, except that each if statement assigns a state value to the signal `next_state`. The state value assigned to `next_state` becomes the new value of `present_state` at the next triggering clock edge, as described by the `state_reg` process.

Example FSMs in this chapter are coded using the template in Listing 10.3.1.

Two- and One-Process FSM Templates

The three-process FSM template has the advantage that it is a one-to-one conceptual mapping to the FSM block diagram in Figure 10.1.2. It also has the advantage, from the coding perspective, that it is very readable and easy to modify and maintain. However, it is certainly not the only template that we can use for an FSM.

One common variation is to use two processes and a set of conditional signal assignment statements. In this approach, the state register and next state logic are implemented using processes as before. The output logic is implemented by the set of conditional signal assignment statements using one statement for each output. The conditions in the signal assignment statement are expressions involving states and, in the case of Mealy FSMs, inputs.

Another FSM template variation uses one process to describe both the state register and next state logic, and uses conditional signal assignment statements for the output logic. This template is often used by state diagram editors that automatically generate VHDL code for an FSM from a graphical representation of its state diagram.

Finally, the entire FSM can be coded as a single process. This is the least readable of the approaches.

Regardless of the approach used, if the same FSM is described, the resulting synthesized logic is the same.

10.4 STATE DIAGRAM DEVELOPMENT

A state diagram can be considered as representing an algorithm. However, at the start of the development of a state diagram the algorithm is often unknown and the state diagram must be developed in a heuristic fashion.

Each state in a state diagram is given a name. Sometimes abstract names are chosen, particularly in the early development of a state diagram. More practically, states are named to reflect the circumstances of the system when it is in each state.

The three-process FSM template in Listing 10.3.1 uses enumerated states. A type is declared that is an enumeration of all of the state names. Signals of this type can then be declared. This is a very powerful technique for the abstract modeling of FSMs.

When it is time to synthesize a FSM description that uses enumerated states, each state's name must be replaced with a unique binary code. We can specify this state assignment or leave it to the synthesizer to make a default assignment.

Moore FSM Positive Edge Detector

Consider the positive edge detector from Section 9.5. It was designed not as a FSM, but instead by conceiving a logic circuit that would perform as desired. Then its VHDL description is written.

In this section, we design a functionally equivalent positive edge detector, first as a Moore FSM and then as a Mealy FSM. Both designs use enumerated states.

Positive Edge Detector State Diagram

A state diagram for a Moore FSM edge detector is shown in Figure 10.4.1. The thought process followed in its development is straightforward. First we need a state for the FSM to be placed in when it is reset by asserting `clr_bar`. This state was abstractly named `state_a`. Since this is a Moore FSM, the output is a function of only the present state. When `clr_bar` is asserted, output `a_pe` is to be a 0. This value for `a_pe` appears below the state name in the state diagram.

Recall that the purpose of the positive edge detector is to detect a 0 to 1 transition on input `a`. There are two conditions that must be met to detect a 0 to 1 transition on `a`. First, `a` must be 0 at a triggering clock edge. Second, `a` must be 1 at the next triggering clock edge.

So, in `state_a`, if `a` is 1 at the next triggering clock edge, nothing of interest has occurred and the FSM transitions back to `state_a`. In effect, the FSM remains in `state_a`. However, if `a` is 0 at the next triggering clock edge, we have met the first condition. The FSM must denote this event by transitioning to a new state. We can name this state `state_b`. Since we have not met all of the conditions for a positive edge on `a`, output `a_pe` is 0 in `state_b`.

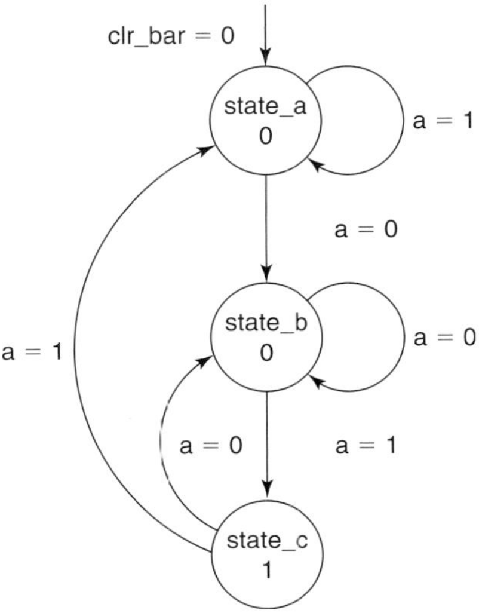

FIGURE 10.4.1
Moore FSM state diagram for a positive edge detector.

In state_b, if a is 0 at the next triggering clock edge, nothing of significance has occurred while in this state, so the transition is back to state_b. However, if a is a 1 at the next triggering clock edge, the FSM has detected a positive edge. This event is denoted by transitioning to a new state, state_c. In state_c, output a_pe is 1.

Since the output is to be 1 for only one clock period after a positive edge on a has been detected, we must transition out of state_c at the next triggering clock edge, regardless of the value of a. If a is 0 at the next triggering clock edge, we have met the first condition for a new positive edge on a, so the transition is to state_b. If instead, a is 1 at the next triggering clock edge, we must wait for it to become 0 before we can detect the next positive edge, so the transition is to state_a.

The state diagram is now complete. The resulting Moore FSM has three states, represented abstractly as state_a, state_b, and state_c. The output for each state is labeled inside the state circle below the state name. It is clear from the state diagram that the output is 1 only in state_c.

At this point, we might conclude that more descriptive, but longer, names for the states might have been waiting_for_0, waiting_for_1, and pos_edge_detected.

From the state diagram, a description of the edge detector can be written. This description, which follows the template in Listing 10.3.1, is given in Listing 10.4.1.

LISTING 10.4.1

Moore FSM description of positive edge detector using enumerated states.

```
library ieee;
use ieee.std_logic_1164.all;

entity pedgefsm is
   port(
      a : in std_logic;
      clr_bar : in std_logic;
      clk : in std_logic;
      a_pe : out std_logic
      );
end pedgefsm;

architecture moore_fsm of pedgefsm is
   type state is (state_a, state_b, state_c);
   signal present_state, next_state : state;
begin

   state_reg: process (clk, clr_bar)
   begin
      if clr_bar = '0' then
```

```
         present_state <= state_a;
      elsif rising_edge(clk) then
         present_state <= next_state;
      end if;
   end process;

   outputs: process (present_state)
   begin
      case present_state is
         when state_c => a_pe <= '1';
         when others => a_pe <= '0';
      end case;
   end process;

   nxt_state: process (present_state, a)
   begin
      case present_state is
         when state_a =>
         if a = '0' then
            next_state <= state_b;
         else
            next_state <= state_a;
         end if;

         when state_b =>
         if a = '1' then
            next_state <= state_c;
         else
            next_state <= state_b;
         end if;

         when others =>
         if a = '0' then
            next_state <= state_b;
         else
            next_state <= state_a;
         end if;
      end case;
   end process;
end moore_fsm;
```

In the declarative part of the architecture, type `state` is declared as an enumeration type consisting of `state_a`, `state_b`, and `state_c`. Signals `present_state` and `next_state` are then declared as being of type `state`.

In the statement part of the architecture body, the first process is the state register process, `state_reg`. If `clr_bar` is asserted, `present_state` is assigned the enumeration value `state_a`. Otherwise, at a triggering clock edge, the `next_state` value becomes the `present_state` value.

The second process is the output process. Since a Moore FSM's output is a function of only its present state, this process has only `present_state` in its sensitivity list. A case statement is used to branch based on the value of `present_state`. When `present_state` is `state_c`, output `a_pe` is assigned `'1'`. The next branch is for all other states. For these states, `a_pe` is assigned `'0'`.

The third process is the next state process. This process determines `next_state` from the `present_state` and `a`. Both of these signals are included in the sensitivity list. One of the branches of the case statement is selected based on `present_state`. Within each case branch an if statement assigns a value to `next_state` based on the value of `a`.

Verifying That a FSM Description Matches a State Diagram

Software tools, such as Aldec's *Code2Graphics*, can convert VHDL code to graphics. Use of such a tool on a description of a FSM is very helpful in verifying the description. For example, we can develop a state diagram by hand. Then, from the hand-drawn state diagram, we can write a three-process VHDL description of the FSM. We can then use the Code2Graphics tool to automatically generate a state diagram from the VHDL description. The automatically drawn state diagram can then be compared to our hand-drawn state diagram to verify the correctness of our VHDL description.

Explicit Version of Moore FSM Edge Detector

For conciseness, the description in Listing 10.4.1 uses a when others branch in the case statements, in both the output and next state processes, to cover those states not explicitly listed. In output process `outputs`, the when others branch covers `state_a` and `state_b`. In the next state process `nxt_state`, the when others branch covers `state_c`.

While Listing 10.4.1 is syntactically correct and will synthesize, it has a drawback. A code to graphics conversion tool may not be able to covert such a description correctly. Outputs of states covered by the when others branch in the output process and state transitions from states covered by the when others branch in the next state process may be missing from the automatically generated state diagram, because they did not appear explicitly in the case statements.

For example, when Listing 10.4.1 is converted to a state diagram, Code2Graphics generates the state diagram in Figure 10.4.2, which is not complete. The outputs associated with `state_a` and `state_b` are missing, and the transition from `state_c` is missing.

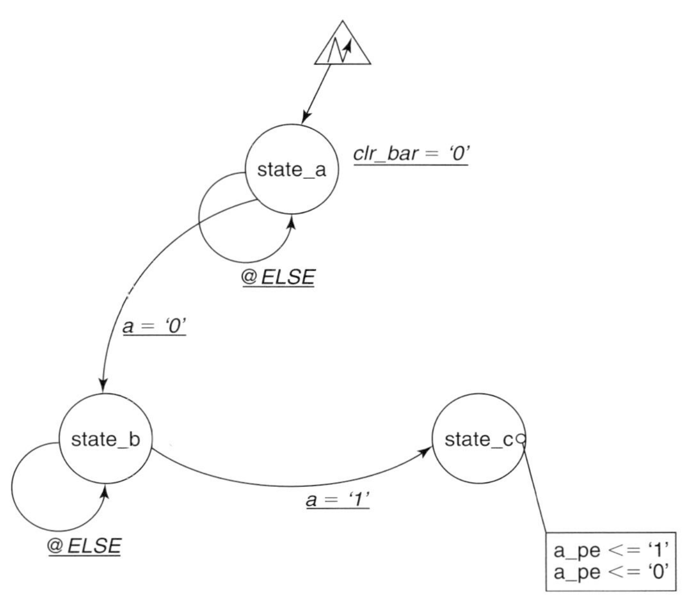

FIGURE 10.4.2
State diagram from Code2Graphics conversion in Listing 10.4.1.

To obtain the complete state diagram using Code2Graphics conversion, the description must indicate each state explicitly in the branches of the case statements. The description in Listing 10.4.2 does this. Recall that, when enumerated states are used, a when others branch is not required at the end of the case statement if each value of the enumeration is listed explicitly.

LISTING 10.4.2
Moore FSM description of positive edge detector with all enumerated states explicitly listed in case statements.

```
library ieee;
use ieee.std_logic_1164.all;

entity pedgefsm is
   port(
      a : in std_logic;
      clr_bar : in std_logic;
      clk : in std_logic;
```

(Cont.)

LISTING 10.4.2 *(Cont.)*

```
      a_pe : out std_logic
      );
end pedgefsm;

architecture moore_fsm of pedgefsm is
   type state is (state_a, state_b, state_c);
   signal present_state, next_state : state;
begin

   state_reg: process (clk, clr_bar)
   begin
      if clr_bar = '0' then
         present_state <= state_a;
      elsif rising_edge(clk) then
         present_state <= next_state;
      end if;
   end process;

   outputs: process (present_state)
   begin
      case present_state is
         when state_c => a_pe <= '1';
         when state_a | state_b => a_pe <= '0';
      end case;
   end process;

   nxt_state: process (present_state, a)
   begin
      case present_state is
         when state_a =>
            if a = '0' then
               next_state <= state_b;
            else
               next_state <= state_a;
            end if;
         when state_b =>
            if a = '1' then
               next_state <= state_c;
            else
               next_state <= state_b;
            end if;
         when state_c =>
            if a = '0' then
               next_state <= state_b;
            else
               next_state <= state_a;
```

```
            end if;
         end case;
      end process;
end moore_fsm;
```

The state diagram resulting from converting Listing 10.4.2 is given in Figure 10.4.3.

Mealy FSM Edge Detector

The edge detector can, alternatively, be implemented as a Mealy FSM. The Mealy FSM state diagram is given in Figure 10.4.4. It requires one less state than the Moore FSM.

A three-process description of the Mealy FSM edge detector is given in Listing 10.4.3.

The Mealy output process includes an if statement in each branch of the case. These if statements assign the output for that state based on the value of input a.

Output Changes in a Mealy FSM

In our Moore FSM, a_pe is a 1 for the clock cycle immediately after a positive edge on a is detected. This is true because in a Moore FSM the output is a function of only

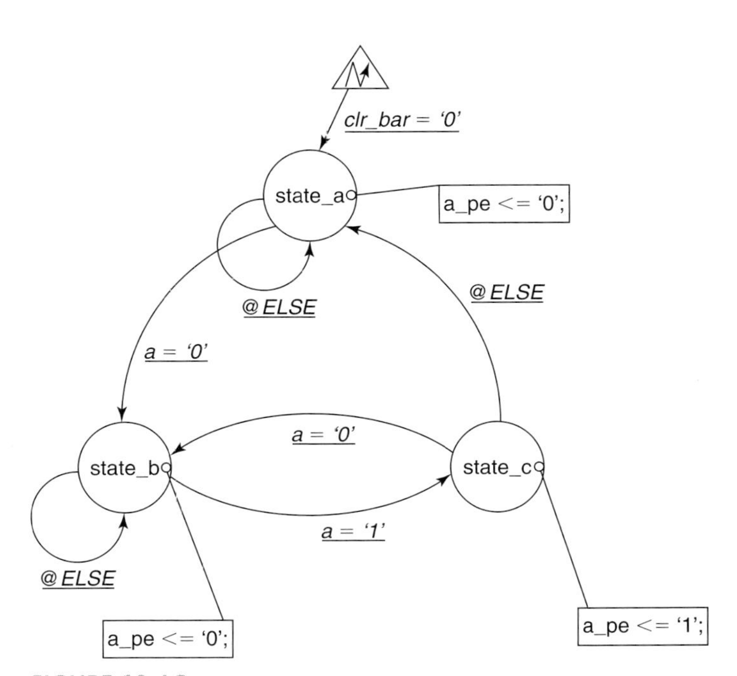

FIGURE 10.4.3
State diagram from Code2Graphics conversion in Listing 10.4.2.

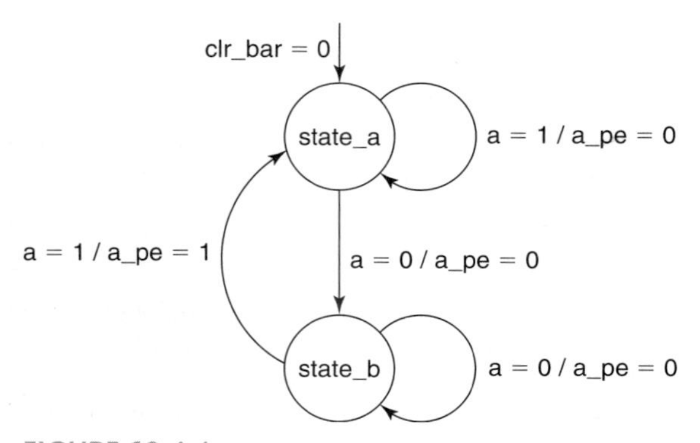

FIGURE 10.4.4
Mealy FSM state diagram for a positive edge detector.

the state, and our FSM remains in `state_c` for one clock cycle after a positive edge on `a` is detected (Figure 10.4.5(a)). In the waveforms in Figure 10.4.5, some state names have been edited to a single letter where the full state name will not fit in the scaled waveform.

LISTING 10.4.3
Mealy FSM description of edge detector using enumerated states.

```
library ieee;
use ieee.std_logic_1164.all;

entity pedgefsm is
   port(
      a : in std_logic;
      clr_bar : in std_logic;
      clk : in std_logic;
      aposedge : out std_logic
      );
end pedgefsm;

architecture mealy_fsm of pedgefsm is
   type state is (state_a, state_b);
   signal present_state, next_state : state;
begin

   state_reg: process (clk, clr_bar)
   begin
      if clr_bar = '0' then
         present_state <= state_a;
      elsif rising_edge(clk) then
         present_state <= next_state;
```

```
      end if;
   end process;

   outputs: process (present_state, a)
   begin
      case present_state is
         when state_a =>
         aposedge <= '0';

         when state_b =>
         if a = '1' then
            aposedge <= '1';
         else
            aposedge <= '0';
         end if;
      end case;
   end process;

   nxt_state: process (present_state, a)
   begin
      case present_state is
         when state_a =>
         if a = '0' then
            next_state <= state_b;
         else
            next_state <= state_a;
         end if;

         when state_b =>
         if a = '1' then
            next_state <= state_a;
         else
            next_state <= state_b;
         end if;
      end case;
   end process;
end mealy_fsm;
```

The waveforms in Figure 10.4.5(b) are for the Mealy FSM. Input a is the same for both Figures 10.4.5(a) and (b). The input shown is asynchronous, since it changes at times other than at a triggering clock edge. Three positive edges occur on input a in the waveform. However, output a_pe is asserted only twice in the waveforms for the Moore FSM (Figure 10.4.5(a)). Since the output of a Moore FSM can only change in response to a state change and its state can only change at a triggering clock edge, the positive edge of a at 750 ns was not detected because it was part of a pulse on a that occurred between triggering clock edges.

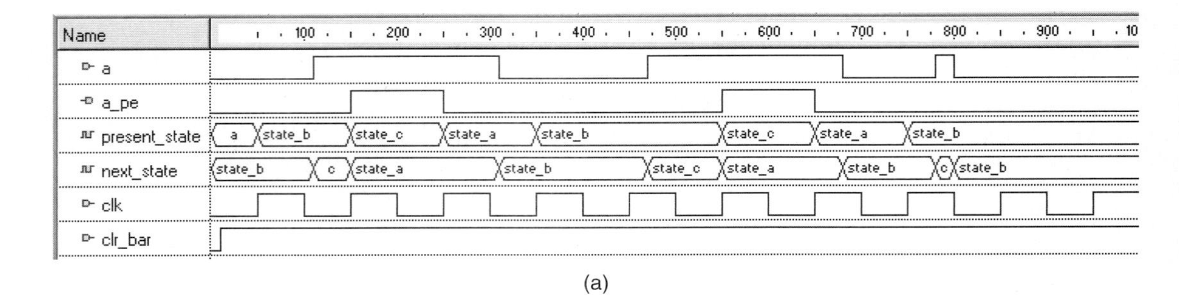

(a)

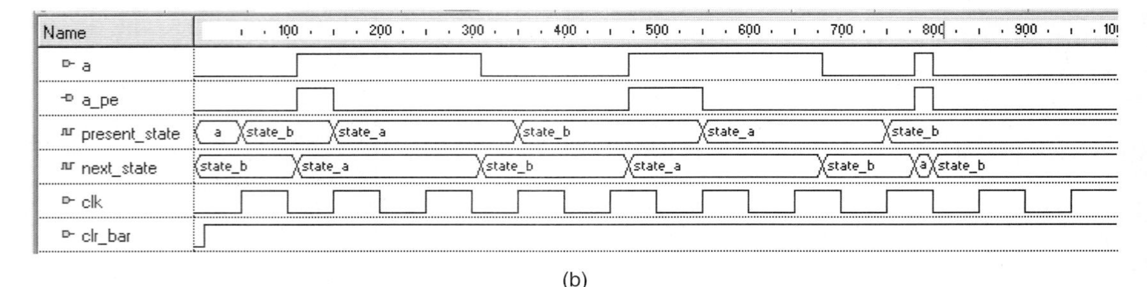

(b)

FIGURE 10.4.5
Waveforms for positive edge detectors: (a) Moore FSM, (b) Mealy FSM.

In our Mealy FSM, a_pe is a 1 if we are in state_b and a is a 1. While we are in state_b, if a changes value between triggering clock edges, then output a_pe changes value. In Figure 10.4.5(b) the positive edge on a at 750 ns is detected, because output a_pe can change at any time a changes.

Since a is an asynchronous input, output a_pe can be 1 for less than a clock period in the Mealy FSM. If input a were synchronous (synchronized to clk), then its value could not change between triggering clock edges and a_pe, when asserted, would be 1 for a full clock period. Thus, if a were a synchronous input, the outputs of the two FSMs would be identical.

Registered Outputs

While in a Mealy FSM input signal changes can be passed through to the output independently of the clock, in some applications we don't want outputs to be asserted by input signal changes that occur between triggering clock edges. Such input changes could occur if the inputs to the FSM are asynchronous or if the system providing the inputs has hazards. One way to eliminate these erroneous output values between triggering clock edges is to synchronize the output to the clock. This can be done by storing the output in a register. Figure 10.4.6 shows the addition of an output register to a Mealy FSM to synchronize its outputs.

Since the clock for the output register is the same as that for the state register, the registered output is delayed by one clock period.

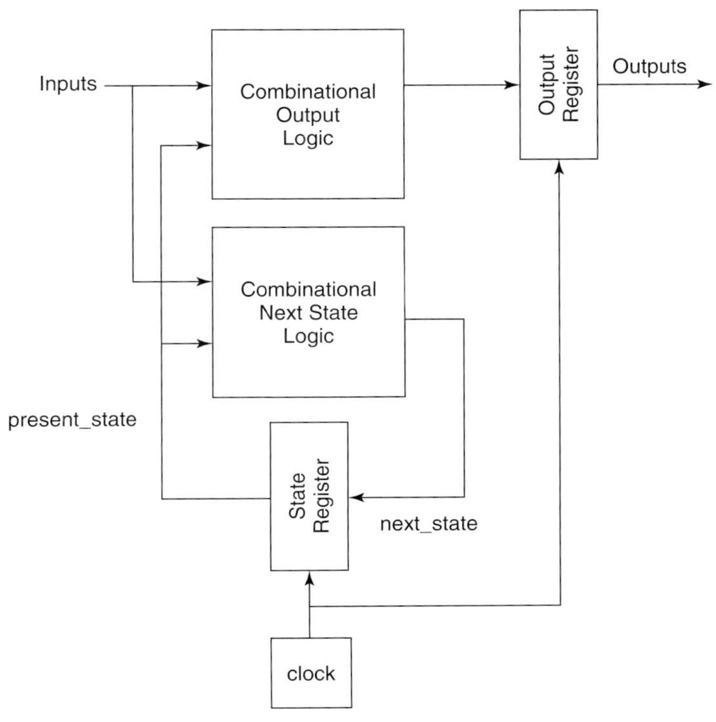

FIGURE 10.4.6
Register outputs added to a Mealy FSM.

10.5 DECODER FOR AN OPTICAL SHAFT ENCODER

Before continuing our discussion of FSMs, we introduce an input device that we will use in some examples in this and later chapters.

Optical Shaft Encoder (OSE)

An *optical shaft encoder* (OSE) converts rotary mechanical motion into a digital output. It does this by first converting the rotation of its shaft into interruptions of a light beam. It then converts these light beam interruptions into electrical pulses. An OSE designed for applications where a hand-operated rotary input device is desired is shown in Figure 10.5.1.

One common use of an OSE is on the front panel of an electronic instrument where it allows an operator to increment or decrement a displayed parameter. For example, clockwise (CW) rotation might cause the parameter to be incremented and counter clockwise (CCW) rotation cause it to be decremented.

FIGURE 10.5.1
An optical shaft encoder (OSE). *Courtesy of US Digital.*

Signals Having a Quadrature Phase Relationship

When its shaft is rotated, the OSE in Figure 10.5.1 produces two digital output waveforms, designated channel A and channel B. These waveforms have a quadrature phase relationship (signal B either lags or leads signal A by one quarter of a period). When the OSE is turned CW, signal A makes a 0 to 1 transition 90 degrees before signal B's 0 to 1 transition (A leads B). When the OSE is turned CCW, signal B makes a 0 to 1 transition 90 degrees before signal A's 0 to 1 transition (A lags B) (Figure 10.5.2).

The quadrature phase relationship between A and B causes A and B to have four different combinations of values in one period of either signal. For example, in the first period of B in Figure 10.5.2, the values of A, B are 00, 10, 11, and 01.

OSEs are commonly available that provide from 16 to 256 pulses on each channel for each revolution of the shaft.

Counting Pulses from an OSE

Assume that we want to use an OSE that generates 32 pulses per revolution to increment or decrement a counter. When the OSE is rotated CW, the counter should

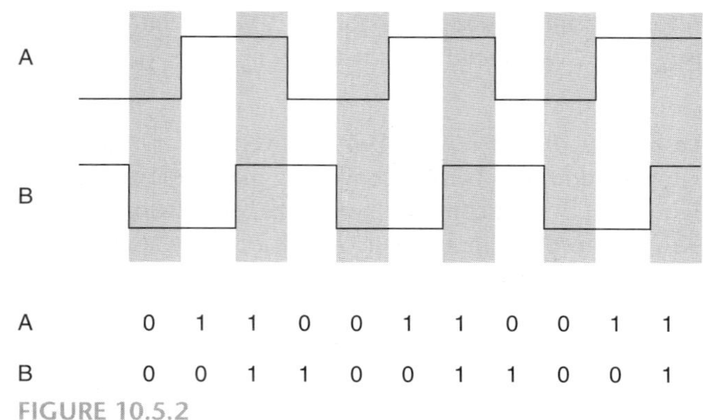

FIGURE 10.5.2
Quadrature signals from an optical shaft encoder.

increment; when it is rotated CCW, the counter should decrement. The counter has count enable, count direction, and system clock inputs. Its count enable input enables the counter to count triggering clock edges.

OSE Output Counted Directly

A straightforward approach might be to connect OSE output A to the clock input of the counter and invert OSE output B and connect it to the counter's direction input. The counter's enable input would be hardwired to always be asserted. With this approach, the counter directly counts the positive edges of A.

The drawback of this approach is that, if we want to use the OSE and counter in a larger sequential system, we will end up with two clock signals: signal A from the OSE and the system clock. The resulting system would not be fully synchronous. A fully synchronous approach would be to derive a signal from A to enable the counter and use the system clock to clock the counter. Each time a positive edge is detected on A, the count enable signal is asserted for one clock cycle. At the next triggering clock edge, the counter either increments or decrements.

Decoder for an OSE

We can design a decoder to generate the count enable and direction control signals to the counter, so that we can have a fully synchronous system (Figure 10.5.3).

Simple OSE Decoder

A simple OSE decoder could be implemented using a positive edge detector and an inverter. The positive edge detector would have A as an input and its output would be connected to the counter's count enable input. The inverter would have B as its input and its output would be connected to the counter's direction control. If the OSE is turned CW one complete revolution, the counter increments 32 times. If the OSE is turned CCW one complete revolution, the counter decrements 32 times.

Four Times Decoder for an OSE

We can increase the effective resolution of the OSE by taking advantage of the quadrature nature of its A and B outputs (Figure 10.5.2). We can design a four times decoder (4X decoder) that enables the counter to count once for each combination of A and B. This increases the effective resolution of the OSE and decoder combination by a factor of four, compared to simply counting positive edges of A (or B) directly.

A state diagram of a Mealy FSM 4X decoder is shown in Figure 10.5.4. Each directed arc is labeled with the inputs and corresponding outputs separated by a slash, (`b, a / up, cnten`).

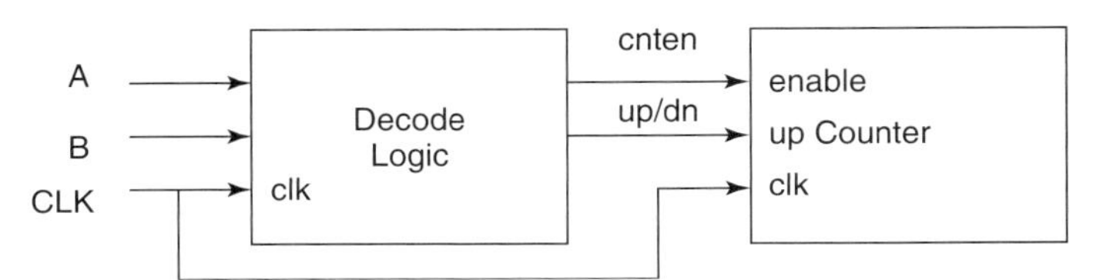

FIGURE 10.5.3
OSE decoder and counter block diagram.

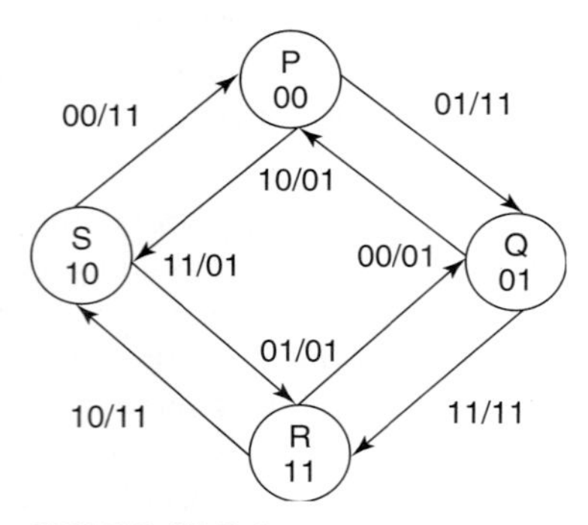

FIGURE 10.5.4
Four times decoder logic state diagram.

The states are named `state_p`, `state_q`, `state_r`, and `state_s`. To be complete, a state diagram must specify the next state for every possible input. For simplicity, the state diagram in Figure 10.5.4 does not show transitions from a state back to itself. For all of these transitions, `up` = `'-'` and `cnten` = `'0'`. For example, in `state_p`, explicit transitions are shown for input combinations (`b`,`a`) of 01 and 10. For all other input combinations the FSM remains in `state_p`.

Multiple Initial States

Normally, a FSM has only one initial state. This design is unique in that any of its states can be the initial state, depending on the values of `b` and `a` (the shaft's initial position) when the system is reset. When reset, the system is placed in whichever state corresponds to the initial values of `b` and `a`. This approach simplifies the design, but requires that the assignment of binary values to encode each state be explicitly specified. The state assignments must correspond to the different combinations of `b` and `a`.

A description of the 4X decoder is given in Listing 10.5.1.

LISTING 10.5.1
Four times OSE decoder.

```
library ieee;
use ieee.std_logic_1164.all;

entity fx_decode is
   port (a, b, clk, rst_bar: in std_logic;
      cnten, up: out std_logic);
end;

architecture behavior of fx_decode is
   subtype state is std_logic_vector(1 downto 0);
   signal present_state, next_state : state;
   constant state_p: state := "00";
```

```
    constant state_q: state := "01";
    constant state_r: state := "11";
    constant state_s: state := "10";

begin

    state_reg: process (clk)
    begin
       if rising_edge(clk) then
          if ( rst_bar = '0' ) then
             present_state <= (b,a);
          else
             present_state <= next_state;
          end if;
       end if;
    end process;

    outputs: process (a, b, present_state)
       variable input : std_logic_vector (1 downto 0);
    begin
       input := (b,a);

       case present_state is

          when state_p =>
          if input = "01" then up <= '1'; cnten <= '1';
          elsif input = "10" then up <= '0'; cnten <= '1';
          else up <= '-'; cnten <= '0';
          end if;

          when state_q =>
          if input = "11" then up <= '1'; cnten  <= '1';
          elsif input = "00" then up <= '0'; cnten <= '1';
          else up <= '-'; cnten <= '0';
          end if;

          when state_r =>
          if input = "10" then up <= '1'; cnten <= '1';
          elsif input = "01" then up <= '0'; cnten <= '1';
          else up <= '-'; cnten <= '0';
          end if;

          when state_s =>
          if input = "00" then up <= '1'; cnten <= '1';
          elsif input = "11" then up <= '0'; cnten <= '1';
          else up <= '-'; cnten <= '0';
          end if;
```

(Cont.)

LISTING 10.5.1 *(Cont.)*

```
      end case;
   end process;

   nxt_state: process (a, b, present_state)
      variable input : std_logic_vector (1 downto 0);
   begin
      input := (b,a);

      case present_state is

         when state_p =>
         if input = "01" then next_state <= state_q;
         elsif input = "10" then next_state <= state_s;
         else next_state <= state_p;
         end if;

         when state_q =>
         if input = "11" then next_state <= state_r;
         elsif input = "00" then next_state <= state_p;
         else next_state <= state_q;
         end if;

         when state_r =>
         if input = "10" then next_state <= state_s;
         elsif input = "01" then next_state <= state_q;
         else next_state <= state_r;
         end if;

         when state_s =>
         if input = "00" then next_state <= state_p;
         elsif input = "11" then next_state <= state_r;
         else next_state <= state_s;
         end if;

      end case;
   end process;
end behavior;
```

To allow us to explicitly specify the binary value for each state's encoding, constants are used. These constants are assigned the desired literal values in the declarative part of the architecture body. In Listing 10.5.1 this was accomplished by declaring a `std_logic_vector` subtype for the state's type and then declaring each state's name and value as a constant of that subtype.

The architecture consists of three processes. The first is the state register process. If `reset_bar` is `'0'` at a triggering clock edge, the system resets so that the present state is equal to the initial input combination `b,a`. This means that it is possible for any of the four states to be the initial state. Otherwise, the next state is loaded into the state register.

The second process is the output process. This process uses a case statement with a branch for each state. In each branch, an if statement assigns a value to `up` and to `cnten` based on the values of `b` and `a`. When `cnten` is assigned a `'0'`, `up` is assigned a don't care (`'-'`). This is appropriate because, when the counter is not enabled to count, the value of its direction control does not matter. By assigning `up` don't care values, we allow the synthesizer to use these values to simplify the combinational logic that generates `up`.

The third process is the next state process. Again, a case statement with a branch for each state is used. In each branch, an if statement assigns a next state value based on the values of `b` and `a`.

10.6 STATE ENCODING AND STATE ASSIGNMENT

To synthesize a FSM, its states must be encoded using a *binary code*. A binary code is a set of binary words. The particular code chosen and the *state assignment* (the assignment of words from the code to particular states) can significantly affect the area and speed of the synthesized logic.

There are several different encoding schemes from which to choose. If states are enumerated in a design description, the choice of encoding scheme can be left to the synthesizer. Or, the synthesizer can be told which scheme to use. Alternatively, we can explicitly specify the encoding of each state, to give us complete control over state assignment, as was done in the 4X OSE decoder.

State Encoding Choices

A *state encoding* maps an abstract state representation (the enumerated states) into a binary encoding. Each bit in the binary encoding requires one flip-flop in the state register. There are several common encoding schemes:

- Sequential (natural binary)
- Random
- Gray
- Johnson
- One hot
- One cold
- Default

Sequential Encoding

Sequential encoding simply assigns each state a natural binary value. This is done based on the position value of each state name in the type declaration enumeration. The first state listed in the enumeration is assigned binary 0, the next state is assigned binary 1, the next state binary 2, and so on. The number of bits in the encoding is n, where n is the smallest value such that 2^n is greater than or equal to the number of states.

The Moore FSM positive edge detector (Figure 10.4.1) has three states, so a sequential encoding requires 2 bits. The sequential assignment is given in Table 10.6.1.

The synthesizer must synthesize the combinational next state logic and the combinational output logic. If the order of the state names in the declaration of the enumeration type is changed, a different state assignment results. A different state assignment may require combinational logic that is more or less complex.

For example, with the state assignment in Figure 10.6.1, output `a_pe` is simply the output of the most significant bit of the state register. This is possible because this bit is 1 only in `state_c` and `a_pe` is also a 1 only in `state_c`. Accordingly, no output combinational logic is required for this assignment.

Number of Possible State Assignments

Unfortunately, the number of possible state assignments to consider can be large. For a code that has 2^n code words and a FSM with s states, the number of possible state assignments is:

$$\frac{(2^n)!}{(2^n - s)!}$$

For example, for the Moore FSM positive edge detector there are 24 different possible state assignments.

Order of States in an Enumeration

If all the flip-flops in a PLD reset to 0s, then whichever state the FSM is to be in when it is reset should be listed first in the enumeration. This state will then be assigned the code of all 0s by the synthesizer.

Random Encoding

Random code is the same as sequential code, except the order of the code words is random.

Gray Encoding

In going from one code to the next code in sequence in a Gray code, only one bit changes. Using Gray encoding, a number of states less than or equal to 2^n can be

Table 10.6.1
Sequential state encoding.

state	encoding
state_a	00
state_b	01
state_c	10

Table 10.6.2
Gray code state encoding.

state	encoding
state_a	00
state_b	01
state_c	11

encoded in n bits. Gray encoding is advantageous for controllers whose state diagrams have long paths without branching. Successive Gray codes are assigned to the states of the long paths. Since only one bit changes in going from one code to the next code in sequence, hazards are also minimized. A Gray code encoding of the positive edge detector Moore FSM is given in Table 10.6.2.

In this particular encoding assignment, the transition from the last state to the first state requires more than one bit to change. This is because the number of code words used is less than 2^n.

Johnson Encoding

Johnson encoding has properties similar to Gray encoding. The Johnson code starts with all 0s. For each subsequent code value a 1 is shifted in from the right, until a code word with all 1s is produced. Then, each subsequent code word has a 0 shifted in from the right, until we have the original code word with all 0s; see "Johnson Counter" on page 346.

The Johnson code also has the property that only one bit changes in going from one code word to the next successive code word. A Johnson code of n bits has $2 \times n$ codewords. For only three states, the Johnson encoding happens to be the same as the Gray code in Table 10.6.2.

One-hot Encoding

In *one-hot encoding*, each code value contains a single 1 and all other bits are 0's (one bit "asserted" per state). As a result, n bits are required to encode n states. This encoding is suitable for speed optimization and for target PLD architectures that have a large number of flip-flops, such as FPGAs. A one-hot encoding for the positive edge detector Moore FSM is given in Table 10.6.3.

One-cold Encoding

An encoding that has only one bit unasserted and all other bits asserted is called a *one-cold encoding*. The properties of a one-cold encoding are similar to those of a one-hot encoding.

Table 10.6.3
One-hot state encoding.

state	encoding
state_a	001
state_b	010
state_c	100

Modified One-hot Encoding

This encoding is a slight variation of one-hot encoding. The all zeros state is a valid state in this encoding and is normally used as the reset state. So, valid states either have only one flip-flop set or no flip-flops set. Accordingly, n-1 flip-flops are required to encode n states. When one-hot encoding is specified, some synthesizers use this code.

Default Encoding

If we don't specify an encoding, the synthesizer chooses a default encoding. This default encoding may be chosen based on the number of states and the target PLD architecture. For example, one synthesizer uses sequential encoding if there are less than five states, one hot encoding for 5–24 states, and Gray encoding for more than 25 states.

An *optimal encoding* is one that results in the least area (smallest number of gates and flip-flops) or the least delay (fastest circuit), or both, depending on the design objectives.

Specifying an Encoding

The synthesizers's default encoding may not be optimal for a particular FSM. Accordingly, the synthesizer allows us to specify a more desirable encoding.

Specification of an encoding may be accomplished by either menu selection or through inclusion of attributes in the design description. If attributes are used, they are defined by the synthesis tool and require a context statement specifying the library and package in which the attributes are declared.

States are assigned code words from left to right. If the number of states in the enumeration type is not a power of 2, the remaining code words may be processed as don't cares by the synthesizer.

Soft Encoding versus Hard Encoding

Using enumerated states in a FSM description is referred to as *soft encoding*. It allows us to use the synthesizer to try different state assignments to find an optimal one without modifying the FSM's description. In contrast, using constants, as in Listing 10.5.1, to specify a state assignment is referred to as *hard encoding*. To try a different state assignment when using hard encoding requires that the constant state values be changed in the FSM's description.

FSM Logic Complexity versus State Encoding

Next state logic complexity is related to the number of state bits that must be decoded to determine the present state and the number of state bits that must be changed in going from the present state to the next state.

Output logic complexity is dependent on how much decoding is required to generate the outputs from the present state and input values. For Moore FSMs, where the output is a function of only the present state, it is often possible to choose a state encoding that is identical to the output values, so no combinational output logic is required. This was the case when a sequential state assignment was chosen for the positive edge detector.

State Bits Used Directly as Outputs

The approach used for the positive edge detector can be extended to any Moore FSM. One bit of the state encoding can be assigned to each output. Each of these bit's values are defined for each state, depending on the output requirements as defined by either a state diagram or a state table. As few additional bits as necessary are then added to the state encoding to make each state unique. A number of state bits larger than the minimum number necessary to encode the required number of states is the result. This may, in turn, increase the complexity of the next state logic. The resulting state assignment is then hard encoded.

FSM Logic Complexity versus Target PLD Architecture

The logic complexity resulting from a given state assignment is also related to the architecture of the target PLD.

SPLDs and CPLDs generally consist of a relatively small number of logic blocks. Each block contains AND/OR logic feeding a flip-flop. The result is that a relatively large amount of logic capacity is associated with each flip-flop. The AND gates in each block have a large number of inputs, so a large number of state bits can be included in each product term. Because a SPLD or CPLD has a relatively small number of flip-flops, use of a highly encoded state assignment such as sequential or Gray is preferable.

In contrast, an FPGA generally consists of a relatively large number of logic blocks. Each block has a small amount of logic that feeds a flip-flop. If a highly-encoded state assignment is used with an FPGA, the amount of logic needed to decode the next state and outputs may require multiple blocks to feed a single flip-flop or to decode a single output.

One-hot encoding has the advantage that only one bit must be decoded to determine the present state. This can reduce the complexity of the next state combinational logic, but does so at the cost of additional bits in the state register. FPGA architectures, with their large number of small logic blocks, each with a flip-flop, efficiently implement one-hot encodings. Typically, the number of blocks required to implement a FSM with s states using an FPGA and one-hot encoding is s blocks.

Another advantage of one-hot encoding is that any one state assignment is as optimal as any other. So, no time is spent trying to find the optimal encoding from among all the possible one-hot state assignments.

As is clear from this discussion, state assignment involves tradeoffs. Accordingly, for complex designs it may be advantageous to try several alternative state assignments and compare the place and route results for the target PLD.

Synthesizer Features

Synthesizers from different vendors have a range of tool-specific features that help in optimizing the synthesized logic for a design. These features may include tools for automatically encoding FSMs.

FSM Compiler

Some synthesis tools are able to detect FSMs in the VHDL source code and optimize them. For example, Synplicity's Synplify Pro synthesizer has a feature called FSM Compiler. When enabled, FSM Compiler searches the code for FSMs. It looks for registers with feedback that are controlled by the current value of the register (for example, case or if statements that test the current value of a state register).

When FSM Compiler finds a FSM, it converts it to a symbolic graph form and performs logic optimizations. These logic optimizations include removing unreachable states, and their associated next state logic, and re-encoding state representations. The result is a synthesized FSM optimized for area and speed. This optimization is based primarily on the number of states in the FSM.

However, if we enable the FSM Compiler when synthesizing a FSM where we have explicitly specified a state assignment, such as the hard encoding in the 4X OSE decoder, we can end up with a completely different state assignment in the synthesized logic and not the expected result. For explicitly specified state assignments, the FSM Compiler must be disabled to get the expected result.

Automatic Determination of Encoding Style

Some synthesizers provide a feature that automatically evaluates different encoding styles for a FSM and selects the one that is best based on overall design constraints. For example, Synplicity's Synplify Pro contains a tool called FSM Explorer that does this. FSM Explorer is run after FSM Compiler and takes considerably more time to run. FSM Explorer is not available for all PLD vendor technologies.

10.7 SUPPOSEDLY SAFE FSMS

Reachable States

Normally, when we draw a state diagram during the development of a FSM, every state in the diagram is *reachable*. That is, there is at least one transition into each state from another state, and at least one transition out of each state to another state. Logically, it would not make sense to have a state that has no transition into it from some other state. This would mean that the state could never be entered during the operation of the FSM and, therefore, is not needed. If we had a state that had no transition out of it to another state, the FSM could never leave the state once it is entered. In effect, the FSM would hang up or lock up in that state.

Logical and Physical States in a FSM

We must distinguish between the logical states and physical states in a FSM. The logical states are the states abstractly represented on the FSM's state diagram. The physical states are the 2^n possible combinations of the n flip-flops in the FSM's state register. In general not all of the physical states are used to encode the logical states. Any physical states not used to encode logical states in the state diagram are *unreachable* in the normal operation of the FSM.

When the number of logical states s in a FSM is a power of 2 and a binary encoding (such as sequential or gray) is used, the number of logical states and physical states is the same. The state register consists of $n = \log_2(s)$ flip-flops and all of the physical states are reachable.

When the number of states s in a FSM is not a power of 2 or when an encoding such as one-hot is used, the state register consists of n flip-flops, where n is the smallest value such that $n > \log_2(s)$. In this situation, one or more of the physical states are not used to encode a logical state and are unreachable. That is, under normal operation (normal flow through the state diagram) the FSM's state register will never have a value corresponding to an unreachable state.

Unsafe States

Due to electrical or other reasons, the contents of a FSM's state register may be disturbed so that its value corresponds to an unreachable physical state. It is clear now that an unreachable physical state is "unreachable" only in terms of the normal operation of a FSM. Once a normal FSM has entered an unreachable state its operation is no longer predictable. Unreachable states in a FSM implementation are also referred to as *unsafe states* or *invalid states*. A FSM with unsafe states is considered an *unsafe FSM*. On a subsequent triggering clock edge, after having entered an unreachable state, the FSM may transition to a safe state and then return to normal operation or it may transition from one unsafe state to another and *lock up*, never transitioning back to a valid state.

Number of Unsafe States versus Encoding

The number of unreachable states is a function of the number of logical states and the state encoding chosen. Consider a FSM with three states that uses a binary code or a gray code. The number of unreachable states is $2^2 - 3 = 1$. For a one-hot encoding of the same FSM, the number of unreachable states is $2^3 - 3 = 5$. In this sense, the one-hot code could be said to be inherently less safe.

Attempting to Make a FSM Safe

How unsafe states are dealt with depends on the application. If the FSM is part of a system that does not require high reliability, then it may be appropriate to ignore them. This is particularly so when, in addition, the probability of the FSM entering an unsafe state is extremely small or the nature of the application is such that, if the FSM enters an unsafe state, the system cannot recover on its own in an acceptable way.

When a system must be highly reliable an attempt to modify the FSM so that it is a "safe" FSM may be appropriate. The term "safe" FSM is commonly used to mean a FSM designed so that, if it transitions into an unsafe state, it will subsequently transition to a safe state that is part of the normal operation of the FSM (usually the initial state) or transition to a special error state and then to a safe state.

Before discussing this approach further, it is important to emphasize that making a FSM safe in this regard is not necessarily sufficient to ensure that the system containing the FSM is a safe or highly reliable system in the general sense. To achieve a highly reliable system may require substantial efforts at the system level involving error correction and detection. Such considerations are beyond the scope of this text.

Creating a Safe FSM at the VHDL Level

When describing a FSM using an enumeration data type for the states, there are no unsafe states at the VHDL description level. The possible states of the FSM are only those in the enumeration, and all of these are reachable in a proper description of a

FSM. The unsafe states don't come into existence until the FSM is synthesized. At that point, an encoding is chosen and, if the number of possible codes in the encoding is not equal to the number of states in the enumeration, unsafe states are created.

When Others in Enumerated States FSMs

It might seem that using a when others clause at the end of the case statement in the next state process solves the problem. We could write:

```
when others =>
        next_state <= initial_state
```

or

```
when others =>
        next_state <= error_state
```

Using when others is syntactically correct, even if all of the enumerated states have explicitly been covered previously in the case statement. However, the when others only applies to the logical states in the VHDL enumeration, not to the states of the encoding. Usually, by default, the synthesizer treats all unreachable states as don't cares to optimize the next state logic. Thus, the when others does not have the desired effect. The synthesizer may give a warning, such as "OTHERS clause is not synthesized."

When using an enumerated data type for states, there is no way to directly create a safe FSM in VHDL if the number of logical states is not equal to the number of possible physical states based on the encoding chosen.

The logic created by the Synplify Pro synthesizer using sequential (binary) encoding for the Moore FSM version of the positive edge detector from Listing 10.4.1 is shown in Figure 10.7.1.

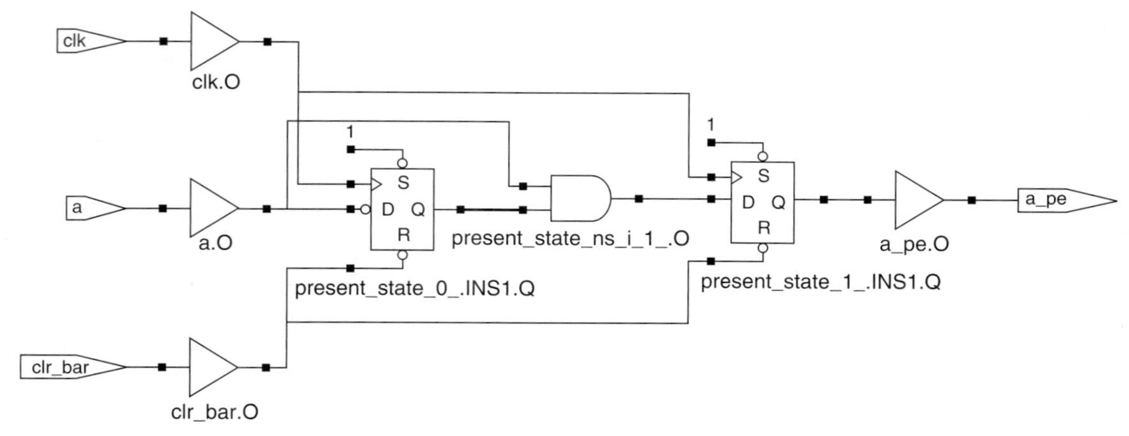

FIGURE 10.7.1
Synthesized logic for Moore version of positive edge detector, Listing 10.4.1.

Using syn_encoding for a Safe FSM

A synthesizer may provide attributes that can be used with FSMs whose states are enumerated types to implement a safe FSM. Consult the synthesizer's documentation to see if that is the case.

For example, for the Synplify Pro synthesizer, the value *safe* of the *syn_encoding attribute* can be used with the default encoding style value or with any of the other encoding style values to add reset logic to force a FSM into the reset state, if it enters an unsafe state. The syn_encoding attribute tells the synthesizer to generate logic that will reset the FSM if it enters the unsafe state. If the two attributes below are added to Listing 10.4.1,

```
type state is (state_a, state_b, state_c);
   signal present_state, next_state : state;
attribute syn_encoding : string;
attribute syn_encoding of present_state : signal is
           "safe,default";
```

the *recovery logic* generated by the synthesizer in response to these syn_encoding attributes is shown in Figure 10.7.2.

In response to the safe attribute, combinational logic is synthesized to detect invalid states. If an invalid state is detected, the flip-flop on the left is set on the next rising clock edge. On the falling edge of that same clock, the flip-flop on the right is set.

The output of the flip-flop on the right is "ORed" with the original reset signal `clr_bar`. The resulting signal drives the clear/preset pins of the state bits, forcing the circuit to its reset state. Once this valid state is reached, the next rising edge of the clock will clear the flip-flop on the left, the next falling edge of the clock will clear the flip-flop on the right, and normal operation will resume.

Note that the flip-flop on the right is triggered on the falling edge of the clock. This is done to prevent any hazardous conditions that could result from removing the reset signal too close to the active clock edge of the state register. One result of this is that this logic cannot be placed and routed to a target PLD that does not have

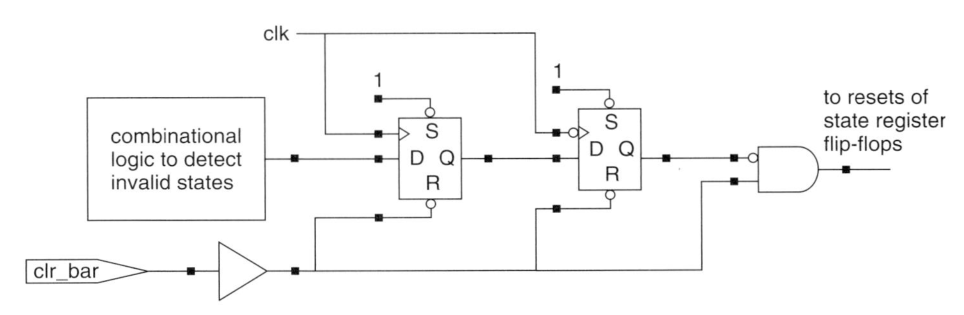

FIGURE 10.7.2

Automatically synthesized recovery logic to force a FSM into its reset state from an unsafe state.

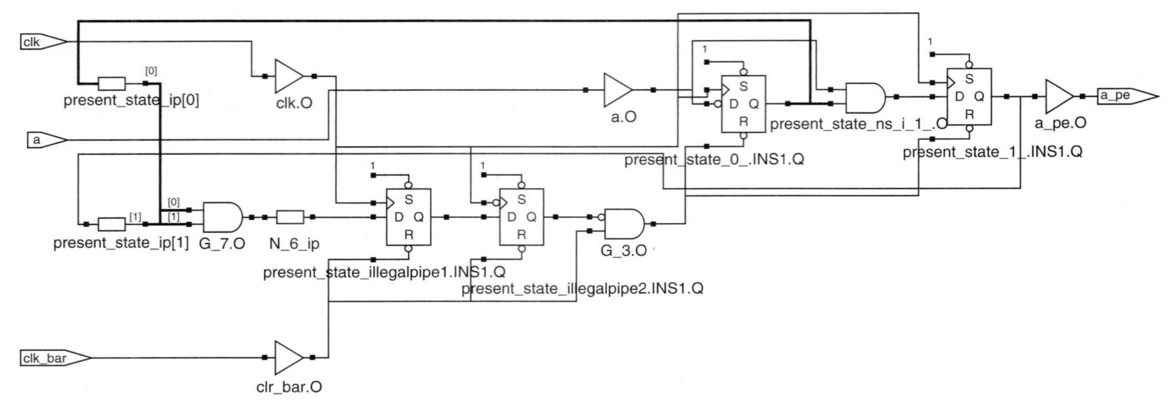

FIGURE 10.7.3
Complete FSM with recovery logic.

memory elements that can be configured to trigger on both the rising edge and falling edge of the clock.

This implementation eliminates the possibility of the FSM getting "stuck" in an unsafe state and not returning to a safe state. This solution has minimal impact on the timing of the system.

Note that the transition out of an unsafe state is not implemented exactly as described by the when others branch of the source code. This deviation from the defined when others branch behavior only occurs for unsafe states. If the when others branch contained any safe state transitions, they would be implemented as described in the source code.

The complete FSM including the recovery logic is shown in Figure 10.7.3. The single unsafe state is the encoding 11. When this state is entered, the recovery logic clears the flip-flops in the state register, forcing the FSM back to its reset state.

Exact Implementation Using a Hard Encoding

It is possible to get an implementation of the circuit that fully implements the when others branch, if it is necessary to do so. This requires disabling the reachability analysis of the state machine, which is done by turning off the FSM compiler, and using a hard encoding instead of an enumerated type. This can have a significant adverse effect on the area and timing of the circuit.

10.8 INHIBIT LOGIC FSM EXAMPLE

Double Buffering

Consider a design where we need to be able to read a counter's output while it is counting ("read on the fly"). If the counter's output were read directly and the read occurred while the counter was in the process of changing value, the value read might be incorrect. To eliminate this problem, we can *double buffer* the counter's outputs.

In Figure 10.8.1, a 12-bit counter's output is to be read by a microprocessor with an 8-bit data bus. The microprocessor must execute two read bus cycles to read the 12-bit value, one byte at a time. If its output were not double buffered, the time required to execute the two bus cycles would substantially increase the likelihood of reading an incorrect value, if the counter were counting.

A 12-bit register is used to double buffer the counter's output. This buffer register has an enable input that is controlled by an Inhibit Logic FSM. The Inhibit Logic FSM has four inputs: `sel`, `oe_bar`, `clk`, and `reset_bar`. Its output is the signal `inhibit`, which drives the buffer register's enable input.

When enabled, the buffer register stores the counter's output at each triggering clock edge. Thus, the buffer register's contents are a copy of the counter's contents delayed by one clock cycle. As long as the counter is not being read by the microprocessor, the buffer register remains enabled.

To read the first byte from the 12-bit counter, the microprocessor makes the `sel` input `'0'` to select the first byte and asserts the `oe_bar` input to turn on the three-state output buffers. When this happens, the Inhibit Logic FSM must make its `inhibit` output a `'1'` to disable the buffer register. This freezes the buffer register's contents. The microprocessor then reads the first byte. Bits 3 down to 0 of this byte contain the most significant four bits of the count. Bits 7 down to 4 of this byte are all 0's.

The buffer register must be kept disabled until after the microprocessor completes reading the second byte from the buffer register. The second byte contains the least significant 8 bits of the 12-bit count. To read the second byte, the microprocessor changes `sel` to `'1'` and again asserts `oe_bar`.

When `oe_bar` is unasserted after the second byte is read, the buffer register is again enabled and stores the value from the 12-bit counter at the next triggering clock edge.

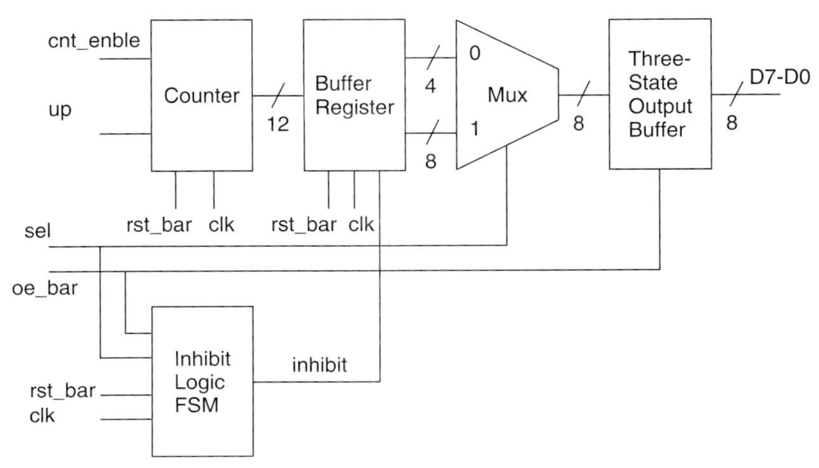

FIGURE 10.8.1
Block diagram of Inhibit Logic FSM controlling a double buffer register.

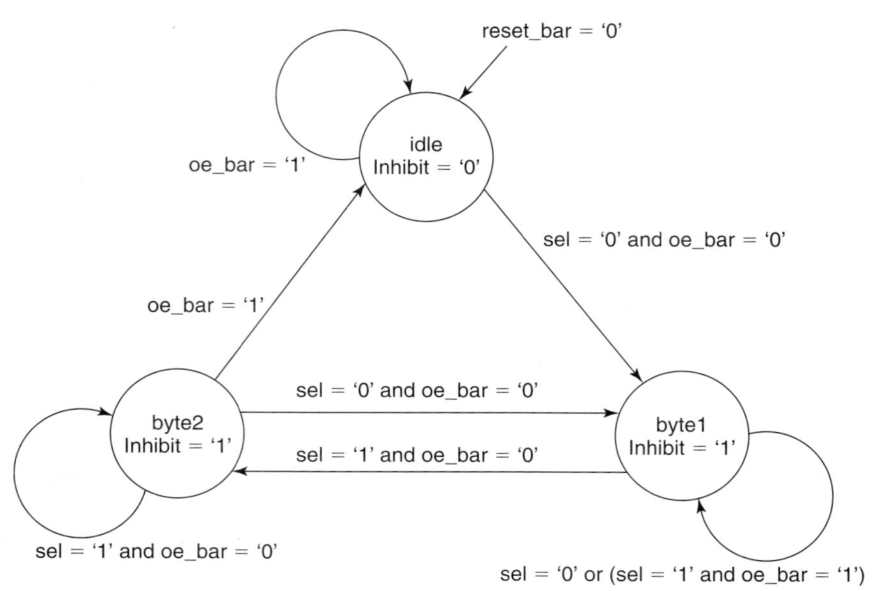

FIGURE 10.8.2
State diagram for Inhibit Logic FSM.

A state diagram for the Inhibit Logic FSM is given in Figure 10.8.2. The states are named to represent the associated operation. In the `idle` state, the counter is not being read and `inhibit` is `'0'`. In state `byte1`, the first byte is being read from the buffer register. In state `byte2`, the second byte is being read. In both these latter two states `inhibit` is `'1'`, causing the buffer register to be disabled so that its contents cannot change value. Since output `inhibit` is a function of only the present state, this is a Moore FSM.

Note that the inhibit logic cannot simply be a combinational function of `oe_bar` and `clk`. That is, we cannot simply have `inhibit` synchronized to the clock and asserted whenever `oe_bar` is asserted. The reason for this is that two bus cycles are required to read the two data bytes. During each bus cycle `oe_bar` is asserted and then unasserted by the microprocessor. However, to ensure that the value read during the second bus cycle is correct, `inhibit` must remain asserted for the duration of both bus cycles.

A design description of the Inhibit Logic FSM in Figure 10.8.1 is given in Listing 10.8.1.

LISTING 10.8.1
Inhibit Logic FSM using enumerated states.

```
library ieee;
use ieee.std_logic_1164.all;

entity inhibit_fsm is
```

```
   port (
      rst_bar: in std_logic;
      clk: in std_logic;
      sel: in std_logic;
      inhibit : out std_logic;
      oe_bar: in std_logic
      );
end inhibit_fsm;

architecture behav of inhibit_fsm is
   type inhib_state is (idle, byte1, byte2);
   signal present_state, next_state : inhib_state;

begin

   inhib_sreg: process (clk)
   begin
      if rising_edge(clk) then
         if rst_bar = '0' then
            present_state <= idle;
         else
            present_state <= next_state;
         end if;
      end if;
   end process;

   inhib_output: process (present_state)
   begin
      case present_state is
         when byte1 | byte2 =>
         inhibit <= '1';
         when others=>
         inhibit <= '0';
      end case;
   end process;

   inhib_nxt_state: process (present_state, sel, oe_bar)
   begin
      case present_state is

         when idle =>
         if sel = '0' and oe_bar = '0' then next_state <= byte1;
         else next_state <= idle;
         end if;

         when byte1 =>
         if sel = '1' and oe_bar = '0' then next_state <= byte2;
```

(Cont.)

LISTING 10.8.1 *(Cont.)*

```
          else next_state <= byte1;
          end if;

          when byte2 =>
          if oe_bar = '1' then next_state <= idle;
          elsif sel = '0' and oe_bar = '0' then next_state <= byte1;
          else next_state <= byte2;
          end if;

      end case;
   end process;

end behav;
```

The Inhibit Logic FSM is partitioned into three processes: `inhib_sreg`, `inhib_output`, and `inhibit_nxt_state`.

The Inhibit Logic FSM uses enumerated states. A type is declared for the enumerated states, and signals for the present state and next state are declared to be of this type:

```
type inhib_state is (idle, byte1, byte2);
signal present_state, next_state : inhib_state;
```

At this point, the states are abstract, which is very convenient for functional simulation. The signals `present_state` and `next_state` can be displayed as part of the output waveforms of a functional simulation. These signals will be displayed having the values `idle`, `byte1`, or `byte2`. This makes it convenient to verify that state transitions occur in the correct order. However, before we can proceed with synthesis, we must either choose a state assignment ourselves or let the synthesizer choose.

10.9 COUNTERS AS MOORE FSMs

A simple counter is a special case of a Moore FSM where there are no inputs (other than the clock and reset) and the outputs are taken directly from the state register. Figure 10.1.5 is modified to represent a simple counter as a Moore FSM in Figure 10.9.1. This is accomplished by eliminating the inputs and the combinational logic that computes the outputs.

2-bit Binary Counter FSM

For example, consider a 2-bit binary counter described as a Moore FSM. If the counter resets to 0, it can be represented by the state diagram in Figure 10.9.2.

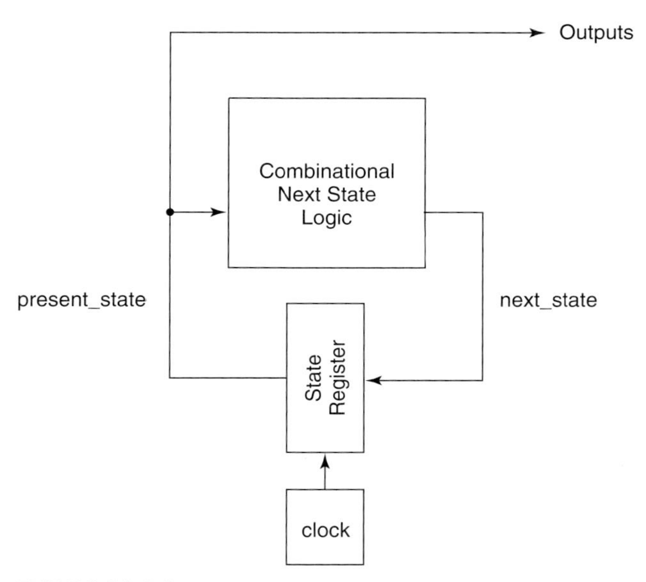

FIGURE 10.9.1
A simple counter as a Moore FSM that requires no combinational output logic.

When the counter is reset, it is placed in its initial state and its outputs are 00. At each triggering clock edge, the counter transitions to its next state. Since there are no inputs and the outputs are taken directly from the state register, the directed arcs are not labeled. As previously stated, an unlabeled arc indicates an unconditional transition.

Table 10.9.1 is the state table for the 2-bit counter. This table provides, in tabular form, the same information as the state diagram.

A description of the 2-bit counter as a Moore FSM is given in Listing 10.9.1.

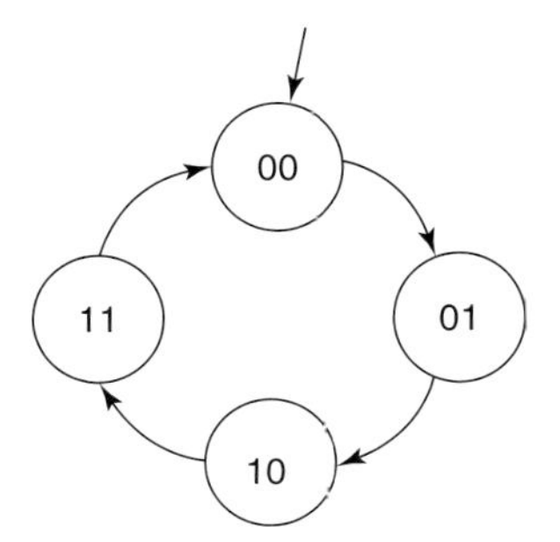

FIGURE 10.9.2
State diagram for a 2-bit binary counter.

Table 10.9.1
State table for a 2-bit counter.

Present State	Next State
00	01
01	10
10	11
11	00

LISTING 10.9.1
Counter with state encoding defined by constants.

```
library ieee;
use ieee.std_logic_1164.all;

entity cnt2bitc is
   port (clk, reset: in std_logic;
   count: out std_logic_vector(1 downto 0));
end;

architecture behavior of cnt2bitc is
   subtype state is std_logic_vector (1 downto 0);
   signal present_state, next_state : state;
   constant state0: state := "00";
   constant state1: state := "01";
   constant state2: state := "10";
   constant state3: state := "11";

begin
   process (clk)
   begin
      if rising_edge(clk) then
         if ( reset='1' ) then
            present_state <= state0;
         else
            present_state <= next_state;
         end if;
      end if;
   end process;

   process (present_state)
   begin
      case present_state is
         when state0 =>
            next_state <= state1;
         when state1 =>
```

```
            next_state <= state2;
         when state2 =>
            next_state <= state3;
         when state3 =>
            next_state <= state0;
         when others =>
            next_state <= state0;
      end case;
      count <= present_state;
   end process;
end behavior;
```

Counter State Assignment

To allow us to explicitly specify the binary value for each state of the counter, hard encoding is used. These constants are assigned the desired literal values in the declarative part of the architecture body. We can easily change the count sequence by changing the values assigned to the constants in their declarations.

Counter FSM Description

The description in Listing 10.9.1 follows the FSM template in Listing 10.3.1. However, since this is a Moore FSM whose outputs are directly equal to its present state, only two processes are needed. The first infers, resets, and clocks the state register. The second describes the combinational logic that determines the next state value.

Waveforms for a functional simulation of this FSM, using a 100 ns clock, are shown in Figure 10.9.3.

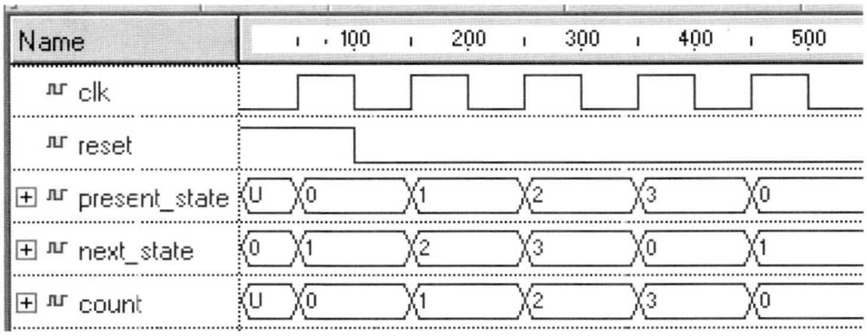

FIGURE 10.9.3
Functional simulation of 2-bit counter FSM.

PROBLEMS

10.1 What information does the state of a FSM represent? Where is this information stored in the FSM?

10.2 Briefly describe the difference(s) between an asynchronous sequential system and a synchronous sequential system. What is the criteria for a synchronous system to be fully synchronous?

10.3 What determines the minimum length of time a fully synchronous FSM remains in a state? How can such a FSM remain in the same state for a longer period of time?

10.4 Briefly describe the difference(s) between a Mealy and a Moore FSM.

10.5 Draw a state diagram for each of the following types of flip-flops. Also specify which are Mealy FSMs and which are Moore FSMs.
(a) D flip-flop
(b) S-R flip-flop
(c) J-K flip-flop
(d) T flip-flop

10.6 Draw a state diagram for a 2-bit right shift register, which is otherwise identical to the 4-bit shift register in Listing 9.2.1.

10.7 Draw a state diagram for a 3-bit counter, which is otherwise identical to the 12-bit counter in Listing 9.4.4.

10.8 What is the operation of the FSM represented by the state diagram in Figure 10.2.1(a)? In other words, what does this FSM do? What is the operation of the FSM represented by the state diagram in Figure 10.2.1(b)?

10.9 Draw a state/output table for the FSMs in Figures 10.2.1(a) and 10.2.1(b).

10.10 What is the purpose of each of the three processes in a three-process template for a FSM description? Which of these processes are sequential and which are combinational? What is the advantage of using the three-process approach to describing a FSM?

10.11 In which process of the three process template does the FSM's reset appear? Under what condition would the reset signal's name appear in this process's sensitivity list?

10.12 Rewrite the process `state_reg` in the template in Listing 10.3.1 to implement an asynchronous reset.

10.13 Write a three-process description of the FSM represented by the state diagram in Figure 10.2.1(a).

10.14 Write a three-process description of the FSM represented by the state diagram in Figure 10.2.1(b).

10.15 Write a two-process description of the FSM represented by the state diagram in Figure 10.2.1(b).

10.16 Write a description for an alternative architecture for the BCD synchronous up/down counter in Problem 9.21. The entity declaration remains the same. Describe the BCD counter as a FSM. The state register is to be written as a process. The next state logic and the output logic are to each be written as conditional signal assignment statements. Use an unsigned signal to represent the count.

When considered as an FSM, is this a Mealy or Moore machine? Explain your reasoning.

10.17 A 2-bit sequencer (counter) produces the output sequence 00, 11, 10, 01, and then repeats. The sequencer has a synchronous reset input, `reset_bar`, which, when asserted, causes the output to be 00.

(a) Draw a state diagram for the sequencer.

(b) Write a complete design description of the sequencer. The description must consist of two or three processes that form a FSM.

(c) Is this FSM a Mealy or Moore FSM? Justify your answer.

10.18 Draw a state diagram, using a minimal number of states, to describe a Moore FSM implementation of the event detector in Problem 9.22. Write an architecture body that describes the event_detector as a Moore FSM. The state register and next state logic are to be described using processes. The output logic is to be described using a conditional signal assignment statement. If the input is a square wave, what is its maximum frequency relative to the system clock frequency, so that each input event is represented by a separate pulse at the system's output? Explain your reasoning.

10.19 A periodic waveform generator system is required that generates three outputs: w, x, and y. The system has inputs `clk` and `rst_bar`. The output waveform repeats every 80 ns. One period of the required output is represented by the following waveforms:

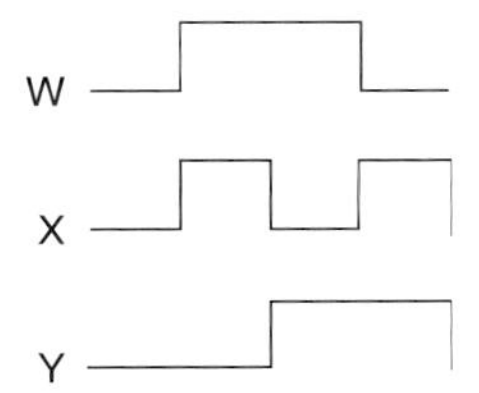

(a) Write a complete VHDL description of a FSM that implements the waveform generator. The description must follow the three-process template and use enumerated states. The system is reset by the asynchronous reset signal `rst_bar`.

(b) Specify the state assignment that you would expect to lead to the simplest hardware. Justify your answer.

10.20 A sequence recognizer generates a 1 output whenever it receives the input sequence 101.

(a) Draw a state diagram for a Moore FSM implementation this sequence recognizer. Use as few states as possible.

(b) Write a complete description of a Moore FSM implementation of this sequence recognizer. Use a three-process format with enumerated states and a synchronous reset.

10.21 A sequence recognizer generates a 1 output whenever it receives the input sequence 101.

(a) Draw a state diagram for a Mealy FSM implementation of this sequence recognizer. Use as few states as possible.

(b) Write a complete description of a Mealy FSM implementation of this sequence recognizer. The implementation must have a registered output. Use a three-process format with enumerated states and a synchronous reset.

10.22 A clocked synchronous system is described by the following state diagram:

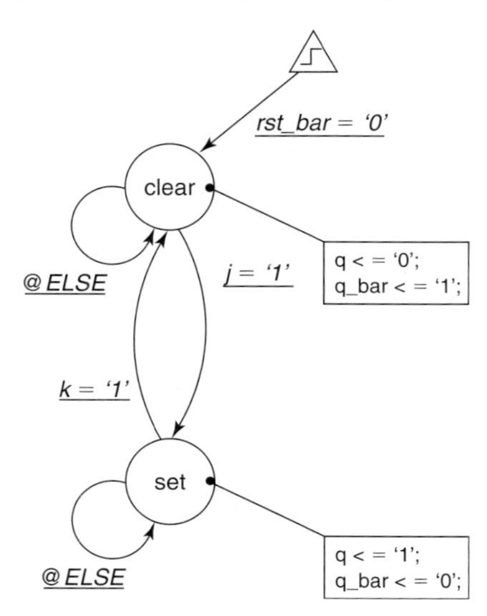

(a) Write a design description of the synchronous system describing it as a FSM with enumerated states and following the three-process template given in this chapter. Use the entity declaration given below:

```
entity jkfsm is
   port(
      j : in std_logic;
      k : in std_logic;
      rst_bar : in std_logic;-- synchronous reset
      clk : in std_logic;
      q : out std_logic;
      q_bar : out std_logic
      );
end jkfsm;
```

(b) Is the machine as described by its state diagram a Mealy or Moore FSM? Explain your answer.

10.23 A clocked synchronous sequence detector that detects the input sequence, or subsequence, 1101 is to be designed. The sequence detector has a data input `x` and a data output `y`. The input sequence is sampled on the rising edge of the sequence detector's clock input, `clk`. When the last 1 in a sequence 1101 is clocked in, output `y` should become and remain a 1 until the next triggering clock edge.

(a) Draw a state diagram of a Mealy FSM with the minimum number of states that realizes the sequence detector.

(b) Write a complete two-process VHDL FSM design description of the sequence detector. One process handles the state change and the other handles combinational functions. Write this description in a fashion that allows the synthesizer to do the state assignment.

10.24 A programmable delay is created using a 10-bit counter, a 10-to-1 multiplexer, and a FSM. A block diagram of the system follows. The system is able to produce any one of 10 possible delays based on the value of its `select_counter_bit` input. For each increase of the `select_counter_bit` value by 1, the delay doubles.

The idea behind the design is that the `select_counter_bit` value selects one of the 10 output bits of the 10-bit counter. When the `start_delay` input is asserted, the `end_of_delay` output becomes a 0 and the counter begins counting. When the selected counter bit changes from 0 to 1, the counter stops counting and `end_of_delay` becomes a 1.

(a) Draw a state diagram for a Moore type FSM that implements the `prog_delay_fsm` component of the block diagram. Use as few states in the diagram as possible.

(b) Write a three-process description of the `prog_delay_fsm` entity.

(c) Write an expression that specifies the length of time that `end_of_delay` is a 0.

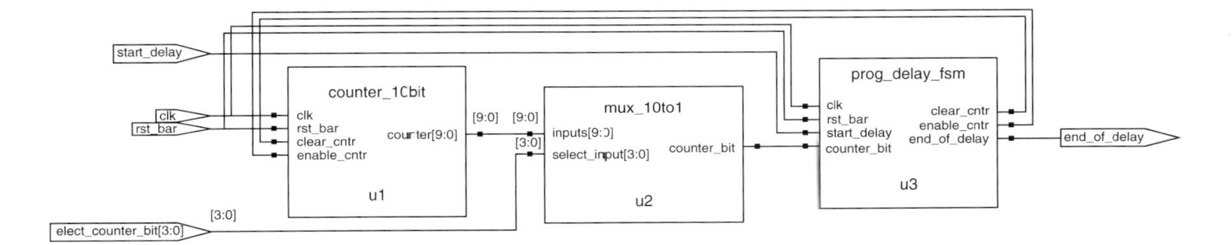

10.25 The state diagram in Figure 10.5.4 is for a Mealy FSM that implements a 4X decoder. Draw a state diagram for a Moore FSM that implements the same functionality.

10.26 What is a one-hot state assignment and what advantages does it have? With what type of PLD is a one-hot state assignment most likely to be advantageous and why?

10.27 What is the difference between a sequential and a Gray code state assignment? If a FSM has eight states, which of these two assignments will require the most bits in its state register?

10.28 Assuming two different state assignments require the same number of bits in their state registers, is it possible for the choice of assignments to have any impact on the logic synthesized? Explain your answer.

10.29 If you are using explicit state encoding for a FSM, should you enable or disable the "FSM Compiler" feature in Synplify? Explain your answer.

10.30 Write a complete Moore FSM VHDL design description of a 3-bit Gray code counter. The counter has an active-low synchronous clear input (`clr`). The 3-bit Gray code count sequence is 000, 001, 011, 010, 110, 111, 101, 100, and 000 (and then repeats).

10.31 A sequencer (counter) has a 3-bit output and is implemented as a FSM. The sequencer's states are constants named `state_a`, `state_b`, and so on. The sequencer starts in `state_a` and sequences through five unique states, returning to `state_a`. The state assignment is defined in the design description by the initial value assigned to each constant. Write the sequencer design description and assign initial values to the states so that the output sequences in binary from "000" to "100" and then back to "000".

10.32 A counter is used to generate a sequence of outputs that are control signals, thus implementing a controller. The control signals are directly encoded as the counter's outputs. The control signals are labeled A, B, C, and D. Their sequence of values is 0000, 0001, 0110, 1100, 1000 and 0000 (sequence repeats).

(a) Write a complete VHDL design description of the controller as a FSM using either two or three processes as is appropriate. The outputs are to be declared as scalars. The counter has a synchronous reset that is active low.
(b) Is this controller a Mealy or Moore FSM? Explain your answer.

10.33 A sequence recognizer has a single input and a single output. The output of the sequence recognizer is a 1 if and only if the last four input values were 0101. The sequence recognizer has a synchronous active low reset input.

(a) Draw a state diagram for the sequence recognizer.
(b) Write a complete VHDL design description of the sequence recognizer as a two- or three- (whichever is most appropriate) process FSM.

Chapter 11

ASM Charts and RTL Design

An alternative way to graphically represent a FSM is to use an algorithmic state machine (ASM) chart. ASM charts are similar to traditional flow charts, except that they also contain implicit timing information in a manner similar to state diagrams.

All digital systems can be thought of as implementing algorithms. In a complex system, each task in an algorithm might be accomplished by a separate component. A task involves either the accomplishment of some action or a data manipulation. A FSM can be used to control when each component performs its task to accomplish the algorithm.

A system that performs a data manipulation can be viewed as consisting of a data path and a control. The data path consists of the components that manipulate and store the data. The control consists of the component that sequences when specific manipulations take place and when specific data is stored. The control is implemented by one or more FSMs. Design at this level of abstraction is called register transfer level (RTL) design.

11.1 ALGORITHMIC STATE MACHINE CHARTS

Algorithms

An *algorithm* is a finite sequence of steps for performing a task or solving a logical or mathematical problem. All digital systems can be viewed as implementing algorithms. In some systems, this view is obvious; for example, when a system implements a computational algorithm. In others, the algorithmic nature of the system may not be as readily apparent.

In the previous chapter we used state diagrams as a graphical way of conceptualizing and representing FSMs. A state diagram is also a graphical way of representing an algorithm.

ASM Chart

Another way of graphically representing a FSM or an algorithm, one that is an alternative to a state diagram, is an *algorithmic state machine chart* (ASM chart). A FSM represented by an ASM chart is often referred to as an *algorithmic state machine* (ASM). This method is preferred by some designers, particularly when conceptualizing a complex system's implementation in terms of its algorithmic nature. ASM charts are also easily translated to VHDL code.

State diagrams, state tables, and ASM charts are different forms of FSM representations that provide equivalent information. The choice of which to use is often simply based on a designer's preference.

Basic ASM Chart Elements

The basic elements of an ASM chart are the state box, decision box, and conditional output box (Figure 11.1.1). In this section, we consider these elements and their use in low-level ASM charts that directly describe FSMs.

State Box

Each state in an ASM chart is represented by a *state box*. The symbol for a state box is a rectangle (Figure 11.1.1(a)). A state box has one *entrance path* and one *exit path*. The state's name is listed above the state box to the left. If a state assignment has been made, the state code is listed above the box to the right. A state box is equivalent to a state circle in a state diagram.

Only one state in an ASM chart is active at a time. Like the FSM it represents, an ASM chart is in one state for the duration of a clock cycle. At the end of the clock cycle it transitions to its next state.

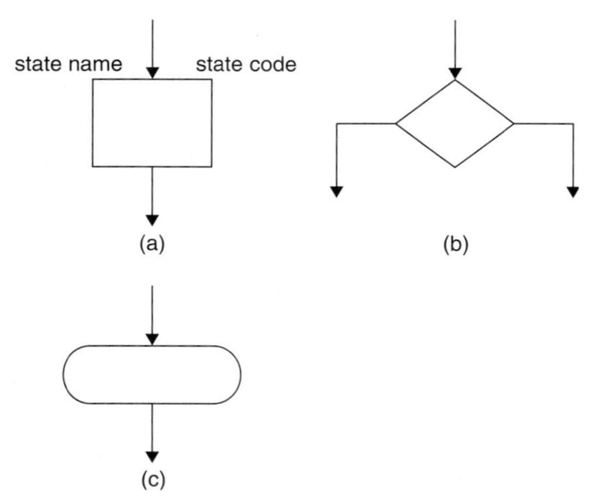

FIGURE 11.1.1
Basic elements of an ASM chart: (a) state box, (b) decision box, and (c) conditional output box.

Inside a state box are listed signal assignments that are made unconditionally while in that state. These are assignments that depend only on the state of the FSM (Moore assignments).

Immediate Assignments

Assignments made in a state box can be either immediate or delayed. An *immediate assignment* takes effect immediately upon entering the state.

Most low-level ASM charts deal with only binary-valued signals. For conciseness, only assignments that assert a signal are listed in a state box. The asserted signal's name is written in the state box without writing an assignment symbol and value. If a signal's name does not appear in a state box, it is assumed that the signal is assigned its unasserted value in that state.

Alternatively, if a style of ASM chart is used where an assignment symbol is used to represent an immediate assignment, then an equal sign (=) is used and the assigned value is specified.

Delayed Assignments

A *delayed assignment* does not take effect until the state following the one in which the assignment is made. On a low-level ASM chart, we simply write in the state box the name of the signal that is asserted to cause the delayed assignment to occur. For example, consider a counter whose output is named `count` and that has an enable input named `en_count`. If the value of `count` is to be incremented in a particular state, then `en_count` appears in that state box. This means that `en_count` is immediately asserted in that state. At the next triggering clock edge, `count` is incremented. Therefore, the newly incremented value of `count` is not available until the next state time.

ASM charts can be used to represent systems at different levels of abstraction. An ASM chart representing an algorithm at a high level of abstraction may use an assignment symbol to represent a delayed assignment. Either a left or a right pointing arrow ($\leftarrow$) or ($\rightarrow$) is used. This assignment symbol is called the *delay operator*. For example, if the counter is to be incremented in a particular state, we could write in the state box:

$count \leftarrow count + 1$

The new counter value (count + 1) is computed in the current state, but the counter is not loaded with this new value until the next triggering clock edge. So, the new count value is not available at the counter's output until the next state. This is an abstract representation of the operation to be performed on the counter and would likely be used in an ASM chart representing an algorithm at a high level of abstraction. Later, in a lower-level ASM chart, this operation is represented by writing only the name of the counter's enable input in the state box.

Note that signals assigned a value with a delay operator are not limited to scalar signals and could be vectors. In addition, signals assigned a value using the delay operator keep their previous values until a new value is assigned.

Decision Box

A *decision box* (or condition box) is a diamond-shaped box (Figure 11.1.1(b)) that contains a conditional expression involving the FSM's inputs. This condition determines the exit path to the next state. A decision box is used in conjunction with a conditional output box to determine the conditions under which Mealy output assignments are made.

A decision box has one entrance path and two exit paths. One path is taken when the condition is true and the other path is taken when it is false. Exit paths are usually labeled 1 and 0 or T and F, for true and false, respectively.

Conditional Output Box

A *conditional output box* specifies assignments that are made when all the conditions in the decision boxes that provide the path from a state box to the conditional output box are true. These are Mealy assignments. An output specified in a conditional output box can change value during the state time, if the inputs in the decision boxes leading to the conditional output box change value. A conditional output box has rounded corners (Figure 11.1.1(c)) to differentiate it from a state box. It has one entrance path and one exit path.

ASM Block

An *ASM block* consists of one state box and the tree of decision boxes and conditional output boxes that may optionally emanate from that state box up to, but not including, the state boxes that terminate the tree. An ASM block has one entrance path and one or more exit paths.

Each ASM block describes operations executed in a single state. As previously stated, the state box represents the state just as a circle does on a state diagram. The decision boxes and conditional output boxes represent the state transitions and Mealy outputs that are represented by labeled directed arcs in a state diagram.

Decision boxes and conditional output boxes belong to a single state box and cannot be shared by other state boxes.

The operations specified in the state box and all decision boxes and conditional output boxes in an ASM block execute concurrently while in that state.

ASM Block Example

A simple example of an ASM block is given in Figure 11.1.2. The portion of the ASM chart that represents the single ASM block `s2` is enclosed by a dashed rectangle. This ASM block has one entrance path and three exit paths. The state represented by this ASM block is state `s2`. State boxes `s3`, `s4`, and `s5` are not part of this ASM block. Output `v` is unconditionally asserted in state `s2`. Thus, `v` is a Moore output. Output `w` is asserted only if input `a` is asserted. Output `x` is asserted only if inputs `a` and `b` are both asserted. Outputs `w` and `x` are Mealy outputs.

Parallel versus Series Decision Boxes in an ASM Block

Two equivalent ASM blocks are shown in Figure 11.1.3. These ASM blocks are equivalent, because of the property that all operations within an ASM block are performed concurrently, not sequentially.

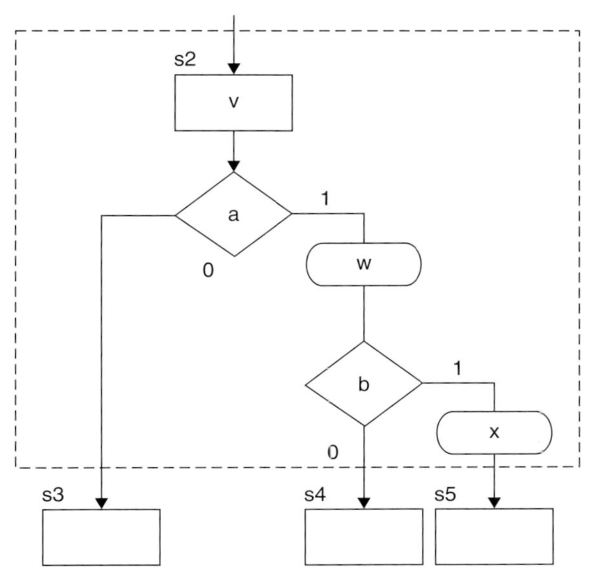

FIGURE 11.1.2
An ASM block.

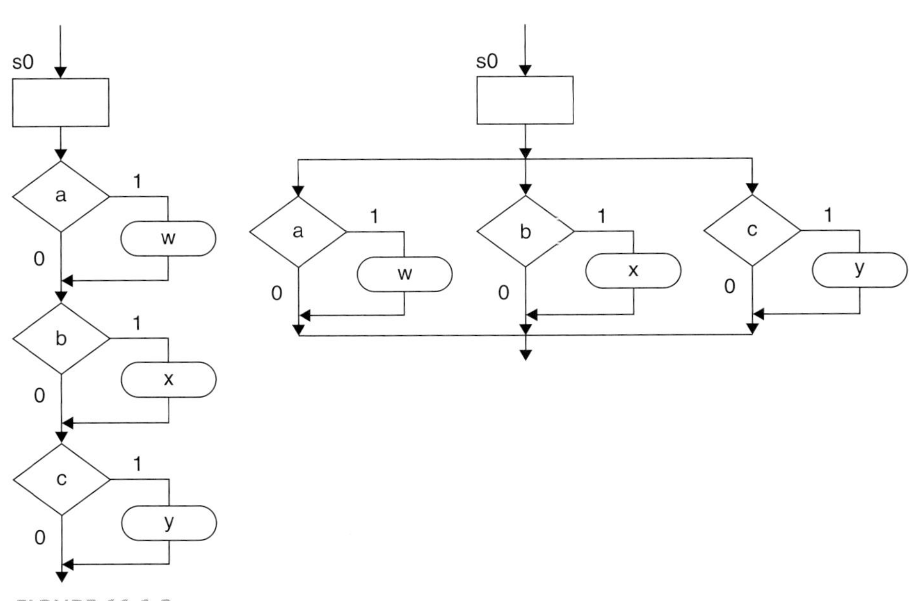

FIGURE 11.1.3
Equivalent ASM blocks.

In each case the ASM block indicates that, in state `s0`, `w` is asserted only if `a` is 1, `x` is asserted only if `b` is 1, and `y` is asserted only if `c` is 1. While the series representation is often preferred, they are equally valid.

General Rules for Constructing an ASM Chart

An ASM chart consists of one or more interconnected ASM blocks. For an ASM chart to be well defined, there must be a unique next state for each state and set of conditions. And, each exit path from an ASM block must connect to a state box.

A decision box in an ASM block must follow and be associated with a state box. A conditional output box must follow a decision box.

Moore FSM Positive Edge Detector ASM Chart

An ASM chart for the Moore FSM positive edge detector, equivalent to the state diagram in Figure 10.4.1, is given in Figure 11.1.4.

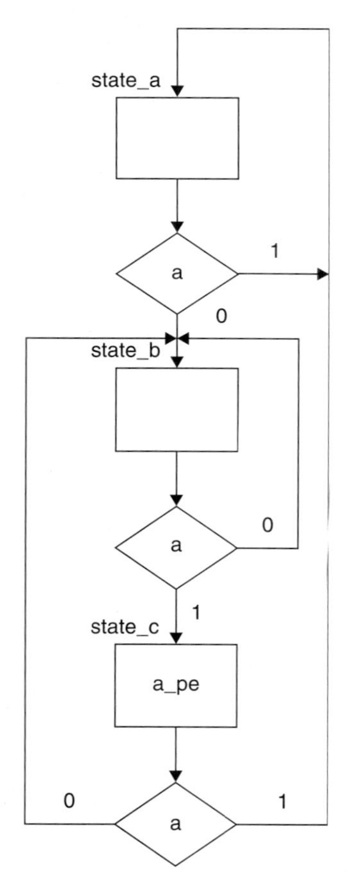

FIGURE 11.1.4
ASM chart for Moore FSM positive edge detector.

There are three ASM blocks in this chart. Note that, in an ASM chart, the ASM blocks are usually not delineated by dashed lines, as was done for conceptual purposes in Figure 11.1.2. Each ASM block in Figure 11.1.4 consists of a state box and a single decision box. For example, the state box for `state_a` is followed by a decision box that determines, based on the value of `a`, whether the next state transition is back to `state_a` or is to `state_b`. The conditional expression in the decision box is simply the input name, `a`. This expression is equivalent to `a` = 1.

Output `a_pe` is asserted only in state `state_c`. The assertion of `a_pe` is indicated by simply writing the name `a_pe` in the state box `state_c`. This means `a_pe` is asserted in this state. Since `a_pe` is asserted high, this is equivalent to writing `a_pe` = 1 in state box `state_c`. The other states do not list `a_pe`, so `a_pe` is not asserted in those states. This is equivalent to writing `a_pe` = 0 in state boxes `state_a` and `state_b`.

With this shorthand notation in mind, the correspondence between this ASM state chart and the state diagram in Figure 10.4.1. is clear.

Mealy FSM Positive Edge Detector ASM Chart

An ASM chart for the Mealy FSM positive edge detector, equivalent to the state diagram in Figure 10.4.4, is given in Figure 11.1.5.

There are two ASM blocks in this chart. The first consists of the state box for `state_a` and a decision box. The second consists of a state box for `state_b`, a decision box, and a conditional output box.

In ASM block `state_b`, output `a_pe` is listed in a conditional output box. Since the path to the conditional output box is through the decision box's exit path corresponding to `a` being asserted, the assertion of `a_pe` is conditioned on `a` being 1. This assignment is conditional and immediate.

The correspondence between this ASM chart and the state diagram in Figure 10.4.4 is also clear.

ASM Charts versus State Diagrams

While state diagrams and ASM charts are equivalent, there are some practical differences that may make one or the other preferable.

A state diagram is usually drawn to represent a completely specified FSM. Thus, for every state, a next state transition is shown for each possible input combination. In addition, the value of every output is indicated for each possible input condition while in that state. The resulting extensive labeling of arcs for Mealy outputs and output values inside, or associated with, state symbols for Moore outputs can make state diagrams for practical FSMs difficult to interpret. This extensive labeling also implies that in each state each input and each output is equally important to the understanding of the operation of the FSM. For a practical FSM, this is usually not the case.

Clearly, state diagrams can be simplified by following certain conventions like not labeling transitions back to the same state in Moore machines, as was done in some examples in the previous chapter.

The state boxes in ASM charts only list the outputs that are asserted in each state. These are the outputs that are important when in that state. The decision boxes in an

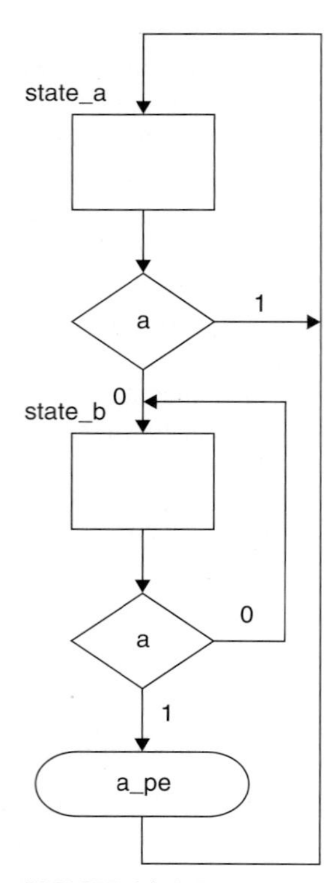

FIGURE 11.1.5
ASM chart for Mealy FSM positive edge detector.

ASM block only list the inputs that are important in determining the next state transitions and the conditional outputs when in that state. These are the inputs that are important in that state. This simplification often makes it easier to interpret the operation of an FSM represented by an ASM chart.

11.2 CONVERTING ASM CHARTS TO VHDL

If the three-process FSM template in Listing 10.3.1 is followed, writing a VHDL description of a FSM from an ASM chart is straightforward. The state register and next state processes are written in exactly the same fashion as before for state diagrams. For the output process, some additional interpretation is required to write the VHDL code from the ASM chart.

For the next state process, the contents of the case statement's when branches are determined directly from the ASM blocks. For example, Figure 11.2.1 shows the ASM chart for the Mealy positive edge detector. The middle column in the figure

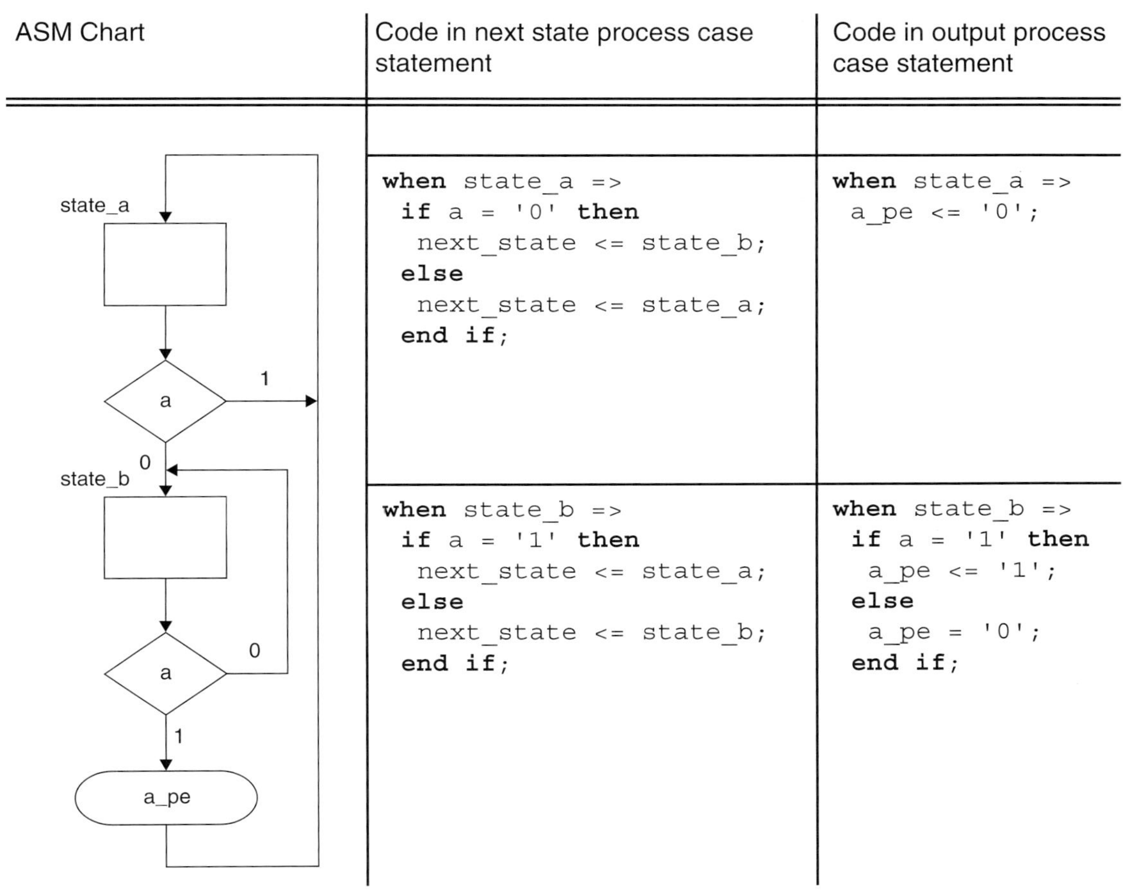

FIGURE 11.2.1
VHDL when branch code corresponding to ASM blocks for Mealy FSM positive edge detector.

gives the when branches of the case statement in the next state process. Each when branch is associated with the ASM block to its left.

Missing Output Assignments in ASM Charts

Since the output and next state processes in the three-process FSM template describe combinational logic, each time one of these processes executes, each of its outputs must be assigned a value.

For the output process, only assignments that assert outputs are listed in the ASM chart. This means that, when writing code for a when branch that corresponds to an ASM block where no assignment to an output is indicated, the appropriate assignment must be added.

For example, in the ASM block for state_a in Figure 11.2.1, there is no assignment to a_pe listed. Accordingly, a_pe is not asserted in that state and must be assigned '0'. In the rightmost column, the when branch for state_a includes this assignment.

In state_b, when a = '1', the conditional output box indicates that a_pe is asserted. Accordingly, a_pe is assigned a '1' by the if statement in the when branch corresponding to the ASM block for state_b, in the rightmost column in Figure 11.2.1.

In state_b, when a = '0', no conditional assignment to a_pe is listed. Accordingly, a_pe is assumed to be unasserted for this condition. The if statement in the rightmost column requires an else branch that assigns a_pe a '0', since that is its unasserted value. Note that we want to assign a value to a_pe equal to its unasserted value. We do not want to remember the value of a_pe, which would be the result if we left out the else branch.

When the code for the when branches is placed in the case statements of the appropriate processes and the state register process is added, the description is complete. The resulting VHDL code is the same as in Listing 10.4.3.

A Simpler Output Process

When writing VHDL code for a FSM directly from an ASM chart, there is a simpler approach that can be used for the output process. Since the ASM chart only indicates when a signal is asserted, we can simplify writing the code from the chart by including statements at the beginning of the output process to assign all outputs their unasserted values.

These default assignment statements are followed by the case statement. The when branches of the case statement then only contain assignments to the outputs that are to be asserted. If the assertion of an output is conditional (Mealy output), the assignment statement is contained in an appropriate if statement. Code following the simpler form of output process is given in Listing 11.2.1.

LISTING 11.2.1
Simplified output process for positive edge detector written directly from ASM chart.

```
outputs: process (present_state, a)
begin
   a_pe <= '0';    -- default unasserted value

   case present_state is
      when state_a =>
      null;

      when state_b =>
      if a = '1' then
         a_pe <= '1';
```

```
        end if;
    end case;
end process;
```

Using Concurrent Statements for FSM Outputs

It is sometimes simpler to replace the output process with conditional signal assignment statements. The ASM chart can be examined and a statement written to assign the asserted value to an output for all the conditions where such an assignment is warranted and default to the unasserted value for all other conditions.

For example, from examination of the ASM chart in Figure 11.2.1, we could write the conditional signal assignment statement:

```
a_pe <= '1' when present_state = state_b and a = '1'
            else '0':
```

This single statement replaces the output process in Listing 10.4.3.

11.3 SYSTEM ARCHITECTURE

In previous chapters we have considered the design of combinational systems and simple sequential systems. We have also considered the FSM approach to designing sequential systems.

In theory, any synchronous sequential system can be implemented as a single FSM. However, for a complex system with many inputs, outputs, and states, implementation as a single FSM is impractical. It is preferable to decompose such a system into simpler components.

Once the operation of a system has been specified in terms of an algorithm, the tasks that must be accomplished to perform the algorithm are known. For each task one or more well-defined components and their interconnection to accomplish the task are determined. Each task must be performed at the appropriate time, as determined by the algorithm, to accomplish the system's overall task. In effect, the system is decomposed into *functional units* (combinational and sequential components) that implement tasks, and one or more FSMs that control when each task is performed.

The interconnection of functional units and FSMs comprises the architecture of the complex system. Conceptually, two common architectural approaches for a complex sequential system are: functional units and control, and data path and control.

These architectures are conceptually represented in Figure 11.3.1 for an algorithm that consists of tasks A, B, and C. In both architectures, the control part of the system is implemented by one or more FSMs.

Figure 11.3.1(a) represents an architecture where each functional unit is enabled by the control unit to carry out a particular task. When a functional unit completes its task, it signals the control unit via a status signal. The control unit sequences the operation of the functional units in the appropriate order, to achieve the overall

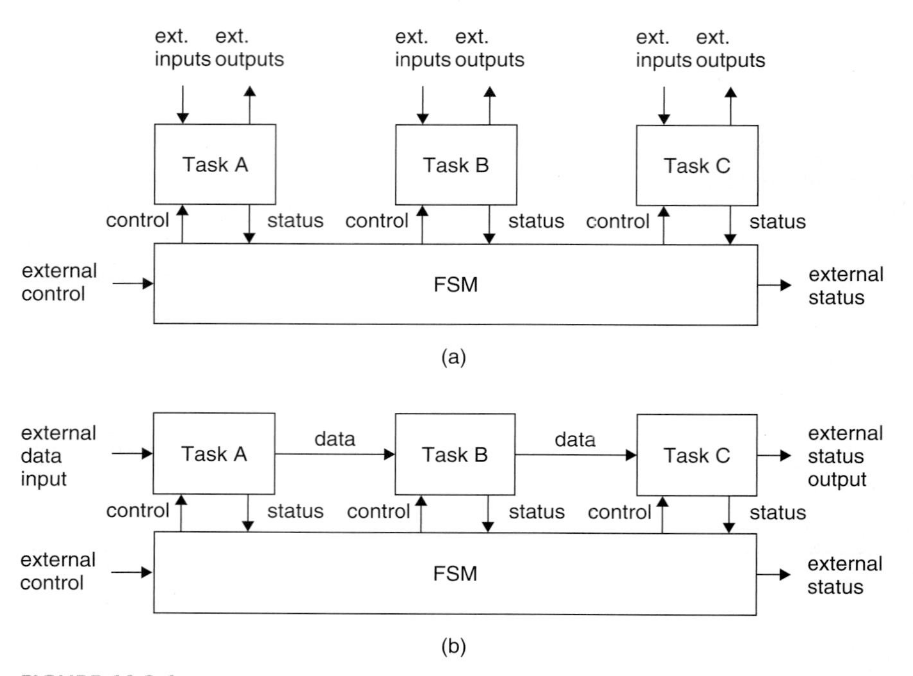

FIGURE 11.3.1
Common architectures: (a) functional units and control, (b) data path and control.

system function. In a simple system the tasks may be executed in sequence. Alternatively, multiple functional units may be active at one time or a single functional unit may be active repeatedly.

Figure 11.3.1(b) represents an architecture where there is a data path and control. The system is decomposed into a collection of functional units through which data flows (a data path). Each functional unit consists of combinational logic and a register. The control enables each functional unit to carry out a particular task (data manipulation) on the data as it passes along the data path. The sequence of data manipulations performed accomplishes the desired overall transformation of data. The data output from the data path is the computed result.

The main distinction between these two approaches is whether the system is viewed primarily as carrying out a sequence of actions or a sequence of data manipulations. In both approaches the functional units are accomplishing the tasks and the control unit (FSM) is sequencing task execution to perform the algorithm.

Register Transfer Level (RTL) Design

The methodology used to design the architecture in Figure 11.3.1(b) is called register transfer level (RTL) design. The data path stores, routes, and processes data. Its operation is described in terms of data transfers from one register to another register and data manipulations performed by combinational logic that exists between the registers.

The control subsystem controls the functions performed in the data path. The control system is a FSM and the architecture is also referred to as a *finite state machine with data path (FSMD)*. Registers in the data path and the control unit (FSM) use the same system clock signal.

Data Path

The *data path* is used to perform data manipulations or numerical computations. It consists of the registers that store data and the combinational logic used to transform the data. Input data flows into the data path and the result flows out. Input and output data are often vectors and the data path connections implemented as buses.

The data path receives control signals from the control subsystem and provides status signals to the control subsystem. The control inputs to the data path determine what functions are performed by the combinational logic and when data path registers are loaded.

For example, in the Inhibit Logic System in Figure 10.8.1, repeated here in Figure 11.3.2, the counter, buffer register, multiplexer, and three-state output buffer comprise the data path. The Inhibit Logic FSM is the control.

When viewed in the context of Figure 11.3.1(b), there is no external data input for this data path, but it does have an external output, `D7-D0`. The external control inputs are `cnt_enable`, `up`, `sel`, and `oe_bar`. In this example two of the external control inputs go to the FSM, `sel` and `oe_bar`. The only control signal from the FSM to the data path is `inhibit`. There is no status signal from the data path back to the FSM.

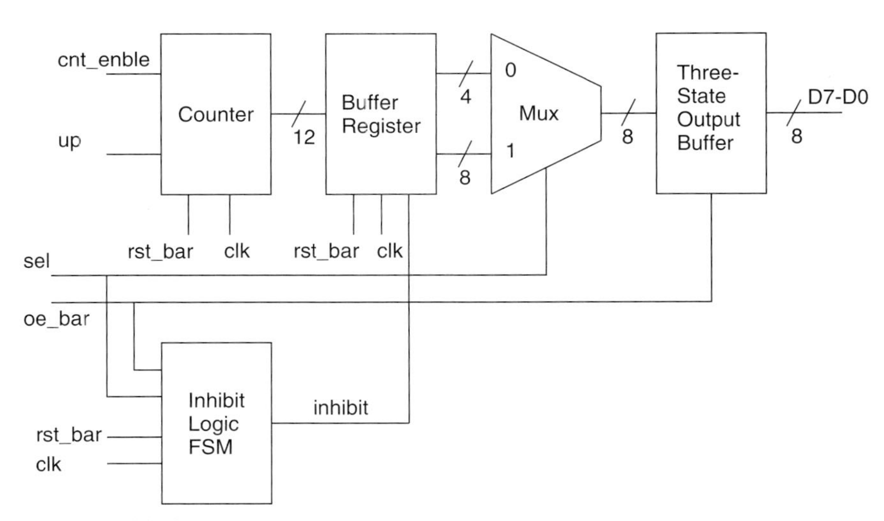

FIGURE 11.3.2
Counter to microprocessor interface architected as a data path and control.

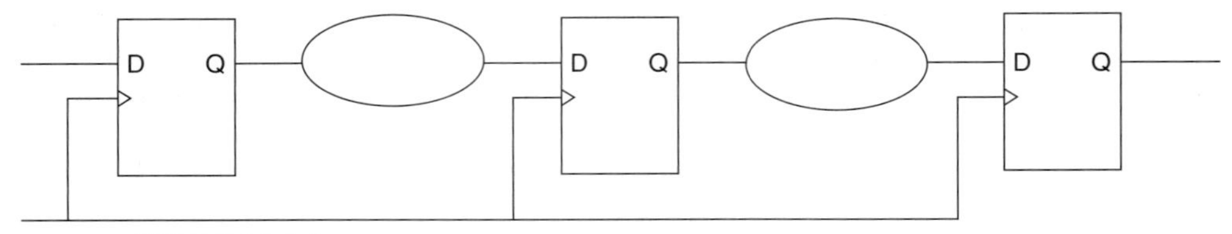

FIGURE 11.3.3
The cloud representation of a data path.

Generic View of a Data Path

A more generic view of a data path represents it as a sequence of simple storage registers separated by combinational logic (Figure 11.3.3). This view is particularly appropriate for a pipelined data path.

Combinational Logic "Cloud"

As shown in Figure 11.3.3, the combinational logic between registers is often conceptually represented by a "cloud" symbol. Each combinational logic "cloud" either transforms or routes the output of the storage register to its left and provides the transformed or routed value as input to the storage register on its right. Often, a simple storage register and the combinational logic associated with it combine to produce an operational register such as a shift register, counter, or arithmetic and logic unit (ALU). Alternatively, the combinational logic might provide data routing between registers, such as demultiplexing or multiplexing.

Data in registers represent variables manipulated by the data path. The combinational logic in the data path manipulates the input data stream, as specified by control inputs from the control subsystem. Outputs of registers become valid after a triggering clock edge. Register outputs are transformed by the combinational logic during the time between triggering clock edges. Results computed by the combinational logic are stored in a register at the next triggering clock edge.

Control

The control subsystem produces control signals based on its external inputs, status inputs from the data path, and its present state. Control signals direct data routing in the data path by selecting multiplexer and demultiplexer channels and enabling or disabling registers. In addition to their clock inputs, registers typically have one or more enable inputs that are controlled to determine whether or not a register stores its input data at the next triggering clock edge. Combinational logic blocks may also have control inputs that determine what function they perform at any given time. For example, the select inputs to demultiplexers and multiplexers are controlled to route data.

Control Word

The values of all the control inputs to the data path are sometimes aggregated to form a *control word*. The bits of the control word determine the source and destination registers for data transfers and the operations performed in the data path during each clock cycle. To cause the data path to accomplish its algorithm on the input data, the control word must have the appropriate value for each clock cycle.

Scheduling

The sequence of control words from the FSM is said to schedule the use of the resources (registers and combinational logic) in the data path.

11.4 SUCCESSIVE APPROXIMATION REGISTER DESIGN EXAMPLE

As an example of the partition of a system into a data path and control, this section describes the design of a *successive approximation register (SAR)*. A SAR is an operational register that is a component of a successive approximation analog-to-digital converter.

Analog-to-Digital Converter

An *analog-to-digital converter (ADC)* converts its analog input voltage to a digital output. An ADC's analog input range is constrained and only voltages within its range can be converted. For example, an ADC might have a 0 to 2.5V input range. Accordingly, the full-scale (FS) value of this input range is 2.5 V.

ADC Output as a Binary Fraction

An n-bit binary ADC produces an n-bit binary output that is proportional to its analog input voltage. The n-bit output is a binary fraction that represents the analog input voltage as a fraction of the ADC's FS input value. The binary point is assumed to be to the left of the most significant bit of the output. For example, if its input voltage is 1.875 V, a 4-bit ADC with a 0 to 2.5 V input range would produce an output of 1100. This output represents the binary fraction 0.1100 or 12/16 (75%) of FS. Thus, from the binary output, the analog input voltage can be computed to be 0.75×2.5 V = 1.875 V.

Successive Approximation ADC

One popular type of ADC is a successive approximation ADC. It performs a conversion by generating a sequence of approximations that ultimately converge to the final binary output. A successive approximation ADC is constructed using a SAR, a digital-to-analog converter (DAC), and an analog comparator (Figure 11.4.1).

SAR Operation

The purpose of the SAR is to generate the sequence of approximations that are output to the DAC during a conversion. At power on, the SAR's `reset_bar` input is asserted for a short time, putting the SAR in its initial state.

When a conversion is desired, the SAR's start of conversion (`soc`) input is asserted. The SAR then executes its successive approximation algorithm to determine the binary result. During execution of this algorithm, the SAR uses the logic value at its `comp_in` input to determine the next approximation value to output to the DAC. When the conversion is complete, the end of conversion (`eoc`) status output is asserted and the last approximation is the binary result. This value is available at output `result`.

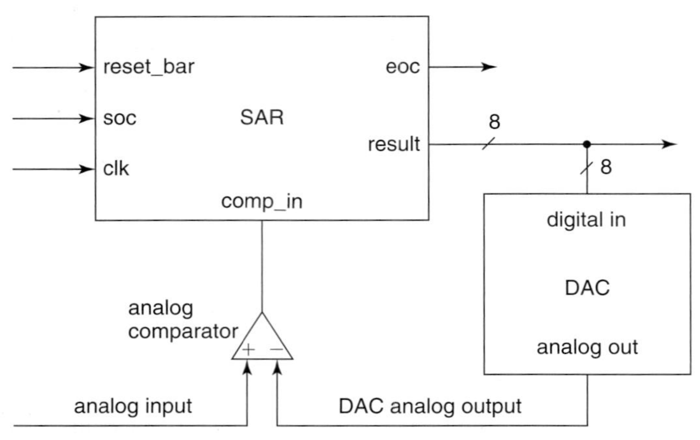

FIGURE 11.4.1
Successive approximation ADC block diagram.

Successive Approximation Algorithm

The successive approximation algorithm is illustrated in Figure 11.4.2. This flowchart implies that the content of a single register is being manipulated, bit by bit, to produce the result.

The algorithm is a binary search that requires one iteration (clock cycle) to determine each bit of the result. During each iteration, an approximation to the final value is used as input to the DAC to generate an analog voltage (Figure 11.4.1). The analog comparator compares the DAC's analog output voltage with the analog input voltage being converted.

During the first iteration, the most significant bit of the approximation is a 1 and all of its other bits are 0s. This approximation value is output to the DAC to determine the value of the most significant bit of the result.

If the DAC analog output voltage produced by this approximation is less than or equal to the input analog voltage, the comparator output is a 1. In this case, the most significant bit in the final result must be a 1, and is, therefore, left a 1 in the approximation.

If, instead, the DAC voltage produced by the first approximation is greater than the input analog voltage, the comparator output is a 0. In this case, the most significant bit in the final result must be a 0, and this bit is changed to a 0 in the approximation.

To determine the next bit in the result, the corresponding bit in the approximation is made a 1, with all other more significant bits of the approximation keeping their previously determined values. The output of the comparator is used to determine whether the bit currently being determined must be a 1 in the final result and, therefore, must be left a 1 in the approximation, or whether it must be changed back to 0. The steps described in this paragraph are repeated for each bit until the least significant bit has been determined, completing the conversion.

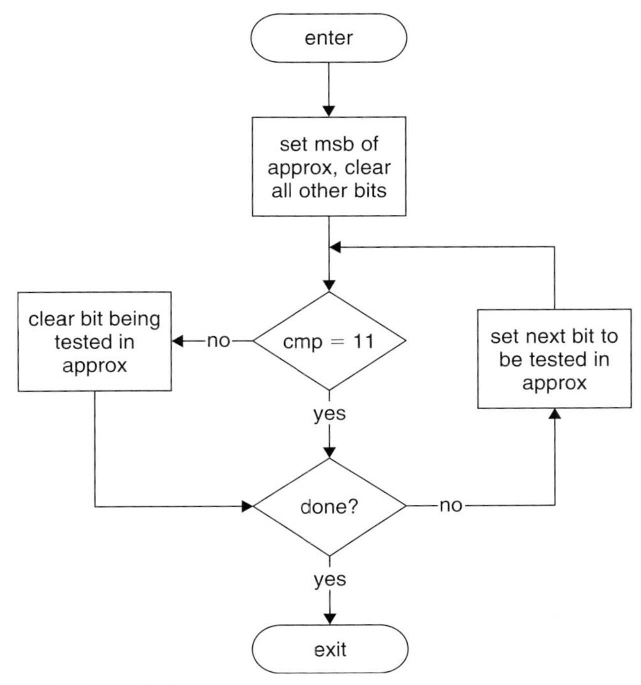

FIGURE 11.4.2
Flowchart of the successive approximation algorithm.

Figure 11.4.3 shows the sequence of approximations created by a 4-bit successive approximation ADC with a range of 0 to 2.5 V and an input voltage of 1.875 during a conversion. The figure shows the approximations output during the four iterations required for the conversion. The fifth interval shows the final result at the end of the conversion.

VHDL Representation of SAR Algorithm

The successive approximation algorithm represented in Figure 11.4.2 could, alternatively, have been described directly in VHDL, as shown in Listing 11.4.1, for an 8-bit conversion.

This behavioral description is not synthesizable because the synthesizer requires all of the wait statements in a process to be identical. However, it is a concise and unambiguous way to describe the algorithm to someone who knows VHDL.

RTL Design of SAR

The RTL design of the SAR involves creating a hardware architecture to implement the successive approximation algorithm. When creating such an architecture, it may be that a direct one-to-one mapping of the algorithm to hardware is not conceptually the simplest approach. For example, in the architecture to be described, two registers are actually used, rather than the single register implied in Figure 11.4.2 and Listing 11.4.1.

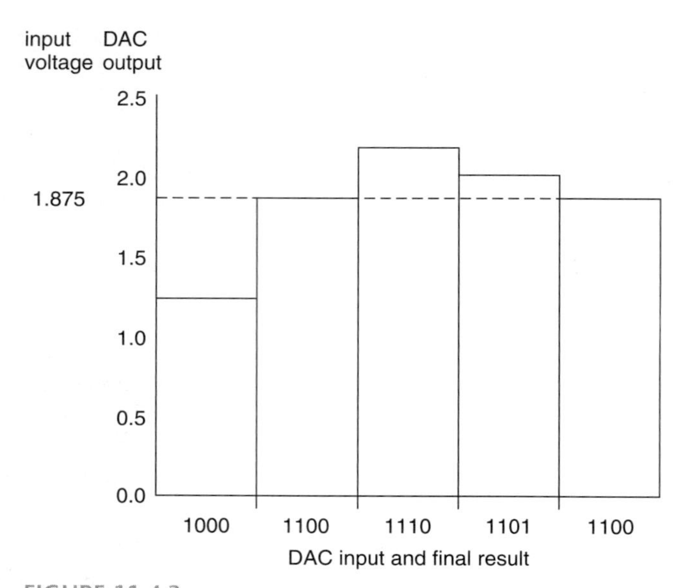

FIGURE 11.4.3
Approximations for a 4-bit conversion of an input voltage of 1.875 V by a 0 to 2.5 V ADC.

LISTING 11.4.1
Nonsynthesizable process describing successive approximation algorithm.

```
sac_alg : process
   variable approx : std_logic_vector(7 downto 0);
begin
   wait until soc = '1';        -- wait to start conversion
   eoc <= '0';                  -- unassert eoc
   approx := (others => '0');  -- generate approximations
   for i in 7 downto 0 loop
      approx(i) := '1';
      result <= approx;         -- output approximation
      wait until clk = '1';
      wait for tpda;            -- DAC and comparator progation delay
      if comp_in = '0' then     -- read comparator and clear bit if necessary
         approx(i) := '0';
      else
         approx(i) := '1';
      end if;
   end loop;
   result <= approx;            -- output result
   eoc <= '1';                  -- assert eoc
   wait until clk = '1';
end process;
```

SAR RTL Architecture

In our two-register approach, register `approx_reg` is used to hold the value of the previous approximation during each iteration of the conversion and to hold the final result at the end of the conversion.

Another register, `shftreg`, holds a single 1 in the position of the bit of the result being determined during a particular iteration of the algorithm. This 1 is shifted right one bit position after each bit of the result is determined.

The logical OR of the contents of these two registers provides the current approximation to the DAC's inputs.

A RTL block diagram of the SAR is given in Figure 11.4.4. It shows the four components that comprise the system: two registers, an OR gate, and a FSM. This diagram is for an 8-bit conversion, but the design is easily scaled to any number of bits.

Component approx_reg

Component `approx_reg` is an 8-bit register that is cleared when reset. If its `load` input is asserted, `approx_reg` loads `"00000000"` at the next triggering clock edge. If its `load` input is not asserted and both its enable inputs, `en1` and `en2`, are simultaneously asserted, `approx_reg` stores its 8-bit input data at the next triggering clock edge.

Component shiftreg

Component `shiftreg` is an 8-bit right shift register that is cleared when reset. If its `load` input is asserted, `shiftreg` loads `"10000000"` at the next triggering clock edge. If its `load` input is not asserted and its enable input `en` is asserted, `shiftreg` shifts its contents to the right at the next triggering clock edge.

Component or_function

The outputs of `approx_reg` and `shiftreg` are bitwise ORed by component `or_function`. It is the output of `or_function` that is the current approximation value during the conversion and the final result when the conversion is complete.

SAR Data Path

When viewed in terms of a data path and a control, the SAR's data path consists of `approx_reg`, `shft_reg`, and `or_function`. This data path has no external data inputs. The output of `or_function` is the data path's output. Control inputs to the data path are the external input `comp_in` and outputs from the control FSM to `en`, `en2`, and `load`. The status output from the data path to the control FSM is `last_bit_sig` (`result(0)`).

SAR Control

Control `saadc_fsm` is a Moore FSM. To start a conversion, its external control input `soc` must be asserted. In response, `saadc_fsm` asserts the load inputs of both registers. At the next triggering clock edge component, `approx_reg` is loaded with `"00000000"` and `shiftreg` is loaded with `"10000000"`. The OR of these two register's outputs is the first approximation, `"10000000"`.

Output `result` from the SAR is input to the DAC (Figure 11.4.1). The DAC produces an analog output voltage proportional to its binary input value. The analog

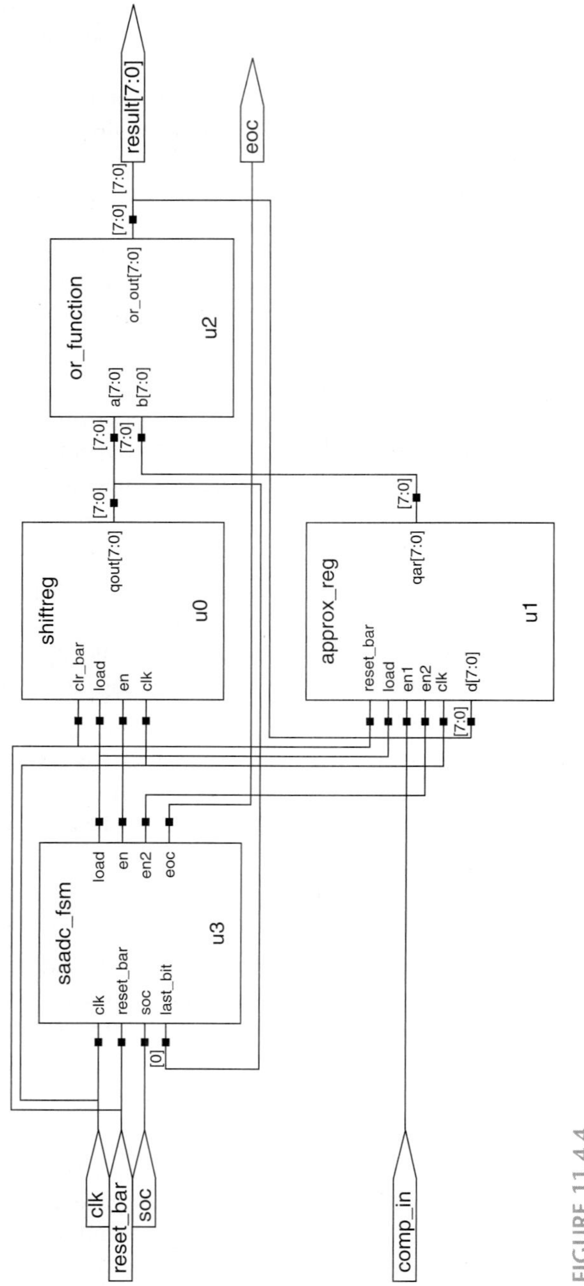

FIGURE 11.4.4
Block diagram of SAR.

comparator compares the DAC output voltage with the analog input voltage being converted. If the DAC output voltage is less than or equal to the analog input voltage being converted, the comparator output is '1'; otherwise, it is '0'.

The comparator output being a '1' means that the bit value being determined during this iteration must be a '1' in the remaining approximations and in the final result.

The SAR reads the comparator output at its comp_in input. The value at comp_in directly enables input en1 of approx_reg. Enable input en2 is controlled by output signal reg_en from the FSM. The FSM keeps en2 asserted from the start of a conversion until its end. As a result, during a conversion, if the comparator output is a '1', approx_reg is loaded with the current approximation at the next triggering clock edge, making the final value of the bit that was being determined a '1' in approx_reg.

In contrast, if the comparator output is a '0', then approx_reg is not loaded with the current approximation at the next triggering clock edge. This leaves approx_reg with its previous value, which has a 0 in the position of the bit that was just determined.

At each triggering clock edge, the shift register contents are shifted to the right. When the '1' in the shift register is in the least significant bit position, the last bit is being determined and the conversion is complete at the next triggering clock edge.

SAR Control FSM State Diagram

With the data path determined and a knowledge of how the successive approximation algorithm maps to the data path, we can develop a state diagram for the control FSM. For a simple FSM we can draw the state diagram directly. For a more complicated FSM it may be helpful to create a table listing the operations that must take place in the data path and the control signal values (control word) for these operations. Operations that can take place simultaneously are grouped together and the corresponding control word asserts all the signals to cause these operations.

Since this is to be a Moore FSM, each operation or set of operations that take place simultaneously requires a separate state. For the SAR, the operations are do nothing, initialize the registers, and shift the shift register. These operations are listed in Table 11.4.1.

Table 11.4.1
Operations for SAR FSM.

Operation	State	en	en1	en2	eoc	load
no op	idle	0	comp_in	0	1	0
initialize registers	initial	0	comp_in	0	0	1
shift register	shift	1	comp_in	1	0	0

The outputs of the FSM are en, en2, eoc, and load. Of these, en, en2, and load are control signals to the data path. Output eoc is an external status output. Also listed in the table is en1, which is not an output of the FSM, but an input of the approx_reg that is driven by comp_in. For approx_reg to be loaded, both en1 and en2 must be asserted. For the shift register to be loaded, only en must be asserted.

The "no operation" corresponds to the ADC being idle; this is the state of the system before and after a conversion. State idle is introduced to control this operation. The only output asserted in this state is eoc, to indicate the end of a prior conversion.

At the beginning of a conversion, in response to soc being asserted, both shiftreg and approx_reg must be loaded with their initial values. This is accomplished by asserting load. The state introduced to control this operation is named initial.

After the initial state, the shift register must be shifted at each triggering clock edge until its least significant bit is a 1. In addition, if comp_in is a 1, approx_reg must be loaded with the output of or_function. State shift is added to control this operation. In state shift, if the least significant bit of the shiftreg is a 0, the next state is again shift. If the least significant bit of the shiftreg is a 1, the conversion is complete and the next state is idle.

The completed state diagram is shown in Figure 11.4.5.

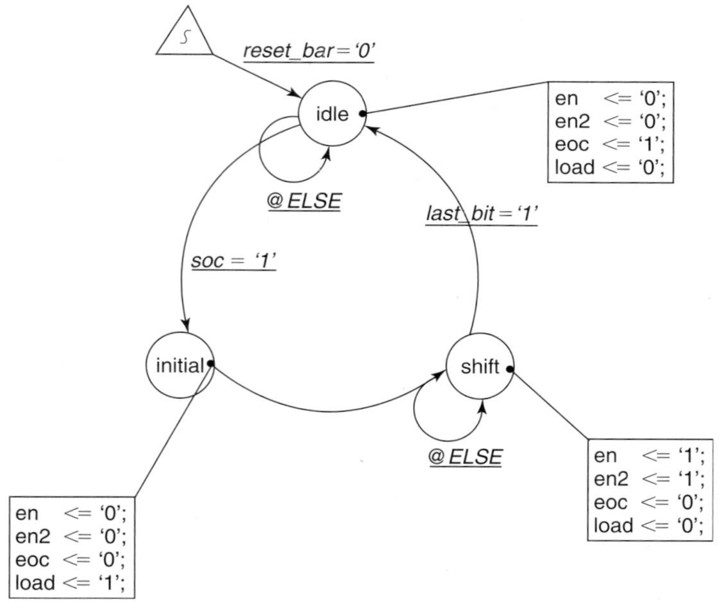

FIGURE 11.4.5
SAR FSM state diagram.

Starting from the `idle` state, if `soc` is asserted, the first triggering clock edge (clock 1) puts the FSM in the `initial` state. In this state, a control word is output by the FSM that enables both registers to be loaded with their initial values at the next triggering clock edge. The next triggering clock edge (clock 2) loads the registers and places the FSM in the `shift` state.

The next seven triggering clock edges (clocks 3–9) determine in sequence, starting with the most significant bit, 7 bits of the result. At each of these triggering clock edges, the state transition is from the `shift` state back to the `shift` state.

The ninth triggering clock edge causes the 1 that is being shifted to the right in the shift register to be shifted into bit position 0. This corresponds to the condition `last_bit = '1'`. At the last triggering clock edge (clock 10), determination of the least significant bit of the result is complete and the FSM transitions back to its `idle` state. In the `idle` state `eoc` is a `'1'`.

SAR ASM Chart

Instead of conceptualizing the SAR control using a state diagram, an ASM chart could be used (Figure 11.4.6).

By only listing the signals that are asserted, the ASM chart is simpler than the previous state diagram.

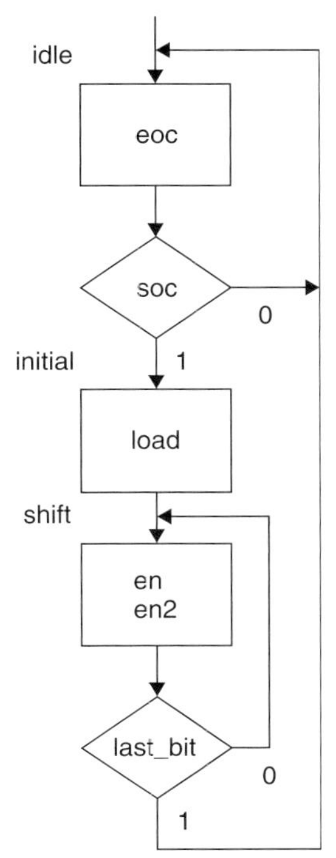

FIGURE 11.4.6
ASM chart for SAR FSM.

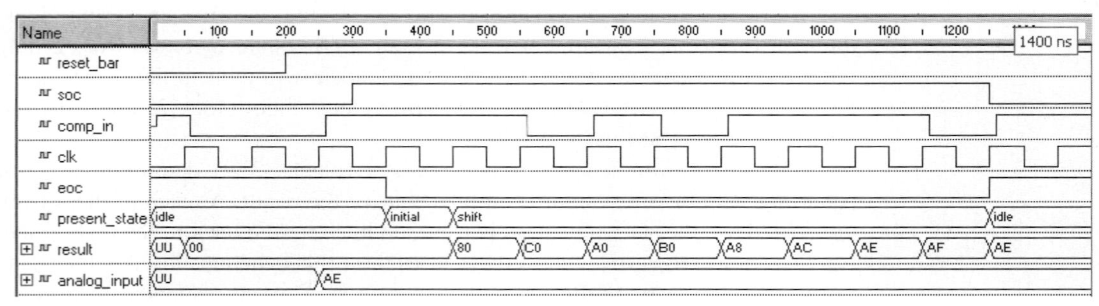

FIGURE 11.4.7
Simulation waveforms from successive approximation conversion.

Expanding the Number of Bits in the ADC

This design can be modified to produce an n-bit successive approximation converter, for any n, by simply changing `approx_reg` and `shiftreg` to n-bit registers, `or_function` to an n-bit OR, and using an n-bit DAC. Control FSM `saadc_fsm` does not need to be modified for a different value of n. The conversion time is always $n + 2$ clock periods.

Simulation of Successive Approximation Conversion

Simulation waveforms from a conversion are shown in Figure 11.4.7. In this simulation run, the value of the analog input voltage was set to 68% of FS input range. Since 68% corresponds to 174/256, the expected conversion result is 174 or 10101110 in binary.

This simulation was performed by creating a VHDL model of a digital equivalent of the DAC and analog comparator combination. This model is used with the SAR to form a complete ADC. The ADC was then used as the UUT in a testbench. Details of the combined DAC and comparator model and the testbench are given in Section 14.5.

The waveform for `present_state` shows the sequence of states in a conversion. The fact that the states are enumerated makes it easy to observe the present state as the simulation progresses, which is very helpful for debugging. The waveform for `result` shows the sequence of approximations from the initial approximation to the final result.

A description of the SAR control `saadc_fsm` as a Moore FSM is given in Listing 11.4.2.

The SAR components are interconnected by the structural description `saadc` in Listing 11.4.3.

LISTING 11.4.2
Description of SAR FSM.

```
library ieee;
use ieee.std_logic_1164.all;

entity saadc_fsm is
   port (
      clk : in std_logic;

      reset_bar : in std_logic; -- syn. reset
      soc : in std_logic; -- start of conversion
      last_bit : in std_logic; -- from lsb of shift register
      load : out std_logic; -- shift reg <= x"80", approx_reg <= x"00"
      en : out std_logic;  -- enable shift reg to shift
      en2 : out std_logic; -- enable approx_reg
      eoc : out std_logic); -- end of conversion
end;

architecture behavior of saadc_fsm is

   type state is (idle, initial, shift);
   signal present_state, next_state : state;

begin

   state_reg: process (clk)-- state register process
   begin
      if rising_edge(clk) then
         if reset_bar = '0' then
            present_state <= idle;
         else
            present_state <= next_state;
         end if;
      end if;
   end process;

   output: process (present_state)-- output process
   begin
      case present_state is
         when idle => en <= '0'; en2 <= '0'; eoc <= '1'; load <= '0';
```

(Cont.)

LISTING 11.4.2 *(Cont.)*

```
            when initial => en <= '0'; en2 <= '0'; eoc <= '0'; load <= '1';
            when shift => en <= '1'; en2 <= '1'; eoc <= '0'; load <= '0';
            when others => en <= '0'; en2 <= '0'; eoc <= '0'; load <= '0';
        end case;
    end process;

    nx_state: process (present_state, soc, last_bit)-- next state process
    begin
        case present_state is
            when idle =>
            if soc = '1' then next_state <= initial;
            else next_state <= idle;
            end if;

            when initial => next_state <= shift;

            when shift =>
            if last_bit = '1' then next_state <= idle;
            else next_state <= shift;
            end if;

            when others => next_state <= idle;
        end case;
    end process;
end behavior;
```

LISTING 11.4.3
Top-level structural description of SAR.

```
library ieee;
use ieee.std_logic_1164.all;

entity saadc is
    port (
        reset_bar: in std_logic;
        soc: in std_logic;  -- start conversion input
        comp_in: in std_logic;  -- from comparator output
        clk: in std_logic;
        eoc: out std_logic; -- end of conversion
        result: out std_logic_vector (7 downto 0)
        );
end saadc;

architecture structural of saadc is

    signal shift_en_sig, reg_en_sig : std_logic;
```

```
    signal load_sig: std_logic;
    signal sr_sig, sar_sig, result_sig : std_logic_vector(7 downto 0);

begin

    u0: entity shiftreg port map (clr_bar => reset_bar, load => load_sig,
        en => shift_en_sig, clk => clk, qout => sr_sig);

    u1: entity approx_reg port map (d => result_sig, reset_bar => reset_bar,
        load => load_sig,
    en1 => comp_in, en2 => reg_en_sig, clk => clk, qar => sar_sig);

    u2: entity or_function port map (a => sr_sig, b => sar_sig,
        or_out => result_sig);

    u3: entity saadc_fsm port map (clk => clk, reset_bar => reset_bar,
        soc => soc,last_bit => sr_sig(0), load => load_sig, en => shift_en_sig,
        en2 => reg_en_sig, eoc => eoc);

    result <= result_sig;

end structural;
```

11.5 SEQUENTIAL MULTIPLIER DESIGN

When we perform a binary multiplication by hand, we examine each bit of the multiplier in succession, starting with its least significant bit (Figure 11.5.1). If the multiplier bit being examined is a 1, we write down a copy of the multiplicand as the partial product resulting from that multiplier bit. If the multiplier bit is a 0, we write down 0, as the corresponding partial product. The partial product resulting from each multiplier bit is written shifted left one bit position relative to the previous partial product. This reflects the fact that each successive multiplier bit has a weight that is twice that of the bit to its right. After all the multiplier bits have

```
    1110    multiplicand
    1010    multiplier
    ----
    0000    1st partial product
   1110     2nd partial product
  0000      3rd partial product
 1110       4th partial product
--------
10001100    product
```

FIGURE 11.5.1
Example of binary multiplication performed by hand.

been examined and all partial products recorded, the partial products are added to produce the product.

If we create an architecture that directly maps this algoritm to hardware for a n-bit multiplier and a n-bit multiplicand, it would require two n-bit registers to hold the operands, n n-bit registers to hold the partial products, $n - 1$ n-bit adders, and a $2n$-bit register to hold the product.

This hardware can be significantly reduced if the algorithm is modified so that, as soon as each partial product is determined, it is added to a single register that accumulates the product. This allows the n n-bit registers used to hold the partial products and the $n - 1$ n-bit adders to be replaced by a single n-bit adder. However, logic that provides the relative 1-bit displacement to the left for each partial product's addition to the accumulated product is required.

This 1-bit displacement to the left for the addition of each partial product to the accumulated product can be efficiently accomplished by making the register that accumulates the product a right shift register. Shifting the accumulated product to the right relative to the partial product after each addition is equivalent to shifting the partial product to the left relative to the accumulated product (Figure 11.5.2). The result is the classic add and shift multiplication algorithm.

A block diagram of a data path for the modified multiplication algorithm using an 8-bit multiplier and an 8-bit multiplicand is given in Figure 11.5.3.

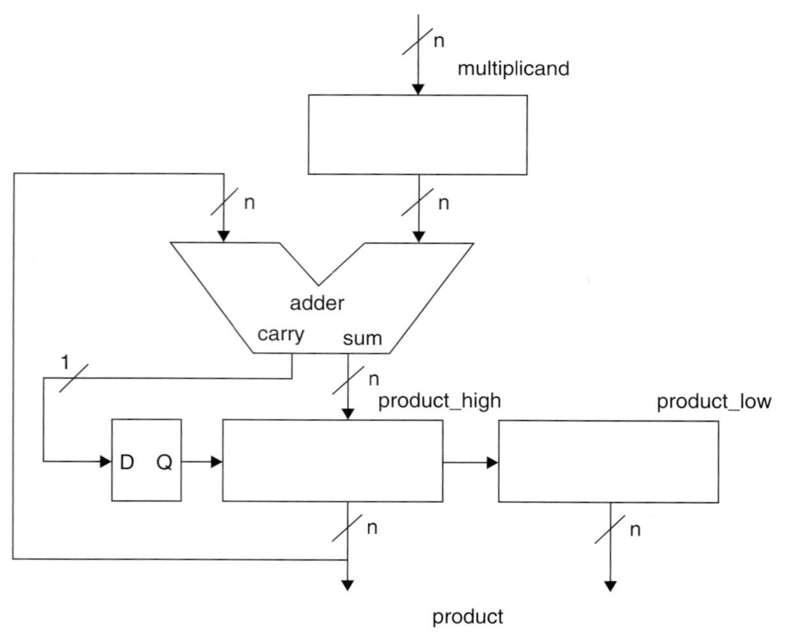

FIGURE 11.5.2
Shifting the accumulated product right relative to the partial product.

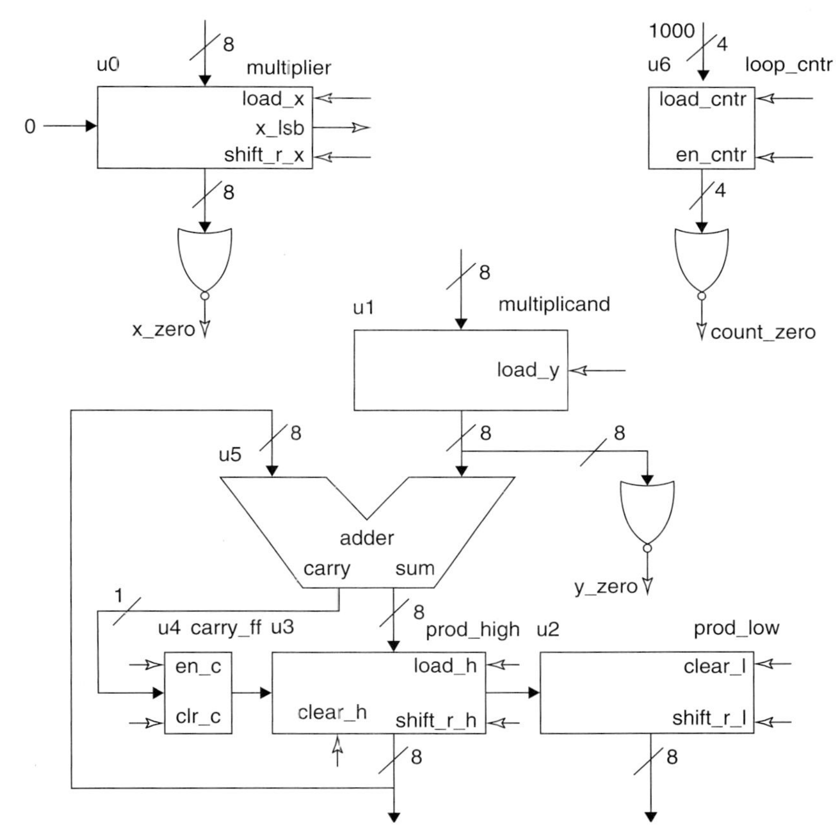

FIGURE 11.5.3
Data path for 8 x 8 multiplier.

Multiplier Register

First we specify each register and the combinational logic of the data path. Before starting a multiplication the operands are loaded. The multiplier is loaded into the 8-bit register `multiplier`. During the alogrithm we will need to examine each bit of the multiplier in turn, starting with the least significant bit. This is facilitated by making this register a right shift register. The least significant bit of this register is available as the data path status signal `x_lsb`.

If the multiplier is zero at the start of a multiplication, the product will be zero and there is no need to go through all the steps of the algorithm. Accordingly, the outputs of register `multiplier` or NORed to generate the status signal `x_zero`. This signal is asserted whenever the content of `multiplier` is zero.

Multiplicand Register

The multiplicand is loaded into the 8-bit register `multiplicand`. If the multiplicand is zero at the start of a multiplication, the product will be zero and there is no need to go through all the steps of the algorithm. Accordingly, the outputs of register multiplicand or NORed to generate the status signal y_zero. This signal is asserted whenever the content of multiplicand is zero.

Product Accumulator Registers

For an 8-bit multiplier and an 8-bit multiplicand a 16-bit register is needed to hold the product. In Figure 11.5.3, this 16-bit register is comprised of two 8-bit registers: prod_high and prod_low. At the start of a multiplication we must clear both of these registers. Thus, each register has a clear input, clear_h and clear_l, respectively.

If the multiplier bit being examined is a 1, we must add the multiplicand to the accumulated product. If it is a 0, we add zero to the accumulated product. This is equivalent to not performing an addition. Each successive addition of either the multiplicand or 0 must be made one bit position to the left relative to the accumulated product. This is accomplished by making registers prod_high and prod_low right shift registers.

Adder

An 8-bit adder is required to add the multiplicand to the accumulated product. The adder adder is a purely combinational circuit. Its inputs are the output from register prod_high and the output from register multiplicand. The sum from adder is the input to register prod_high. The adder's carry is the input to carry_ff.

Carry Flip-flop

The carry from an addition is part of the accumulated product. This carry must be shifted into prod_high when the accumulated product is shifted right. Accordingly, flip-flop carry_ff is needed to hold the carry.

Loop Counter

Finally, the algorithm contains a loop and we need to count its iterations. This could be done as part of the FSM or as part of the data path. We have chosen to do this in the data path by including a 4-bit counter named loop_cntr. We want to count eight interations. At the start of a multiplication, loop_cntr is loaded with the value 8 and subsequently counted down to 0. The outputs of loop_cntr or NORed to generate the status signal count_zero. This signal is asserted whenever the content of loop_cntr is zero.

First Multiplier FSM Block Symbol

Each of the components in the data path was individually specified. The inputs to the block symbol for the FSM in Figure 11.5.4 are the status outputs from the data path, shown with open arrows, and the external inputs to the multiplier, shown with solid arrows. The control outputs from the FSM to the data path are shown with open arrows. These outputs are labeled with the name of the data path component input to which they connect. The single external status output from the FSM is the signal done and it is shown with a solid arrow.

First Multiplier ASM Chart

An ASM chart for the multiplier's FSM is given in Figure 11.5.5. The states are idle, start, ck_oprnds, ck_lsb, add, shift_r, and ck_count. This ASM chart is for a Moore FSM.

After system reset, the FSM stays in the idle state until the start_multiply external input is asserted. When this input is asserted, the FSM transitions to the start state at the next triggering clock edge.

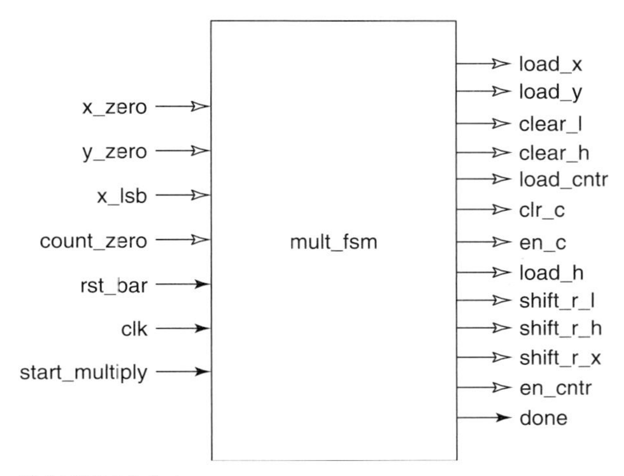

FIGURE 11.5.4
Multiplier FSM with data path control outputs labeled corresponding to data path component input labels.

In state `start`, the signals to enable loading the `multiplier`, `multiplicand`, and `loop_cntr` and clearing the `prod_high`, `prod_low`, and `carry_ff` registers are asserted. On the next triggering clock edge these registers are loaded and cleared, respectively, and the FSM transitions to the `ck_oprnds` state.

In state `ck_oprnds`, if either `multiplier` or `multiplicand` is zero, `x_zero` or `y_zero` is asserted, the FSM transitions back to state `idle` on the next triggering clock edge and the multiplication is complete. If neither `multiplier` or `multi-plicand` is zero, then the next state is `ck_lsb`.

In state `ck_lsb`, the next state transition is based on the value of the multiplier bit being examined. If `x_lsb` is 1, the transition is to the `add` state. Otherwise, the transition is to the `shift_r` state.

In state `add`, signals to enable the `carry_ff` to store the carry and the `prod_h` register to load the sum from the adder are asserted. On the next triggering clock edge these actions occur and the FSM transitions to the `shift_r` state.

In state `shift_r`, signals to enable `prod_high` and `prod_low` to shift right, the `carry_ff` to be cleared, and the `loop_cntr` to be decremented are asserted. On the next triggering clock edge these actions occur and the FSM transitions to the `ck_count` state.

In state `ck_count`, the `count_zero` status signal is checked. If it is asserted the multiplication is complete and the transition is back to the `idle` state. Otherwise, the transition is back to the `ck_oprnds` state and the multiplication continues.

Simpler Multiplication ASM Chart

In examining the FSM outputs asserted in each state of the ASM chart, it can be seen that some outputs are asserted in only a single state. All of the outputs asserted in only a single state can be replaced by a single output that is connected to all the component inputs associated with the replaced output signals. For example, outputs

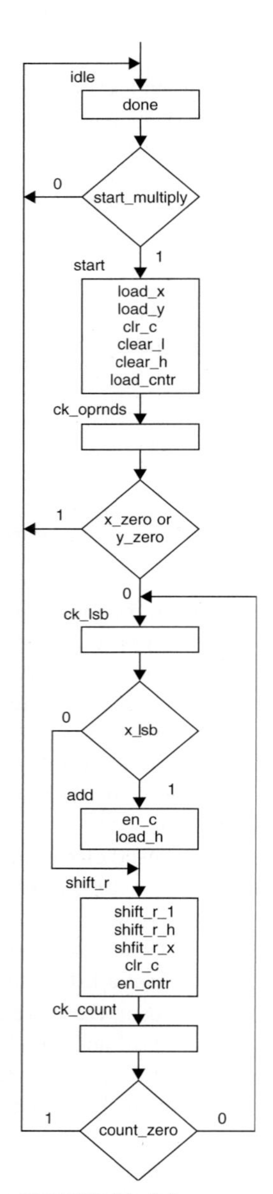

FIGURE 11.5.5
First ASM chart for multiplier FSM.

`load_x`, `load_y`, `clear_l`, `clear_h`, and `load_cntr` are only asserted in state `start`. All of these FSM outputs can be replaced by a single output named `init`. This output signal must be connected to the `load_x`, `load_y`, `clear_l`, `clear_h`, and `load_cntr` inputs of the appropriate data path components.

If two or more FSM outputs are asserted only in the same states, they can also be combined. A simplified ASM chart that reflects these kinds of simplifications is given in Figure 11.5.6

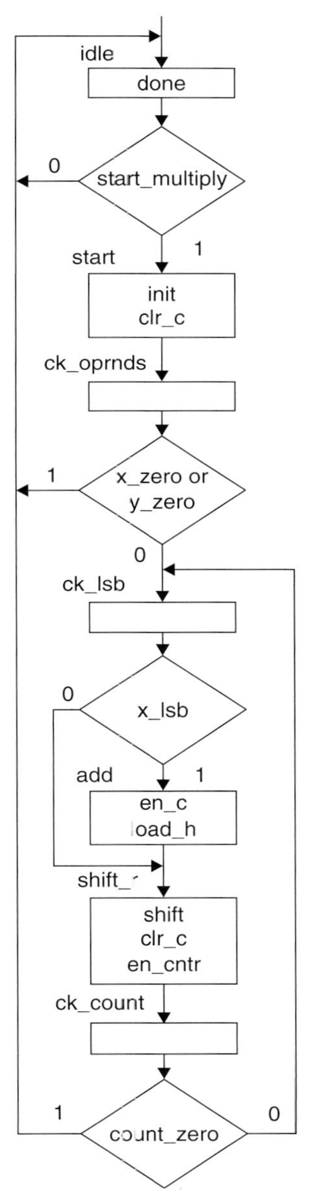

FIGURE 11.5.6
Simplified ASM chart of multiplier FSM.

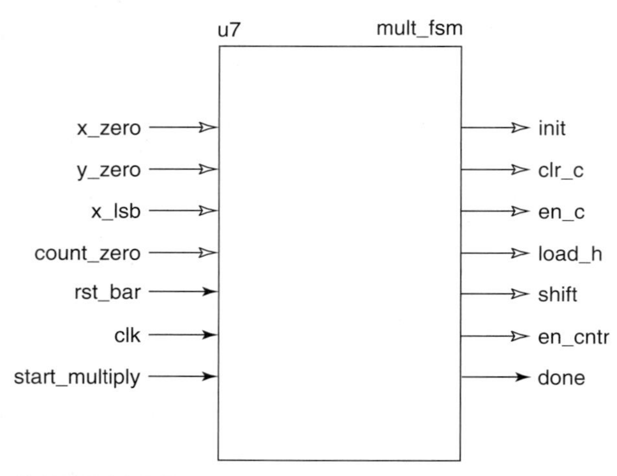

FIGURE 11.5.7
Simplified block symbol for multiplier FSM.

Multiplier FSM Code

Based on the simplified ASM chart, the block symbol for the multiplier's FSM is given in Figure 11.5.7.

The code for the multiplier FSM design entity `mult_fsm` is given in Listing 11.5.1.

LISTING 11.5.1
Design entity mult_fsm.

```
library ieee;
use ieee.std_logic_1164.all;

entity mult_fsm is
    port(
        start_multiply : in std_logic;
        x_lsb : in std_logic;
        x_zero : in std_logic;
        y_zero : in std_logic;
        count_zero : in std_logic;
        rst_bar : in std_logic;
        clk : in std_logic;
        init : out std_logic;
        clr_c : out std_logic;
        en_c : out std_logic;
        load_h : out std_logic;
        shift : out std_logic;
        en_cntr : out std_logic;
        done : out std_logic
        );
end mult_fsm;
```

```
architecture behavioral of mult_fsm is
    type state is (idle, start, ck_oprnds, ck_lsb, add, shift_r, ck_count);
    signal present_state, next_state : state;
begin

    state_reg: process (clk, rst_bar)
    begin
        if rst_bar = '0' then
            present_state <= idle;
        elsif rising_edge(clk) then
            present_state <= next_state;
        end if;
    end process;

    output_process: process (present_state)
    begin
        -- set all outputs to unasserted values
        init <= '0'; shift <= '0'; clr_c <= '0'; en_c <= '0';
        load_h <= '0'; en_cntr <= '0'; done <= '0';

        case present_state is
            when idle => done <= '1';
            when start => init <= '1'; clr_c <= '1';
            when ck_oprnds => null;
            when ck_lsb => null;
            when add => en_c <= '1'; load_h <= '1';
            when shift_r => shift <= '1'; clr_c <= '1'; en_cntr <= '1';
            when ck_count => null;
        end case;
    end process;

    nxt_state: process (present_state, start_multiply, x_lsb, x_zero,
        y_zero, count_zero)
    begin
        case present_state is
            when idle =>
                if start_multiply = '0' then
                    next_state <= idle;
                else
                    next_state <= start;
                end if;

            when start => next_state <= ck_oprnds;

            when ck_oprnds =>
                if (x_zero = '1' or y_zero = '1') then
                    next_state <= idle;
```

(Cont.)

LISTING 11.5.1 *(Cont.)*

```
                else
                    next_state <= ck_lsb;
                end if;

            when ck_lsb =>
                if x_lsb = '0' then
                    next_state <= shift_r;
                else
                    next_state <= add;
                end if;

            when add => next_state <= shift_r;

            when shift_r => next_state <= ck_count;

            when ck_count =>
            if count_zero = '1' then
                next_state <= idle;
            else
                next_state <= ck_lsb;
            end if;

        end case;
    end process;
end behavioral;
```

Multiplier Top Level The code for the multiplier top-level design entity `seq_multiplier` is given in Listing 11.5.2.

LISTING 11.5.2
Top-level design entity for the sequential multiplier.

```
library ieee;
use ieee.std_logic_1164.all;

entity seq_multiplier is
    port(
        start_multiply : in std_logic;
        rst_bar : in std_logic;
        clk : in std_logic;
        multiplier : in std_logic_vector(7 downto 0);
        multiplicand : in std_logic_vector(7 downto 0);
        done : out std_logic;
```

```
        product : out std_logic_vector(15 downto 0)
        );
end seq_multiplier;

architecture behavioral of seq_multiplier is
    signal init, shift, clr_c, en_c, load_h, en_cntr : std_logic;
    signal x_lsb, x_zero, y_zero : std_logic;
    signal cy_reg, carry, count_zero : std_logic;
    signal multiplicand_reg, product_l_sig, product_h_sig,
    sum : std_logic_vector(7 downto 0);
begin


    u0: entity multiplier port map (d => multiplier, load_x => init,
         shift_r_x => shift,
        clk => clk, rst_bar => rst_bar, x_lsb => x_lsb, x_zero => x_zero);

    u1: entity multiplicand port map (d => multiplicand, load_y => init,
         clk => clk,
        rst_bar => rst_bar, q => multiplicand_reg, y_zero => y_zero);

    u2: entity prod_low port map (si => product_h_sig(0), clear_l => init,
         shift_r_l => shift,
        clk => clk, rst_bar => rst_bar, q => product_l_sig(7 downto 0));

    u3: entity prod_high port map (d => sum, load_h => load_h, si => cy_reg,
        shift_r_h => shift, clear_h => init, clk => clk, rst_bar => rst_bar,
        q => product_h_sig);

    u4: entity carry_ff port map (d => carry, enff => en_c, clrff => clr_c,
        clk => clk, rst_bar => rst_bar, q => cy_reg);

    u5: entity adder port map (a => product_h_sig, b => multiplicand_reg,
        sum => sum, carry => carry);

    u6: entity loop_cntr port map (load_cntr => init, en_cntr => en_cntr,
        clk => clk, rst_bar => rst_bar, count_zero => count_zero);

    u7: entity mult_fsm port map (start_multiply => start_multiply,
        x_lsb => x_lsb, x_zero => x_zero, y_zero => y_zero,
        count_zero => count_zero, rst_bar => rst_bar, clk => clk, init => init,
        clr_c => clr_c, en_c => en_c, load_h => load_h, shift => shift,
        en_cntr => en_cntr, done => done);

    product(7 downto 0) <= product_l_sig;
    product(15 downto 8) <= product_h_sig;

end behavioral;
```

PROBLEMS

11.1 List each of the basic ASM chart elements and state its purpose.

11.2 For each of the state diagrams in Figure 10.2.1 draw an equivalent ASM chart.

11.3 For the state diagram in Figure 10.5.4 draw an equivalent ASM chart.

11.4 For the state diagram in Figure 10.8.2 draw an equivalent ASM chart.

11.5 For the state diagram in Problem 10.22 draw an equivalent ASM chart.

11.6 Write a VHDL description of the component `approx_reg`. See "Component approx_reg" on page 449.

11.7 Write a VHDL description of the component `shiftreg`. See "Component shiftreg" on page 449.

11.8 Write a VHDL description of the component `or_function`. See "Component or_function" on page 449.

11.9 Compile and simulate the successive approximation algorithm in Listing 11.4.1. Attempt to synthesize this description and describe the results of this attempt.

11.10 Create a state diagram and an ASM chart for a Mealy version of the SA FSM corresponding to Figure 11.4.6. Does the Mealy version have fewer states? If so, explain qualitatively how this affects the performance of the ADC as its number of bits is increased.

11.11 Create a RTL architecture for a direct (one-to-one) implementation of the successive approximation algorithm as described in Figure 11.4.2 and Listing 11.4.1. That is, use a single register with appropriate combinational logic and a FSM. Compare the efficiency and performance of the architecture with that in Figure 11.4.4.

11.12 Draw a version of Figure 11.5.1 that illustrates a multiplication using the same operand values but using an accumulated product that is shifted to the right after each addition.

11.13 Write the code for each of the design entities in Figure 11.5.3. Use information from Listing 11.5.2 to determine the names of the ports for each design entity. Note that the NOR gates are part of their associated design entities and do not represent separate design entities.

11.14 Create an ASM chart for a Mealy FSM for the sequential multiplier. Does this ASM chart have fewer states than the Moore FSM chart in Figure 11.5.4?

11.15 Write the code for the Mealy FSM corresponding to the ASM chart from Problem 11.14.

Chapter 12

Subprograms

VHDL subprograms are similar to subprograms in conventional programming languages. A subprogram is an encapsulated sequence of sequential statements that define an algorithm. The algorithm uses the values of input parameters, passed to the subprogram when it is called, to compute results or cause some desired effect.

The actual code for a subprogram appears only once in the text of a program. However, the subprogram can be executed by calling it from anywhere in the program. Unlike subprograms in conventional programming languages, VHDL subprograms can also be executed as concurrent statements.

12.1 SUBPROGRAMS

Subprograms simplify the organization of descriptions by providing information hiding. Designs that use subprograms are more modular and compact. As a result, they are easier to read and understand. Writing a description is further simplified by reusing previously written and verified subprograms. This not only reduces the time required to write a description, but also reduces errors.

The main use of subprograms is to perform common and frequently repeated operations. While this use makes a program more modular, subprograms are not the primary mechanism for producing hierarchy in synthesizable VHDL descriptions. Instead, design entities serve that purpose. Subprograms are not design units and, therefore, they cannot be separately compiled.

Kinds of Subprograms

VHDL provides two kinds of subprograms; functions and procedures. A *function* computes and returns a single value. This value is computed using the values of parameters passed to the function when it is called. A function is called from an expression. The value returned by a function is used in the expression that contains the call. An expression containing a function call can be part of either a sequential or concurrent statement. However, all statements within a function are sequential.

A *procedure* can return one or more values, or it may return no values and be used only for its effect. A procedure call is a statement. It can appear in a process or another subprogram as a sequential statement. Or, it can appear alone in an architecture body as a concurrent statement. However, all statements within a procedure are sequential.

Concurrent Subprogram Calls Example

Before considering functions and procedures in detail, we look at an example that includes a function and a procedure that are called using concurrent statements. In this example, these subprograms are used like components in a mixed architecture.

The logic diagram for a positive-level D latch comprised of NAND gates is shown in Figure 12.1.1.

For instructional purposes, the D latch description in Listing 12.1.1 has each NAND component implemented in a different way:

u1: concurrent signal assignment statement (`nand_2csa`)
u2: component instantiation statement (`nand_2c`)
u3: process statement (`nand_2`)
u4: function in a concurrent signal assignment statement (`nand_2f`)
u5: concurrent procedure call statement (`nand_2p`)

While this is certainly not an efficient way to describe a D latch, it illustrates how a number of different kinds of concurrent statements can each implement an identical operation. It also illustrates how functions and procedures use signals to communicate with other concurrent statements.

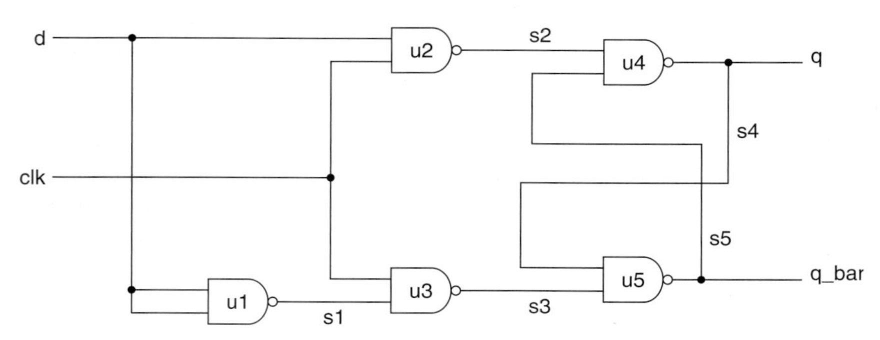

FIGURE 12.1.1
Logic diagram for a D latch composed of two-input NAND gates.

LISTING 12.1.1
Description of a D latch that includes using a function and a procedure.

```
library ieee;
use ieee.std_logic_1164.all;

entity nand_2c is  -- Entity and architecture for 2-input NAND component
   port (in1, in2 : in std_logic;
      out1 : out std_logic);
end nand_2c;

architecture dataflow of nand_2c is
   begin
      out1 <= in1 nand in2;
end dataflow;


library ieee;
use ieee.std_logic_1164.all;

entity dlatch is  -- Entity and architecture for positive level D latch
   port (d, clk : in std_logic;
      q, q_bar : out std_logic);
end dlatch;

architecture mixed of dlatch is

              -- Function body for NAND gate function
function nand_2f (signal in1, in2 : in std_logic) return std_logic is
   begin
      return (in1 nand in2);
end nand_2f;

              -- Procedure body for NAND gate procedure
procedure nand_2p (signal in1, in2 : in std_logic; signal out1 : out std_logic)
is
   begin
      out1 <= (in1 nand in2);
   end nand_2p;

signal s1, s2, s3, s4, s5 : std_logic;  -- Signals to connect gates

begin

   u1: s1 <= d nand d; -- concurrent signal assignment NAND

   u2: entity nand_2c port map (in1 => d, in2 => clk, out1 => s2); -- component
```

(Cont.)

LISTING 12.1.1 *(Cont.)*

```
   u3: process (clk, s1) -- process NAND
     begin
        s3 <= clk nand s1;
     end process u3;

   u4: s4 <= nand_2f(s2, s5); -- function NAND

   u5: nand_2p(s4, s3, s5); -- procedure NAND

   q <= s4;        -- assignment of signals to
   q_bar <= s5;    -- output pins

end mixed;
```

The description starts with the entity declaration and architecture body for the design entity nand_2c, which is later instantiated in the top-level design entity.

Next is the entity declaration for the top-level design entity dlatch.

nand_2f Function

In the declarative part of the architecture body for dlatch is the body of the function nand_2f. This function corresponds to component u4 in the logic diagram. This function's body also serves as its declaration.

```
function nand_2f (signal in1, in2 : in std_logic)
   return std_logic is
      begin
            return (in1 nand in2);
end nand_2f;
```

Following keyword **function** is the name of the function, nand_2f, followed by a formal parameter list. Two parameters, in1 and in2, are passed to the function. Both parameters are std_logic signals. This function interfaces to the other concurrent statements in the architecture body through the signals in its parameter list and its return value.

Function nand_2f is like a component in the sense that it can be instantiated multiple times and its internal operations are not visible to the instantiating architecture (information hiding).

Functions always return a value. Function nand_2f returns a value that is type std_logic. The body of this function consists of a single **return** statement that returns the NAND of parameters in1 and in2.

nand_2p Procedure

Following the function body for nand_2f is the procedure body for procedure nand_2p:

```
procedure nand_2p (signal in1, in2 : in std_logic;
                   signal out1 : out std_logic) is
    begin
        out1 <= (in1 nand in2);
    end nand_2p;
```

Following keyword **procedure** is the name of the procedure, nand_2p. The formal parameter list specifies three parameters, all std_logic signals. Parameters in1 and in2 are inputs and parameter out1 is an output. Procedure nand_2p contains a single sequential signal assignment statement that assigns out1 the NAND of parameters in1 and in2.

Also declared in the architecture body's declarative part are signals s1 through s5. These signals provide the communications between "components" u1 through u5.

In the statement part of the top-level architecture are seven concurrent statements. The first five correspond to the "components" u1 through u5. The last two assign the values of signals s4 and s5 to output ports q and q_bar, respectively.

u1, u2, and u3 are, respectively, a concurrent signal assignment statement, a component instantiation statement, and a process statement.

Calling a function or procedure in this example is analogous to instantiating a component. A function can be used to instantiate a "component" that has a single output. The function call must be part of an expression. Statement u4 is a concurrent signal assignment that assigns the value returned by function nand_2f to signal s4. In this function call, actual parameters, signals s2 and s5, replace formal parameters in1 and in2, respectively. This function is called to update the value of s4 whenever there is an event on either s2 or s5.

Concurrent procedure call statement u5 calls procedure nand_2p. Actual parameter signals s4, s3, and s5 replace formal parameters in1, in2, and out1, respectively, of the procedure body. The procedure is called whenever there is an event on input parameters s4 or s3. When called, nand_2p updates the value of s5.

12.2 FUNCTIONS

A function can be considered a generalized form of an operator and used to define a new operator. A function is called in an expression. When called, a function calculates and returns a result. The definition of a function can be given in two parts: a function declaration and a function body. The *function declaration* defines the function's calling conventions and is optional.

```
function _declaration ::=
function designator [ ( formal_parameter_list ) ] return type_mark;

designator ::= identifier | operator_symbol
```

FIGURE 12.2.1
Simplified syntax for a function declaration.

A *function body* (*function definition*) defines a function's execution. When a function is defined in the declarative part of an architecture body or process, usually only the function body is provided. In such a case, the function body also serves as the function's declaration.

Function Declaration

A function declaration describes the interface to a function and the type of the value it returns. It provides no information on the function's realization.

The syntax for a function declaration is given in Figure 12.2.1

In the syntax, the term designator represents the name of the function. A designator can be either an identifier or one of the predefined operator symbols.

Visibility of a Function

A function can be declared in the declarative part of an architecture body, in the declarative part of a process, or in a package. A function can also be declared in another subprogram or in a package body. Where a function is declared determines its visibility (from where in the description it can be called). A function declared in an architecture body can be called only from the statement part of that architecture. A function declared in a process can be called only from within that process. A function declared in a package can be called from any design entity preceded by a library clause and use clause that make that package visible.

Function Body

A function body provides the same interface and type information as a function declaration. Its syntax is given in Figure 12.2.2. The first part of a function body is identical to a function declaration. In addition, a function body includes the sequential statements that implement the function.

```
function_body ::=
function designator [ ( formal_parameter_list ) ] return type_mark is
    { subprogram_declarative_item }
begin
    { sequential_statement }
end [ designator ] ;
```

FIGURE 12.2.2
Syntax for a function body (definition).

In Listing 12.1.1, function `nand_2f` is defined in the declarative part of architecture `mixed`. This function body serves as both the declaration and body for this function. Function `nand_2f` is visible only in architecture `mixed`.

A function's designator is either an identifier or one of the predefined VHDL operator symbols. A VHDL operator symbol is used when the purpose of the function is to overload the operator. Overloading of operators is discussed in Section 12.5.

Formal Parameters

As seen in Listing 12.1.1, a formal parameter list is somewhat akin to an entity's port interface list. *Formal parameters* are simply placeholders. The names of the formal parameters are used in the statements inside a function to refer to the values that are passed as actual parameters (arguments) when the function is called.

As seen in Figure 12.2.1, the parameter list for a function is optional. However, the use of parameters makes a function more general, so that it can perform its algorithm using different data each time it is called.

The *formal*_parameter_list is the list of parameters passed to the function. It has the simplified form given in Figure 12.2.3.

Formal Parameter Class

Interface declarations in the formal parameter list are separated by semicolons. A formal parameter's class and mode determine how access to that parameter is implemented.

A formal parameter's class indicates the class of the parameter ***within*** the function. That is, how the parameter can be referenced in the statement part of the function body. In general, the class of a formal parameter of a function may be either constant, signal, or file. Parameters of class file have no mode. Since files are not used as parameters in functions that are synthesized, they are not discussed further in this chapter.

The default class of a parameter passed to a function is constant. However, the actual parameter that replaces a formal constant parameter when a function is called can be an expression, constant, variable, or signal.

Formal Parameter Mode

The mode of a formal parameter specifies how the parameter may be accessed ***within*** the statement part of the function body. The syntax for the formal parameter list indicates that the *mode* of a parameter used in a function must be **`in`**, and that its specification is optional. Because their mode is **`in`**, function parameter values can only be read in a function body; they cannot be written.

formal_parameter_list ::=
interface_declaration { ; interface_declaration }

interface_declaration ::=
[**constant**] identifier_list : [**in**] subtype_indication [:= *static*_expression]
| [**signal**] identifier_list : [**in**] subtype_indication [:= *static*_expression]

FIGURE 12.2.3
Simplified formal parameter list for a function.

Taking advantage of the defaults, the body of the function in Listing 12.1.1 could be written as:

```
function nand_2f (in1, in2 :std_logic)return std_logic is
    begin
        return (in1 nand in2);
    end nand_2f;
```

Function Parameter Default Values

For formal parameters of class constant and mode in, optional default values can be specified in the formal parameter list. If a default is specified, an actual value can be omitted in the call. This can be done by using the keyword **open** for the actual parameter in the call. Another method of omitting the actual value is simply to not associate any value with the formal. In this case, any formal parameter for which no value is specified must appear in the parameter list after all formals for which values are specified.

Declarative Part of a Function

Since within a function's body formal parameters cannot be modified, variables may need to be declared in a function to store intermediate values. ***Variables declared in a function are initialized on each call to the function. Values of variables do not persist from one call to the next.*** Functions only use variables to store intermediate values. Since a variable is updated immediately after assignment, its value can be returned to the function's caller. Neither shared variables nor signals can be declared in a function.

Other items that can be declared in a function are types, subtypes, constants, and nested subprograms.

Statement Part of a Function

Within the statement part of a function body all statements must be sequential. With the exception of wait statements, any of the sequential statements that we have previously considered can be used. ***Functions cannot contain wait statements.*** If a function calls a procedure, the procedure cannot contain a wait statement. All functions execute in zero simulation time.

Return Statement

A function can have only one return value and it must be of the type specified in the function's declaration. A function must contain at least one *return statement.* A return statement has the form given in Figure 12.2.4.

```
return_statement ::=
return [ expression ] ;
```

FIGURE 12.2.4
Syntax for a return statement.

The value represented by the expression is the value returned by the function. ***The value returned can be assigned to a signal or variable of the return type, or used to build an expression of that type.***

A function can have more than one return statement. This often occurs when return statements are used in different branches of an if statement or case statement. Multiple return statements may also appear in a loop, to break out of the loop and then after the loop, to handle the situation where the loop is completed without a return statement within the loop being executed.

When a return statement is executed, the flow of control continues with completion of the evaluation of the statement containing the function call.

Function Call

The caller passes parameters to a function as part of a *function call*. A function call consists of the function name followed by a list of actual parameters. The function call can appear alone on the right-hand side of an assignment statement, or as part of a more complex expression.

A function declared in the declarative part of an architecture body can be called in the statement part of the architecture body using a concurrent statement, or can be called by a sequential statement in a process in the architecture body. A function declared in the declarative part of a process can only be called from within that process.

When a function call appears in a concurrent signal assignment statement, it is executed whenever there is an event on a signal in its parameter list. In contrast, a function in a process is executed only when the statement containing the function call is reached during the sequential execution of the process's statements.

Actual Parameters

Actual parameters are the names of the actual data objects whose values are mapped to the formal parameters when a function is called. The function uses the values of the actual parameters in its execution.

Parameter Matching

The class of the actual parameters must match the class of the formal parameters, except when the formal parameter's class is constant. ***A constant class formal parameter can be replaced by an actual parameter that is either an expression, constant, variable, or signal. The type of the actual parameter must match the type of the formal parameter.***

Formal parameters are specified with a type. An actual parameter that is a subtype of the type of a formal parameter can also be passed. This allows a function written in terms of base type formal parameters to handle actuals that are the base type or any of its subtypes.

Parameter Association

Actual parameters are associated with formal parameters using either *positional association* or *named association*. Association is expressed in the same way as in the port map of a component instantiation.

The single value returned from a function called as an operand from within an expression replaces the function call itself within the expression.

An example of a function declared in an architecture body and called by a concurrent signal assignment statement is given in Listing 12.2.1. The entity has an eight-element std_logic_vector input and a std_logic output that indicates whether the parity of the input vector is even or not. Parity is even if the vector has an even number of bits that are 1s, or if none of its bits are 1s.

The body of the function `parity_even` appears in the declarative part of the architecture. This function has a single formal parameter `s` that is a std_logic_vector whose class is constant and mode is, by default, in. A variable named `result` is declared in the declarative part of the function and is initialized to `'1'`.

A loop is used to sequence through each element of the vector passed to the function. If the element being examined is a `'1'`, the value of `result` is complemented. Once the loop is completed, a return statement returns the value of `result`.

The architecture body contains a single concurrent call to the function `parity_even`. The value returned by the function is assigned to output `even`.

LISTING 12.2.1
Parity generator using a function.

```
library ieee;
use ieee.std_logic_1164.all;

entity parity_even_entity is
   port (
      in_vector : in std_logic_vector(7 downto 0);
      even : out std_logic
      );
end parity_even_entity;

architecture behavioral of parity_even_entity is

   function parity_even (s: std_logic_vector(7 downto 0)) return std_logic is
      variable result : std_logic := '1';
      begin
      for i in 7 downto 0 loop
         if s(i) = '1' then
            result := not result;
         end if;
      end loop;
      return result;
   end parity_even;

   begin

   even <= parity_even(in_vector);

end behavioral;
```

Variables in Functions

Unlike in processes, variables declared in functions are created and initialized each time the function is called. As a result, variables in functions do not preserve their values between executions of the function. Thus, they can only be used to store values for the duration of a single execution of the function. For example, this means that variables in functions cannot be used to describe latches or flip-flops. Since the values of variables are initialized each time a function is called, a variable can be initialized with the value of one of the function's formal parameters.

Function Return Values

After its execution completes, a function's return value replaces the function call in the expression in which the call appears.

Synthesizable Functions

Functions are synthesizable if they are statically determinable. *Statically determinable* means determinable when the description is compiled. Where loops are involved, this requires that the number of iterations be fixed. The number of loop iterations when the function `parity_even` is called is known to be eight.

Summary of Function and Procedure Characteristics

The characteristics of functions are summarized in column two of Table 12.2.1. For comparison, column three lists the corresponding characteristics of procedures, which are discussed in the next section.

Table 12.2.1
Subprograms: Functions and procedures.

Characteristic	Function	Procedure
Purpose	compute a value	compute a value or cause an effect
Call is a(n)	expression	sequential statement concurrent statement
Values returned	one and only one, replaces call in expression	zero, one, or more
Return statement in body	required	not required
Formal parameter list	not syntactically required, but practically necessary	not required
Formal parameter classes	constant signal file[a]	constant variable signal file[a]
Formal parameter modes allowed and (default class)	in (constant)	in (constant) inout (variable) out (variable)
Mode **in** formal parameters allowed default values for synthesis	constant	constant

(Cont.)

Table 12.2.1 *(Cont.)*

Characteristic	Function	Procedure
Formal parameter and associated actual parameters formal => actual	constant => expression[b] signal => signal file => file[a]	constant => expression variable => variable signal => signal file => file[a]
Declaration of signals in subprogram body	not allowed	not allowed
Signal assignments in subprogram body	not allowed	allowed, including visible signals not passed as parameters
wait statements	not allowed at all	not allowed if procedure is called by a function or a process with a sensitivity list[a]
synthesizable?	if statically determinable and no file parameters	if statically determinable and no file parameters or wait statements

[a]Files are not supported for synthesis.

[b]Since the actual parameter can be an expression, it can be a variable, signal, or constant.

12.3 PROCEDURES

Procedures are like functions, in that they are encapsulated sequences of sequential statements. However, there are several important differences.

Procedure Declaration

Like a function, the definition of a procedure can be given in two parts: a procedure declaration and a procedure body. A *procedure declaration* defines a procedure's calling convention and is optional. It provides no information on the procedure's implementation. A procedure declaration has the form shown in Figure 12.3.1.

Procedure Body

A *procedure body* (*procedure definition*) provides the same interface and type information as a procedure declaration. In addition, a procedure body includes the sequential statements that define the procedure's implementation.

When a procedure is defined in the declarative part of an architecture or a process, usually only the procedure body is provided. In such cases, the procedure body also serves as the procedure's declaration. The syntax for a procedure body is given in Figure 12.3.2. A procedure's designator is always an identifier; no operator symbols are allowed.

```
procedure_declaration ::=
    procedure identifier [ ( formal_parameter_list ) ]
```

FIGURE 12.3.1
Syntax for a procedure declaration.

```
procedure_body ::=
        procedure identifier [ ( formal_parameter_list ) ] is
                { subprogram_declarative_item }
        begin
                { sequential_statement }
        end [ designator ] ;

formal_parameter_list ::=
interface_declaration { ; interface_declaration }

interface_declaration ::=
[ constant ] identifier_list : [ in ] subtype_indication [ := static_expression ]
| [signal] identifier_list : [ mode ] subtype_indication [ := static_expression ]
| [variable] identifier_list : [ mode ] subtype_indication [ := static_expression ]
```

FIGURE 12.3.2
Simplified syntax for a procedure body.

Procedure Formal Parameters

In addition to constants, signals, and files, variables can also be passed to procedures.

In addition to mode `in`, procedure parameters can also be mode `out` or `inout`. Thus, procedures can return values by modifying the value of parameters that are mode out or inout.

Unlike a function, a procedure's declaration and body do not specify a return type. Use of parameters is the only way values can be returned from a procedure. Parameters of mode out or inout are used like a return value in a function, except there can be any number of them. If a class is not specified, a formal parameter of mode in defaults to constant class and a formal parameter of mode out or inout defaults to the variable class.

Parameters of mode out cannot be read in the procedure body. They are only assigned values to transfer information back to the caller. Parameters of mode in or inout can be read in the procedure body. Attributes of formal parameters of any mode can be read, with the exception of the signal-valued attributes `'stable,'` `'quiet,'` `'transaction,'` and `'delayed.'`

Variables in Procedures

Variables can be declared in the declarative part of a procedure. Unlike variables declared in processes, procedure local variables are created and initialized each time the procedure is called. Values of a variable declared in a procedure do not persist between calls to the procedure.

Procedure Actual Parameters

Procedure actual parameters are the names of the actual data objects whose values are mapped to the formal parameters when a procedure is called.

Signal Actual Parameters

Signals cannot be declared in a procedure. However, signals can be passed to a procedure as parameters and assigned new values within the procedure. A signal formal parameter can be mode in, out, or inout.

Pass by Value

When a scalar constant or variable is passed to a procedure during a procedure call, it is passed by value. A copy of the value of the actual data object is passed to the formal parameter and used by statements in the procedure body. The value of a formal variable parameter of mode out or inout is passed back to the corresponding actual parameter when the procedure returns. Arrays or records can be passed either by value or by reference; the IEEE Std 1076 standard allows either method of implementation.

Pass by Reference

The passing of signal parameters is more complicated. Instead of passing a value, a reference to the signal, the driver of the signal, or both are passed based on the mode of the signal.

When a caller passes a signal parameter of mode in, instead of passing the value of the signal, ***a reference to the signal object itself is passed***. Any reference to the formal parameter within the procedure is like a reference to the actual signal.

A consequence of this is that, if a procedure executes a wait statement, a signal's value may be different after the wait statement completes and the procedure resumes. This fact is often used to advantage by procedures in testbenches. This behavior differs from that of a constant parameter of mode in, which has the same value throughout the execution of a procedure. When a procedure executes a wait statement it is the calling process that actually suspends.

When a caller passes a signal of mode out, the procedure is passed ***a reference to the driver of the signal.*** When the procedure performs a signal assignment to the formal parameter, the transaction is scheduled on the driver for the actual signal.

When a caller passes a signal of mode inout, both a reference to the signal and a reference to its driver are passed to the procedure.

Side Effects

It is possible for a procedure to read or write signals or variables that are not in its parameter list. A procedure's reading or writing a signal or variable that is not in its parameter list is called a *side effect*. A procedure can have a read side effect on any signal or variable visible to the procedure. Whether a procedure can have a write side effect on a signal or variable depends on where the procedure is declared. A procedure declared in the declarative part of a process can have a write side effect on any signal or variable visible to the procedure. A procedure declared in an architecture's declarative part or in a package cannot have a write side affect on a signal.

The reason for creating a procedure with side effects is to limit the number of parameters in its formal parameter list. However, using procedures with side effects is not considered good programming practice. Such procedures are not very readable and can be difficult to reuse, because one or more of the procedure's side effects may be overlooked.

Sequential Procedure Call

A procedure call is a statement. A procedure can be called from within a process or another subprogram. Such a call is a *sequential procedure call*. In a process, a procedure is called when the call statement is reached in the sequential execution of statements in the process. A sequential procedure call is equivalent to inserting inline, at the call statement, the code in the procedure's body.

A procedure that uses mode out or inout parameters that are variables can only be called by a sequential procedure call. Furthermore, these parameters cannot be used to assign values to signals.

Return Statements in Procedures

When the last statement in a procedure is completed, the procedure returns. If the procedure call was sequential, the flow of control returns to the statement following the call. A return statement is not required.

However, a return statement can be used to return from the middle of a procedure. When a return statement is used in a procedure, it does not specify a return value. Execution of the return statement immediately terminates the procedure and returns control to the caller.

When a procedure returns, the value of any formal parameters of mode out or inout are copied back to the corresponding actual parameters, transferring information back to the caller.

Concurrent Procedure Call

A procedure call can also exist alone in an architecture body as a *concurrent procedure call*. As a shorthand for component instantiation, procedures can represent components with multiple outputs. The procedure in a concurrent procedure call is called whenever there is an event on a signal that is an input parameter (mode in or inout) to the procedure. ***The parameter list of a concurrent procedure cannot contain a variable, since a variable cannot exist outside of a process***.

Synthesizeable procedures cannot have wait statements. In contrast, nonsynthesizeable procedures can. However, since a process cannot have both a sensitivity list and a wait statement, a process that calls a procedure that has a wait statement cannot have a sensitivity list.

Procedure Parameter Default Values

For formal parameters of class constant, signal, or variable and mode in, optional default values can be specified in the formal parameter list. However, for this class signal or variable this value is ignored for synthesis. If a default is specified, an actual value can be omitted in the call. This can be done by using the keyword `open` for the actual parameter in the call. Another method of omitting the actual value is simply to not associate any value with the formal. In this case, any formal parameter for which no value is specified must appear in the parameter list after all formals for which values are specified.

When a formal parameter that is a signal is to be left unconnected in a procedure call, the keyword `open` is used as the actual parameter.

12.4 ARRAY ATTRIBUTES AND UNCONSTRAINED ARRAYS

To maximize their reusability, subprograms should be written to be as widely applicable as possible. When a subprogram has an array parameter, we would like to be able to use the same subprogram to handle arrays of any length. Use of unconstrained arrays and array attributes when writing subprograms allows this to be accomplished.

A subprogram that uses unconstrained arrays as formal parameters can be written to operate on actual array parameters of any size. This eliminates the need to write a different subprogram for each different-sized actual array.

The predefined VHDL *array attributes* provide information about the index range of an array. The array attributes are listed in Table 12.4.1. In this table, A is an array type object and N is an integer between 1 and the number of dimensions in A. Attributes are written by following the array name with a quote mark (') and the attribute name.

If an array has only one dimension, or if we wish to refer to only the first dimension, N can be omitted from the array attribute.

Attributes of array bounds do not return the array element; rather, the element's index is returned. The range and reverse_range attributes can be used anyplace a range specification is required, as an alternative to specifying the left and right bounds and the range direction.

For example, if `d_bus` is declared as

```
signal d_bus : std_logic_vector(7 downto 0);
```

its array attributes would have the values:

```
d_bus'left = 7
d_bus'right = 0
d_bus'high = 7
d_bus'low = 0
d_bus'range = 7 downto 0
d_bus'reverse_range = 0 to 7
d_bus'length = 8
d_bus'ascending = false
```

Unconstrained Arrays as Formal Parameters

Using unconstrained arrays as formal parameters and using array attributes within a subprogram allows us to write a subprogram that manipulates arrays in a form that maximizes its reuse. Within the subprogram body, attributes of the formal parameters are used to set the size of local unconstrained arrays.

Where a formal parameter is an unconstrained array, the values of the attributes of the actual constrained array passed to the subprogram set the size of the array manipulated. Attributes of the actual are never passed into a subprogram. A reference to an attribute of a formal parameter is legal only if the formal has such an attribute. Such a reference retrieves the value of the attribute associated with the formal.

Table 12.4.1
Array attributes.

Attribute	Value Returned
A'left[N]	left bound of index range of Nth dimension of A
A'right[N]	right bound of index range of Nth dimension of A
A'high[N]	upper bound of index range of Nth dimension of A
A'low[N]	lower bound of index range of Nth dimension of A
A'range[N]	index range of Nth dimension of A
A'reverse_range[N]	reverse of index range of Nth dimension of A
A'length[N]	number of elements in index range of Nth dimension of A
A'ascending[N]	true if index range of Nth dimension of A is defined with an ascending range; otherwise, false

Since the size of an unconstrained array is constrained when the subprogram is called, the subprogram is synthesizable. Each time a subprogram is called in a description, a new copy of the subprogram is created. Each copy is constrained to the size of its actual parameter before synthesis. Each call produces a separate hardware implementation.

The even parity function in Listing 12.2.1 is rewritten in Listing 12.4.1 to take as its input parameter an unconstrained array.

LISTING 12.4.1
Parity generation function written to handle unconstrained arrays.

```
library ieee;
use ieee.std_logic_1164.all;

entity parity_even_entity is
   port (
   byte_in : in std_logic_vector(7 downto 0);
   nibble_in : in std_logic_vector(3 downto 0);
   byte_even : out std_logic;
   nibble_even : out std_logic
      );
end parity_even_entity;

architecture behavioral of parity_even_entity is

function parity_even (s: std_logic_vector) return std_logic is
      variable result : std_logic := '1';
      begin
```

(Cont.)

LISTING 12.4.1 *(Cont.)*

```
      for i in s'range loop
         if s(i) = '1' then
            result := not result;
         end if;
      end loop;
      return result;
   end parity_even;

   begin

   byte_even <= parity_even(byte_in);
   nibble_even <= parity_even(nibble_in);

end behavioral;
```

The design entity in Listing 12.4.1 has two input vectors: one is an eight-element vector, `byte_in`; the other is a 4-element vector, `nibble_in`. Separate outputs indicate whether each input vector has even parity.

The body of the `parity_even` function in Listing 12.4.1 has a formal parameter `s` that is an unconstrained array. The range attribute is used to determine the range for the loop, based on the range of the actual std_logic_vector passed to the function.

In the architecture body, `parity_even` is called twice, once with the actual `byte_in` and once with the actual `nibble_in`. A synthesizer generates two logic blocks, one for each call, as shown in Figure 12.4.1.

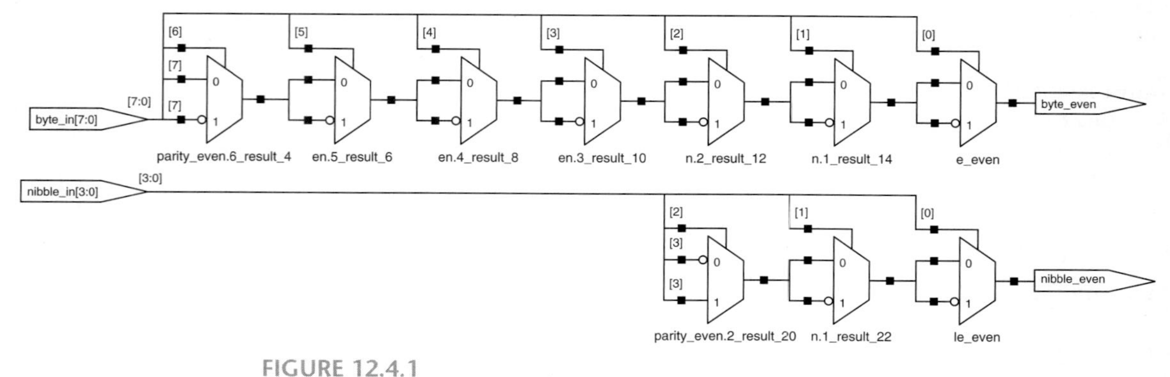

FIGURE 12.4.1
Synthesized logic from two different calls to parity generator function in Listing 12.4.1.

Normalized Unconstrained Arrays

A subprogram with unconstrained array formal parameters should be written to accept input arrays not only of any size, but with any index direction. To accomplish this, the subprogram must be written so that is does not rely on the direction or the range of the actual parameter. This can be accomplished through a technique called *normalization*. Using this technique, input arrays are immediately assigned to local variables whose sizes depend on the sizes of the actual parameters, but that have descending ranges ending in 0. These variables are then used instead of the formal parameters.

For example, the program in Listing 12.4.2 uses a procedure to compare the magnitudes of input vectors `p` and `q`. Three active-low values are computed by the procedure `p_gt_q_bar`, `p_lt_q_bar`, and `p_eq_q_bar`.

LISTING 12.4.2
Magnitude comparison procedure.

```
library ieee;
use ieee.std_logic_1164.all;

entity mag_comp_proc is
   port ( p_in : in std_logic_vector (3 downto 0);
      q_in : in std_logic_vector (0 to 3);
      p_gt_q_bar, p_eq_q_bar, p_lt_q_bar : out std_logic);
end mag_comp_proc;

architecture behavior of mag_comp_proc is

   procedure mag_comp (p : in std_logic_vector; q : in std_logic_vector;
      signal p_gt_q_bar, p_lt_q_bar, p_eq_q_bar : out std_logic) is

      variable p_v : std_logic_vector (p'length - 1 downto 0) := p;
      variable q_v : std_logic_vector (q'length - 1 downto 0) := q;
      variable  p_gt_q_bar_v, p_lt_q_bar_v, p_eq_q_bar_v : std_logic := '1';

   begin
      assert p_v'length = q_v'length
      report "vector parameters must be same length"
      severity failure;

      for i in p_v'range loop
         if ((p_v(i) = '1') and (q_v(i) = '0')) then
            p_gt_q_bar_v := '0';
            exit;
```

(Cont.)

LISTING 12.4.2 *(Cont.)*

```
         elsif ((p_v(i) = '0') and (q_v(i) = '1')) then
            p_lt_q_bar_v := '0';
            exit;
         end if;
   end loop;

      if ((p_gt_q_bar_v = '1') and (p_lt_q_bar_v = '1')) then
         p_eq_q_bar_v := '0';
      end if;

      p_gt_q_bar <=  p_gt_q_bar_v;
      p_lt_q_bar <=  p_lt_q_bar_v;
      p_eq_q_bar <=  p_eq_q_bar_v;

   end mag_comp ;

begin

   mag_comp (p_in, q_in, p_gt_q_bar, p_lt_q_bar, p_eq_q_bar);

end behavior;
```

The constrained actual parameters passed to the procedure are 4-bit vectors. One of these vectors has a descending range and the other has an ascending range. The procedure is called by a concurrent procedure call in the architecture body.

The procedure body has two parameters of mode in that are unconstrained arrays, and three parameters that are signals of mode out. The signal parameters of mode out return the results. These parameters must be signals, because the procedure is called using a concurrent procedure call.

The normalization takes place in the declarative part of the procedure. Two array variables are declared. Each is declared with a length equal to that of its associated formal parameter by using the length attribute. Each variable is assigned an initial value equal to that of its associated formal parameter. As a result, no matter the index direction of the actual arrays passed to the procedure, the procedure is written in terms of array variables with descending ranges.

In the procedure body an assert statement is used to make sure that the arrays passed to the procedure are of equal length. If not, an error message is displayed during simulation. A synthesizer simply ignores the assert statement.

A loop is used to sequence through the elements of the two arrays to determine the three outputs. This part of the procedure is equivalent to the program in Listing 5.7.1.

Unconstrained Array Return Value

An unconstrained array can be used to return an array value from a subprogram. In Listing 7.9.1, a process was used to sequentially compute and apply all input combinations to a half adder UUT in a testbench. A large number of combinational testbenches could use this same kind of stimulus computation. It would be useful if this computation were implemented by a subprogram, so that it could be reused. A procedure that does this is given in Listing 12.4.3.

Procedure `binary_stimulus` has three formal parameters. Parameter `n` is an integer input parameter indicating number of bits in the stimulus (number of UUT inputs to be driven). Parameter `p` specifies the time between generation of each stimulus combination. Parameter `slv` is an unconstrained vector that is the stimulus value.

The procedure uses a loop to produce all possible 2^n combinations of n bits. The time between each combination is p seconds. Since the function `to_unsigned` from package NUMERIC_STD is used, an appropriate use clause is required in the testbench.

As an example of the use of this procedure, consider a testbench for a UUT with std_logic inputs `a, b, c,` and `d`. A vector is declared in the declarative part of the testbench to connect the stimulus from procedure `binary_stimulus` to the UUT:

```
signal bin_stim : std_logic_vector (3 downto 0);
```

The statement part of the testbench's architecture contains the instantiation of the UUT and the following two concurrent statements.

```
binary_stimulus (4, 20 ns, bin_stim);

(a, b, c, d) <= bin_stim;
```

The first of these statements is the concurrent call to procedure `binary_stimulus`. The actual parameters specify that the stimulus generated is four bits wide, is generated every 20 ns, and is returned as vector `bin_stim`. The second concurrent statement applies vector `bin_stim` to the aggregate `(a, b, c, d)`.

During simulation initialization the concurrent procedure `binary_stimulus` is executed once. It generates the sixteen stimulus values, each separated by 20 ns, and then returns.

LISTING 12.4.3
Procedure to generate a binary stimulus.

```
Procedure binary_stimulus (n : integer; p : time ;
    signal slv : out std_Logic_vector) is
begin
    for i in 0 to 2**n - 1 loop
        slv <= std_logic_vector (to_unsigned(i, n));
        wait for p;
    end loop;
end binary_stimulus;
```

Index Range of Unconstrained Return Values

An *unconstrained return value* has the index range of the formal parameter used to return its value. If the vector returned is assigned in its entirety to an equal length vector with a different index range, there is no problem. The assignment is, as always, made from left to right, independently of the index ranges. However, if an element from the returned vector is selected directly, it is the index range of the formal parameter that must be considered in selecting the desired element.

This can be a problem when a subprogram is used and an element is selected directly from a returned vector without knowledge of the index range of the formal parameter used to return its value. In particular, this can be a problem when using subprograms from standard packages or de facto standard packages where we have not examined the subprogram body.

An example to clarify this is given in Listing 12.4.4. This design entity consists of two attempts at selecting bit 0 from the entity's input vector.

Output `outbit_a` is bit 0 taken from the input vector, which has a range of 3 down to 0. Output `outbit_b` is bit 0 taken directly from an identity function, which takes the input vector as its actual parameter and returns this vector. We would expect these two bit values to be the same because they are both bit 0.

LISTING 12.4.4
Design entity to select a bit from a vector.

```
library ieee;
use ieee.std_logic_1164.all;

entity select_bit is
   port(
      invect : in std_logic_vector(3 downto 0);
      outbit_a, outbit_b : out std_logic
      );
end select_bit;

architecture behavioral of select_bit is
   function identity (slv : std_logic_vector) return std_logic_vector is
      variable lcl_vec : std_logic_vector(0 to slv'length - 1);
   begin
      lcl_vec := slv;
      return lcl_vec;
   end identity;

begin
   process (invect)
   begin
      outbit_a <= invect(0);   -- selecting bit 0
      outbit_b <= identity(invect)(0); -- selecting from function_return_value
   end process;
end behavioral;
```

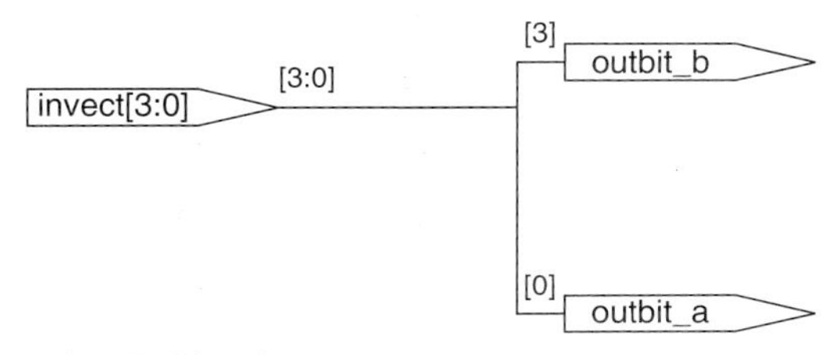

FIGURE 12.4.2
Synthesized logic from the code in Listing 12.4.3.

However, the identity function assigns its input vector to a local variable and this local variable has a index range increasing from 0. As a result the assignment

```
outbit_b <= identity(invect)(0);
```

selects bit 0 of the value returned by the function. This index position is defined by the local variable used to return the vector, not by the input vector passed to the function. Bit 0 is the leftmost bit of the returned vector, not its rightmost bit.

The problem becomes even clearer when the design entity is synthesized. The synthesizer realizes that the two outputs are simply the rightmost and leftmost elements of the input vector and synthesizes the "logic" in Figure 12.4.2.

From the schematic of the synthesized logic it is clear that `output_a = invect(0)` and `output_b = invect(3)`.

12.5 OVERLOADING SUBPROGRAMS AND OPERATORS

Subprogram Overloading

VHDL allows the same subprogram name to be used for different subprograms. This reuse of names is called *overloading*. Most often overloading is used when the same or similar operations are to be performed on parameters of different types. We have repeatedly encountered overloading when using the logical operators on operands of type std_logic. As we will see later in this section, an operator is just a special form of a function.

The logical operators are predefined for type bit and boolean. We have used these same logical operator names on operands of type std_logic. We are able to do so because the functions that overload these logical operator names for operands of type std_logic are provided in package STD_LOGIC_1164.

When a VHDL compiler encounters an overloaded subprogram, it determines which of the subprograms with the same name is intended by examining the number and type of operands used in the subprogram call and, in the case of functions, the return type. This process is called *operator resolution*. As long as each subprogram with the same name is unique in these characteristics, the compiler can determine which one to use. It is an error if a compiler cannot distinguish between overloaded subprograms based on these three factors.

Note that it is the base type of parameters that is used in operator resolution. Parameters of different subtypes that have the same base type are not differentiated. Also, the class of a parameter is not considered in operator resolution. Two overloaded functions that differ in only the class of a parameter cannot be differentiated.

Predefined VHDL Operators

The predefined VHDL operators depend on the type. Table 12.5.1 lists which synthesizable types have which predefined operators.

Bit and boolean are special cases of enumeration types. Only the bit and boolean types have predefined logical operators. Bit and boolean arrays are special cases of the array type. While the concatenation, equality, and ordering operators are predefined for any one-dimensional array, only bit and boolean one-dimensional arrays have the logical and shifting operators predefined.

Std_logic is simply an enumeration type. Like all enumeration types, other than bit and boolean, it has only the equality and ordering operators when it is declared. To add any other operators to std_logic requires operator overloading. This operator overloading is provided in package STD_LOGIC_1164.

Array types such as unsigned and signed, declared in package NUMERIC_STD, have the operators concatenation, equality, and ordering predefined as would any other array type. Logical, shifting, and arithmetic operators for these types are provided by overloading functions in package NUMERIC_STD.

When we declare a new type of our own, the set of predefined operators for that type automatically become available. Any other operators that we would like for this new type we will have to provide as overloaded functions.

Table 12.5.1
Predefined Operators.

Type	Arithmetic	Logical	Concatenation	Equality	Ordering	Shifting
enumeration				✓	✓	
bit		✓		✓	✓	
boolean		✓		✓	✓	
integer	✓			✓	✓	
record				✓		
array			✓	✓	✓	
bit array		✓	✓	✓	✓	✓
boolean array		✓	✓	✓	✓	✓

LISTING 12.5.1

Use of overloaded AND operator for std_logic written in the normal function form.

```
library ieee;
use ieee.std_logic_1164.all;

entity outfix_and is
   port (x, y: in std_logic; r: out std_logic);
end outfix_and;

architecture function_call of outfix_and is
begin

r <= "and" (x,y);

end function_call;
```

Operator Overloading

A common use of functions is to define operators for a type, since operators are just functions.

The rules for operator overloading only allow the predefined VHDL operators to be overloaded. The name of the function is the operator symbol enclosed in quotes. For example, the **and** operator could be declared as:

```
function "and" (l, r : bit) return bit;
```

Because it is a built-in operator, the **and** function can be called using *infix notation* as:

```
r <= x and y;
```

or, alternatively, in the normal function call format as:

```
r <= "and" (x, y);
```

Note that in this form the operator name is enclosed in double quotes.

For example, since the **and** operator is already overloaded for std_logic operands in package STD_LOGIC_1164, we can write the **and** operation in the normal function form-as shown in Listing 12.5.1.

The declaration and body of overloaded subprograms and operators are usually placed in a package and package body, respectively, so that these subprograms can be used in multiple design units. Additional examples of operator overloading are presented in Chapter 13.

12.6 TYPE CONVERSIONS

A *type conversion* converts a value from one type to another. VHDL provides explicit predefined type conversions. Predefined type conversions cannot be overloaded. User-defined type conversions are also allowed. User-defined type conversions are written as functions.

Predefined Type Conversions

Predefined type conversions are available to convert from one *closely related* type to another. Two types are considered "closely related" if they meet either of the following conditions:

1. They are both abstract numeric types (integer or floating point).
2. They are both arrays that have the same dimensions, have the same or closely related index types, and have elements that are the same type. Two array types may be closely related, even if their corresponding index positions have different directions.

An explicit predefined type conversion has the form shown in Figure 12.6.1.

As the syntax indicates, a predefined type conversion uses the name of the target type as if it were a function.

We have previously used predefined type conversions. For example, in the counter in Listing 9.4.2 the unsigned signal `count_us` was used to hold the count. However, the counter's output port `count` was type std_logic_vector. Accordingly, we needed to convert from unsigned to std_logic_vector. The assignment statement:

```
count <= std_logic_vector(count_us);
```

was used to accomplish this conversion. The type_mark specified is `std_logic_vector`. The conversion is possible because the unsigned and std_logic_vector arrays involved are closely related. They have the same dimensions, the same index types, and their elements have the same base type (std_ulogic).

User-Defined Type Conversions

User-defined type conversions are functions that take a value of one type and return a value of another type. Predefined type conversions cannot be overloaded. So, names of types cannot be used to name user-defined type conversion functions. A widely used convention for naming user-defined type conversion functions is to call the function `to_type`, where `type` is the target type of the conversion.

```
type_conversion ::=
type_mark ( expression )
```

FIGURE 12.6.1
Syntax for an explicit type conversion.

Listing 12.6.1
Function to convert a natural number to a std_logic_vector.

```
function to_std_logic_vector(arg: natural; size: positive)
   return std_logic_vector is
      variable result: std_logic_vector (size-1 downto 0);
      variable temp: natural;
   begin
      temp := arg;
      for i in 0 to size-1 loop
   if (temp mod 2) = 1 then
      result(i) := '1';
   else
      result(i) := '0';
   end if;
      temp := temp / 2;
   end loop;
   return result;
end;
```

For example, assume we needed to convert from subtype natural to std_logic_vector. We cannot use a predefined type conversion, because these two types are not closely related, because one is a scalar and the other is an array. Accordingly, we must either find a conversion function in a standard package that accomplishes this, or write our own function. A user-defined function named `to_std_logic_vector` is given in Listing 12.6.1.

This function requires two parameters, the natural number to be converted and the length of the std_logic_vector to be returned. Following common practice, the type to be converted is the first parameter and any information required for the conversion is given as subsequent parameters. The conversion algorithm uses a loop to successively divide the natural number by 2. Each division produces one bit of `result`, starting from the least significant bit.

A mod 2 division is used in the if statement's condition to determine the remainder after division by 2. This value is assigned to the bit being determined.

If we have a natural number that ranges from 0 to 255, we can convert and assign it to a std_logic_vector of length 8 using the following function call:

```
std_lg_vect <= to_std_logic_vector(nat_num, 8);
```

Formal Parameter Default Values

Formal parameters of a subprogram that are of mode **in** and class constant can be assigned default values for synthesis. If a default value is included in a parameter specification, we have the option of omitting the actual value when the subprogram is called. If the actual parameter is at the end of the parameter list, we can simply omit it. Otherwise, keyword **open** is used in place of the actual parameter value.

LISTING 12.6.2
Function using a default parameter.

```
function to_std_logic_vector(arg: natural; size: positive := 8)
   return std_logic_vector is
```

For example, the parameter list for the function in Listing 12.6.1 could be defined using a default value for the parameter `size`, as shown in Listing 12.6.2.

The mode for a function parameter is always in and the class defaults to constant. As a result, formal parameter `size` can be assigned a default value.

The function could then be called as:

```
std_lg_vect <= to_std_logic_vector(nat_num);
```

or as:

```
std_lg_vect <= to_std_logic_vector(nat_num, open);
```

To convert a natural number to a std_logic_vector that has a dimension other than 8, the desired dimension must be specified in the call.

As another example of a conversion function using a default parameter, consider the function `to_bit` that converts a std_ulogic value to a bit value. This function is provided in package STD_LOGIC_1164 and shown in Listing 12.6.3.

This function must map the nine valued std_ulogic type parameter to a two valued type bit result. The function maps values `'0'` and `'L'` to `'0'` and values `'1'` and `'H'` to `'1'`. When the std_ulogic parameter is one of the other five values, a second formal parameter `xmap` with the default assignment of `'0'` is used to map these values. The value of `xmap` is returned as the result when the std_ulogic value to be converted is other than `'0'`, `'L'`, `'1'`, or `'H'`. If we want these five values to be mapped to zero, we can call the function with a single parameter. However, if we would like these other five values mapped to `'1'` instead, we call the function with `'1'` specified as the second parameter.

LISTING 12.6.3
To_bit conversion function using a default parameter.

```
function to_bit ( s : std_ulogic; xmap : bit := '0') return bit is
    begin
      case s is
        when '0' | 'L' => return ('0');
        when '1' | 'H' => return ('1');
        when others => return xmap;
      end case;
    end;
```

LISTING 12.6.4
Function to convert a std_logic type to a character type.

```
function chr(sl: std_logic) return character is
   variable c: character;
begin
   case sl is
      when 'U' => c:= 'U';
      when 'X' => c:= 'X';
      when '0' => c:= '0';
      when '1' => c:= '1';
      when 'Z' => c:= 'Z';
      when 'W' => c:= 'W';
      when 'L' => c:= 'L';
      when 'H' => c:= 'H';
      when '-' => c:= '-';
   end case;
   return c;
end chr;
```

Conversion of a Std_logic_vector to a String

Another useful example of a type conversion is the conversion of a std_logic_vector value to a string. For example, such a function can be used to display the value of a std_logic_vector in an assert or report statement. This conversion can be accomplished using two functions. The first function, `chr`, converts a single std_logic value to the corresponding character type value. This function is given in Listing 12.6.4.

The formal parameter's type for this function is std_logic. The function's return type is character. The function declares a variable `c` of type character. A case statement uses an expression that is the value of the std_logic parameter passed to the function. Each branch of the case has a choice, which is one of the nine std_logic values. The branch whose choice corresponds to the value of the parameter passed to the function assigns the character variable `c` the corresponding character type value. The return statement returns the value of the variable.

The second function, `str`, converts a std_logic_vector to a string using the function `chr`. Function `str` is given in Listing 12.6.5.

A string variable is declared to hold the result. Since the index range for a string is type positive, the index for this variable ranges from 1 to the length of the std_logic_vector passed to the function. An integer variable is also declared for use as the string index.

The conversion is accomplished in a loop. Each time through the loop one element of the std_logic_vector is converted to a character by calling the `chr` function. Note that the loop index cannot also be used to index the element of `result`, because they may have different ranges. After the loop terminates, the string variable's value is returned by the function.

LISTING 12.6.5
Function to convert a std_logic_vector to a string.

```
function str(slv: std_logic_vector) return string is
   variable result : string (1 to slv'length);
   variable r : integer;
begin
   r := 1;
   for i in slv'range loop
      result(r) := chr(slv(i));
      r := r + 1;
   end loop;
   return result;
end str;
```

If these two functions are declared in the architecture of a testbench, the report clause of an assert statement can be used to display a message that gives the stimulus value that causes an assertion violation. For example, if the stimulus is a std_logic_vector named `input`, the report statement

```
report "Output incorrect for input of " & str(input)
```

displays the value of `input` as a string of `0`s and `1`s.

Since these two functions would find widespread use in testbenches, it is more convenient to place them in a package in a library. They would then be available for use in any testbench, as long as the appropriate library and use clauses were included.

PROBLEMS

12.1 List the basic similarities and differences between a function and a procedure in VHDL.

12.2 Write a program that implements the D latch in Figure 12.1.1 using only the function `nand_2f` and local signals.

12.3 Write a program that implements the D latch in Figure 12.1.1 using only the procedure `nand_2p` and local signals.

12.4 What is meant by the class of a formal parameter? What are the possible classes for the formal parameters of a function? What is the default class of a formal parameter of a function? What is meant by the mode of a formal parameter? What are the possible modes for the formal parameters of a function? What is the default mode of a formal parameter for a function?

12.5 What are the differences between variables declared in a function and variables declared in a process?

12.6 Write a function named `nand_fcn` that returns the result of an eight-input NAND operation on its inputs. The function's parameters are all type std_logic. If the function is called with fewer than eight actual input parameters the other unused inputs default to `'1'`s. As a result, this function can be used to implement a NAND gate with one to eight inputs. What could a one-input NAND gate be?

12.7 A two-wide two-input AND-OR-INVERT gate consists of two 2-input AND gates whose outputs are ORed together and the result inverted. Write a function `a0i_2w_2i` that implements an equivalent operation.

12.8 What are the possible classes for the formal parameters of a procedure? What are the possible modes for the formal parameters of a procedure? What is the default class of a formal parameter of a procedure that is mode out?

12.9 Write a procedure `norproc` that implements a two-input NOR gate. Draw a logic diagram of a S-R (set–reset) latch consisting of cross-coupled NOR gates. Write a VHDL description of a S-R latch. The architecture for this latch must be based on concurrent calls to the procedure `norproc`.

12.10 A function named `ones_count` has a single formal parameter that is an unconstrained std_logic_vector. The function returns an integer. The integer specifies the number of elements that are 1s in the actual vector passed to the function. Write the function.

Write a design description of a five-input majority voting circuit. This circuit uses the function `ones_count`. The five-input majority voting circuit has the following entity declaration:

```
entity majority is
   port(
      voters : in std_logic_vector(4 downto 0);
      maj  : out std_logic;
end majority;
```

Output `maj` is a `'1'` whenever the majority (more than half) of the bits of `voters` are `'1'`s.

12.11 An entity named `reorder` has an 8-bit std_logic_vector input and an 8-bit std_logic_vector output. The bits of the entity's output vector have the reverse order of the bits of its input vector. Write a description of the design entity. The description must use a single concurrent call to a function to implement the entity. The function is defined in the declaration region of the architecture body. The function, named `reorder_vec`, returns a std_logic vector whose bits have the reverse order of the bits in the std_logic_vector passed to the function. The function `reorder_vec` must be written to handle a vector of any length. Draw a diagram of the circuit that would be synthesized from this description.

12.12 Write a description of a system that determines whether two 8-bit std_logic_vector inputs are both even or both odd parity. The entity's name is `parity_comp`. The input vectors are `x` and `y`. The output is the std_logic signal `same_par`. Signal `same_par` is a `'1'` if both input vectors are even parity or both are odd parity; otherwise, it is a `'0'`. Use the function `parity_even` from Listing 12.2.1 in this design.

12.13 The function `to_integer`, which converts a std_logic_vector to an integer, has the following declaration:

```
function to_integer (slv: in std_logic_vector) return natural;
```

Use the function `to_integer` to write a function that overloads the + operator to add two std_logic_vectors. Assume the two vectors to be added and their sum are all the same length. The carry from the addition is discarded.

12.14 The VHDL code for a procedure is to be written. This procedure is similar in purpose to the function `to_integer`, whose declaration was given in Problem 12.13. The procedure's declaration is:

```
procedure vector_to_int (slv : in std_logic_vector; x_flag :
out boolean; q : inout integer);
```

The procedure converts the std_logic_vector `slv` to an integer and returns the integer value as `q`. However, if any element of the vector is other than a `'0'` or a `'1'`, then `x_flag` is returned true to indicate this condition.

Chapter 13

Packages

In Chapter 12 we saw how subprograms are used to advantage in designs. As designs become more complex and more subprograms are used, placement of subprogram declarations or bodies in the declarative part of an architecture body or a process can make the overall form of the program less clear. In addition, to reuse these subprograms in other design units requires that their code be duplicated. Subprograms are not design units, so they cannot be placed alone in a separate file and compiled.

In VHDL, a package is used to collect together related commonly used declarations, so that they can easily be made visible for use by multiple design units. These declarations include constants, types, functions, and procedures. Since a package is a design unit, it can be placed alone in a separate file and compiled.

Commonly used subprograms for type conversion, resolution of multiply driven signals, and overloading operators are usually placed in packages. Packages may also contain definitions of constants and additional types. Packages are typically placed in libraries. We have previously used types and functions provided in packages in the IEEE library.

Use of subprograms available in standard packages saves time and is preferable to writing our own equivalent subprograms.

13.1 PACKAGES AND PACKAGE BODIES

A subprogram can be defined locally (within an architecture or process). Its usage is then limited to that architecture or process. The only way a subprogram can be made available to multiple design units is for it to be declared in a package. The subprogram is then also readily available for reuse in later designs.

A package containing subprograms is defined in two parts: a package declaration (or simply package), and a package body. The package contains the declarations of the subprograms. The package body contains the bodies of the subprograms declared in the package.

Package

A *package* is a primary design unit used to organize and collect together related commonly used declarations. These include declarations of constants, types, functions, and procedures. Declarations are organized in a package, so that they can be easily made visible for use by multiple design units. By placing related declarations in a common package, the text of entity declarations and architecture bodies that use items in the package is simplified. Also, if one subprogram calls another, it is more convenient if they are in the same package.

A common use of a package is to declare one or more new types, the functions required to overload the predefined operators to apply to the new types, and any functions to define additional operators.

A package declaration has the simplified syntax given in Figure 13.1.1.

A package declaration defines the visible contents of a package. It tells us how to interface to (use) the declared items. For example, a subprogram declaration in a package provides an interface template for designs that call the subprogram. It provides all of the information necessary to use the functions and procedures, such as a subprogram's name and its formal parameters.

Package Body

A package accomplishes information hiding by separating the external view of the declared items from their realization. The code that realizes a subprogram is given separately in an associated secondary design unit called a *package body*. A package body has the simplified syntax given in Figure 13.1.2.

A package body must have the same name as its associated package. It contains the bodies of all subprograms declared in the package. Thus, the package body provides the hidden implementation details of each subprogram. A package body is required only if a package declares subprograms or deferred constants.

```
package_declaration ::=
        package identifier is
            { package_declarative_item }
        end [ package_simple_name ] ;

package_declarative_item ::=
        subprogram_declaration | type_declaration | subtype_declaration
        | constant_declaration | signal_declaration | component_declaration
        | attribute_declaration | attribute_specification | use_clause
```

FIGURE 13.1.1
Simplified syntax for a package declaration.

package_body ::=
 package body *package*_simple_name **is**
  { package_body_declarative_item }
 end [*package*_simple_name] ;

package_body_declarative_item ::=
 subprogram_declaration | subprogram_body | type_declaration
 | subtype_declaration | constant_declaration | ~~*shared* variable_declaration~~
 | file_declaration | alias_declaration | use_clause | ~~group_template_declaration~~
 | ~~group_declaration~~

FIGURE 13.1.2
Syntax for a package body.

A subprogram body in a package body must have a subprogram declaration portion that is exactly as written in the package declaration. That is, the names, types, modes, and default values of each formal parameter must be repeated exactly.

Package slv_str

Functions `chr` and `str` (see "Conversion of a Std_logic_vector to a String" on page 560) can be placed in a package and compiled to a library. Listing 13.1.1 shows a package and package body containing these two functions. Any design with a context clause referencing the library and the package `slv_str` can call these functions.

LISTING 13.1.1
A package and package body containing two functions.

```
library ieee;
use ieee.std_logic_1164.all;

package slv_str is
   function chr(sl: std_logic) return character;
   function str(slv: std_logic_vector) return string;
end slv_str;

package body slv_str is

   function chr(sl: std_logic) return character is
      variable c: character;
   begin
      case sl is
         when 'U' => c:= 'U';
         when 'X' => c:= 'X';
         when '0' => c:= '0';
         when '1' => c:= '1';
```

(Cont.)

LISTING 13.1.1 *(Cont.)*

```
         when 'Z' => c:= 'Z';
         when 'W' => c:= 'W';
         when 'L' => c:= 'L';
         when 'H' => c:= 'H';
         when '-' => c:= '-';
      end case;
      return c;
   end chr;

   function str(slv: std_logic_vector) return string is
      variable result : string (1 to slv'length);
      variable r : integer;
   begin
      r := 1;
      for i in slv'range loop
         result(r) := chr(slv(i));
         r := r + 1;
      end loop;
      return result;
   end str;

end slv_str;
```

If desired, the package and package body can be placed in separate files and compiled separately. In that case, the file containing the package must be compiled before the file containing the package body.

Deferred Constants

When a constant is declared in a package, its name and type must be specified. However, the assignment of a value to the constant can be done either as part of its declaration in the package, or it can be deferred to the package body. If the assignment of a value to a constant is done in the package, the constant is a *fully specified constant*. In contrast, if the assignment is deferred to the package body, the constant is a *deferred constant*. A deferred constant's declaration in the package is repeated in the package body, along with the addition of the initialization expression for the constant's value.

If a package declares only types, signals, or fully specified constants, no associated package body is necessary.

Declarations in a Package Body

In addition to the bodies of any subprograms or deferred constants declared in a package, a package body may contain full declarations of additional types, subtypes, constants, and subprograms used in the package body to realize the subprograms declared in the package. These items are not visible outside of the package body.

Global Signals

A package body cannot declare any signals. Signals may only be declared in a package. Signals declared in a package are *global signals*. They are visible anywhere the package is used.

Packages in Libraries

Packages and package bodies are compiled to libraries. They may be compiled to the working library or to a resource library. Usually, a package is compiled into the working library (`work`) by default. However, packages that are of general use are typically compiled to a resource library. Different VHDL compilers use different ways to specify into which library a package is compiled. These procedures are delineated in the compiler's documentation.

The order of compilation of packages and package bodies is important. Since a package body depends on information defined in its associated package, the package must be compiled before its package body. A package must also be compiled before an entity declaration or architecture body that uses any items declared in the package.

A more complete discussion of libraries and the compilation of design units is given in Sections 15.2 and 15.3.

13.2 STANDARD AND DE FACTO STANDARD PACKAGES

As we have seen, the predefined types, operators, and functions declared in packages STANDARD and TEXTIO from the library STD are not sufficient for our simulation and synthesis needs. VHDL allows us to define our own data types and operators and place them in packages so that we can use them in our designs.

However, most of the data types and operators that we normally need are already available in IEEE standard packages and in de facto standard packages commonly included in the IEEE library provided by tool vendors with their simulation or synthesis tools. Use of these standard packages or de facto standard packages saves us the time required to develop our own packages and improves code portability. Another advantage is that many tool vendors provide optimized implementations of these packages.

Some of these packages are listed in Table 13.2.1. Listed first are the predefined packages STANDARD and TEXTIO. Packages STD_LOGIC_1164, NUMERIC_BIT, and NUMERIC_STD are defined by IEEE standards and included in the IEEE Library. Although packages STD_LOGIC_UNSIGNED, STD_LOGIC_SIGNED, and STD_LOGIC_ARITH are usually included by tool vendors in their IEEE libraries, they are not actually IEEE standards. However, because of their widespread use, they are de facto standards.

This section provides a brief overview of these packages. Later sections of this chapter provide a more detailed look at some of these packages, to provide the background necessary to allow the actual source code of these packages to be read

Table 13.2.1
Standard packages commonly used in synthesis.

Packages	Library	Standard Number	Standard Title
standard, textio	std	IEEE Std 1076	IEEE Standard VHDL Language Reference Manual
std_logic_1164	ieee	IEEE Std 1164	IEEE Standard Multivalue Logic System for VHDL Model Interoperability (Std_logic_1164)
numeric_bit, numeric_std	ieee	IEEE Std 1076.3	IEEE Standard VHDL Synthesis Packages
std_logic_unsigned, std_logic_signed	ieee	None	None
std_logic_arith	ieee	None	None

and understood and to provide further examples of how subprograms are written. The source code is contained in the actual standards documents for those packages that are IEEE standards. Tool vendors sometimes supply the source code with their IEEE library.

Tool vendors are not allowed to add declarations to or delete declarations from packages that are IEEE standards. However, they are allowed to modify the package bodies to create more efficient implementations. Any modified package body must be semantically identical to the original package body in the standard.

Package STANDARD

VHDL compilers come with a library called Std. *Library Std* is predefined by the VHDL 1076 standard. This library contains two packages: STANDARD and TEXTIO. It is in package STANDARD that the predefined types and operators of VHDL are declared. For example, type bit and the `and` operator.

Because almost every VHDL description needs to use types and operators from package STANDARD, its contents are automatically made visible to all design units.

Package TEXTIO

Package TEXTIO contains predefined types, functions, and procedures used for reading and writing files.

Package STD_LOGIC_1164

Type std_ulogic and subtype std_logic, along with their corresponding vector types std_ulogic_vector and std_logic_vector, are defined in package STD_LOGIC_1164.

The basic logical operators (`and`, `or`, `not`, and so on) are overloaded for these types in this package.

Package NUMERIC_STD

Std_logic_vector, as defined in package STD_LOGIC_1164, is not a numeric representation. However, it can be interpreted as representing an array of logic values, an unsigned value, or a signed value. To treat std_logic_vector as a numeric value, arithmetic operators such as addition, subtraction, and inequality would have to be defined by the user, and their functions written. Many of these functions would have to differ, depending on whether a std_logic_vector was to be interpreted as an unsigned or signed value.

To avoid confusion as to how a std_logic_vector is to be interpreted, separate types can be created. IEEE Std 1076.3 provides two packages for this purpose: NUMERIC_STD and NUMERIC_BIT. NUMERIC_STD defines two vector types whose elements are type std_logic. These two types are named *unsigned* and *signed* because they represent unsigned and signed numeric values, respectively. The related arithmetic operators are also defined in this package.

Package NUMERIC_STD provides essentially the same functionality as an earlier package named STD_LOGIC_ARITH. Package STD_LOGIC_ARITH was developed by EDA tool vendor Synopsys. Although the Synopsys package is provided by most VHDL tool vendors in their implementations of the IEEE library, it is proprietary and not a standard. In fact, package NUMERIC_STD was developed by the IEEE so that there would be a nonproprietary standard numeric package available that could be adopted by all tool vendors.

Package NUMERIC_STD has significantly fewer conversion functions than package STD_LOGIC_ARITH. Instead, many type conversions are done using the inherent type conversion capability of the VHDL language.

Package NUMERIC_BIT

Package NUMERIC_BIT also defines two vector types, also named unsigned and signed. However, their elements are type bit, rather than std_logic.

Packages STD_LOGIC_UNSIGNED and STD_LOGIC_SIGNED

Originally, Synopsys developed two packages, STD_LOGIC_UNSIGNED and STD_LOGIC_SIGNED, that provide arithmetic, conversion, and comparison for type std_logic_vector. These separate packages interpret std_logic_vector as either unsigned or signed, respectively. Package STD_LOGIC_UNSIGNED performs unsigned arithmetic on std_logic types and returns std_logic types. Package STD_LOGIC_SIGNED performs signed arithmetic on std_logic types and returns std_logic types.

A drawback of these packages is that we can only use one or the other in a design. Therefore, our design would have to deal with either unsigned or signed std_logic_vector numbers exclusively. This presented a problem in designs that needed to deal with both unsigned and signed numbers.

Package STD_LOGIC_ARITH

Synopsys later developed package STD_LOGIC_ARITH. This package allows use of both unsigned and signed interpretations of std_logic_vectors in a single design.

Package STD_LOGIC_ARITH declares two types, unsigned and signed, that are vectors of std_logic elements. In this way, the arithmetic operators for these two types are defined appropriately. Arithmetic comparison functions between unsigned and signed and conversion functions between unsigned, signed, or std_logic_vector are also provided.

Table 13.2.2
Selected arithmetic operators in standard and de facto standard packages.

Function	Operand Types	Result Types	Package
+	signed, unsigned integer, natural	signed, unsigned	numeric_std
–	signed, unsigned integer, natural	signed, unsigned	numeric_std
*	signed, unsigned integer, natural	signed, unsigned	numeric_std
/	signed, unsigned integer, natural	signed, unsigned	numeric_std
rem	signed, unsigned integer, natural	signed, unsigned	numeric_std
mod	signed, unsigned integer, natural	signed, unsigned	numeric_std
+	signed, unsigned integer, std_logic	signed, unsigned std_logic_vector	std_logic_arith
–	signed, unsigned integer, std_logic	signed, unsigned std_logic_vector	std_logic_arith
*	signed, unsigned integer, std_logic	signed, unsigned, std_logic_vector	std_logic_arith
+	std_logic, integer, std_logic_vector	std_logic_vector	std_logic_unsigned
–	std_logic, integer, std_logic_vector	std_logic_vector	std_logic_unsigned
*	std_logic, integer, std_logic_vector	std_logic_vector	std_logic_unsigned

Logical operators for type unsigned and signed are not defined in STD_LOGIC_ARITH. Instead, the unsigned or signed type operands must be converted to std_logic_vector, the logical operation performed, and the result converted back to unsigned or signed. The conversions are accomplished by type conversions rather than functions, since the elements of the different types are all std_logic.

A Summary of Selected Package Functions

Tables 13.2.2–13.2.6 list selected functions from the packages previously discussed in this section. To properly use the types and functions in these packages, their definitions need to be clearly understood. Accordingly, realizations of some of the most commonly used functions are discussed in detail in later sections. These later discussions provide additional examples of how functions are written and help us develop the capability to examine and understand function bodies in packages.

Table 13.2.3
Selected relational operators in standard and de facto standard packages.

Function	Operand Types	Result Types	Package
>	signed, unsigned integer, natural	boolean	numeric_std
<	signed, unsigned integer, natural	boolean	numeric_std
=	signed, unsigned integer, natural	boolean	numeric_std
>=	signed, unsigned integer, natural	boolean	numeric_std

Function	Operand Types	Result Types	Package
<=	signed, unsigned integer, natural	boolean	numeric_std
/=	signed, unsigned integer, natural	boolean	numeric_std
>	signed, unsigned integer	boolean	std_logic_arith
<	signed, unsigned integer	boolean	std_logic_arith
=	signed, unsigned integer	boolean	std_logic_arith
>=	signed, unsigned integer	boolean	std_logic_arith
<=	signed, unsigned integer	boolean	std_logic_arith
/=	signed, unsigned integer	boolean	std_logic_arith
>	std_logic_vector, integer	std_logic_vector	std_logic_unsigned
<	std_logic_vector, integer	std_logic_vector	std_logic_unsigned
=	std_logic_vector, integer	std_logic_vector	std_logic_unsigned
>=	std_logic_vector, integer	std_logic_vector	std_logic_unsigned
<=	std_logic_vector, integer	std_logic_vector	std_logic_unsigned
/=	std_logic_vector, integer	std_logic_vector	std_logic_unsigned

Table 13.2.4
Selected shift operators in standard and de facto standard packages.

Function	Operand Types	Result Types	Package
shift_left	signed, unsigned, natural	signed, unsigned	numeric_std
shift_right	signed, unsigned, natural	signed, unsigned	numeric_std
rotate_left	signed, unsigned, natural	signed, unsigned	numeric_std
rotate_right	signed, unsigned, natural	signed, unsigned	numeric_std
sll	signed, unsigned, integer	signed, unsigned	numeric_std
srl	signed, unsigned, integer	signed, unsigned	numeric_std
rol	signed, unsigned, integer	signed, unsigned	numeric_std
ror	signed, unsigned, integer	signed, unsigned	numeric_std
shl	signed, unsigned integer	signed, unsigned	std_logic_arith
shr	signed, unsigned integer	signed, unsigned	std_logic_arith
shl	std_logic_vector, integer	std_logic_vector	std_logic_unsigned
shr	std_logic_vector, integer	std_logic_vector	std_logic_unsigned

Table 13.2.5
Selected type conversion functions in standard and de facto standard packages.

Function	Operand Types	Result Types	Package
to_integer	signed, unsigned	integer, natural	numeric_std
to_unsigned	std_logic_vector, natural	unsigned	numeric_std
to_signed	std_logic_vector, natural	signed	numeric_std
conv_integer	signed, unsigned std_ulogic, integer	integer	std_logic_arith
conv_signed	signed, unsigned std_ulogic, integer	signed	std_logic_arith
conv_unsigned	signed, unsigned std_ulogic, integer	unsigned	std_logic_arith
conv_std_logic_vector	signed, unsigned std_ulogic, integer	std_logic_vector	std_logic_arith
conv_integer	std_logic_vector	integer	std_logic_unsigned

13.3 PACKAGE STD_LOGIC_1164

To eliminate portability problems resulting from the use of nonstandard multivalued logic types, IEEE Std 1164 was developed and was ratified by the IEEE in 1993. This standard contains package STD_LOGIC_1164, which defines types, subtypes, and functions to extend VHDL by adding a nine-value data type called std_ulogic. Package STD_LOGIC_1164 is also called the *standard logic value package,* or *MVL9* (for multivalued logic, nine values). All VHDL tool vendors support this standard. Std_logic, a subtype of std_ulogic, was introduced in Section 3.4 and has been used extensively throughout this book.

Resolution Function for Std_logic

As previously discussed, std_logic is a resolved type. Since a resolved signal can have multiple sources, its declaration must also include a resolution function. The resolution function determines the resultant signal value from the values of a signal's

Table 13.2.6
Selected miscellaneous functions in standard and de facto standard packages.

Function	Operand Types	Result Types	Package
resize	singed, unsigned, natural	signed, unsigned	numeric_std
std_match	signed, unsigned, std_logic_vector, std_ulogic_vector	boolean	numeric_std
ext	std_logic_vector	std_logic_vector	std_logic_arith
sxt	std_logic_vector	std_logic_vector	std_logic_arith
abs	signed	signed	std_logic_arith

multiple sources. Package STD_LOGIC_1164 contains the declaration of a resolution function name `resolved`. The function's declaration is:

```
function resolved ( s : std_ulogic_vector ) return std_ulogic;
```

The function's body, as defined in package body STD_LOGIC_1164, is shown in Listing 13.3.1.

Type `stdlogic_table` is declared as:

```
type stdlogic_table is array(std_ulogic, std_ulogic) of
std_ulogic;
```

LISTING 13.3.1
The resolution function resolved as defined in package body STD_LOGIC_1164.

```
type stdlogic_table is array(std_ulogic, std_ulogic) of std_ulogic;

constant resolution_table : stdlogic_table := (
--      ---------------------------------------------------------
--      | U    X    0    1    Z    W    L    H    -        |   |
--      ---------------------------------------------------------
        ( 'U', 'U', 'U', 'U', 'U', 'U', 'U', 'U', 'U' ), -- | U |
        ( 'U', 'X', 'X', 'X', 'X', 'X', 'X', 'X', 'X' ), -- | X |
        ( 'U', 'X', '0', 'X', '0', '0', '0', '0', 'X' ), -- | 0 |
        ( 'U', 'X', 'X', '1', '1', '1', '1', '1', 'X' ), -- | 1 |
        ( 'U', 'X', '0', '1', 'Z', 'W', 'L', 'H', 'X' ), -- | Z |
        ( 'U', 'X', '0', '1', 'W', 'W', 'W', 'W', 'X' ), -- | W |
        ( 'U', 'X', '0', '1', 'L', 'W', 'L', 'W', 'X' ), -- | L |
        ( 'U', 'X', '0', '1', 'H', 'W', 'W', 'H', 'X' ), -- | H |
        ( 'U', 'X', 'X', 'X', 'X', 'X', 'X', 'X', 'X' )  -- | - |
    );

function resolved ( s : std_ulogic_vector ) return std_ulogic is
    variable result : std_ulogic := 'z';  -- weakest state default
begin
    -- the test for a single driver is essential otherwise the
    -- loop would return 'x' for a single driver of '-' and that
    -- would conflict with the value of a single driver unresolved
    -- signal.
    if    (s'length = 1) then    return s(s'low);
    else
        for i in s'range loop
            result := resolution_table(result, s(i));
        end loop;
    end if;
    return result;
end resolved
```

This statement declares a two-dimensional array of type std_ulogic. The range of each of the array's dimension is the enumerated range of type std_ulogic, or ('U', 'X', '0', '1', 'Z', 'W', 'L', 'H', '-').

Constant resolution_table is then defined. Its type is the previously defined two-dimensional array of type std_ulogic named stdlogic_table. This table of constants concisely defines the rules for resolving a multiply driven std_logic signal. The entries in the table must define a commutative function, so that the values may be passed to the function in any order. The commutative nature of the table entries is apparent because entry values are reflected along a diagonal from the top left to the bottom right.

Unlike other functions we have considered, we never write a call to a resolution function. Instead, the function is automatically called by the simulator each time a transaction is executed for a driver of a multiply driven std_logic signal. Function resolved takes as its parameter a single unconstrained std_ulogic_vector array created and passed to it by the simulator. The simulator uses the new driver transaction value and the values of any other drivers of the signal to form the array that it passes to the function resolved. The result returned by resolved is the new value of the multiply driven signal.

When called, resolved first initializes the variable result to the value 'Z'. The if statement then uses the length attribute to check whether the length of the array passed to the function is 1. If this is the case, the signal has a single driver and its resolved value is the value of that single driver. The low attribute returns the lower bound of the index range of the array passed to the function. So, s(s'low) is simply the sole element of the one element array that was passed to the function.

If the length of the array passed is greater than 1, a loop is entered. This loop takes the loop index i through all the values in the range of the array. Each time through the loop, the value of result is used as the row index into resolution_table and the value of array element i is used as the column index. The table entry at this row and column becomes the new value of result. When the loop is completed, the value of result is the resolved value of all the elements of the array passed to the function. The value of result is then returned by the function.

Std_ulogic Subtypes Without Strength: X01, X01Z, UX01, UX01Z

Package STD_LOGIC_1164 includes the declaration of four additional subtypes that exclude weak strength values. These subtypes are used by functions in the package. Each subtype's name is an acronym formed by the values included in the subtype. Their declarations are:

```
subtype X01 is resolved std_ulogic range 'X' to '1';
-- ('X','0','1')
subtype X01Z is resolved std_ulogic range 'X' to 'Z';
-- ('X','0','1','Z')
subtype UX01 is resolved std_ulogic range 'U' to '1';
-- ('U','X','0','1')
subtype UX01Z is resolved std_ulogic range 'U' to 'Z';
-- ('U','X','0','1','Z')
```

The range of these subtypes is specified in terms of the ordered enumeration that defines std_ulogic. Therefore, **range** 'X' **to** '1' starts with 'X', includes '0', and ends with '1'.

None of these subtypes includes any of the values that represent weak signals: 'W', 'L', 'H'. These subtypes are useful when we don't care about the strength of a signal, only its logical value.

Each of these subtypes is also a *closed subtype*. This means that the result of resolving values of the subtype is a value of the subtype.

Overloaded Logical Operators for Std_logic

Recall that the logical operators defined in IEEE Std 1076 for bit and boolean types are overloaded in package STD_LOGIC_1164 for std_logic. The functions that define these overloaded operators also use table lookups. The tables and functions are declared in the STD_LOGIC_1164 package and the functions are realized in the STD_LOGIC_1164 package body. For example, the table for the **and** function is given in Listing 13.3.2

The entries in the table are restricted to values of subtype UX01. Thus, the result of a table lookup for the **and** operation is a value of subtype UX01.

The **and** function for std_ulogic types, as given the in the package body, is:

```
function "and"  ( l : std_ulogic; r : std_ulogic )
return UX01 is
    begin
        return (and_table(l, r));
    end "and";
```

In the function body, the name of the function is placed in quotes, ("and"). These quotes indicate that the name is that of one of the predefined operators. The parameters are denoted l and r to indicate that these are the operands to the left and right

LISTING 13.3.2
Table of constants used to overload the AND function for std_logic.

```
constant and_table : stdlogic_table := (
--      ---------------------------------------------------
--      | U    X    0    1    Z    W    L    H    -        |   |
--      ---------------------------------------------------
        ( 'U', 'U', '0', 'U', 'U', 'U', '0', 'U', 'U' ),   -- | U |
        ( 'U', 'X', '0', 'X', 'X', 'X', '0', 'X', 'X' ),   -- | X |
        ( '0', '0', '0', '0', '0', '0', '0', '0', '0' ),   -- | 0 |
        ( 'U', 'X', '0', '1', 'X', 'X', '0', '1', 'X' ),   -- | 1 |
        ( 'U', 'X', '0', 'X', 'X', 'X', '0', 'X', 'X' ),   -- | Z |
        ( 'U', 'X', '0', 'X', 'X', 'X', '0', 'X', 'X' ),   -- | W |
        ( '0', '0', '0', '0', '0', '0', '0', '0', '0' ),   -- | L |
        ( 'U', 'X', '0', '1', 'X', 'X', '0', '1', 'X' ),   -- | H |
        ( 'U', 'X', '0', 'X', 'X', 'X', '0', 'X', 'X' )    -- | - |
);
```

```
alias_declaration ::=
alias alias_designator [ : subtype_indication ] is name;

alias_designator ::=
identifier | character_literal | operator_symbol
```

FIGURE 13.3.1
Simplified syntax for an alias declaration.

of the operator, respectively. The return value is indicated to be of subtype `UX01`. The **and** function returns the value obtained from the table lookup using table `and_table` with `l` as the row index and `r` as the column index.

Aliases

Some of the code used to implement functions in STD_LOGIC_1164 takes advantage of the use of aliases. Accordingly, we consider aliases here, prior to looking at an example of the realization of the **and** function for std_logic_vectors. An *alias* in VHDL is an alternate name for a named item. An alias declaration follows the syntax shown in Figure 13.3.1.

Typically, aliases are used to make it easier to read and understand code. For example, a long selected name can be aliased to a simpler name. Any operations performed using the alias actually apply to the original item, with one important exception.

The exception applies to attributes related to the index range of an array that is aliased. If an alias declaration defines an index range for an alias that is different from the index range for the item being aliased, index range attributes using the alias name return values based on the index range specified in the alias declaration. Elements in the alias denote the corresponding elements in the aliased array from left to right. This makes it possible to use an alias to normalize the range of an array parameter passed to subprogram.

Use of an alias for normalizing an array in a subprogram obviates the need to introduce an additional variable in the subprogram for that purpose, as was done in Section 12.4.

Overloaded Logical Operators for Std_logic_vector

Package STD_LOGIC_1164 also provides overloaded functions for logical operations on std_logic_vectors. These functions utilize attributes of the index range of aliases as just discussed. For example, the function for the AND of std_logic_vectors is given in Listing 13.3.3.

LISTING 13.3.3
Function for the AND of two std_logic_vectors.

```
function "and"  ( l, r : std_logic_vector ) return std_logic_vector is
        alias lv : std_logic_vector ( 1 to l'length ) is l;
        alias rv : std_logic_vector ( 1 to r'length ) is r;
        variable result : std_logic_vector ( 1 to l'length );
```

```
begin
    if ( l'length /= r'length ) then
        assert false
        report "arguments of overloaded 'and' operator not same length"
        severity failure;
    else
        for i in result'range loop
            result(i) := and_table (lv(i), rv(i));
        end loop;
    end if;
    return result;
end "and"
```

This function declares two aliases `lv` and `rv`. These aliases are used to make sure the index range is in the same direction for both operands. Each alias specifies a range from 1 to the length of the corresponding operand for which it is an alias. The variable `result` is also declared with a length equal to the length of parameter `l`.

The function first tests whether the lengths of the two parameters are the same. If they are not, an error is asserted. If the lengths are the same, a loop uses the `and_table` to determine, for each of the corresponding pairs of elements from the two parameters, the corresponding element of `result`. These pairs of elements from the two parameters are specified in terms of their aliases.

Strength Stripping Functions

Functions are provided in package STD_LOGIC_1164 that convert from std_ulogic to subtypes without strength values. These functions are also table lookups. For example, the table used to convert from type std_ulogic to type X01 uses the following array type defined in the package:

```
type logic_X01_table is array (std_ulogic'low to
std_ulogic'high) of X01;
```

The range of this array (`std_ulogic'low` to `std_ulogic'high`) is the enumeration of all the std_ulogic values.

The lookup table that serves as the basis for the conversion is:

```
constant cvt_to_X01 : logic_X01_table := (
                        'X',   -- 'U'
                        'X',   -- 'X'
                        '0',   -- '0'
                        '1',   -- '1'
                        'X',   -- 'Z'
                        'X',   -- 'W'
                        '0',   -- 'L'
                        '1',   -- 'H'
                        'X'    -- '-'
                       );
```

This is a one-dimensional array of constants. The comments indicate which std_ulogic value is being converted. The index into the table is the std_logic value to be converted. The table entry is the type `X01` value to which it is converted. The weak values simply map to the normal logic values. Values `'U'`, `'Z'`, and `'-'` map to `'X'`.

The function in package STD_LOGIC_1164 that converts from std_ulogic to `X01` is named `to_X01`. This function's body is:

```
function to_X01  ( s : std_ulogic ) return  X01 is
    begin
        return (cvt_to_X01(s));
    end;
```

This function simply uses the std_ulogic value being converted as the index into table `cvt_to_X01` and returns the value at that index position.

Edge Detection Functions

We have frequently used the `rising_edge` function in previous descriptions. This function is defined in package STD_LOGIC_1164 as:

```
function rising_edge  (signal s : std_ulogic) return
boolean is
    begin
         return (s'event and (to_X01(s) = '1') and
                         (to_X01(s'last_value) = '0'));
    end;
```

This function returns a boolean value. True is returned if three conditions are met:

1. Signal s has changed value during this simulation cycle (s'EVENT = true).
2. The value of s after the change, when converted to type X01, is '1' (to_X01 (s) = '1').
3. The value of s before the change when converted to type X01 was '0' (to_X01 (s'LAST_VALUE) = '0').

As defined, this function returns true only for a transition from `'0'` to `'1'`, or `'L'` to `'1'`, or `'0'` to `'H'`, or `'L'` to `'H'`. Thus, the determination of an edge is irrespective of the driving strength.

The body of the `falling_edge` function simply reverses the values `'0'` and `'1'` in the body of `rising_edge`.

13.4 PACKAGE NUMERIC_STD (IEEE STD 1076.3)

The Need for Numeric Types

Package STD_LOGIC_1164 defines std_ulogic and its subtypes std_logic and std_logic_vector. It also provides logical and other operators. However, it does not define a numeric interpretation of std_logic_vector, nor does it provide any arithmetic operators. Thus, std_logic_vector does not have a defined numeric interpretation.

As discussed in Chapter 3, this problem can be solved in one of two ways. Functions to overload the predefined arithmetic operators to include operations on type std_logic_vector can be written. Alternatively, new vector types can be created specifically for representing unsigned and signed integer values for synthesis and overloaded arithmetic operators can be defined for these new types.

Proprietary Numeric Types

Initially, synthesis tool vendors added their own packages to overload the arithmetic operators for std_logic_vector, or to define additional data types that had a numeric interpretation. One such package, STD_LOGIC_ARITH, is discussed in the next section. However, these packages are not standards.

IEEE Numeric Standard

To provide vector types for unsigned and signed arithmetic and to ensure the widest possible portability between vendors' tools, the IEEE created *IEEE Std 1076.3*. This standard defined two numeric packages. It also interpreted VHDL data types previously defined by IEEE Std 1076 and IEEE Std 1164 as they relate to actual hardware.

IEEE Std 1076.3 is defined in the document *IEEE Standard VHDL Synthesis Packages,* which is often simply referred to as the *Numeric Standard or Synthesis Standard.*

Packages NUMERIC_BIT and NUMERIC_STD

The two packages defined in Std 1076.3 are NUMERIC_BIT and NUMERIC_STD. NUMERIC_BIT is based on vectors with elements that are type bit. NUMERIC_STD is based on vectors with elements that are type std_logic. Outside of these element type differences the two packages provide similar functions. Since we are using std_logic for synthesis, our interest is in package NUMERIC_STD.

Types Unsigned and Signed

The NUMERIC_STD package defines two arithmetic data types, unsigned and signed, along with arithmetic, shift, type conversion, and logical operators. Functions in the package perform arithmetic operations on unsigned and signed types, and return these types. The only operator defined in this package for std_logic_vectors is the `std_match` function, previously introduced in "Std_match Function" on page 141. Function `std_match` is also defined in this package for types unsigned and signed. There are no other operators provided in the NUMERIC_STD package for std_logic or std_logic_vectors.

Package NUMERIC_STD defines the types unsigned and signed as unconstrained arrays:

```
type unsigned is array (natural range <>) of std_logic;
type signed is array (natural range <>) of std_logic;
```

While the elements of both the unsigned and signed types are std_logic, they are two separate types, distinct from each other and from std_logic_vector.

Type unsigned is used to represent unsigned binary integer numbers. The most significant bit is on the left. Type signed is used to represent signed binary integer numbers. The most significant bit (the sign bit) is on the left. Objects of type signed are interpreted as 2's complement numbers. The overloaded operators are written to manipulate each type according to its intended interpretation. Using two separate types precludes any confusion over the interpretation of a given bit pattern.

Range of Unsigned and Signed Numbers

Since unsigned and signed are arrays, the precision of the numbers they represent is limited only by the array length specified when a particular object of either type is declared. Accordingly, arithmetic using unsigned or signed can be performed to any precision.

Operators Provided

Package NUMERIC_STD provides an extensive set of operators: resize, arithmetic, comparison, logical, and concatenation. All vector return values are normalized, so that the direction of the index range is down to and the right bound is 0.

The primary focus in this section is on the resize and arithmetic operators. Additional comments are provided concerning specific interpretations of some of the other operators.

Unsigned Resize Function

The package includes a resize function to extend or truncate an unsigned or signed vector. The resize function declaration for an unsigned vector is:

```
function resize (arg: unsigned; new_size: natural)
return unsigned;
```

Parameter `arg` is the unsigned vector to be resized, and parameter `new_size` specifies the new length of the returned vector. If the new length is greater than the length of the original vector, 0s are added to the left of the vector to increase its length. If the new length is less than the original length, the appropriate number of bits is truncated from the left to decrease the vector's length. If `new_size` is a negative value, a null vector is returned with its range specified as `0 downto 1`.

Signed Resize Function

The signed version of the `resize` function resizes a signed vector and returns a signed result. If the new length is greater than the length of the original vector, copies of the sign bit (leftmost bit) of the original vector are added to the left of the vector to increase its length. This corresponds to the common computer software sign extension operation.

If the new length is less than the original length, the sign bit is kept and the appropriate number of bits immediately to the right of the sign bit is truncated to decrease the vector's length. This is a rather unusual truncation operation. Typically, a truncation operation on a signed quantity simply removes the required number of bits from the left (sign bit included), leaving the possibility that the result's sign is changed.

Arithmetic Operators

Several arithmetic operators are defined in NUMERIC_STD; they are unary –, abs, +, –, *, /, mod, and rem. Unary – and abs are defined only for type signed. The other operators are defined for both types unsigned and signed.

Addition and Subtraction

Addition and subtraction of unsigned or signed vectors are easily performed using NUMERIC_STD's + and – operators. Operands must be either both unsigned, or unsigned and natural, or both signed, or signed and integer. A type unsigned and a type signed cannot be added or subtracted.

Addition and subtraction operands can be of different lengths. The addition and subtraction functions automatically use the resize function to make the operands the same length before adding or subtracting. The result returned has the length of the larger of the two operands. If one operand is type unsigned or type signed and the other is natural or integer, respectively, the result has the length of the vector.

If the result is larger than the length of the larger of the two operands (overflow), the result wraps around. For example, adding one to the most positive unsigned number gives zero. Adding one to the most positive signed number gives the most negative signed number.

Carry or Borrow

Adding or subtracting two n-bit operands gives an n-bit result. There is no carry or borrow. This is an advantage when addition or subtraction is used with an unsigned operand to describe an up counter or down counter, respectively. The counter "rolls over" in the desired fashion.

When it is necessary to produce a carry or borrow, the operands have to be extended by one bit in order to accommodate the carry or borrow. A '0' is concatenated to the left of the operands to produce an $n + 1$ bit result. The least significant n bits of the result is the sum or difference, and the most significant bit is the carry or borrow.

A description of a 4-bit full adder using signed operands is given in Listing 13.4.1. A 4-bit full adder must add together two 4-bit operands and a carry in to produce a 4-bit sum and a carry out.

LISTING 13.4.1
A 4-bit full adder using type signed.

```
library ieee;
use ieee.std_logic_1164.all;
use ieee.numeric_std.all;


entity adder_signed is
    port(
        carry_in : in std_logic;
        augend : in std_logic_vector(3 downto 0);
```

(Cont.)

Listing 13.4.1 (Cont.)

```
        addend : in std_logic_vector(3 downto 0);
        carry_out : out std_logic;
        sum : out std_logic_vector(3 downto 0)
         );
end adder_signed;

architecture dataflow of adder_signed is
signal sum_lcl : signed(4 downto 0);
begin
    sum_lcl <= signed('0' & augend) + signed('0' & addend)
        + signed'('0' & carry_in);
    sum <= std_logic_vector(sum_lcl(3 downto 0));
    carry_out <= std_logic(sum_lcl(4));

end dataflow;
```

The entity is declared to have std_logic_vector and std_logic inputs and outputs. If we tried to create the adder using the signal assignment statement

```
sum <= signed(augend) + signed(addend)
          + carry_in);
```

we would have two problems. First, the sum would be a 4-bit quantity and the carry out would be lost. Second, `carry_in` is type std_logic and the signed addition operation is not defined for an operand of this type.

It might seem that the first problem could be solved by resizing one of the operands to 5 bits and taking this 5th bit as the carry. The addition operation would then automatically resize the second operand, so that both operands would be 5 bits. However, since the operands are signed, resizing is accomplished by sign extension. The 5th bit would then not correspond to the carry, but instead would simply be the sum of the most significant bits of the extended operands.

The correct carry can only be produced if both operands are initially equal length and a 0 is concatenated to the left of each operand, as is done in Listing 13.4.1. If the operands are not of the same length, the shorter operand must first be resized, using the resize function, and then a `'0'` concatenated to the left of both operands.

The second problem is resolved by making the scalar std_logic `carry_in` a std_logic_vector by concatenating a `'0'` to its left and then type qualifying the result as signed. The addition operation sign extends this 2-bit vector using 0s.

Note that, in order to use signed addition, use statements for both NUMERIC_STD and STD_LOGIC_1164 are required prior to the entity declaration.

Unary – and abs

The unary – and abs operators apply only to the signed type. The "–" operator computes the two's complement of a number. The result is the same size as the original vector. Since the two's complement number range is not symmetrical, having one negative value more than positive values, negation of the most negative value causes an overflow and the result is the most negative number itself.

The abs operator returns the absolute value of a parameter. If the parameter is negative, this is accomplished by negating the parameter. If the parameter is the most negative value, negating this value returns the same value. So, for this one special case, the absolute value returned can be negative!

Multiplying Operators

VHDL classifies the operators: *, /, mod, and rem as being multiplying operators. Standard 1076.6 states that all the multiplying operators in package NUMERIC_STD are supported for synthesis by compliant synthesizers. However, other synthesizers may place constraints on the synthesis of the /, mod, and rem operators. These constraints are discussed in the paragraphs that follow.

Multiplication

The multiplication operator is defined for both unsigned and signed operands. The operands can be of different lengths. The length of the product returned is the sum of the lengths of the operands. As a result, there is no overflow. The multiplication operator is synthesizable. For example, given two 8-bit unsigned operands the following statement will cause an 8×8 multiplier to be synthesized:

```
product <= multiplier * multiplicand;
```

Division

A division operator is also defined. However, this operation can only be synthesized under some very strict constraints. Division is synthesizable only when both operands are static, or when the right operand is a static power of 2. In the first case, the result is a constant that is computed during compilation. In the second case, the multiplication is equivalent to a right shift a fixed number of positions.

For division, modulo, or remainder, if the right operand is zero during simulation, an assertion violation with severity level error will occur.

If we want a general-purpose divider synthesized, we have to write a complete design description of one. This is a VHDL description of a conventional combinational or sequential divider.

Mod

The modulus operator mod is often supported for synthesis only when the operands are constants and when the right operand is a power of 2.

Rem

The remainder operator rem is often supported for synthesis only when the operands are constants and when the right operand is a power of 2.

Exponentiation

Exponentiation is, like division, synthesizable only under very strict constraints. Either both operands must be static or the left operand must be a static value of 2.

Comparison Operators

NUMERIC_STD provides a full set of relational operators. Among these operators the inequality operators are overloaded to provide the correct numeric comparisons for type unsigned and type signed operands.

A description of a 4-bit signed comparator is given in Listing 13.4.2.

Boolean Operators

The predefined logical operators: **not**, **and**, **or**, **nand**, **nor**, **xor**, and **xnor** are overloaded in package NUMERIC_STD for types unsigned and signed.

Shift Operators

The predefined shift operators **sll**, **srl**, **rol**, and **ror** are overloaded for unsigned and signed. However, the predefined shift operators **sla** and **sra** are not. Recall that none of these operators are overloaded for std_logic in package STD_LOGIC_1164.

Shift Functions

In addition to overloading the logical shift operators, package NUMERIC_STD provides four shift functions. They are logical shifts and rotates similar to the built-in shift operators. No arithmetic shift functions are provided. The shift functions are:

```
shift_left
shift_right
rotate_left
rotate_right
```

LISTING 13.4.2
A 4-bit signed comparator.

```
library ieee;
use ieee.std_logic_1164.all;
use ieee.numeric_std.all;

entity comparator_signed is
    port(
       p : in std_logic_vector(3 downto 0);
       q : in std_logic_vector(3 downto 0);
       p_lt_q : out std_logic;
       p_eq_q : out std_logic;
       p_gt_q : out std_logic
        );
end comparator_signed;

architecture dataflow of comparator_signed is
begin

    p_lt_q <= '1' when signed(p) < signed(q) else '0';
    p_eq_q <= '1' when signed(p) = signed(q) else '0';
    p_gt_q <= '1' when signed(p) > signed(q) else '0';

end dataflow;
```

These functions all take two parameters. The first is the unsigned or signed value to shift. The second is a type natural value that specifies the shift distance. Since the second parameter is type natural, each shift is in only one direction.

Type Conversions

Functions are provided in package NUMERIC_STD to provide conversions between types unsigned, signed, and integer. These functions were introduced in Chapter 3. (See "Types Unsigned and Signed" on page 107.) All of these functions have the name `to_type`, where *type* is the type being converted to. When converting type unsigned or type signed to integer, the function has a single parameter, the vector being converted. When converting an integer to unsigned or signed, the function has two parameters. The first is the integer being converted, and the second is a type natural that specifies the number of bits in the result vector.

13.5 PACKAGE STD_LOGIC_ARITH

Package STD_LOGIC_ARITH predated package NUMERIC_STD and has found widespread use. It contains arithmetic functions that accept signed and unsigned parameter types and return std_logic types. This package has essentially the same functionality as NUMERIC_STD. While typically included in tool vendors' IEEE libraries, this package is not an IEEE standard.

Many of the operators and functions in package STD_LOGIC_ARITH are extensively overloaded. This can sometimes cause ambiguities that create compiler errors.

Since IEEE Std 1076.6 limits the array types used to represent unsigned and signed numbers to NUMERIC_STD and NUMERIC_BIT, use of package STD_LOGIC_ARITH is precluded if strict adherence to IEEE Std 1076.6 is to be maintained. However, use of this package is encountered in legacy code, so some familiarity with it is useful.

Resize Function

STD_LOGIC_ARITH contains functions `conv_unsigned` and `conv_signed` that are used to resize (truncate or extend) values of the respective types. These functions are similar to the resize function in NUMERIC_STD, except for when `conv_signed` is used to truncate a signed value. In this case, bits are truncated from the left to create a result of the desired size. This includes truncating the sign bit. If the original value is small enough to be represented in the truncated result, there is no change in sign. However, if the original value is too large to fit in the truncated result, the value's sign changes. This corresponds to the conventional approach to signed number truncation. But, it is in contrast to the approach used by the `resize` function in NUMERIC_STD, where the sign bit is kept and then the necessary number of bits following the sign bit is truncated to create a result of the desired size.

Arithmetic Operators

There are fewer arithmetic operators in package STD_LOGIC_ARITH than in package NUMERIC_STD. STD_LOGIC_ARITH provides only the operators sign –,

sign +, abs, +, –, and *. The unary sign operator (–) and the abs operator apply only to signed type.

There are a large number of permutations of the arithmetic operators provided by overloading. These permutations allow types to be mixed without type conversions. For example, an unsigned and a signed can be added and a signed result returned, or an unsigned and a signed can be added and a std_logic_vector result returned.

Comparison Operators

The complete set of comparison operators is provided. Comparison operators for unsigned and signed are defined to give the correct interpretations for their respective numeric representations. In addition, the operands compared do not have to be the same length.

Shift Functions

Because package STD_LOGIC_ARITH predates the 1993 version of the VHDL standard, the shift operators introduced in that version are not supported. Instead, two shift functions, `shl` and `shr`, are provided. Whereas VHDL's built-in shift operators use an integer to specify the number of bit positions to be shifted, STD_LOGIC_ARITH requires the shift distance to be specified by an unsigned type.

Boolean Operators

There are no logical operators provided in package STD_LOGIC_ARITH. Logical operations can be accomplished by converting the unsigned or signed operands to std_logic_vector and using the STD_LOGIC_1164 logical operators. The std_logic_vector result can be converted back to either unsigned or signed.

Type Conversions

Four different types are used in package STD_LOGIC_ARITH: unsigned, signed, std_logic_vector, and integer. Type conversions between std_logic_vector and either unsigned or signed are implicit conversions provided by the built-in conversions between similar array types. Conversion functions are provided to accomplish the other type conversions. The functions are named `conv_`*`type`* where *`type`* is the name of the type being converted to.

Conversion functions that convert to an integer have only one formal parameter, the vector to be converted. Conversion functions that convert to a vector have two formal parameters. The first parameter is the vector to be converted and the second is the number of bits in the result vector. When the vector converted from and the vector converted to are the same type, the conversion function is acting as the previously discussed, resize function.

13.6 PACKAGES FOR VHDL TEXT OUTPUT

Several nonstandard packages are available to provide text output. One such package is Stefan Doll's package `txt_util`. This package is available on the Web and provides a means to output text and manipulate strings. The package includes functions to convert various types to strings, procedures to print messages to the screen, and to read and write strings to text files.

The functions to convert different types to strings are particularly useful in testbenches to display the actual value of a std_logic_vector when it differs from the expected value. For example, in addition to a function `str` equivalent to that in Listing 12.6.5, it also contains a function `hstr`, which converts a std_logic_vector to a hexadecimal string.

PROBLEMS

13.1 Write a package and package body for a package called `gates`. The parameters passed to and returned by the subprograms in this package are all std_logic. This package includes the following subprograms:

(a) A procedure `inverter` that realizes an inverter.

(b) A procedure `and_2` that realizes a two-input AND.

(c) A procedure `or_2` that realizes a two-input OR.

13.2 Assuming the package `gates` from Problem 13.1 has been compiled to a library called `mylib`, write a VHDL design description of the circuit shown below that uses only concurrent calls to procedures in the package `gates`.

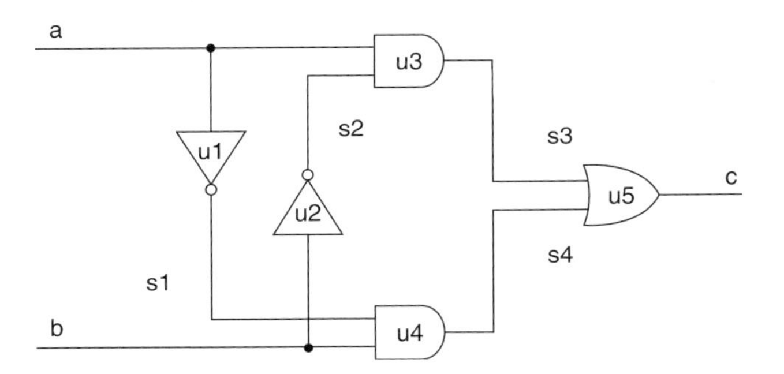

13.3 Using the 4-bit full adder in Listing 13.4.1, create a structural description of an 8-bit full adder. Write a simple testbench to verify that the 8-bit full adder gives the correct sum and carry out.

13.4 Give a set of example values that shows why the following statements cannot be used to create a 4-bit full adder.

```
sum_lcl <= resize(signed(augend),5)
        + resize(signed(addend),5)
        + signed'('0' & carry_in);
sum <= std_logic_vector(sum_lcl(3 downto 0));
carry_out <= std_logic(sum_lcl(4));
```

Chapter 14
Testbenches for Sequential Systems

Exhaustive verification is relatively easy to accomplish for combinational systems. In contrast, for even simple sequential systems it may be difficult. For complex sequential systems, it can be practically impossible. Accordingly, instead of an exhaustive verification, a comprehensive verification, one well short of exhaustive, is often the objective. A comprehensive verification is one that provides a high degree of confidence that a design is correct.

Achieving a comprehensive verification is predicated on how well the stimulus we apply covers the circumstances in which the UUT is expected to operate. Ultimately, our confidence in a verification is limited by the thoroughness and appropriateness of the stimulus.

Testbenches for sequential systems require generation of a clock and a reset signal, in addition to data and control stimulus signals.

Use of procedures simplifies stimulus generation. Each procedure can be written to create the stimulus required to exercise a particular UUT operation or functionality. With stimulus generation for each UUT operation encapsulated into a separate procedure, a complex stimulus is created by simply calling these procedures in the desired order.

14.1 SIMPLE SEQUENTIAL TESTBENCHES

Figure 7.1.1, which shows the constituents of a simple combinational testbench, is modified in Figure 14.1.1 to represent the constituents of a simple sequential testbench by adding a system clock and a reset.

System Clock and System Reset

All synchronous systems require a system clock signal. Accordingly, their testbenches must provide this signal. Generation of an accurate system clock is important, because

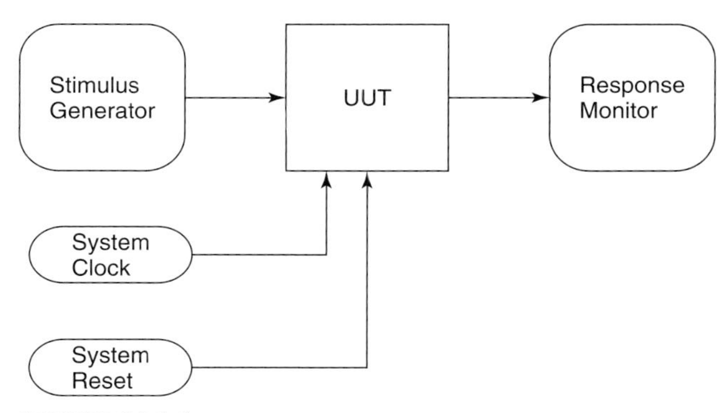

FIGURE 14.1.1
Typical constituents of a simple sequential testbench.

other stimulus signals are ususally synchronized to the system clock. The system clock must accurately model the clock that will exist in the UUT's application environment.

In addition to a system clock, a reset signal is required. The reset signal is asserted at power on to place the sequential system in its initial state. The reset stimulus must accurately model the reset signal in the UUT's application environment.

14.2 GENERATING A SYSTEM CLOCK

Since other stimulus signals in the testbench are usually synchronized to the system clock, its generation is considered first. A system clock can easily be generated using either a concurrent signal assignment statement or a process. Use of a process provides greater flexibility.

50% Duty Cycle Clock

System clocks are often specified as having a 50% duty cycle. An example of a 50% duty cycle clock generated using a concurrent signal assignment statement is

```
clk <= not clk after period/2;
```

where constant `period` specifies the clock's period and is declared in the declarative part of the testbench architecture. Signal `clk` must be given an initial value of either `'0'` or `'1'` when it is declared. Otherwise, since the **not** of `'U'` is `'U'`, the clock would remain `'U'` throughout the simulation.

An example of a process that generates a 50% duty cycle clock is given in Listing 14.2.1.

In Listing 14.2.1, constant `period` is declared in the declarative part of the process. An assignment statement makes the initial phase of the clock `'0'`. Next, an infinite loop repeatedly waits for half the clock period and then complements the clock's value.

LISTING 14.2.1
Process to generate a 50% duty cycle clock.

```
clock_gen: process
constant period : time := 100 ns;
begin
   clk <= '0'; -- Initial phase of the clock
   loop
      wait for period/2;
      clk <= not clk;
   end loop;
end process;
```

Stopping the Clock in a Simulation

As discussed in Chapter 6, a simulation stops when a specified simulation time limit is reached or when there are no more events to simulate. It is sometimes more convenient to run a simulation without having to specify a time limit. We must then depend on the simulation stopping when there are no more events.

When simulating a combinational system without using a time limit, after the last stimulus value has been applied and the events associated with it have been simulated, there are no more events for the simulator to process and the simulation stops.

However, when simulating a sequential system without using a time limit, the system clock generated by either of the previous methods runs continuously. The simulation does not end because the clock is always generating new events, even after the last data or control stimulus has been applied. As a result, if the clock is not stopped, the simulation does not stop.

Using a Boolean Signal to Stop the Clock

A boolean signal can be used to stop the clock. For example, if boolean signal `end_sim` is assigned the value `false` at the beginning of the testbench, the clock in Listing 14.2.2 runs until `end_sim` is assigned the value `true`.

LISTING 14.2.2
Clock that is stopped by an external signal.

```
clock_gen : process
begin
   clk <= '0';
   loop
      wait for period/2;
      clk <= not clk;
      exit when end_sim = true;
   end loop;
   wait;
end process;
```

LISTING 14.2.3
Process to create a clock with a specified duty cycle.

```
clock_gen: process
   constant duty_cycle : real := 0.40;
   constant period : time := 100 ns ;
   constant clk_high : time := duty_cycle * period ;
begin
   loop
      clk <= '0';
      wait for period - clk_high;   -- clock low time
      clk <= '1';
         wait for clk_high;   -- clock high time
      end loop;
   end process;
end behavioral;
```

After the stimulus process has finished applying stimulus to the UUT, it assigns the value `true` to `end_sim`. At that point, the clock process exits its loop and executes an infinite wait. There are then no more events for the simulator to process and the simulation ends.

To stop the system clock, a boolean signal, like `end_sim`, can be added to any clock description.

Nonsymmetrical Clock

A clock's specification may call for a duty cycle other than 50%. In such cases, we need to be able to specify how long the clock is high (or low), in addition to specifying its period. The process in Listing 14.2.3 generates a clock with a specified duty cycle.

This example is for a clock having a 40% duty cycle. The `clk_high` time is specified as a constant whose value is the product of the duty cycle and clock period. The `clk_low` time is simply the clock period minus the `clk_high` time. The resulting waveform is shown in Figure 14.2.1.

Concurrent Procedure Clock

Instead of a concurrent signal assignment or a process, a concurrent procedure can be used to generate a clock. A clock described by a concurrent procedure can be placed in a package for easy reuse in other testbenches.

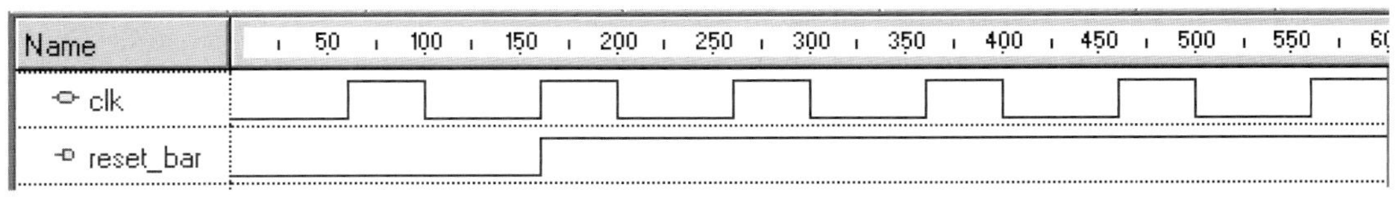

FIGURE 14.2.1
Clock waveform from process in Listing 14.2.3 and reset waveform from process in Listing 14.3.2.

Phase-Related Periodic Waveforms

In addition to the system clock, some data or control inputs to a sequential system may also be in the form of periodic signals. For example, if a system receives inputs from an optical shaft encoder, these inputs can be modeled as two periodic signals with a 90-degree phase shift.

Listing 14.2.4 uses two calls to the concurrent procedure `clock_gen`, declared in the declarative part of the architecture, to generate a two-phase clock corresponding to the outputs of an optical shaft encoder (OSE).

Procedure `clock_gen` has one formal parameter that is a signal of mode out and three formal parameters that are constants of mode in. After waiting for a period of time equal to the desired phase shift, the procedure enters an infinite loop where it repeatedly generates a signal that is a `'1'` for a time equal to `high`, and then is `'0'` for a time equal to `period` minus `high`. Note that there is no actual subtraction operation in this code.

Inside the loop, the assignment to `clk` is a waveform that is `'1'` initially, and then changes to `'0'` after `high` ns. The wait statement following the assignment to `clk` suspends the procedure and causes two transactions to be posted to the driver for `clk`. After `period` ns, the procedure becomes active again and the assignment to `clk` is executed again, causing the clock waveform to repeat with the specified period.

LISTING 14.2.4
Two-phase clock generated by concurrent calls to a procedure.

```
library ieee;
use ieee.std_logic_1164.all;

entity shaft_encoder is
   port(
      clk_a : out std_logic;
      clk_b : out std_logic);
end shaft_encoder;

architecture behavioral of shaft_encoder is
   procedure clock_gen (signal clk: out std_logic;
      constant period, high, phase : in time ) is
   begin
      clk <= '0';
      wait for phase;
      loop
         clk <= '1', '0' after high;
         wait for period;
      end loop;
   end clock_gen;
begin
   clock_gen (clk_a, 100 ns, 50 ns, 0 ns);
   clock_gen (clk_b, 100 ns, 50 ns, 25 ns);
end behavioral;
```

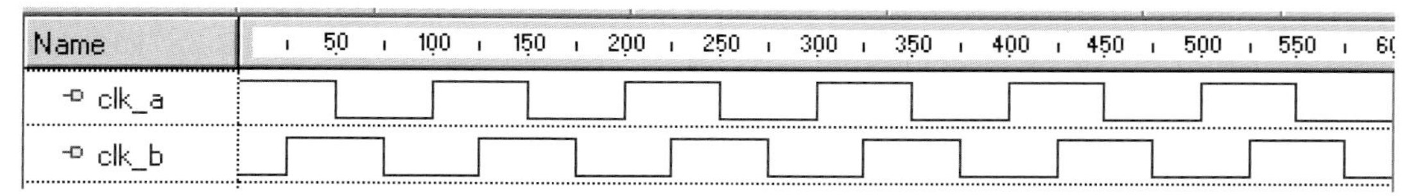

FIGURE 14.2.2
Two-phase waveform generated by concurrent procedure calls in Listing 14.2.4.

In the statement part of the architecture, there are two concurrent procedure calls to generate the two-phase clock representing the OSE's output. The two procedures are invoked during the initialization phase of the simulation. Neither procedure is ever returned from. The resulting waveforms appear in Figure 14.2.2.

14.3 GENERATING THE SYSTEM RESET

The reset signal to a hardware system typically starts in its asserted state at power on, remains in that state for a specified period of time, then changes to its unasserted state and remains there for as long as power continues to be applied to the system. The duration of the assertion of the reset signal is specified as either a fixed time or some multiple of the clock's period, and should be synchronized to the system clock to properly reset the UUT.

Generating a fixed-duration reset pulse is simple. For example, if the reset signal should be `'0'` for 160 ns and then become `'1'`, a concurrent signal assignment can be used:

```
reset_bar <= '0', '1' after 160 ns;
```

Alternatively, but less concisely, a reset process can be written as in Listing 14.3.1.

Synchronizing Reset Duration to the System Clock

If the reset pulse duration is specified as some minimum number of clock periods, then the reset can be written to synchronize its duration to the clock, as shown in Listing 14.3.2.

LISTING 14.3.1
Process for a fixed-duration reset pulse.

```
reset: process
begin
   reset_bar <= '0';
   wait for 160 ns;
   reset_bar <= '1';
   wait;
end process;
```

LISTING 14.3.2
Process for a reset pulse whose duration is a multiple of the clock period.

```
reset: process
begin
   reset_bar <= '0';
   for i in 1 to 2 loop
   wait until clk = '1';
   end loop;
   reset_bar <= '1';
   wait;
end process;
```

This process first assigns a '0' to reset_bar. A loop then counts two rising edges of clk. After these two rising edges, reset_bar is assigned a '1'.

An advantage of this approach is that if the system clock period is changed, the reset pulse is automatically adjusted to provide the minimum required duration.

14.4 SYNCHRONIZING STIMULUS GENERATION AND MONITORING

In this section we consider some issues associated with synchronizing stimulus generation and output verification to the system clock.

Stimulus Timing

A system's specification may include an interface timing specification that details input signal timing. If so, the testbench's stimulus must accurately model the timing specified. We consider generating stimulus that meets detailed interface timing specifications in Section 14.9. In this section, our attention is on generating stimulus when the exact interface timing is not specified. This is appropriate when a system is being designed for generic applications.

Testbench for a Two-Bit Counter

An example of a testbench for a 2-bit counter illustrates some timing issues of concern. Assume that the UUT in Figure 14.1.1 is a 2-bit binary counter with only clock and reset inputs. Also assume that the system clock is generated by the process in Listing 14.2.2 and the reset signal is generated by the process in Listing 14.3.2.

Synchronizing Reset to the System Clock

The reset signal is synchronized to the system clock. Initially, rst_bar is asserted. A loop is then used to wait until the first two triggering clock edges have occurred. When the loop is completed, rst_bar is unasserted. The waveforms produced by this simulation appear on a waveform viewer, as shown in Figure 14.4.1.

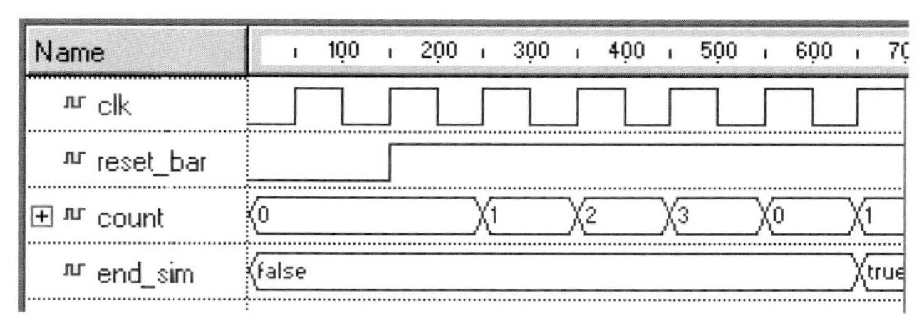

FIGURE 14.4.1
Waveforms for functional simulation of 2-bit counter.

Effect of Delta Delays

Looking at the `clk` and `rst_bar` signals, it might appear from Figure 14.4.1 that, for a functional simulation, the counter will starting counting at the second triggering clock edge. However, by examining the code in Listing 14.3.2, we see that the reset process suspends each time the statement

```
wait until clk = '1':
```

is executed. During the same simulation cycle where `clk` changes from 0 to 1 for the second time, the reset process resumes execution and the statement

```
rst_bar <= '1';
```

is executed, followed by the instruction:

```
wait;
```

When the wait instruction is executed, the reset process suspends and the transaction that changes `rst_bar` to a `'1'` is posted with a time component 1 δ later than the current simulation time. Thus, it is not until the next simulation cycle (a delta cycle) that `rst_bar` is actually changed to a `'1'`. Since the waveform viewer does not show delta cycles, the 1 δ delay is not visible in Figure 14.4.1. The actual relationship between the clock and reset signals is represented in Figure 14.4.2. As a result, the counter does not start counting until the third triggering clock edge. Thus, in a functional simulation, a δ delay can have the same effect as delaying an entire clock cycle. Knowing when the counter actually starts to count is critical for correctly writing the monitor.

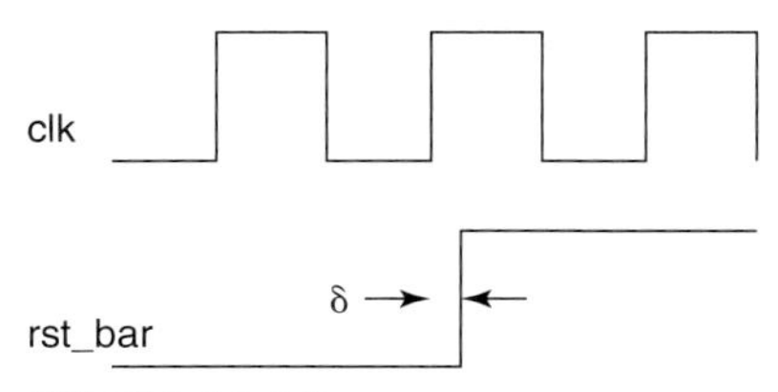

FIGURE 14.4.2
Actual delta delay between the clock triggering edge and synchronized reset.

Synchronizing the Monitor

Output verification must be synchronized to the system clock. Assume that we want to write a process to verify the 2-bit counter's outputs. The monitor process must wait until after the first two triggering clock edges. Then, after each subsequent triggering clock edge, the monitor process can use an assertion statement to verify the counter's output value. For example, if, after the first two triggering clock edges, we wanted to wait for the third triggering clock edge and then check for a count of `"01"`, we might attempt to use the following sequence of statements:

```
wait until clk = '1';
assert count = "01"
report "missed count 01"
severity error;
```

The monitor process containing these statements is resumed during the simulation cycle where the event corresponding to the triggering clock edge occurs. During this same simulation cycle, the counter process is also resumed. The counter process posts a transaction assigning `count` the value `"01"`. This value will not take effect until the next simulation cycle. However, during the current simulation cycle, the assertion statement verifies whether `count` is `"01"`. Since it is not (yet), an error is erroneously reported.

To achieve proper verification, the assertion statement must wait until after `count` actually takes its new value. This can be done by waiting for any small nonzero time after the triggering clock edge. For example, the code

```
wait until clk = '1';
wait for 1 ns;
assert count = "01"
report "missed count 01"
severity error;
```

works correctly, since the additional wait for 1 ns ensures that the counter has taken its new value in response to the triggering clock edge.

The monitor process in Listing 14.4.1 can be used to exhaustively verify the outputs of a *n*-bit binary counter.

LISTING 14.4.1
Monitor process for n-bit binary counter.

```
monitor: process
   constant n : integer := 2; -- number of bits in counter
begin
   for i in 1 to 2 loop     -- skip clocks while counter is reset
      wait until clk = '1';
   end loop;
   wait for 1 ns;

   -- verify counter's output after each clock
```

```
    for i in 0 to 2**n loop    -- count until clock rolls over to 0
      assert count = (std_logic_vector(to_unsigned(i mod 2**n, n)))
      report "Count of " & integer'image(i mod 2**n) & " failed"
      severity error;
      wait until clk = '1';
      wait for 1 ns;
    end loop;

    end_sim <= true;
    wait;
  end process;
```

The first loop in process `monitor` skips the first two triggering clock edges, which occur while the counter is reset. Since we are skipping the clock edges that occur while the counter is reset, a simpler alternative is to replace the first loop with the instruction:

```
wait until rst_bar = '1';
```

The second loop verifies the counter's output prior to the first triggering clock edge that is counted, and then 1 ns after each of the next 2^n triggering clock edges. Thus, the counter's output is checked as it counts from 0 to $2^n - 1$ and then rolls over to 0. This monitor process provides an exhaustive verification for this simple counter.

If the instruction

```
wait for 1 ns;
```

is removed from the second loop, `count` will be erroneously reported as incorrect for all but the initial count after reset.

Note that, in this example, to stop the simulation the monitor process assigns `end_sim` the value `true` after the end of the second loop.

Synchronizing Monitoring for a Timing Simulation

While waiting for any small nonzero delay after the triggering clock edge works for a functional simulation, a timing simulation is more complicated. If the system's specification specifies a maximmum delay, then that is the value that must be used. If not, the delay must be greater the propagation delay of the UUT timing model.

A simple approach that usually allows the same testbench to be used for both functional and timing simulations, when a timing specification is not given, is to synchronize monitoring to the nontriggering clock edge. For a system with a 50% duty cycle clock, this gives half a clock period for output signal transitions due to propagation delay in the UUT to settle.

Synchronizing Stimulus to System Clock

Our simple counter example has no inputs other than its clock and reset. A more complex counter would have additional inputs such as count enable, count direction, load enable, and load data. As synchronous inputs, they would also need to be synchronized to the system clock. Usually, their values would be expected to become valid just after the preceding triggering clock edge. When exact timing is not specified, also synchronizing these stimuli to the nontriggering clock edge is often appropriate.

Testbench for Digital Noise Filter

A testbench for the digital noise filter from Figure 9.2.3 and Listing 9.2.4 illustrates synchronizing stimulus and monitoring to the falling edge of the system clock.

LISTING 14.4.2
Testbench for digital noise filter from Figure 9.2.3 and Listing 9.2.4.

```
library ieee;
use ieee.std_logic_1164.all;

entity filter_tb is
end filter_tb;

architecture tb_architecture of filter_tb is
   -- Stimulus signals - signals mapped to the input and inout ports of UUT
   signal cx : std_logic;
   signal clk : std_logic := '0';
   signal rst_bar : std_logic;
   -- Observed signals - signals mapped to the output ports of UUT
   signal y : std_logic;
   signal end_sim : boolean := false;

   constant period : time := 100 ns;

   -- stimulus and response signal values
   constant inputs : std_logic_vector  := "00110111001100010111";
   constant outputs : std_logic_vector := "00000000011111111000";

begin
   -- Unit Under Test port map
   UUT : entity filter
   port map (cx => cx, clk => clk, rst_bar => rst_bar, y => y );

   -- system clock generation
   cpu_clock_gen : clk <= not clk after period/2 when end_sim = false else unaff
ected;

   -- system reset generation
   reset_gen : rst_bar <= '0', '1' after 2 * period;

   -- stimulus and response process
   stim_respon : process
```

```
    begin
        wait until rst_bar = '1';
        for i in inputs'range loop
            cx <= inputs(i);
            wait until clk = '0';
            assert y = outputs(i)
            report "system failure at " & integer'image(i) & " clocks after reset"
            severity error;
        end loop;
        end_sim <= true;
        wait;
    end process;
end tb_architecture;
```

In Listing 14.4.2 the single process `stim_respon` is used to both generate the stimulus and verify the response. The first time this process is executed, it suspends and waits for `rst_bar` to be unasserted. After `rst_bar` is unasserted, it resumes and enters the loop. The first time through the loop, the process applies the first input value from a sequence of input values represented by the vector `inputs` and then suspends waiting for the falling edge of the clock.

Subsequently, on the falling edge of each clock pulse, the process verifies the output by comparing its value to the corresponding element in the vector `outputs` of expected output values. It then applies the next input value and suspends waiting for the next falling edge of the clock. After all of the input values have been applied and their outputs verified, the loop terminates. Signal `end_sim` is assigned the value `true` and the simulation ends. The waveforms from this simulation are given in Figure 14.4.3.

The verification provided by this testbench is clearly not exhaustive. It can be made more comprehensive by adding additional carefully selected values to the vectors `inputs` and `outputs`.

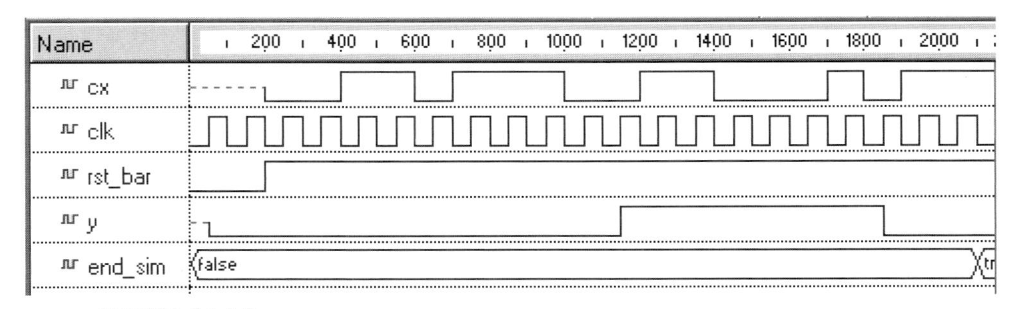

FIGURE 14.4.3
Waveforms for simulation of digital noise filter using testbench in Listing 14.4.2.

14.5 TESTBENCH FOR SUCCESSIVE APPROXIMATION REGISTER

A testbench for the successive approximation register (SAR) design from Section 11.4 illustrates the complete design of a sequential testbench and its use of a system clock and a system reset. It also demonstrates the use of a procedure to synchronize actions in the stimulus and response process to the operation of the UUT. The testbench structure is shown in Figure 14.5.1.

The testbench's architecture contains five concurrent statements:

- Instantiation of SAR as the UUT
- Process to generate system clock
- Concurrent signal assignment to generate system reset
- Concurrent signal assignment to model DAC and comparator
- Process to provide stimulus and to monitor response

The stimulus-and-response process generates a value corresponding to the analog input voltage to be converted, and generates the start of conversion signal `soc` to the SAR. At the completion of a conversion, this same process compares the conversion result to the analog input value that was converted. The testbench code is given in Listing 14.5.1.

DAC and Analog Comparator Model

The DAC has digital inputs and an analog output. The analog comparator has analog inputs and a digital output. At first thought, it might seem that, due to the analog nature

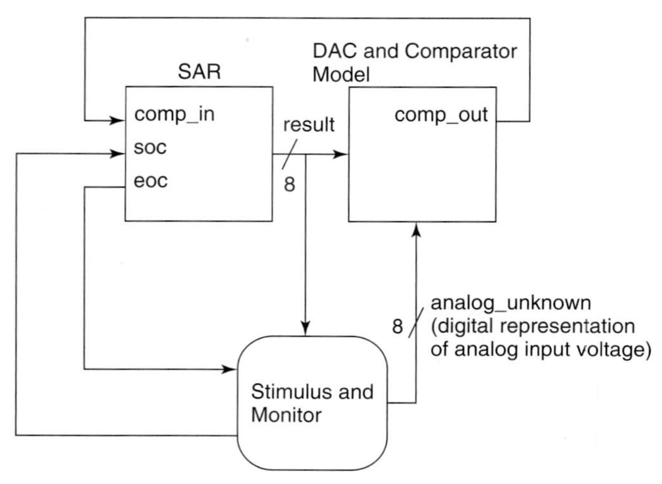

FIGURE 14.5.1
Structure of successive approximation register testbench.

LISTING 14.5.1
Successive approximation register (SAR) testbench.

```
library ieee;
use ieee.std_logic_1164.all;
use ieee.numeric_std.all;

entity saadc_tb is
end saadc_tb;

architecture tb_architecture of saadc_tb is

   -- stimulus signals
   signal reset_bar : std_logic;
   signal soc : std_logic;
   signal comp_in : std_logic;
   signal clk : std_logic := '0';
   -- observed signals
   signal eoc : std_logic;
   signal result : std_logic_vector(7 downto 0);
   signal analog_unknown : std_logic_vector(7 downto 0);

   constant period : time := 100 ns;
   signal end_sim : boolean := false;

begin
   -- unit under test instantiation
   uut : entity saadc
   port map (reset_bar => reset_bar, soc => soc, comp_in => comp_in,
      clk => clk, eoc => eoc, result => result );

   -- system clock generation
   cpu_clock_gen : process
   begin
      clk <= '0';
      wait for period/2;
      loop
      clk <= not clk;
      wait for period/2;
      exit when end_sim = true;
   end loop;
   wait;
end process;

   -- system reset generation
   reset_gen : reset_bar <= '0', '1' after 2 * period;
```

(Cont.)

LISTING 14.5.1 *(Cont.)*

```
    -- DAC and analog comparator model instantiation
  comp_in <= '1' after 10 ns when analog_unknown >= result else '0' after 10 ns;

    -- stimulus and response
    stim_resp: Process
       -- definition of procedure to control soc for a conversion
       procedure soc2eoc is
       begin
          wait until falling_edge(clk);
          soc <= '1';
          wait until eoc = '1';
          soc <= '0';
          wait until rising_edge(clk);
       end procedure ;

    begin
       -- intialize soc and wait until reset is complete
       soc <= '0';
       wait until reset_bar = '1';
       wait until rising_edge(clk);

       -- apply every possible analog unknown value
       for i in 0 to 255 loop
          analog_unknown <= std_logic_vector(to_unsigned(i,8));
          soc2eoc;    -- start conversion and wait for its end
          assert result = analog_unknown
          report "conversion failed for analog unknown equal to " &
          integer'image(i)
          severity error;
       end loop;
end_sim <= true;
       wait;
    end process;
end tb_architecture;
```

of these devices, they would be difficult to model in VHDL. However, modeling the DAC and comparator together as a single entity is surprisingly simple.

In the successive approximation ADC, the analog output of the DAC provides one of the inputs to the analog comparator (Figure 11.4.1). When combined, the DAC and analog comparator have only one analog input signal, the analog voltage to be converted. If this signal is represented in the simulation as a digital value (`analog_unknown`), the DAC and analog comparator combination is then simply equivalent to a digital comparator (Figure 14.5.1). One input to this digital comparator is the result from the SAR's output and the other is the digital representation of the analog input voltage to be converted. The digital comparator's output corresponds to the analog comparator's output. As a result, the DAC and analog comparator combination is modeled by the concurrent signal assignment statement:

```
comp_in <= '1' after 10 ns
     when analog_unknown >= result
     else '0' after 10 ns;
```

The 10 ns delays specified by the after clauses model the combined delay of the DAC and analog comparator. Typically, this value is much greater than 10 ns for an actual DAC and analog comparator. For a timing simulation, the actual DAC and analog comparator combined delay must be used.

SAR Testbench Stimulus-and-Response Process

The stimulus-and-response process (`stim_resp`) uses procedure `soc2eoc`, declared in the declarative part of the process, to assert `soc` to start a conversion. This procedure then waits until the conversion is complete (`eoc = '1'`) to unassert `soc`. Use of this procedure simplifies the body of the process.

Exhaustive SAR Verification

In the statement part of the `stim_resp` process, `soc` is unasserted and then the process waits for the completion of the system reset. A loop then successively generates the digital representation of all possible values for the analog input voltage. Since this is an 8-bit converter, the digital values corresponding to the range of the analog input voltage are from 0 to 255.

Each time through the loop, the next value in its range is assigned to `analog_unknown`. The `soc2eoc` procedure is then called to initiate a conversion and to wait until the conversion ends. After this procedure returns, an assert statement compares the conversion result with `analog_unknown` and reports an error if they are not equal. When the loop is completed, the `end_sim` signal is assigned the value `true`, and the simulation terminates.

This simulation is exhaustive, since the functionality of the system is verified for every possible analog input value. An exhaustive verification is feasible for this sequential system because, while its internal operation is sequential, the result from a conversion is a function of only the current analog input value to be converted. Figure 14.5.2 shows the waveforms for the first four loop iterations.

Comprehensive SAR Verification

As discussed in Chapter 11, the number of bits in the SAR is easily increased to create a larger ADC. The same testbench is easily modified to handle a SAR with a larger number of bits. For a larger number of bits, the time required to simulate each conversion as well as the number of conversions that must be simulated for exhaustive verification increases exponentially. This raises the question of whether an exhaustive functional verification is required or whether a nonexhaustive and less time-consuming comprehensive verification might be adequate.

For example, we might decide that a simulation that includes conversion of values for `analog_unknown` corresponding to the endpoints of the ADC input range (0 and

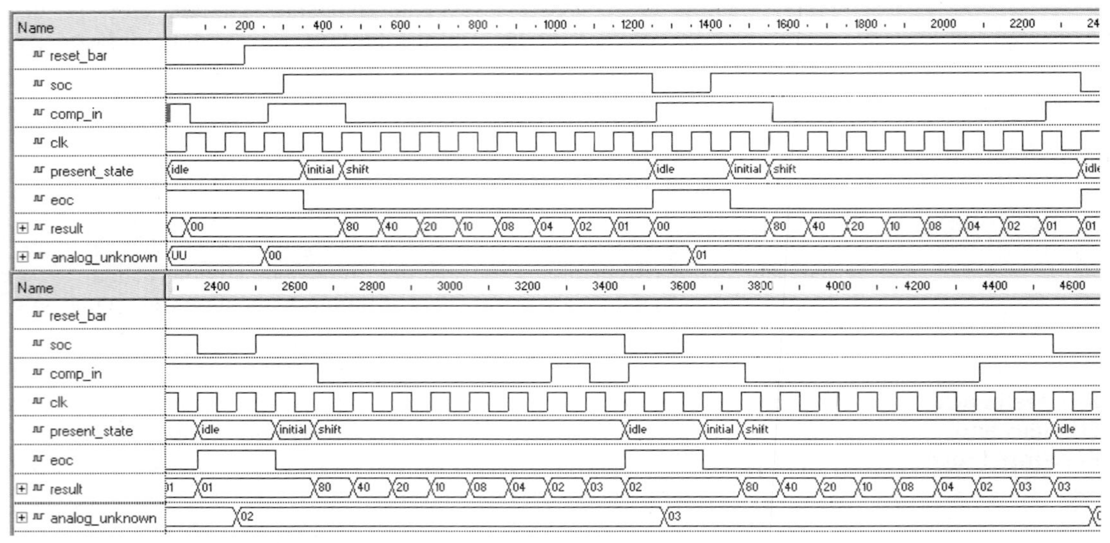

FIGURE 14.5.2
Simulation waveforms for first four analog unknown values used in the SAR testbench.

full scale), values on both sides of the midpoint of the input range, and the two values with alternating 0s and 1s is sufficient to verify the functional operation of the SAR.

14.6 DETERMINING A TESTBENCH STIMULUS FOR A SEQUENTIAL SYSTEM

Since an exhaustive verification is usually not possible for a sequential system, we must determine an appropriate stimulus to comprehensively verify the system's functionality.

Consider the task of determining a stimulus for the D flip-flop with synchronous set and clear described in Listing 8.4.5. A logic symbol for the flip-flop is given in Figure 14.6.1. Simply treating its inputs as a 4-bit vector and sequencing through all

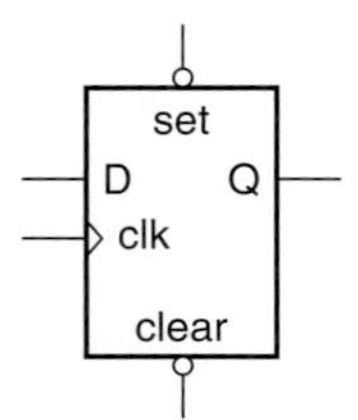

FIGURE 14.6.1
Logic symbol for D flip-flop with synchronous set and clear.

16 combinations in binary is not meaningful with respect to the flip-flop's functionality, and it represents only one of many possible input sequences.

Functionality Based Stimulus Sequence

To provide a comprehensive verification, we need to determine a stimulus that appropriately exercises the system's functionality.

Column one in Table 14.6.1 lists an appropriate sequence of D flip-flop operations to use as a basis for a comprehensive verification. Each operation is chosen to change the state of the flip-flop using the most appropriate values on all of its inputs. For example, when we verify the clear operation, we make the D input a 1, so that we can be sure the effect of the operation was a clear, and not a load of 0 from input D.

First in the sequence of operations is the clear operation. If this operation is successful, the flip-flop's output is 0. The next operation is to load a value into the flip-flop using its D input. The value we need to load is a 1. If we load a 0, we will not be able to tell if the load operation was successful or if it had no effect and we are simply observing the 0 previously stored as a result of the clear operation.

The last column of the table lists the expected output value after each operation. The expected value for q is the value after a triggering clock edge, assuming the corresponding inputs are stable before the triggering clock edge.

The desired stimulus-and-response waveforms are given in Figure 14.6.2.

Setup Time and Hold Time Considerations

An important consideration is when stimulus values should be changed relative to the clock. If we consider only what is required for a functional simulation, inputs can be changed any time prior to the clock, even as late as simultaneously with the clock edge. Inputs can be changed to the next combination as early as one delta after the clock edge.

It is advantageous to create a stimulus that is appropriate for both the functional simulation and the timing simulation, so the same testbench can be used.

When a system is being designed for use in a specific application environment, the timing of stimulus signals should model that environment. For example, if system inputs will be from a microprocessor's system bus, then the bus timing specification determines when stimulus values change.

For proper operation of a PLD, synchronous input values must be stable prior to the triggering clock edge for a time equal to or exceeding the PLD's setup time and

Table 14.6.1
A sequence of operations as stimulus to a D flip-flop.

Operation	d	set_bar	clr_bar	Expected q
clear	1	1	0	0
load 1	1	1	1	1
load 0	0	1	1	0
set	0	0	1	1
load 0	0	1	1	0
load 1	1	1	1	1

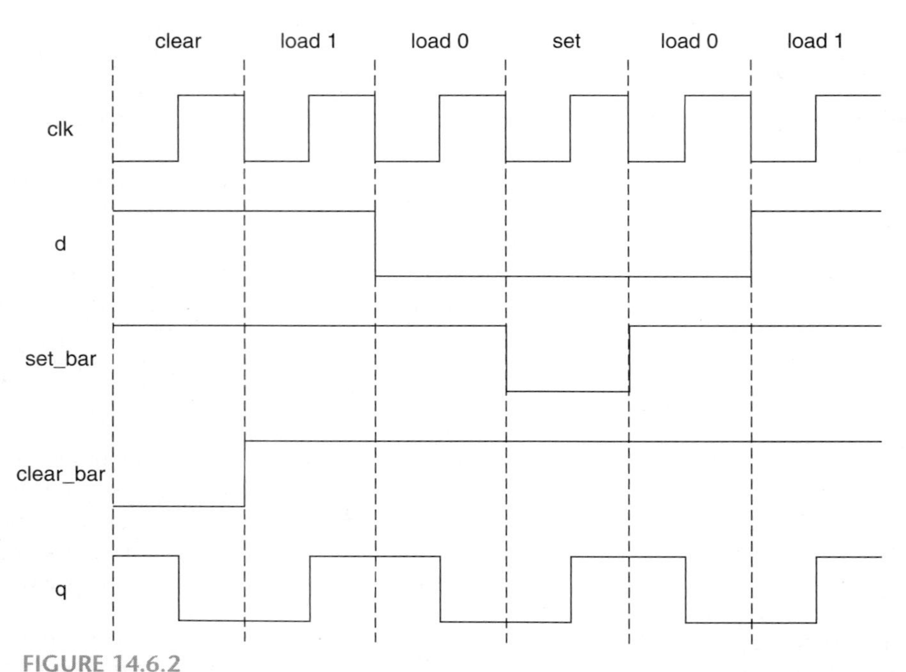

FIGURE 14.6.2
D flip-flop stimulus (and response) waveforms using operation sequence in Table 14.6.1.

must remain stable after the triggering clock edge for a time equal to or exceeding the PLD's hold time.

A set of stimulus signals for the D flip-flop, which can also be used for a timing simulation, are given in Figure 14.6.3.

These signals model the timing of the signals in the application environment relative to the system clock. When the stimulus signals are not stable they are assigned the unknown value, `'X'`. This makes it easy to detect whether the design will meet its timing requirements. Leaving stimulus values constant from the end of the hold time to the beginning of the next setup time would not detect whether the setup and hold timing requirements are met in the application environment.

Drawbacks of Previous Stimulus-Generation Approaches

One approach used for generating stimulus for a combinational system was to write a concurrent signal assignment statement with multiple after clauses for each signal. Writing the four statements needed to generate the waveform in Figure 14.6.3 is possible, but unwieldy. It is also difficult when using this approach to add assertion statements to check for expected output values.

Another approach, used with combinational systems, was to use multiple wait for statements in a process. This approach can also be unwieldy, although it is easier to add assertion statements.

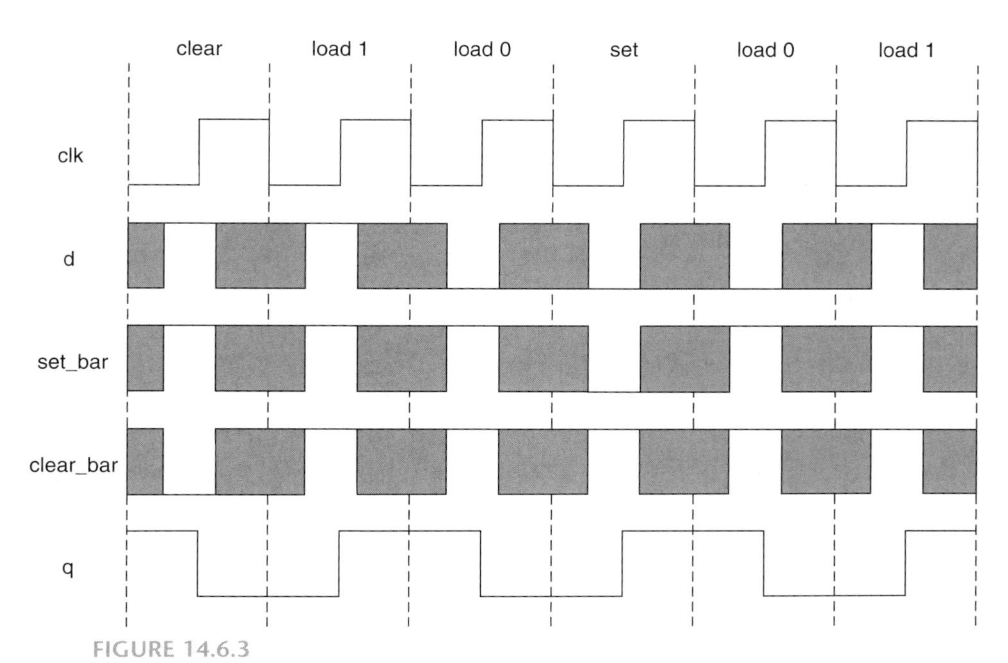

FIGURE 14.6.3
D flip-flop stimulus (and response) waveforms for timing or functional simulation.

Finally, using test vectors read from a table or a file is also unwieldy. These vectors are difficult to write correctly and hard to interpret when applied to sequential systems.

In contrast to the previous approaches, use of multiple processes and procedures are an effective way to generate stimulus. Examples of these approaches are presented in the sections that follow.

14.7 USING PROCEDURES FOR STIMULUS GENERATION

A useful approach to generating stimulus waveforms is to focus on the operations that can be performed by a UUT and write procedures to encapsulate the stimulus to activate those operations. These procedures are written so that they synchronize themselves to the system clock. In the testbench, these procedures are simply called in the appropriate order to produce the desired sequence of UUT operations. Using this approach, the testbench stimulus is easier to code and maintain.

For example, operations for a D flip-flop require only three procedures: a set procedure, a clear procedure, and a load procedure. The load procedure uses a parameter to specify whether a `'0'` or `'1'` is applied to the D input. The load procedure is shown in Listing 14.7.1.

LISTING 14.7.1
Procedure that generates stimulus to load a D flip-flop with synchronous set and clear.

```
procedure load(
   din : in std_logic;
   signal clk : in std_logic;
   signal d : out std_logic;
   signal set_bar : out std_logic;
   signal clr_bar : out std_logic) is
begin
   d <= din;
   set_bar <= '1';
   clr_bar <= '1';
   wait on clk until clk = '1'; -- wait for rising edge of clock
   wait for thold;
   d <= 'X';
   set_bar <= 'X';
   clr_bar <= 'X';
   wait for period - thold - tsetup;
end load;
```

Procedure `load` has five parameters. Parameter `din` is class **`constant`** and mode **`in`**. Using this parameter we can pass to the procedure the value to be loaded into the flip-flop. The other mode **`in`** parameter is the signal `clk`. All other parameters are signals of mode **`out`**. These are the stimulus signals that are assigned values by the procedure. All stimulus parameters in a procedure must be of class signal.

When procedure `load` is called, it immediately assigns values to `d`, `set_bar`, and `clr_bar`, then waits for the rising edge of the clock signal. After the rising edge of the clock, the procedure waits for a time equal to the hold time, `thold`. After the hold time has elapsed, the procedure assigns the unknown value `'X'` to its `d`, `set_bar`, and `clr_bar`, outputs. Finally, the procedure waits for an additional time equal to the clock period minus the setup time and hold time, and then returns. Note that, if the D flip-flop had asynchronous (instead of synchronous) set and clear inputs, the `set_bar` and `clear_bar` outputs from the load procedure would have to maintain their `'1'` values.

The waveforms created by a single `clear` procedure call followed by a `load` procedure call (with `d` = 1) are shown in Figure 14.7.1.

Waiting after the positive clock edge for a time equal to the clock period minus the setup time and hold time leaves the simulation at the point in simulated time where new input values must be assigned in order to meet the setup time requirement for the next operation. The next procedure is called immediately and assigns new values to the stimulus signals. Thus, the required setup time relative to the clock is met.

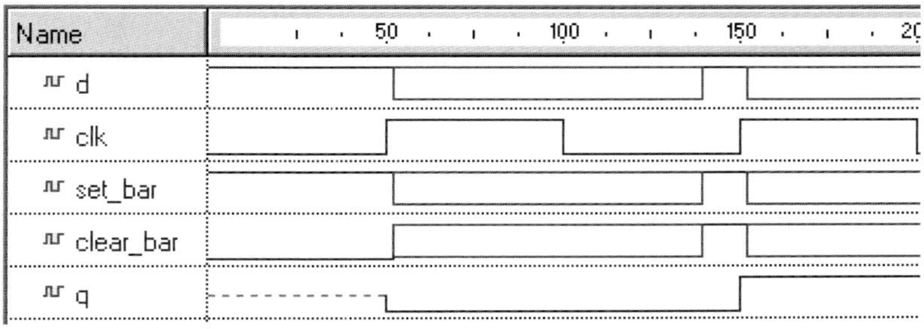

FIGURE 14.7.1
Waveforms generated by a clear procedure followed by a load (d = 1) procedure.

By calling these operation procedures in the appropriate order, the desired input stimulus is generated. For example, the sequence of operations in Table 14.6.1 is created by the following sequence of procedure calls:

```
clear(clk, d, set_bar, clear_bar);
load('1', clk, d, set_bar, clear_bar);
load('0', clk, d, set_bar, clear_bar);
set(clk, d, set_bar, clear_bar);
load('0', clk, d, set_bar, clear_bar);
load('1', clk, d, set_bar, clear_bar);
```

The complete testbench is given in Listing 14.7.2.

LISTING 14.7.2
Testbench for D flip-flop using procedures.

```
library ieee;
use ieee.std_logic_1164.all;

entity d_ff_pe_ssc_tb is
end d_ff_pe_ssc_tb;

architecture tb_architecture of d_ff_pe_ssc_tb is

   -- Signals mapped to the input ports of tested entity
   signal d : std_logic;
   signal clk : std_logic := '0';
   signal set_bar : std_logic;
   signal clear_bar : std_logic;
   -- Signals mapped to the output ports of tested entity
   signal q : std_logic;
```

(Cont.)

LISTING 14.7.2 *(Cont.)*

```
-- Constants
constant period : time := 100 ns;
constant tsetup : time := 10 ns;
constant thold : time := 2 ns;

-- Proceures to encapsulate operations

procedure clear(
   signal clk : in std_logic;
   signal d : out std_logic;
   signal set_bar : out std_logic;
   signal clr_bar : out std_logic) is
begin
   d <= '1';
   set_bar <= '1';
   clr_bar <= '0';
   wait on clk until clk = '1';
   wait for thold;
   d <= 'X';
   set_bar <= 'X';
   clr_bar <= 'X';
   wait for period - thold - tsetup;
end clear;

procedure set(
   signal clk : in std_logic;
   signal d : out std_logic;
   signal set_bar : out std_logic;
   signal clr_bar : out std_logic) is
begin
   d <= '0';
   set_bar <= '0';
   clr_bar <= '1';
   wait on clk until clk = '1';
   wait for thold;
   d <= 'X';
   set_bar <= 'X';
   clr_bar <= 'X';
   wait for period - thold - tsetup;
end set;

procedure load(
   din : in std_logic;
   signal clk : in std_logic;
```

```
      signal d : out std_logic;
      signal set_bar : out std_logic;
      signal clr_bar : out std_logic) is
   begin
      d <= din;
      set_bar <= '1';
      clr_bar <= '1';
      wait on clk until clk = '1';
      wait for thold;
      d <= 'X';
      set_bar <= 'X';
      clr_bar <= 'X';
      wait for period - thold - tsetup;
   end load;

begin

   -- Unit Under Test port map
   UUT : entity d_ff_pe_ssc
   port map
      (d => d,
      clk => clk,
      set_bar => set_bar,
      clear_bar => clear_bar,
      q => q );

   -- Clock generation
   clock_gen: clk <= not clk after period/2;

   -- Test case
   operations : process
   begin
      clear(clk, d, set_bar, clear_bar);
      load('1', clk, d, set_bar, clear_bar);
      load('0', clk, d, set_bar, clear_bar);
      set(clk, d, set_bar, clear_bar);
      load('0', clk, d, set_bar, clear_bar);
      load('1', clk, d, set_bar, clear_bar);
      wait;
   end process;

end tb_architecture;
```

Signals to connect the clock generator and `operations` process to the UUT are declared in the declarative part of the architecture. Also defined in the declarative part of the architecture body are the procedures.

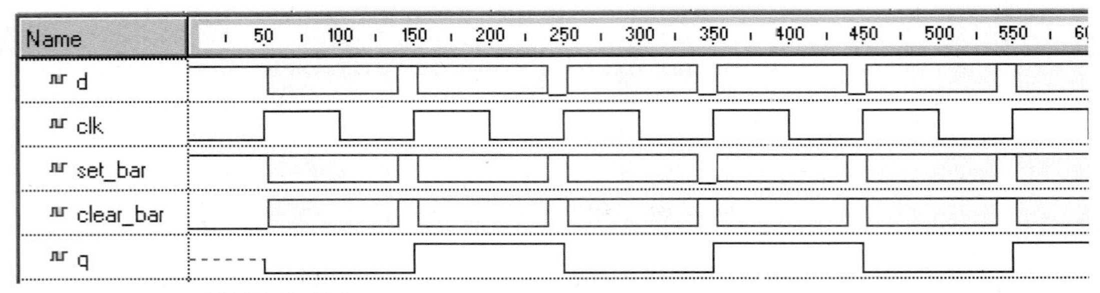

FIGURE 14.7.2
Stimulus-and-response waveforms for D flip-flop testbench.

In the statement part of the architecture are:

- An instantiation statement to instantiate the flip-flop
- A concurrent signal assignment statement to generate the clock
- A process statement with a sequence of calls to the procedures

The waveforms generated are shown in Figure 14.7.2. Stimulus values are unknown until a time equal to the setup time before the clock's rising edge. The stimulus values become unknown again, after a time equal to the hold time from the clock's rising edge. The clock's period, setup time, and hold time are specified by constants in the testbench.

14.8 OUTPUT VERIFICATION IN STIMULUS PROCEDURES

This section discusses using assertion statements within stimulus procedures to verify outputs.

In Listing 14.8.1 the load procedure in Listing 14.7.1 is modified to include assertion statements to verify the output.

LISTING 14.8.1
Using assertion statements in a stimulus procedure to verify outputs.

```
procedure load(
   din : in std_logic;
   signal clk : in std_logic;
   signal d : out std_logic;
   signal set_bar : out std_logic;
   signal clr_bar : out std_logic) is
begin
   d <= din;
   set_bar <= '1';
```

```
    clr_bar <= '1';
    wait on clk until clk = '1';
    assert q'stable(period - tco)   -- Check previous output remained stable
    report "previous output not stable";
    wait for thold;
    d <= 'X';
    set_bar <= 'X';
    clr_bar <= 'X';
    wait for tco - thold;
    assert q = din                  -- Check new output is correct
    report "load operation failed";
    wait for period - tco - tsetup;
end load;
```

First, lets consider the last assertion statement:

```
wait for tco - thold;
assert q = din        -- Check new output is correct
report "load operation failed";
```

This statement checks that the output `q` of the flip-flop is equal to the value of the stimulus `tco` ns after the positive clock edge. Timing parameter `tco` is the clock to output delay, which specifies when the flip-flop's `q` output should become stable after the positive clock edge. Assuming `tco` > `thold`, since a previous **`wait for`** `thold` statement was executed after the rising clock edge, statement **`wait for`** `tco - thold` makes the total time elapsed after the rising clock edge equal to `tco`.

If this last assertion evaluates true, we know that the output is valid at the point in time when it should first become valid. But we don't know if the output remains stable at this valid value until the next clock edge.

Now lets consider the first assertion statement. This statement verifies that the output remains stable from `tco` ns after the positive clock edge until the next positive clock edge. However, it verifies this for the previous clock period, the one that is terminated by the positive clock edge.

```
wait on clk until clk = '1';
-- Check previous output remained stable
assert q'stable(period - tco)
report "previous output not stable";
```

The signal attribute `stable` returns a `true` value when the referenced signal has no events in the preceding time specified by its parameter. Since the assertion statement with the `stable` attribute as its condition follows the statement that waits for the positive clock edge, the assertion is true if `q` remained stable for `period - tco` ns before the rising clock edge. Thus, this assertion statement checks the stability of the output following the positive clock edge used to synchronize the procedure that was executed prior to this procedure.

The next procedure called must contain a copy of the same assertion statement to check the stability of the output caused by the execution of this procedure.

14.9 BUS FUNCTIONAL MODELS

If a UUT interfaces to a parallel or serial bus, the required stimulus signals representing the bus interface can be quite complex. For example, we may be designing a system that interfaces to a central processing unit's (CPU's) system bus.

Fully Functional Model

One approach to generating complex bus stimulus is to use a fully functional RTL model of the CPU. A fully functional model contains all of the CPU's internal registers, pipelines, caching, instruction set architecture, and timing and control. Such a model may be commercially available. If not, or if it is prohibitively expensive, the alternative of creating our own fully functional RTL model is such a complex task that it is not justifiable.

Bus Functional Model

All that is usually needed to verify the operation of a UUT connected to a CPU's system bus is a set of signals that exhibit behavior equivalent to that of the CPU bus signals. Thus, a simpler approach is to create a bus functional model (Figure 14.9.1). A *bus functional model (BFM)* of a device models only the functionality and timing of the device's bus transactions (bus cycles).

The output of a CPU BFM is a timed sequence of bus signal transitions that model the timing of signals on the CPU's bus. A CPU BFM does not contain any of the internal functionality or architecture associated with the CPU's instruction set. In addition to its relative simplicity, use of a BFM results in much faster simulations than use of a fully functional RTL CPU model.

A BFM provides the generation of bus signal stimulus at a higher level of abstraction than dealing with every transition of each bus signal. This higher level of abstraction is referred to as the *transaction level* (bus cycle level).

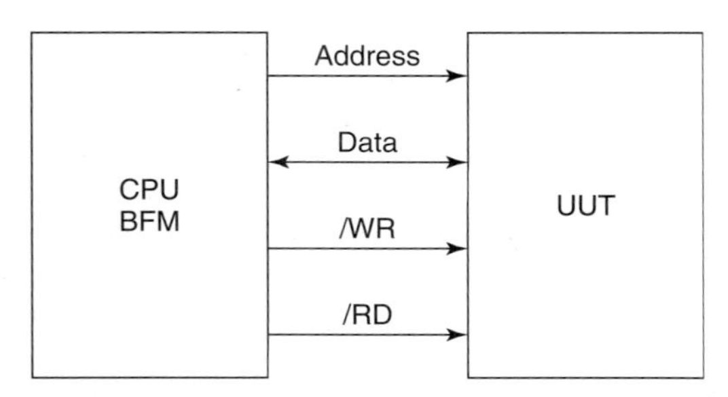

FIGURE 14.9.1
A CPU bus functional model providing stimulus to a UUT.

For a particular UUT verification, it may not be necessary to model the full range of all possible CPU bus cycles. For example, if the UUT does not generate interrupts, it is not necessary to model the interrupt bus cycle. However, if a number of different systems are to be designed that all interface to the same CPU, the effort of writing a complete BFM that supports all possible bus cycles for the CPU may be justified.

VHDL BFMs for some CPUs are commercially available. They allow a CPU vendor to provide their customers a bus accurate model of the CPU, without revealing any CPU architectural implementation details.

Simple CPU Bus Functional Model

While a CPU has other bus cycles that must be modeled in a complete BFM, for verifying the interface to a simple I/O device, read and write bus cycles are usually sufficient. These bus cycles transfer data from and to the UUT.

Timing diagrams for read and write bus cycles for a CPU consisting of an Intel 80C188EB microprocessor with a demultiplexed address and data bus are provided in Figures 14.9.2 and 14.9.3. These are the waveforms that the BFM must generate to model these bus cycles.

A BFM can be implemented as a design entity and instantiated and connected to the UUT in a testbench. However, to change the sequence of bus cycles to generate different test cases, the BFM's design file would have to be modified or the BFM would have to include inputs that command the sequence of bus cycles to be executed and the parameters to be used.

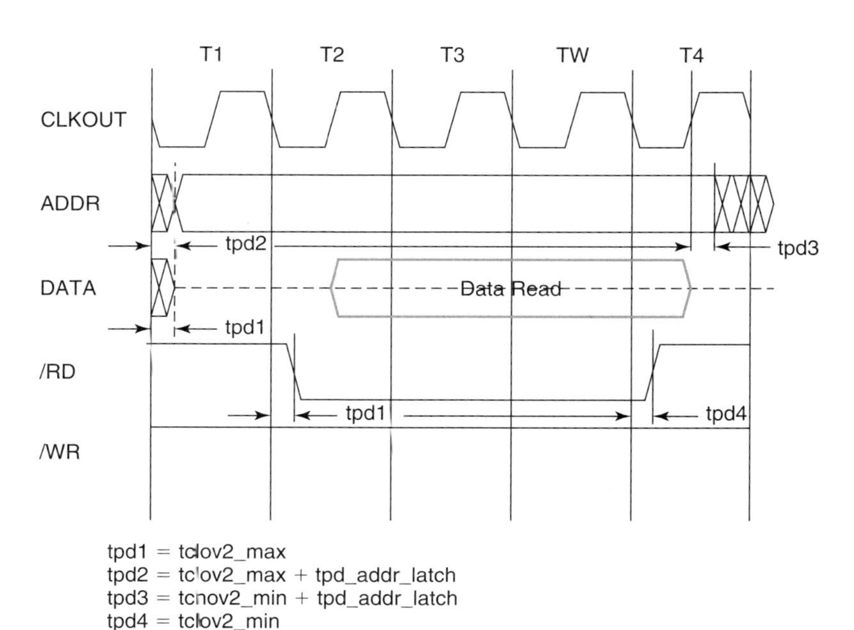

FIGURE 14.9.2
Read bus cycle with one WAIT state for an 80C188EB microprocessor.

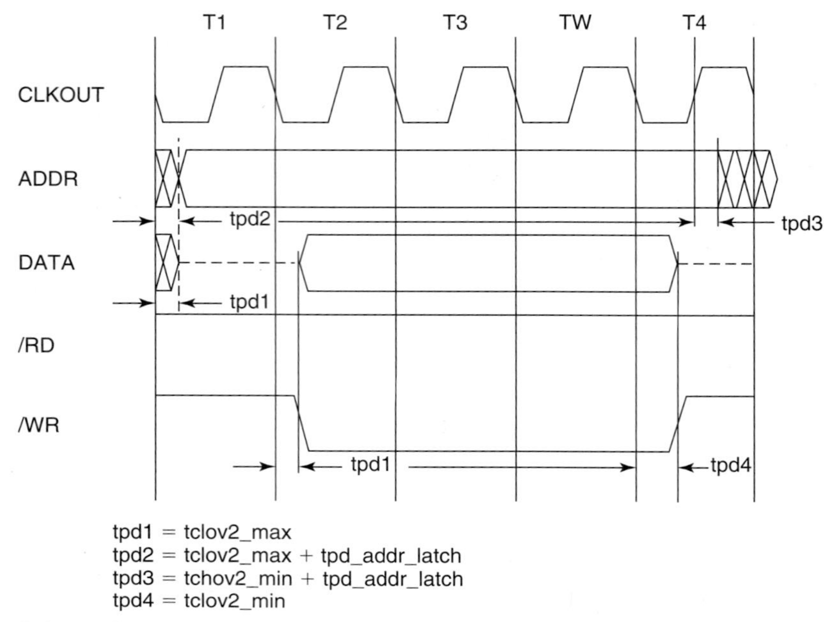

FIGURE 14.9.3
Write bus cycle with one WAIT state for an 80C188EB.

Procedure-Based BFM

A more flexible approach is to implement the BFM as a set of procedures. A separate procedure is written for each kind of bus cycle to be modeled. During a simulation, the sequence of bus cycles executed is determined by the order in which these procedures are called. The procedures are placed in a package. This package and a process in the testbench that calls procedures from it constitute the BFM represented in Figure 14.9.1. A particular test case is created by a sequence of procedure calls with the appropriate parameters.

The discussion that follows of the design of BFM procedures for an 80C188EB is generally applicable to any CPU. For a different CPU, we must use its timing diagrams and timing parameters as the basis.

The package consists of a procedure for each kind of bus cycle modeled: `read_cycle`, `write_cycle`, and `bus_idle`. The bus idle cycle represents a bus cycle where no data is transferred.

To create the proper timing for bus signal transitions, each procedure is synchronized to the `clkout` signal from the CPU. Normally, an 80C188EB bus cycle consists of four periods of the CPU's `clkout` signal. Each clock period starts at the falling edge of `clkout` and ends at the next falling edge. Each of these periods is called a T state and they are numbered from T1 to T4.

WAIT States

Sometimes the duration of a bus cycle must be extended by the insertion of one or more states between states T3 and T4. The purpose of extending a bus cycle is to make the bus cycle's timing compatible with the timing of a slower device connected to the CPU's system bus.

States inserted between T3 and T4 are called WAIT states or TW. During a WAIT state, the value of each bus signal remains as it was in T3. An 80C188EB can be programmed to introduce from 0 to 15 WAIT states in a bus cycle. When addressed, an external device can also request any number of WAIT states by unasserting the 80C188EB's READY input.

Bus Idle State

Usually, completion of one bus cycle is followed immediately by the start of the next bus cycle. Occasionally, when the CPU is not immediately ready to make another data transfer, a read or write bus cycle is followed by one or more bus idle states, or TI. During a bus idle state, bus signals remain essentially the same as they were at the end of the previous bus cycle.

Read Bus Cycle

The `read_cycle` procedure is the most complicated of the three and is described here in detail. What happens during a read bus cycle is shown in Figure 14.9.2. In state T1, at the beginning of the bus cycle, the CPU drives an address onto the address bus. This is the address from which data is read during the bus cycle. At the beginning of the next state (T2), the CPU asserts control signal `rd_bar`.

In response to the assertion of `rd_bar`, the addressed device (the UUT) must drive the data that is being read onto the data bus. At the beginning of T4, the CPU latches the data on the data bus into one if its internal registers. The CPU then unasserts `rd_bar`. The bus cycle ends at the end of T4.

Bus Cycle Package

The bus cycle procedures, placed in a package, are as shown in Listing 14.9.1. The procedures are declared in package `bus_cycle` and defined in package body `bus_cycle`.

LISTING 14.9.1
Package containing bus cycle procedures.

```
library ieee;
use ieee.std_logic_1164.all;

package bus_cycle is
   --constant clk_period : time := 125 ns;
   constant tclov2_min : time := 3 ns;    -- for 13 MHz speed grade
   constant tclov2_max : time := 30 ns;    -- for 13 MHz speed grade
   constant tpd_addr_latch : time := 7 ns;    -- for address latch

   procedure read_cycle (
   addr : in std_logic_vector(19 downto 0); -- address specified for bus cycle
   signal data_rd : out std_logic_vector(7 downto 0);    -- data read
```

(Cont.)

LISTING 14.9.1 *(Cont.)*

```
   wait_states : in integer;-- number of wait states
   signal clk : in std_logic;
   signal addr_out : out std_logic_vector(19 downto 0); -- cpu address generated
   signal data : inout std_logic_vector(7 downto 0);  -- data generated or read
   signal rd_bar : out std_logic) ;  -- cpu read control signal

   procedure write_cycle (
   addr : in std_logic_vector(19 downto 0);  -- address specified for bus cycle
   data_wr : in std_logic_vector(7 downto 0);  -- data to write during bus cycle
   wait_states : in integer;  -- number of wait states
   signal clk : in std_logic;
   signal addr_out : out std_logic_vector(19 downto 0);  -- address generated
   signal data : inout std_logic_vector(7 downto 0);  -- data generated or read
   signal wr_bar : out std_logic) ;  -- cpu write control signal

   procedure bus_idle (
   signal clk : in std_logic;
   signal rd_bar : out std_logic;  -- cpu read control signal
   signal wr_bar : out std_logic);  -- cpu write control signal

end bus_cycle;


package body bus_cycle is

   procedure read_cycle (
      addr : in std_logic_vector(19 downto 0);  -- address specified
      signal data_rd : out std_logic_vector(7 downto 0);  -- data read
      wait_states : in integer;  -- number of wait states
      signal clk : in std_logic;
      signal addr_out : out std_logic_vector(19 downto 0);  -- address generated
      signal data : inout std_logic_vector(7 downto 0); -- data written or read
      signal rd_bar : out std_logic)  -- cpu read control signal
      is
   begin
      wait on clk until clk = '0';  -- T1
      addr_out <= addr after tclov2_max + tpd_addr_latch;
      data <= (others => 'Z') after tclov2_max;
      wait on clk until clk = '0';  -- T2
      rd_bar <= '0' after tclov2_max;
      wait on clk until clk = '0';  -- T3
      for i in 1 to wait_states loop  -- TW
         wait on clk until clk = '0';
      end loop;
      wait on clk until clk = '0';  -- T4
```

```
      data_rd <= data;
      rd_bar <= '1' after tclov2_min;
      wait on clk until clk = '1';
      addr_out <= (others => 'X') after tclov2_min + tpd_addr_latch;
   end read_cycle;


   procedure write_cycle (
      addr : in std_logic_vector(19 downto 0);  -- address specified
      data_wr : in std_logic_vector(7 downto 0);  -- data to write
      wait_states : in integer;  -- number of wait states
      signal clk : in std_logic;
      signal addr_out : out std_logic_vector(19 downto 0);  -- address generated
      signal data : inout std_logic_vector(7 downto 0); -- data writen or read
      signal wr_bar : out std_logic)  -- cpu write control signal
      is
   begin
      wait on clk until clk = '0';  -- T1
      addr_out <= addr after tclov2_max + tpd_addr_latch;
      data <= (others => 'Z') after tclov2_max;
      wait on clk until clk = '0';  -- T2
      data <= data_wr after tclov2_max;
      wr_bar <= '0' after tclov2_max;
      wait on clk until clk = '0';  -- T3
      for i in 1 to wait_states loop  -- TW
         wait on clk until clk = '0';
      end loop;
      wait on clk until clk = '0';  -- T4
      wr_bar <= '1' after tclov2_min;
      wait on clk until clk = '1';
      addr_out <= (others => 'X') after tclov2_min + tpd_addr_latch;
      data <= (others => 'Z') after tclov2_min;
   end write_cycle;

   procedure bus_idle (
      signal clk : in std_logic;
      signal rd_bar : out std_logic;  -- cpu read control signal
      signal wr_bar : out std_logic)  -- cpu write control signal
      is
   begin
      wait on clk until clk = '0';  -- T idle
      rd_bar <= '1' after tclov2_max;
      wr_bar <= '1' after tclov2_max;
      wait on clk until clk = '1';
   end bus_idle;

end bus_cycle
```

Read Bus Cycle Procedure Details

The procedure `read_cycle` has a number of parameters (Figure 14.9.4). Parameters on the left side of the symbol for the `read_cycle` procedure transfer information to and from the procedure by way of procedure calls in the testbench. Parameters on the right side of the symbol are the simulated CPU bus signals that connect to the UUT.

Input parameter `addr` is the 20-bit address passed to the procedure. This is the address from which we want the data to be read. Output parameter `data_rd` is used to return the data that was read. This is the data provided by the UUT during the read bus cycle. Input parameter `wait_states` is an integer that specifies how many wait states are inserted between states T3 and T4. Input parameter `clk` is the system clock that the procedure uses for synchronization. Output parameter `addr_out` is the address signal from the BFM. Parameter `data` is mode **`inout`** and represents the BFM data bus, which is bidirectional. Output parameter `rd_bar` is the BFM control signal that is asserted during a read bus cycle.

Examination of Figure 14.9.2 shows that the CPU's bus signal transitions are synchronized to the falling edge of `clkout`. However, there is a slight delay between the falling edge of `clkout` and the actual change in a bus signal's value. This delay is specified by the 80C188EB timing parameter tclov (time from clock low to output valid). There are two different sets of values for tclov in the 80C188EB data sheet. For the signals of interest here, use of the set tclov2 is appropriate. Both a minimum (`tclov2_min`) and a maximum (`tclov2_max`) value for tclov2 are specified. To stress the UUT, we want to use the worst-case value at each point in time. This is typically the value that requires the UUT to have the fastest response.

State T1

Code to sequence the signals for a bus cycle is straightforward. A wait on statement synchronizes each T state to the falling edge of the clock. Signal changes synchronized to that falling clock edge for a particular T state are then assigned. For state T1 the code is:

```
wait on clk until clk = '0';   -- T1
addr_out <= addr after tclov2_max + tpd_addr_latch;
data <= (others => 'Z') after tclov2_max;
```

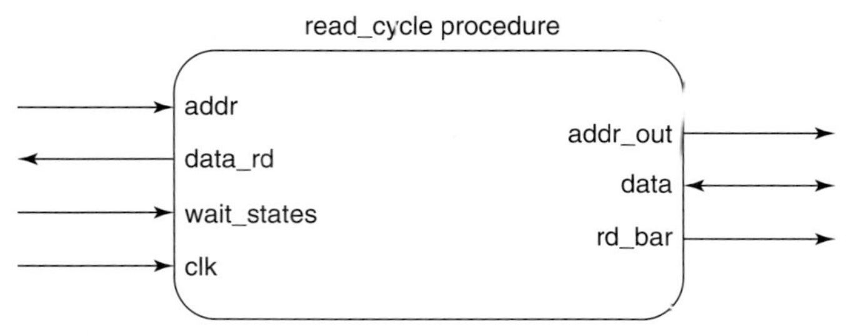

FIGURE 14.9.4
Read_cycle procedure parameters.

After the falling edge of the clock in T1, addr_out and data are assigned values. Signal addr_out is assigned addr, the address value passed to the procedure in the call statement. This assignment is made with a delay of tclov2_max + tpd_addr_latch. Choosing tclov2_max causes the address to appear as late as possible in the bus cycle, which is the worst case. Delay tpd_addr_latch is the delay of the address latch used to demultiplex the address/data bus in an 80C188EB CPU design.

State T2

After the falling edge of the clock in T2, rd_bar is asserted:

```
wait on clk until clk = '0';   -- T2
rd_bar <= '0' after tclov2_max;
```

States T3 and TW

After the falling edge of the clock in T3, the procedure uses a loop to insert the specified number of wait states. This number is passed to the procedure in the call statement as integer parameter wait_states:

```
wait on clk until clk = '0';   -- T3
for i in 1 to wait_states loop   -- TW
      wait on clk until clk = '0';
end loop;
```

If a 0 is passed as the actual value for the number of WAIT states, no WAIT states are inserted in the bus cycle. No signals driven by the BFM change their values during either T3 or TW.

State T4

After the falling edge of the clock in T4, the procedure reads the data and rd_bar is unasserted. In the second half of T4 at the rising edge of the clock, the address is assigned the unknown value:

```
wait on clk until clk = '0';-- T4
data_rd <= data;
rd_bar <= '1' after tclov2_min;
wait on clk until clk = '1';
addr_out <= (others => 'X') after tclov2_min + tpd_addr_latch;
```

Write Bus Cycle BFM Waveforms

The overall design of the write_cycle procedure is similar to that of the read_cycle procedure. The major difference is that the BFM drives the data bus with a data value passed as a parameter. This data is driven on the bus at the beginning of T2 and removed at the beginning of T4. In addition, for a write bus cycle, wr_bar is asserted rather than read_bar.

The waveforms generated by a single call to the procedure write_cycle are shown in Figure 14.9.5.

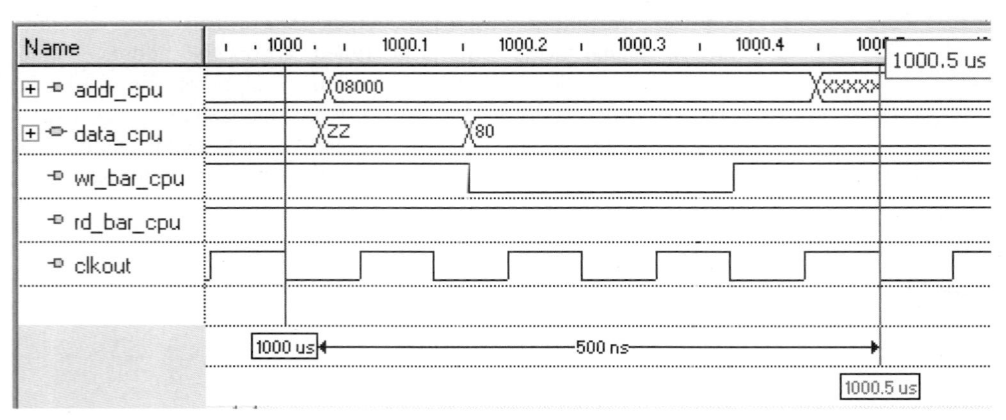

FIGURE 14.9.5
Waveforms produced by a call to the write_cycle procedure at time = 1000 us.

This call was made with an address parameter of 08000 hexadecimal and a data value of 80 hexadecimal. The write cycle in this figure is bounded by the waveform editor cursors.

Sequencing Bus Cycles

Another advantage of a BFM model over a fully functional RTL model is that it is easy to directly create any desired sequence of bus cycles by simply calling bus cycle procedures in the desired order. In contrast, using a fully functional RTL model requires writing and compiling CPU code to produce the desired bus cycles and loading this code into a memory model also connected to the CPU bus in the testbench.

An example of a sequence of calls to the bus cycle procedures to verify the pulse width modulated signal generator UUT is given in the next section.

14.10 RESPONSE MONITORS

Testbench for PWMSG

A separate process can be used to monitor output responses from a UUT. A diagram of a testbench that uses a CPU BFM to write duty cycle values to the microprocessor-compatible PWMSG UUT in Section 9.6 is given in Figure 14.10.1. The testbench components, processes, and procedures are linked by signals.

The reset and clock are described by concurrent signal assignment statements. The CPU BFM is implemented using the package `bus_cycle` from the previous section. Responses from the UUT are monitored by a separate response monitor process. The testbench is given in Listing 14.10.1.

The stimulus process contains a loop that calls procedure `write_cycle`. Each call loads a new value to produce a different duty cycle. The PWMSG is allowed to generate each duty cycle for 10 ms before the next duty cycle value is loaded.

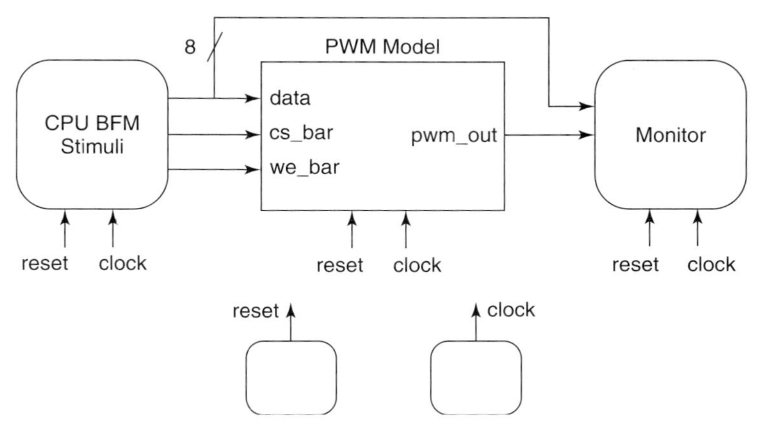

FIGURE 14.10.1
Testbench structure for PWMSG using CPU BFM.

The process stimuli was written to show that it is not difficult to write an exhaustive verification for this UUT. However, running the exhaustive verification takes a relatively long time. It may be preferable to simply have a limited number of calls to procedure `write_cycle` to verify a select number of duty cycle values. For example, a sufficient set of load values might be `x"00"`, `x"01"`, `x"08"`, `x"7F"`, `x"80"`, `x"FE"`, and `x"FF"`.

LISTING 14.10.1
Testbench for pulse width modulated signal generator.

```
library ieee;
use ieee.std_logic_1164.all;
use ieee.numeric_std.all;

library bus_cycle;
use bus_cycle.bus_cycle.all;

entity pwm_tb is
end pwm_tb;

architecture tb_architecture of pwm_tb is
   constant cpu_clk_period : time := 125 ns;

   -- Stimulus signals
   signal pwm_out: std_logic;

   -- Observed signals
   signal d : std_logic_vector(7 downto 0);
```

(Cont.)

LISTING 14.10.1 *(Cont.)*

```
   signaladdr_cpu:  std_logic_vector (19 downto 0);
   signaldata_cpu:  std_logic_vector (7 downto 0);
   signalwr_bar_cpu:  std_logic;
   signalrd_bar_cpu:  std_logic;
   signalclkout:  std_logic := '0';

   signal cpu_reset_bar : std_logic;
   signal cpu_reset : std_logic;
   signal duty_cyc_ld : std_logic_vector(7 downto 0);

begin
   -- Unit Under Test port map
   UUT : entity pwm
   port map (
      data => data_cpu,
      cs_bar => '0',
      we_bar => wr_bar_cpu,
      clk => clkout,
      reset => cpu_reset,
      pwm_out => pwm_out);

   cpu_clock_gen : clkout <= not clkout after cpu_clk_period/2;

   cpu_reset_gen : cpu_reset_bar <= '0', '1' after 2 * cpu_clk_period;

   cpu_reset <= not cpu_reset_bar;


   stimulus : process
   begin
      rd_bar_cpu <= '1';
      wr_bar_cpu <= '1';
      wait for 1 ms;
      for i in 0 to 255 loop
         duty_cyc_ld <= std_logic_vector(to_unsigned(i,8));
         write_cycle(x"08000", duty_cyc_ld, 0, clkout, addr_cpu, data_cpu,
         wr_bar_cpu);
         wait for 10 ms;
      end loop;
      wait;
   end process;

   monitor: process
      variable leading_edge, high_time, pulse_period: time;
```

```
      variable data_written : unsigned(7 downto 0);
      variable duty_cycle_set, duty_cycle_meas : integer;
   begin
      wait until rising_edge(wr_bar_cpu);
      data_written := unsigned(data_cpu);
      wait until pwm_out = '1'; -- skip first pulse
      wait until pwm_out = '1';
      leading_edge := now;
      wait until pwm_out = '0';
      high_time := now - leading_edge;
      wait until pwm_out = '1';
      pulse_period := now - leading_edge;
      duty_cycle_set := 100 * (to_integer(data_written))/255;
      duty_cycle_meas := (100 * high_time)/pulse_period;
      if duty_cycle_set > 1 and duty_cycle_set < 100 then
         assert  duty_cycle_meas <= duty_cycle_set + 1
         and duty_cycle_meas >= duty_cycle_set - 1
         report "duty cycle error" & CR & LF &
         "iteration = " & integer'image(to_integer(unsigned(duty_cyc_ld))) &CR
            & LF &
         "duty cycle set = " & integer'image(duty_cycle_set) &CR & LF &
         "duty cycle measured = " & integer'image(duty_cycle_meas);
      end if;
   end process;
end tb_architecture;
```

The monitor process waits for the rising edge of the `wr_bar_cpu` signal, which indicates that the BFM has written a new value to the PWMSG. The monitor process then waits until the second positive edge of the PWMSG `pwm_out` output. The first transition of `pwm_out` is skipped, because writing a new value to the PWMSG by the microprocessor is not synchronized with the pulse currently being generated by the PWMSG and may, therefore, not have the required duty cycle.

Now

The second output pulse's high-time duration and period are measured using VHDL's `now` function. Function `now` is provided by package STANDARD. This function returns the current simulation time, Tc. Current simulation time values are of subtype `delay_length`.

From the values returned by `now`, the measured duty cycle is computed. The specified duty cycle is computed from the value that was written to the PWMSG by the microprocessor. An assertion statement compares the measured duty cycle with the specified duty cycle. The measured duty cycle has to be within ± 1 percent of the specified duty cycle or a report of the difference is made.

PROBLEMS

14.1 What two stimulus signals are always required in a testbench for a synchronous system? What is the basic criterion for these signals?

14.2 If `period` is an odd number, what is the effect on the actual period of a clock generated by the statement:

```
clk <= not clk after period/2;
```

In other words, is the result from the division truncated, resulting in a shorter period, or rounded up, resulting in a longer period? If period is 15 ns, can the above statement be rewritten to produce a more accurate clock model? What effect does simulator resolution have on this accuracy?

14.3 Can a clock with other than a 50% duty cycle be generated using a concurrent signal assignment statement? If so, write a concurrent signal assignment statement to generate a clock with a 40% duty cycle. If not, explain why.

14.4 Can a concurrent signal assignment statement describe a clock that is stopped by a boolean signal? If so, write such a statement that describes a clock with a 50% duty cycle that is stopped when the boolean signal `end_sim` becomes `true`. If not, explain why.

14.5 The examples of clocks generated by processes in Section 14.2 used loops. Consider the following attempt to describe a clock:

```
clock : process
   begin
      clk <= '0';
      wait for period/2;
      clk <= '1';
      wait for period/2
end process;
```

(a) Explain why this code either does or does not successfully describe a clock waveform.

(b) If the code does successfully describe a clock waveform, can an `end_sim` signal be added to the process to end the simulation? If so, write the modified code.

14.6 Write a procedure named `clk_gen` that has the following declaration:

```
procedure clk_gen (constant freq, duty_cycle : in real;
      signal clk : out std_logic);
```

where the frequency is specified in MHz and the duty cycle is a value between 0.0 and 1.0. This procedure must be written to allow any frequency and duty cycle to be specified when the procedure is called.

14.7 A UUT requires a reset pulse named `rst` that is asserted high for 250 ns at power up and is then unasserted. Write a concurrent signal assignment statement to generate `rst`.

14.8 It is often useful to have code for a reset and a clock that can be reused when we want to quickly verify UUTs, such as simple counters, by examining their output waveforms. Write a procedure that can be called concurrently that has five parameters passed to it: clock period, asserted state of the reset, duration of the reset as a number of clock cycles, initial level of the clock, and number of clock cycles to be generated after the reset is unasserted. The values returned by the procedure are the signals `reset` and `clk`.

14.9 Explain why the following statement is valid for the positive-edge-triggered flip-flop in Listing 8.4.5. For a functional simulation, the D input can be changed to the next input value as late as simultaneously with the clock edge. That is, if a new value is assigned to D in the same delta cycle that the clock makes its 0 to 1 transition, the value stored in the D flip-flop will be the new value assigned to D, not its old value. Use the diagram in Figure 6.6.1 as the basis for your explanation.

14.10 Write a testbench for the 4-bit right shift register in Listing 9.2.1. Use vectors to provide the serial input data stimulus and the expected serial output value.

14.11 Write a testbench for the 4-bit binary counter in Listing 9.4.1.

14.12 Write a testbench for the 12-bit binary counter in Listing 9.4.4.

14.13 Write a testbench for the modulo 32 BCD counter in Listing 9.4.7.

14.14 Write a testbench for the positive edge detector in Listing 9.5.1.

14.15 Write a testbench for the single shot in Listing 9.5.2.

14.16 Generate the stimulus in Figure 14.6.2 using four concurrent signal assignment statements each with multiple after clauses.

14.17 Generate the stimulus in Figure 14.6.2 using a process with signal assignment statements and wait for statements.

14.18 Write a package named `signal_probe` and its associated package body. This package contains a single procedure named `period_probe`, which has the following procedure declaration:

```
procedure period_probe (signal a : in std_logic; signal period : out time);
```

Procedure `period_probe` computes the period of the signal passed to it as parameter `a`. This procedure is to be called as a concurrent procedure. Once called, this procedure never returns. However, its parameter `period` provides a running measurement of the period of `a`. Write this procedure to take advantage of the predefined function `now` in making its computation.

Procedure `period_probe` could be used during functional simulation of a structural style design. Multiple concurrent calls to procedure `period_probe` would provide simultaneous measurements of the periods of multiple signals in the design. The signals that appear as the actual parameters replacing parameter `period` in the calls could be viewed on the simulation waveforms to see the period values. The procedure calls would be removed before the design is synthesized.

Chapter 15
Modular Design and Hierarchy

A large complex system is impossible to comprehend, design, or verify as a single entity. Traditionally, in both hardware and software design, a complex system is partitioned into modules for the purposes of documentation, design, coding, verification, and maintenance. During the partitioning process, each module's interface and function are completely defined, so that when the modules are interconnected they accomplish the system's overall function.

With each module's interface and function completely defined, each can be independently designed and verified. This reduces the overall complexity of system design and verification. In VHDL, modules are implemented as design entities. They are stored in libraries for use in current designs and reuse in future designs.

VHDL provides constructs that support modular design of complex systems. These constructs and their use are discussed in this chapter.

15.1 MODULAR DESIGN, PARTITIONING, AND HIERARCHY

Modular Design

Modular design is a long-established methodology for designing complex hardware or software systems. In modular design, a complex system is partitioned into simpler modules. These modules are interconnected to accomplish the complex system's overall function.

Each module's particular task or function is defined in a way that allows the module to be independently designed and verified, before being combined with the other modules. Designing and verifying each module independently allows us to focus on one module at a time, simplifying the overall design effort.

For software modules to be independently designed and verified, they must be capable of being independently compiled. The same is true for hardware modules described in VHDL. Once verified and properly documented, a module can be placed in a library and used in the original and subsequent designs.

Flat Partitions

In order to take advantage of VHDL's constructs that support modular design, a design must first be partitioned into modules. If a complex system is partitioned into a set of interconnected modules and none of these modules is further partitioned, the result is a *flat* or *single-level partition*.

Hierarchical Partitions

For many complex designs a flat partition is not sufficient, because some of the modules are still too complex. Accordingly, each complex module is further partitioned into simpler modules. This creates a *hierarchical* (multilevel) *partition* of the system.

Top-Down Design

Hierarchical partitioning is the basis of *top-down design*. In top-down design, a system's total function is repeatedly partitioned into simpler modules from the top level down.

The *top-level partition* (first-level partition) in a hierarchy constrains all subsequent (lower-level) partitions (Figure 15.1.1). As a result, the top-level partition is particularly important, since it defines and constrains the system's overall architecture and implementation.

Each module in the top-level partition may itself be partitioned to further reduce its complexity, creating the next lower-level partition. The resulting hierarchy of modules can be represented by a hierarchical block diagram or a hierarchy tree (Figure 15.1.1). A *hierarchy tree* clearly represents the hierarchical relationship of a system's modules. The modules at the bottom of each branch of a hierarchy tree are called *leaf nodes*.

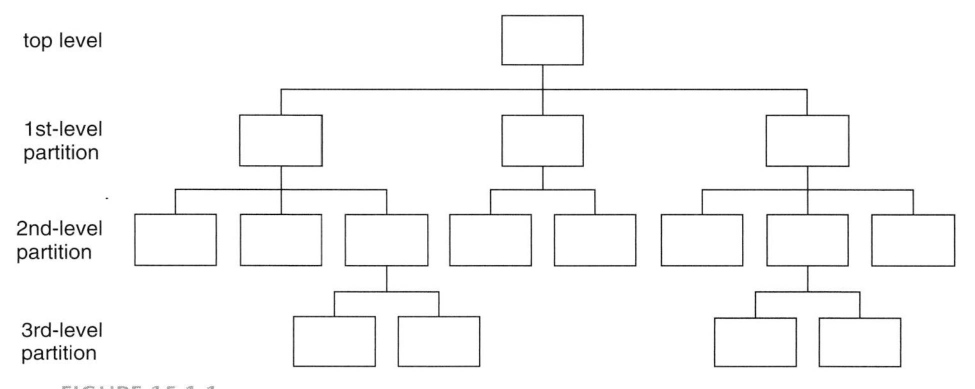

FIGURE 15.1.1
Hierarchy tree for the modular partition of a complex system.

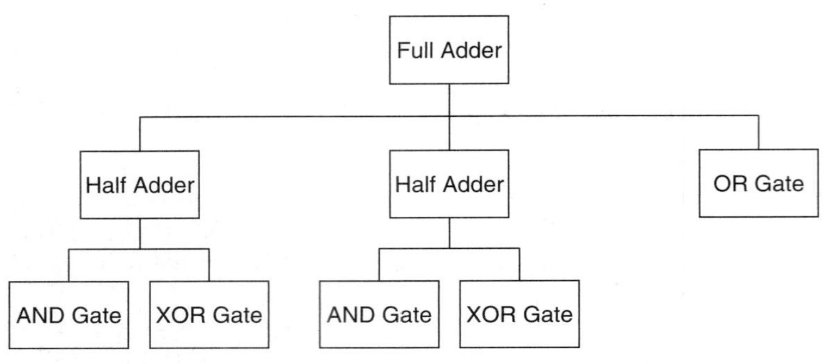

FIGURE 15.1.2
Hierarchy tree for a full adder.

As a very simple example of a hierarchical partition, consider the partition of a full adder. A hierarchy tree representing such a partition is given in Figure 15.1.2.

Typically, the top-level partition consists of the modules that would comprise the blocks in a top-level block diagram of the system. For example, compare the modules in Figure 15.1.2 to the blocks in the block diagram in Figure 2.9.1. It is obvious from both the block diagram and hierarchy tree that some of the same modules are used multiple times in the partition.

The partitioning process is continued until each leaf node represents a relatively simple module that we can easily comprehend and directly code. For example, the XOR gate function can be simplified by further partitioning, as in Figure 15.1.3.

For more experienced VHDL designers, what is considered simple enough to directly code is a lot more complex than it is for beginning designers. So, as we gain more experience in coding VHDL, we are likely to have more complex leaf nodes. However, there are other tradeoffs in deciding the complexity of leaf nodes, such as code readability and maintainability, module reuse, and synthesis efficiency.

Bottom-Up Implementation

Modules corresponding to leaf nodes are coded and verified first. These modules are then interconnected using structural style descriptions to form modules on the next

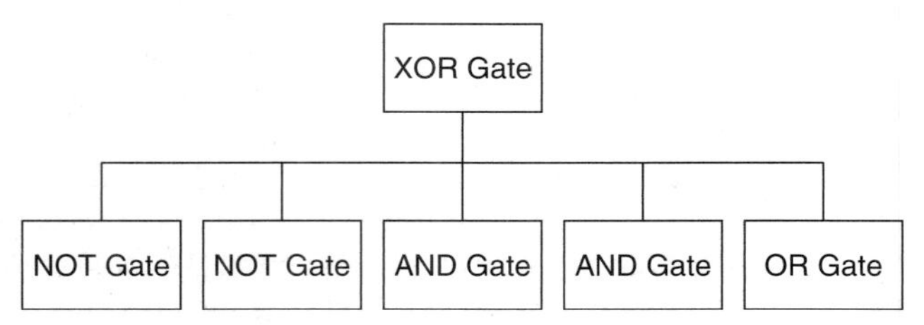

FIGURE 15.1.3
Hierarchy tree for XOR gate.

higher level. This process is repeated for each higher-level partition until the top-level partition is implemented. This approach comprises a *bottom-up implementation.*

Balanced Hierarchy

Modules at the same level in a hierarchy should have a relatively comparable degree of complexity. The number of modules in each hierarchical level should strike a balance between producing an overly horizontal or an overly vertical partition. A *horizontal partition* is one that has a large number of modules in the top-level partition and few levels. A *vertical partition* has few modules at each level and a large number of levels. Either extreme is more difficult to comprehend than is a balanced hierarchy.

Advantages of a Good Design Partition

Effort expended in creating a good design partition is well rewarded. A good partition provides several advantages:

- Design management is easier
- Modules can be designed and verified by different individuals
- Maximum reuse of modules is made possible
- The design description is more readable and easier to comprehend
- Verification is simplified
- Better design results are likely
- The portability of the design is enhanced

Criteria for Delineating Modules

The function(s) realized by a module should be logically coherent. Only related functionality should be contained in a single module.

Each module should have a relatively small number of well-defined inputs and outputs. Definition of a module should also consider its likelihood of being reusable, either in the current design or subsequent designs. Reusability is important, since module reuse provides significant design and verification productivity gains.

While we want to limit the complexity of the leaf node modules so they are simple enough to directly code, we don't want to make them too simple. The simpler they are, the more of them that will be required. Since modules are interconnected using a structural style description; use of very low logical complexity modules results in an inefficient and overly complex structural description. The ultimate example of this inefficiency is the implementation of a complex design using only modules that are two-input NAND gates.

Bottom-Up Verification

Each module should be defined so that it is individually verifiable. Each module is independently verified before being combined with other modules to form a higher-level module. Higher-level modules are independently verified before being combined to form the complete system. The complete system is then verified. This approach is called *bottom-up verification*.

Creating a Partition

Partitioning is an art. The ability to create good partitions improves with experience. In the past, when medium-scale integration (MSI) was the most complex level of IC available, designers partitioned their designs to take maximum advantage of available MSI functions. At that time, this approach minimized system cost by reducing package count and increasing the reliability.

As a result, decoder, encoder, multiplexer, register, shift register, counter, adder, and other common MSI functions became the elemental functions used to implement complex systems.

Designers using PLDs aren't restricted to the same functions that existed when using MSI. However, functions similar to the "elemental functions" from MSI design are still often the basis for lower-level partitions.

Partitioning and Design Reuse

If we can partition a system to make use of existing modules, design time is significantly reduced. Reused modules may be ones that we have created for prior designs or that we have obtained from others, possibly in the form of intellectual property (IP). Reuse of modules increases a system's reliability, assuming a reused module is one that has been verified and proven in a previous design.

Today, the significant advantages of design reuse often lead to defining each module of a partition in a somewhat more general and flexible way than is required for a specific design. This improves the likelihood that modules are reusable in other designs.

Partitions Based on Data Path and Control

A top-level partition of hardware into one or more data paths and control modules ("System Architecture" on page 441) is often made. Recall that a data path consists of registers separated by combinational arithmetic or logic processing modules. A sequential *control* module (a FSM) provides control signals to enable registers and to control the operation of the arithmetic and logic modules.

Partitions Based on a PLD Vendor's Libraries

Ideally, we want our VHDL design descriptions to be portable, so that any target PLD with sufficient logic capacity can be used to realize a system. As a result, we generally want our partitions to be independent of special architectural features of a particular PLD.

Nevertheless, to achieve very high system performance, it is sometimes necessary to take advantage of optimized vendor library modules associated with a specific target PLD's architecture. In such cases, we must create a partition that includes modules from the vendor's library. The result is a design that is not portable.

However, if these vendor modules are isolated and limited to the higher-level partitions, they can be replaced by equivalent modules if we must change the target PLD. Thus, we can still maintain a degree of portability.

Design Entities as Modules

In VHDL, design entities serve the purpose of modules. Structural style descriptions describe how particular design entities are interconnected to form a more complex system.

15.2 DESIGN UNITS AND LIBRARY UNITS

VHDL provides many features that support creating and managing modular designs. Important among these are library units.

As discussed in Section 2.9, each design entity can be described in a separate file. Using separate files provides a degree of modularity that makes design management easier because each file is smaller, improving the readability of its code. Using separate files also allows separate compilation and verification of each design entity. Beyond simply allowing a complex design to be divided into multiple files, VHDL provides more extensive features for facilitating design modularity.

Design Units

One feature associated with independent compilation is that of a design unit. A *design unit* is a VHDL construct that can be independently compiled and stored in a design library. Design units provide modularity for the *design management* of complex systems.

Design Files

A *design file* is a source file containing one or more design units (Figure 15.2.1). A design file is the input to a VHDL compiler. Design units in a design file are compiled in the same order as their textual order in the file.

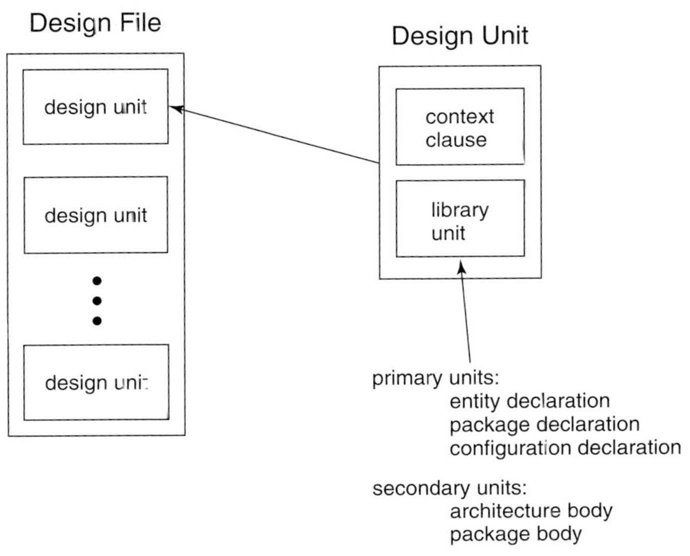

FIGURE 15.2.1
Design file and design units.

Library Units

A design unit is comprised of a context clause and a library unit, as shown in Figure 15.2.1. Compilation of a design unit defines the corresponding library unit that resides in a design library. That is, a library unit is the compiled and stored form of a design unit.

There are five kinds of *library units*. They are divided into two groups, primary units and secondary units. The *primary units* are:

- Entity declaration
- Package declaration
- Configuration declaration

A primary unit describes an external view or interface to either an entity, package, or configuration.

The *secondary units* are:

- Architecture body
- Package body

Note that while the LRM defines entity declarations, package declarations, configuration declarations, architecture bodies, and package bodies syntactically as library units, much of the literature on VHDL refers to them, informally, as design units. In effect, the terms library unit and design unit are often used interchangeably.

A secondary unit describes the behavior or structure of its associated primary unit. Of the secondary units, only architecture bodies have unique names. A package body has the same name as its associated package declaration. A primary unit must be compiled before any associated secondary unit is compiled. A secondary unit must be compiled into the same library as its associated primary unit.

The relationship between a package declaration and its associated package body is similar to the relationship between an entity declaration and its associated architecture body. The primary unit describes the interface, and the secondary unit describes the behavior or structure. However, there can be only one package body for each package declaration, whereas an entity declaration can have multiple architecture bodies.

An entity declaration and an associated architecture body are the only two library units that must exist to form a design entity. Package declarations, package bodies, and configuration declarations are optional. All of the library units, except configuration declarations, have been discussed in previous chapters. Configuration declarations are discussed later in this chapter.

Library Unit Names

The names of primary units must be unique within a given library. Names of alternative architectures associated with a given entity declaration must be unique. However, two architectures associated with different entities can have the same name. A package body's name must be the same as its corresponding package.

15.3 DESIGN LIBRARIES

A simulator can only simulate programs that have been successfully compiled and stored in a library. Accordingly, we have used libraries routinely in previous designs. Libraries also provide features that support design management of complex systems and facilitate design reuse.

A complex design will consist of a number of design units. Many of these design units may need to declare the same constants, data types, and subprograms. Instead of having each design unit repeat the text for these shared declarations and subprogram bodies, they can be placed in packages and package bodies and compiled to a library. Each design unit can then reference the packages in this library to make use of their contents.

Each design entity can be separately compiled and stored in a library. Other design entities that need to instantiate previously compiled design entities as components can simply reference their libraries to do so.

Design Library

A *design library* is a logical storage area in the host computer environment for compiled design units (library units). Its physical structure is implementation dependent and not defined in the VHDL standard. For most host computer implementations, a library is simply a directory. Each VHDL compiler's documentation must be consulted to determine the format and implementation of its libraries.

Library Logical Names

A design library is identified by its *logical name.* We use this name to reference the library in our VHDL descriptions. A path must be assigned to each logical library name to point to the physical directory that contains the library. The way this is done depends on the compiler. With some tools, this is done in an initialization file that we must create. In most tools having an IDE, this assignment is made in a transparent fashion when we set up a new design project.

Working and Resource Libraries

There are two kinds of design libraries: working library and resource library. The *working library* is the library into which the library unit resulting from compilation of a design unit is placed. We can specify into which library we want a design unit to be compiled. If we do not specify a library, the design unit is compiled into the default library named `work`.

A *resource library* is a library containing library units that are referenced within the design unit being compiled. The logical names of the referenced libraries must be listed in a library clause(s) preceding each design unit.

While only one library can be the working library during a compilation, an arbitrary number of resource libraries, including the working library itself, may be used.

Library STD

All VHDL compilers come with the *Library STD* included. This built-in library is provided by IEEE Std 1076. Library STD contains two packages: STANDARD and TEXTIO. It is in *package STANDARD* that the predefined types and operators of VHDL are defined. This package was discussed in Chapter 13. *Package TEXTIO* contains predefined types, functions, and procedures used for reading and writing files.

Library IEEE

VHDL compilers also include the library IEEE. This library contains packages defined by VHDL's supporting standards, such as packages STD_LOGIC_1164 and NUMERIC_STD. Most compiler vendors also include in this library other de facto standard packages such as STD_LOGIC_ARITH. These packages were also discussed in Chapter 13.

User-Defined Libraries

Of course, we can write our own packages and place them in libraries that we create (*user-defined libraries*). We can also place design entities that we have created in our user-defined libraries. Placing subprograms and design entities in user libraries allows us to easily reuse them in subsequent designs.

Intellectual Property Libraries

Third-party intellectual property providers sell libraries containing complex design entities that we can use as modules in our designs.

PLD Vendor Libraries

PLD vendors provide precompiled design entities optimized for the architecture of a particular target PLD family. Entity and package declaration source code is provided with these libraries. However, source code for the architecture bodies and package bodies is usually not provided, only precompiled versions.

Also included in the VHDL compiler's system libraries are libraries containing the primitive component models used to functionally and physically model a particular vendor's PLD families. These models are used in post-synthesis and timing simulations.

15.4 USING LIBRARY UNITS

To use a library unit in another design unit, we must specify where the library unit is to be found. Both the logical name of the resource library containing the library unit and the library unit's name must be made visible to the design unit. This visibility can be provided directly or by selection.

Direct visibility of a library unit and its resource library is made possible by use of a context clause.

Context Clause

A design unit was defined as consisting of a context clause and a library unit. The *context clause,* at the head of a design unit, defines the name environment in which a design unit is compiled.

A context clause consists of a library clause and/or a use clause. For example, most of the design units in previous examples used signals and variables of type std_logic and were preceded by the context clause:

```
library ieee;
use ieee.std_logic_1164.all;
```

Library Clause

A *library clause* defines the logical names of design libraries that are visible to a given design unit. The scope of a library clause starts immediately after the library clause and extends to the end of the declarative region of the design unit in which the library clause appears. Within this scope, each library name defined by the library clause is directly visible.

Use Clause

A *use clause* makes particular declarations in a library directly visible to the library unit it precedes. In the previous example, the keyword **all** makes all the declarations in package STD_LOGIC_1164, in the library ieee, directly visible to the library unit it precedes.

A name made directly visible to a primary library unit by a context clause is automatically visible in any associated secondary library unit. As a result, a design library made visible to an entity declaration is also visible to all its associated architecture bodies.

For example, in Listing 2.6.1 there are two design units, the entity declaration half_adder and its associated architecture body dataflow2. The context clause at the beginning of the program is associated with the entity declaration library unit. The architecture body does not require its own context clause, because it is associated with the entity declaration half_adder, and every name made directly visible to half_adder is automatically visible to the architecture dataflow2, associated with half_adder.

In contrast, in Listing 2.9.1 the context clause preceding the entity declaration half_adder only applies to that entity declaration and its associated architecture body dataflow. Entity declaration or_2 in the design file requires its own context clause.

Library Units Made Visible by Selection

Library units can also be made visible by selection; that is, by using a selected name to reference the library unit instead of using a use clause. The selected name specifies the library name and the library unit's name. An appropriate library clause is still required. An example of this is discussed in "Specifying the Library as Part of an Entity Name" on page 579.

Implicit Context Clause

Because almost every VHDL description needs to use types and operators from package STANDARD, the contents of this package are automatically made visible to all library units. This is done by the *implicit context clause.* Every library unit (except package STANDARD itself) is assumed to be preceded by the implicit context clause:

```
library std, work;
use std.standard.all;
```

This implicit context clause defines two libraries that can be referenced in any library unit without requiring the use of an explicit context clause. The use clause makes all of the declarations in package STANDARD, in the library STD, directly visible to the design unit.

Library Work

The second library defined by the implicit context clause is the working library. This is the library to which any design unit is, by default, compiled. The logical name `work` can be used to reference the working library, regardless of the actual logical name of this library. Tools that allow us to create designs as projects may use the project name as an alias for the library `work`. When this is the case, we can refer to the working library by either name.

Organizing Design Units into Design Files

If an entity declaration and its associated architecture body are in separate design files, each requires a copy of the context clause. This allows each library unit to be independently compiled. The order of compilation is important; we must compile the entity declaration (primary unit) before its associated architecture body (secondary unit).

Since multiple design units in a single design file are compiled in the order of their textual occurrence, design units must be properly ordered in the file. Each primary design unit must precede any associated secondary design unit.

There are two rules that govern the order of compilation of a design unit:

1. A primary unit whose name is referenced within a given design unit must be compiled prior to compiling the given design unit.
2. A primary unit must be compiled prior to compiling any associated secondary units.

User written packages and other design units can be precompiled and placed in a resource library. Design entities in a resource library can be used as modules by design entities in a design file. The appropriate context clauses for any resource libraries used must be included prior to each library unit in a design file.

Compilation Order for Separate Files

An organization of design units that is appropriate to complex designs and to designs developed by groups of engineers uses multiple design files. As seen previously, the full-adder design could be separated into three design files: one for the half adder, one for the OR gate, and one for the top-level entity.

These separate design files have to be compiled in the appropriate order. If entity declarations and architecture bodies are placed in separate design files and there are multiple architecture bodies for a single entity declaration, then the desired architecture body must be compiled last, or a configuration declaration must be used.

15.5 DIRECT DESIGN ENTITY INSTANTIATION

Modular and hierarchical designs are implemented in VHDL using structural style architectures. A structural architecture is, ultimately, a collection of design entities interconnected by signals.

Closer Look at Structural Design

In a structural style description, component instantiation statements are used to create an instance of each design entity. A simplified syntax for a component instantiation statement was introduced in Chapter 2 in Figure 2.6.2.

The syntax in Figure 2.6.2 provides two forms for a component instantiation statement. The first form is for direct instantiation of a design entity. The second form is for indirect instantiation. Indirect instantiation instantiates a component, which serves as a placeholder for a design entity rather than directly instantiating a design entity. The binding of each component to an actual design entity is then accomplished using either default binding or an explicit binding indication.

Direct Design Entity Instantiation

Using direct design entity instantiation, a component instantiation statement directly specifies the name of the design entity and, optionally, the name of the associated architecture body. The statement's port map specifies how each port of this instance of the design entity is either connected to a signal in the enclosing architecture body or an expression, or is left unconnected (using the **`open`** keyword).

Consider the following example. Assume that we have created separate design files for each of the following entity declarations and architecture bodies and compiled them into our working library in the order listed:

- Entity `and_2`
- Architecture `dataflow` of `and_2`
- Architecture `behavioral` of `and_2`
- Entity `or_2`
- Architecture `dataflow` of `or_2`

We now want to use these entity declarations and architecture bodies in a structural style description of the simple combinational circuit in Figure 15.5.1. The description is given in Listing 15.5.1.

This description contains three component instantiation statements. Each one specifies the direct instantiation of a design entity. Two copies of the `and_2` design entity and one copy of the `or_2` design entity are instantiated.

The port maps specify how each design entity's ports are connected to the signals of architecture `structural_1`. Each association element in the port map consists of a design entity port name to the left of the arrow symbol (`=>`) and a signal of the enclosing architecure to the right. Port maps were previously discussed in the section "Port Map" on page 61.

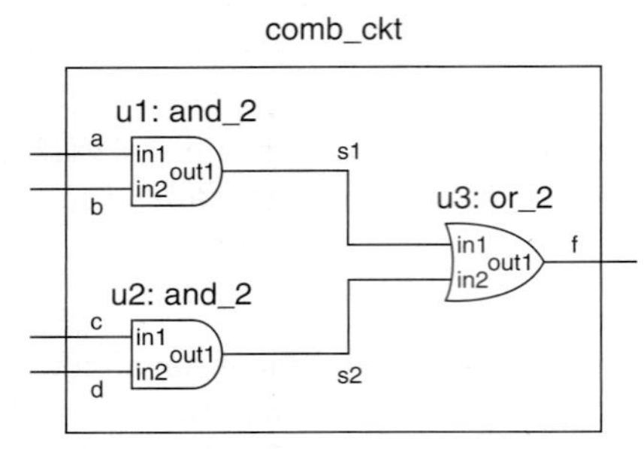

FIGURE 15.5.1
Logic diagram of a simple combinational circuit.

Binding

Binding is the process of associating a design entity and, optionally, a specific architecture body with an instance of a component. To complete the elaboration of a design, for simulation or synthesis, each component must be bound.

Using direct design entity instantiation, the component instantiation statement directly specifies the design entity's name. This provides only part of the information needed to complete the binding.

In Listing 15.5.1, none of the component instantiation statements explicitly specifies the library in which to find the entity declaration and architecture body that comprise each design entity. Nor is the specific architecture body for each design entity specified.

LISTING 15.5.1
Structural description of the combinational circuit in Figure 15.5.1 using direct instantiation of design entities.

```
library ieee;
use ieee.std_logic_1164.all;

entity comb_ckt is
   port( a, b, c, d : in std_logic; f : out std_logic );
end comb_ckt;

architecture structural_1 of comb_ckt is
signal s1, s2 : std_logic;
begin

   u1: entity and_2 port map (in1 => a, in2 => b, out1 => s1);
   u2: entity and_2 port map (in1 => c, in2 => d, out1 => s2);
   u3: entity or_2 port map (in1 => s1, in2 => s2, out1 => f);

end structural_1;
```

Default Binding Rules

The compiler can follow a set of *default binding rules,* defined in the LRM, to accomplish bindings. For example, since no library is specified in Listing 15.5.1, by default the compiler looks in the working library (work) for an entity declaration whose entity name and interface match those specified in the component instantiation statement. If, in the working library, there is more than one architecture body associated with that entity declaration, then the compiler uses the one most recently compiled.

Matching Interfaces

For the purposes of default binding, the interface of a design entity in a library matches that of the design entity in a component instantiation statement, if the generics and ports have the same names and the modes and types of these generics and ports are appropriate for the association.

For entity and_2 in Listing 15.5.1, there are two possible architectures: dataflow and behavioral. Assuming that behavioral was the most recently compiled, instances u1 and u2 will both use that architecture body. There is only one architecture in the working library for or_2, so there is no choice to be made.

Specifying an Architecture as Part of an Entity Name

If we want a binding other than the default binding, we can specify the desired architecture with the entity name in the component instantiation statement. For example, if we want to use architecture dataflow for u1 and architecture behavioral for u2 we can write:

```
u1: entity and_2(dataflow) port map ( ... );
u2: entity and_2(behavioral) port map ( ... );
```

The desired architecture is specified in parentheses following the entity name.

Specifying the Library as Part of an Entity Name

The library that holds the design entity can be specified in the component instantiation statement by using a selected name. Since the default library is work, the first part of the previous component instantiation for u1 is equivalent to:

```
u1: entity work.and_2(dataflow) port map ( ... );
```

The library is specified in the selected name as a prefix (preceding the dot) to the entity name. If the desired design entity is in a library other than work, we simply specify that library.

If for u1 we wanted to use an entity declaration and_2 and an architecture dataflow2 that are both in a library named parts, we could write:

```
u1: entity parts.and_2(dataflow2) port map (.. );
```

In this example, we would have to include an appropriate library clause at the beginning of Listing 15.5.1:

```
library parts;
```

Alternatively, instead of specifying the library name in the component instantiation statement, we can include a use clause with the library clause:

```
library parts;
use parts.and_2;
```

A disadvantage when using direct design entity instantiation is that, if we want to change the architecture body associated with an entity declaration or change the design entity itself, we must edit the component instantiation statement and recompile the file. This editing becomes bothersome in complex designs, particularly when we wish to evaluate a number of alternative architectures in a design.

A more flexible, but less concise, approach is indirect design entity instantiation using components.

15.6 COMPONENTS AND INDIRECT DESIGN ENTITY INSTANTIATION

Indirect design entity instantiation uses a placeholder, called a component, to stand for the design entity in a component instantiation statement. Thus, a component can be viewed as a virtual design entity.

Using components can provide many advantages for design management. For example, in top-down design, we can write a structural description of the first-level partition of the design using components. These components represent the ideal design entities of which we would like to compose the design. They can then serve as a basis for the specifications of the actual design entities that will be developed.

The syntax for the form of a component instantiation statement that uses components is the second form in Figure 2.6.2.

A structural description using components describes how the components are interconnected. For example, a structural description of the combinational circuit in Figure 15.5.1 using components is given in Listing 15.6.1.

LISTING 15.6.1

Structural description of the combinational circuit in Figure 15.5.1 using indirect instantiation of design entities.

```
library ieee;
use ieee.std_logic_1164.all;

entity comb_ckt is
   port( a, b, c, d : in std_logic; f : out std_logic );
end comb_ckt;


architecture structural_2 of comb_ckt is
```

```
component and_2 is    -- component declaration
   port (in1, in2 : in std_logic; out1 : out std_logic);
end component;

component or_2 is    -- component declaration
   port (in1, in2 : in std_logic; out1 : out std_logic);
end component;

signal s1, s2 : std_logic;
begin
      -- component instantiations
   u1: component and_2 port map (in1 => a, in2 => b, out1 => s1);
   u2: component and_2 port map (in1 => c, in2 => d, out1 => s2);
   u3: component or_2 port map (in1 => s1, in2 => s2, out1 => f);

end structural_2;
```

There are two fundamental differences in Listing 15.6.1 and Listing 15.5.1. First, each component in Listing 15.6.1 must be declared before it can be used. Second, each instantiation statement is an instance of a component, not an instance of a design entity.

Component Declarations

A component declaration is placed in the declarative part of the architecture in which the component is used. Alternatively, if a component is likely to be used in multiple designs, its declaration can be placed in a package in a library. Using a package eliminates the need for us to rewrite a component's declaration in each architecture that uses the component.

The simplified syntax for a component declaration was given in Chapter 2 in Figure 2.6.4. As seen in Listing 15.6.1, component declarations are very similar to entity declarations, in that they specify the component's or entity's interface, respectively. There are two components declared in Listing 15.6.1, component `and_2` and component `or_2`.

Instantiations (Instances) of Components

The three component instantiation statements in Listing 15.6.1 use the two components declared in the declarative part of the architecture. Component `and_2` is instantiated twice. The instantiation statements include the keyword **`component`** to make it clear that a component is being instantiated rather than a design entity. This optional keyword is usually omitted from component instantiation statements.

Components as Indirect Specifications of Design Entities

Use of components can be viewed as using indirection in specifying design entities in a structural description. As shown in Figure 15.6.1, the top-level design entity `comb_ckt` is represented by the outer rectangle. Its interface is defined by its entity declaration. Its architecture body consists of three components, represented by the three inner rectangles. Each component's interface is defined by its declaration. The

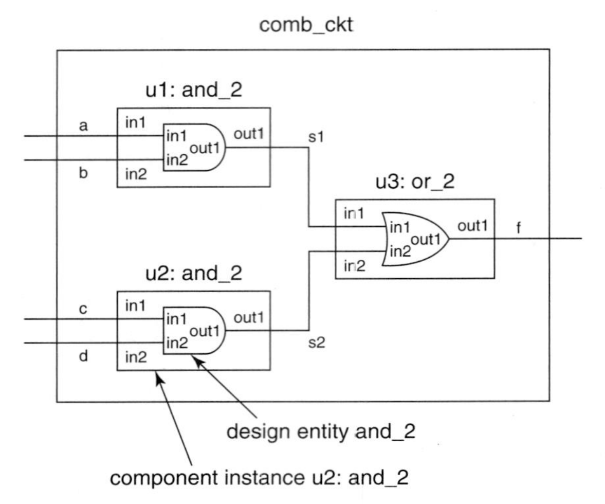

FIGURE 15.6.1
Structural diagram of the combinational circuit illustrating the interconnection of design entities via components.

interconnection of the components is defined by the component instantiation statements. Finally, the binding of components to design entities is accomplished by the default binding. In this example, the design entities are represented by distinctive-shape logic symbols.

Binding Alternatives for Components

Each component instance must, ultimately, be bound to a design entity to complete a design. For components, there are three ways that binding can be accomplished:

- Default binding
- Configuration specification
- Configuration declaration

In contrast to default binding, use of a configuration specification or a configuration declaration allows us to explicitly specify bindings. Configuration specifications are placed in the declarative part of the architecture in which the components are instantiated. Thus, the binding information is provided in the architecture. Configuration specifications are discussed in this section.

A configuration declaration is a design unit. Thus, when using a configuration declaration, the bindings are specified separately from the architecture containing the components. A single configuration declaration can be used to specify the bindings of all of the components in a multilevel hierarchical structural design. Configuration declarations are discussed in the next section.

A single design might bind different components using all three ways.

Default Bindings for Component Instances

When the binding of design entities to components is not explicitly specified, the compiler follows VHDL's default binding rules. The compiler looks for the entity declaration and architecture body in the visible libraries. The compiler first looks in the working library for an entity declaration with the same name as the component name and a matching interface. Where more than one architecture is available for the entity declaration, the most recently compiled architecture is used. If the entity declaration is not found in the working library, then the resource libraries are searched in order. If a design is created so that components and their corresponding design entities have matching names and interfaces, then it is not necessary to use either configuration specifications nor configuration declarations.

Configuration Specifications

A *configuration specification* explicitly specifies the binding of a component to a design entity. It associates *binding information* with *labels* representing specific instances of a given component. It specifies for a particular component instance where the entity to be bound to is to be found, its name, architecture, and how the component's ports are to be associated with the design entity's ports. The configuration specification is put in the declarative part of the architecture body.

The simplified syntax of a configuration specification is given in Figure 15.6.2.

A configuration specification starts with the keyword `for`. Following this keyword are a component specification and a binding indication. The *component specification* specifies the instances of a particular component that the configuration applies to. These instances are specified as either a list of component labels, the keyword `all`, or the keyword `others`, followed by a colon and then the component's name.

Binding Indication

The *binding indication* in a configuration specification specifies any or all of the design entity, generic map, and port map. The entity_aspect of a binding indication allows either an entity or a configuration to be specified. Generics are discussed later in this chapter.

A design description of the combinational circuit using configuration specifications is given in Listing 15.6.2.

configuration_specification ::=
for component_specification binding_indication ;

component_specification ::= instantiation_list : *component*_name
instantiation_list ::= *instantiation*_label { , *instantiation*_label } | **others** | **all**
binding_indication ::= [**use** entity_aspect] [generic_map_aspect] [port_map_aspect]
entity_aspect ::= **entity** *entity*_name [(*architecture*_identifier)]
| **configuration** *configuration*_name | **open**

FIGURE 15.6.2
Simplified syntax for a configuration specification.

LISTING 15.6.2
Structural design of the combinational circuit using components and configuration specifications.

```
library ieee;
use ieee.std_logic_1164.all;

entity comb_ckt is
   port( a, b, c, d : in std_logic; f : out std_logic );
end comb_ckt;


architecture structural_3 of comb_ckt is

   component and_2 is
      port (in1, in2 : in std_logic; out1 : out std_logic);
   end component;

   component or_2 is
      port (in1, in2 : in std_logic; out1 : out std_logic);
   end component;

   signal s1, s2 : std_logic;

   -- configuration specifications
   for u1 : and_2 use entity work.and_2(dataflow);
   for u2 : and_2 use entity work.and_2(behavioral);
   for u3 : or_2 use entity work.or_2(dataflow);

begin

   u1: and_2 port map (in1 => a, in2 => b, out1 => s1);
   u2: and_2 port map (in1 => c, in2 => d, out1 => s2);
   u3: or_2 port map (in1 => s1, in2 => s2, out1 => f);

end structural_3;
```

Three configuration specifications appear in the declarative part of the architecture.

```
for u1 : and_2 use entity work.and_2(dataflow);
for u2 : and_2 use entity work.and_2(behavioral);
for u3 : or_2 use entity work.or_2(dataflow);
```

For each component instance, a configuration specification specifies the entity to which it is to be bound.

If we wanted to bind component instances u1 and u2 to the same design entity, say work.and_2(behavioral), we could write the configuration specifications as:

```
for u1,u2 : and_2 use entity work.and_2(behavioral);
for u3 : or_2 use entity work.or_2(dataflow);
```

or since all occurrences of component and_2 would now be the same we could write:

```
for all : and_2 use entity work.and_2(behavioral);
for u3 : or_2 use entity work.or_2(dataflow);
```

The keyword **all** in this example means all instances of and_2 components.

Let us carefully consider the following example. The configuration specification for u3 could have been written as

```
for u3 : or_2 use entity work.and_2(dataflow);
```

and no compilation error would occur. However, this says that for the u3 instance of component or_2, we want to bind an and_2 design entity. Since the and_2 design entity and the or_2 component declaration have matching interfaces, there is no syntax error. But, the function of the resulting combinational circuit has been changed; instead of being a two-level AND/OR circuit, it is now functionally a two-level AND/AND circuit (logically equivalent to a single four-input AND gate). The effect of using this configuration specification is shown in Figure 15.6.3.

This example illustrates that, when using components, it is easy to change not only the implementation of a function by changing the architecture, but change the function itself.

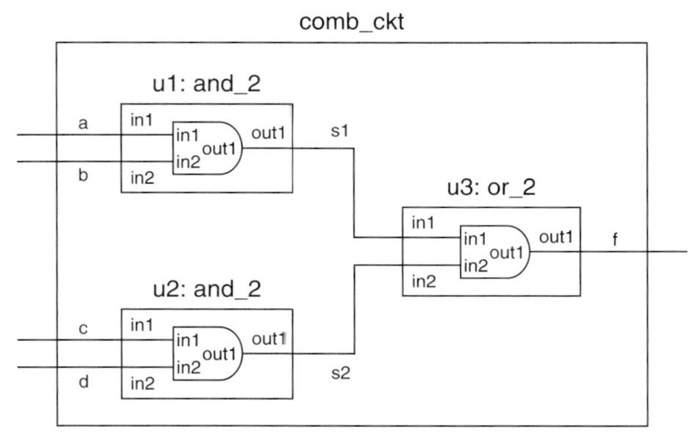

FIGURE 15.6.3
Logic circuit with and_2 design entity bound to u3 or_2 component instance by a configuration specification.

Port Maps in Configuration Specifications

A port map can be included in a configuration specification to configure a design entity to be compatible with a component declaration even though their interfaces do not exactly match. For example, assume that we did not have a design entity `or_2` to bind to the component `or_2`, but we did have a three-input OR design entity in the library `parts`. Assume that the entity declaration for the three-input OR `or_3` is:

```
entity or_3 is
      port( inp1, inp2, inp3 : in std_logic;
            outp1 : out std_logic );
end or_3;
```

We can change the configuration specification for `u3` in Listing 15.6.2 to:

```
for u3 : or_2 use entity parts.or_3(dataflow)
port map (inp1 => in1, inp2 => in2, inp3 => '0',
            outp1 => f);
```

This configuration specification includes a port map that maps the ports of the design entity `or_3` to the ports of the `or_2` component. The third input of the `or_3` entity is mapped to the expression `'0'` (Figure 15.6.4).

A library clause that makes the library `parts` visible is required. However, a use clause is not required, since that is taken care of in the configuration specification.

The ability to use a port map in a configuration specification to bind a component instance to a design entity whose interface does not exactly match is very useful. It allows us to maximize the reuse of some design entities without modifying their original code.

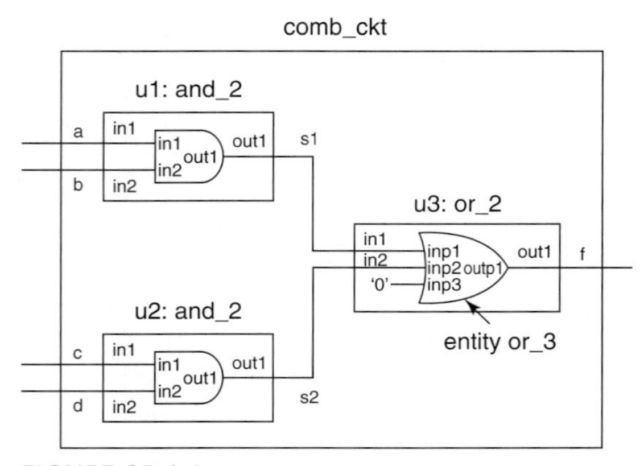

FIGURE 15.6.4
Logic circuit with or_3 design entity bound to u3 or_2 component instance by a configuration specification containing a port map.

15.7 CONFIGURATION DECLARATIONS

Using configuration specifications, as in the previous section, is advantageous when we don't intend to change our component-instance/design-entity bindings. Greater flexibility in specifying bindings can be achieved using a configuration declaration.

A configuration declaration is a design unit that allows bindings of architecture bodies to entity declarations, as well as bindings of components to design entities, to be specified. Since a configuration declaration is a separate design unit, these bindings are specified outside of the architecture body.

Configuration Declaration Syntax

The simplified syntax for a configuration declaration is given in Figure 15.7.1. Configuration declarations can be quite complex. We will consider some straightforward examples.

Specifying an Architecture to Be Bound

The simplest use of a configuration declaration is to specify the binding of an architecture to a design entity when the architecture is not structural (does not contain any components). Suppose that we had a second architecture for `comb_ckt` that was a dataflow architecture named `dataflow`. We now have two different architectures for design entity `comb_ckt`. If we simulated `comb_ckt`, by default, the most recently compiled architecture would be used.

Suppose that we wanted to simulate `comb_ckt` using the `dataflow` architecture and it was not the most recently compiled. We could create a configuration declaration to specify that architecture `dataflow` be used. This configuration declaration is given in Listing 15.7.1.

The first line of the configuration declaration in Listing 15.7.1 consists of keyword **`configuration`**, followed by the name `config1`, which we have chosen to give this configuration. Following keyword **`of`** is the name of the entity being configured.

The second line specifies the name of the architecture, `dataflow`, that is bound to the design entity.

```
configuration_declaration ::=
configuration identifier of entity_name is
    for architecture_name
            { for component_specification binding_indication ;
              end for ; }
    end for ;
end [ configuration ] [ configuration_simple_name ] ;
```

FIGURE 15.7.1
Simplified syntax for a configuration declaration.

LISTING 15.7.1
Configuration declaration for design entity comb_ckt specifying architecture dataflow.

```
configuration config1 of comb_ckt is
   for dataflow       --block configuration
   end for;
end config1;
```

Since a configuration declaration is a design unit, we can place the configuration declaration by itself in a design file. After this file has been compiled, if we want to simulate the description of comb_ckt that uses the dataflow architecture, we simply simulate the configuration declaration config1.

Component Configurations

To write a configuration declaration for design entity comb_ckt that uses architecture structural_2 in Listing 15.6.1, we must not only specify the architecture, but also the bindings of components in the architecture to design entities. This is accomplished using *component configurations* inside the configuration declaration. A component configuration inside of a component declaration allows us to configure the internal structure of the architecture of an entity to which a component is bound; a configuration specification does not allow this.

The simplified syntax of a component configuration is like that of a configuration specification, except for the addition of the keywords **end for** (Figure 15.7.2).

The architecture in Listing 15.6.1 for structural_2 contained no configuration specifications, and its components were bound to design entities using default binding. Configuration declaration config2 in Listing 15.7.2 explicitly configures this architecture to produce the same bindings that were produced using configuration specifications in Listing 15.6.2.

Configuration declaration config2 consists of a block configuration specifying the architecture structural_2. Inside this block configuration are three component configurations, one for each component instance.

Configuration Declarations for a Testbench

Configurations are most often used in simulation as opposed to synthesis. To illustrate this, let's return to our simple half-adder example. Assume that we have written

component_configuration ::=
for component_specification binding_indication ;
end for ;

binding_indication ::= [**use** entity_aspect] [generic_map_aspect] [port_map_aspect]

FIGURE 15.7.2
Simplified syntax for a component configuration.

LISTING 15.7.2
Configuration declaration for structural architecture for comb_ckt.

```
configuration config2 of comb_ckt is
   for structural_2       -- block configuration
      for u1 : and_2      -- component configuration
         use entity work.and_2(dataflow);
      end for;
      for u2 : and_2      -- component configuration
         use entity work.and_2(behavioral);
      end for;
      for u3 : or_2       -- component configuration
         use entity work.or_2(dataflow);
      end for;
   end for;
end config2;
```

a dataflow description of a half-adder, and also have written a testbench named `testbench` that is suitable for performing either a functional, post-synthesis, or timing simulation. The architecture for the entity `testbench` is named `behavior`. In its declarative part, we declare a component named `half_adder`. In the architecture body's statement part, component `half_adder` is instantiated having the label `UUT`.

Configuration for Functional Simulation

When we compile the half-adder description, we get the design entity `half_adder(dataflow)`. We can write a configuration declaration named `functional_sim` for design entity `testbench` that binds the instance `UUT` of component `half_adder` to design entity `half_adder(dataflow)` (Listing 15.7.3). We can then simulate configuration declaration `functional_sim` to perform a functional simulation. Note that, when using a configuration declaration to simulate the UUT, we are not actually simulating the testbench, but rather the configuration declaration for the testbench.

Configuration for Post-Synthesis Simulation

When we synthesize the half-adder description, we get a gate-level model of the design entity. If we examine the text of this model, we can see the name that the synthesizer has given the resulting gate-level architecture. Assume that this architecture

LISTING 15.7.3
Configuration declaration for a functional simulation of a half adder.

```
configuration functional_sim of testbench is
   for behavior
      for UUT : half_adder
         use entity work.half_adder(dataflow);
      end for;
   end for;
end functional_sim
```

LISTING 15.7.4
Configuration declaration for a post-synthesis simulation of a half adder.

```
configuration postsyn_sim of testbench is
   for behavior
      for UUT : half_adder
         use entity work.half_adder(beh);
      end for;
   end for;
end postsyn_sim
```

is named `beh`. We can then write a configuration declaration named `postsyn_sim` for design entity `testbench` that binds instance `UUT` of component `half_adder` to the gate-level design entity `half_adder(beh)` (Listing 15.7.4). We can then simulate configuration declaration `postsyn_sim` to perform a post-synthesis simulation.

Configuration for Timing Simulation

When we place and route the half-adder's EDIF netlist that resulted from the synthesis, we get a timing-model design entity for the half adder placed and routed to the target PLD. If we examine the text of this model, we can see the name that the place-and-route tool has given the resulting timing model architecture. Assume that this architecture is named `structure`. We can write a configuration declaration named `timing_sim` for design entity `testbench` that binds instance `UUT` of component `half_adder` to timing model design entity `half_adder(structure)` (Listing 15.7.5). We can then simulate configuration declaration `timing_sim` to perform a timing simulation.

In summary, with three different half-adder architectures—a functional model, a gate-level model, and a timing model—we can use one of the three different configurations to bind component `UUT` in the testbench to the desired half-adder design entity and architecture.

Using this approach, the configuration declaration becomes the top level in the simulation. Depending on which simulation we wish to perform, we simply simulate the appropriate configuration declaration. The instantiation statement for the UUT in the testbench does not need to be modified, saving us time.

LISTING 15.7.5
Configuration declaration for a timing simulation of a half adder.

```
configuration timing_sim of testbench is
   for behavior
      for UUT : half_adder
         use entity work.half_adder(structure);
      end for;
   end for;
end timing_sim
```

LISTING 15.7.6
Multilevel structural architecture for a half adder.

```
use work.gates.all;
architecture structural of half_adder is  -- Architecture body for half adder

   component xor_2
      port (in1, in2 : in std_logic;
         out1: out std_logic);
   end component;

begin
   u1: component xor_2 port map (in1 => a, in2 => b, out1 => sum);
   u2: component and_2 port map (i1 => a, i2 => b, o1 => carry_out);

end structural;
```

Configuring Multilevel Hierarchies

In the previous half-adder example, the original description of half_adder was bound to a behavioral architecture. As a result, the testbench had a single level of partitioning and the UUT was instantiated at that level. If, instead, the half adder were bound to a structural architecture, there would be an additional level of partitioning, resulting in a multilevel hierarchy.

For example, assume that components and_2, or_2, and invert are declared in a package named gates. A structural architecture for the half adder is given in Listing 15.7.6.

This architecture instantiates two components. Component xor_2 is declared in the declarative part of the architecture. Component and_2 is declared in package gates.

Assume that the only behavioral design entities we have are named and_2x, or_2x, and invert2x, and they are compiled to library work. Component and_2 can be bound to design entity and_2x, but this cannot be done by default, because the component and entity names are different.

For the XOR function, assume we have the structural description in Listing 15.7.7.

LISTING 15.7.7
Structural description of an XOR function.

```
library ieee;
use ieee.std_logic_1164.all;

entity xor_2x is
   port(
      in1, in2 : in std_logic;
      out1 : out std_logic
      );
end xor_2x;
```

(Cont.)

LISTING 15.7.7 *(Cont.)*

```
use work.gates.all;
architecture structural of xor_2x is
   signal s1, s2, s3, s4 : std_logic;
begin

   u1: component invert port map (i => in2, o => s1);
   u2: component invert port map (i => in1, o => s2);
   u3: component and_2 port map (i1 => in1, i2 => s1, o1 => s3);
   u4: component and_2 port map (i1 => s2, i2 => in2, o1 => s4);
   u5: component or_2 port map (i1 => s3, i2 => s4, o1 => out1);

end structural;
```

The components in this description can each be explicitly bound to one of the existing behavioral design entities, `and_2x`, `or_2x`, and `invert2x`, in library `work`.

A more complex version of the syntax for a configuration declaration is given in Figure 15.7.3.

This version of the syntax for a configuration declaration has a recursive form with respect to the block configuration.

The configuration declaration in Listing 15.7.8 uses a single configuration declaration to configure the multilevel hierarchy of `half_adder` to use the behavioral design entities `and_2x`, `or_2x`, and `invert2x`.

Separate Configuration Declarations for Multilevel Hierarchies

Rather than use the recursive capability of a single configuration declaration, a simpler approach to configuring a multilevel architecture is to use separate configuration declarations. This approach makes the overall configuration of a multilevel hierarchy easier to write and comprehend.

configuration_declaration ::=
configuration identifier **of** *entity*_name **is**
    block_configuration
end [**configuration**] [*configuration*_simple_name] ;

block_configuration ::=
    for *architecture*_name
        { **for** component_specification [binding_indication ;]
            [block_configuration]
        end for ;}
end for ;

FIGURE 15.7.3
Syntax for configuration declaration.

LISTING 15.7.8
Configuration for multilevel structural half-adder architecture.

```
configuration functional_sim2 of testbench is
   for behavior   -- architecture of testbench
      for UUT : half_adder use entity work.half_adder(structural);
         for structural    -- architecture of half adder
            for u2: and_2 use entity work.and_2x(dataflow);
            end for;
            for u1 : xor_2 use entity work.xor_2x(structural);
               for structural    -- architecture of xor_2x
                  for u1, u2 : invert use entity work.invertx(dataflow);
                  end for;
                  for u3, u4 : and_2 use entity work.and_2x(dataflow);
                  end for;
                  for u5 : or_2 use entity work.or_2x(dataflow);
                  end for;
               end for;
            end for;
         end for;
      end for;
   end for;
end functional_sim2
```

We can write separate configuration declarations for each structural architecture. For example, a configuration for entity `xor_2x` is given in Listing 15.7.9.

We can then write a configuration declaration for `half_adder` that takes advantage of the existing configuration `xor_2x_config` for `xor_2x`. This is possible because a binding indication can use a configuration. A configuration declaration for `half_adder` that uses the configuration declaration for `xor_2x` to specify the binding of component `u1` of the `half_adder` is given in Listing 15.7.10.

LISTING 15.7.9
Configuration declaration for entity xor_2x.

```
--Entity-Architecture Pair Configuration
configuration xor_2x_config of xor_2x is
   for structural    -- architecture of xor_2x
      for u1, u2 : invert use entity work.invertx(dataflow);
      end for;
      for u3, u4 : and_2 use entity work.and_2x(dataflow);
      end for;
      for u5 : or_2 use entity work.or_2x(dataflow);
      end for;
   end for;
end xor_2x_config;
```

LISTING 15.7.10
Configuration declaration for half_adder that uses an existing configuration declaration for xor_2x.

```
configuration functional_sim3 of testbench is
   for behavior    -- architecture of testbench
      for UUT : half_adder use entity work.half_adder(structural);
         for structural    -- architecture of half adder
            for u2: and_2 use entity work.and_2x(dataflow);
            end for;
            for u1 : xor_2 use configuration work.xor_2x_config;
            end for;
         end for;
      end for;
   end for;
end functional_sim3
```

Using separate configurations in this fashion makes the overall configuration of a multilevel design easier to understand.

15.8 COMPONENT CONNECTIONS

One of the purposes of a component instantiation statement is to specify the mappings between the formal signals of an instance of a component and the actual signals of the encompassing architecture. In this section we consider the interconnection of components in greater detail. While the discussion is in terms of component instantiation statements that use direct design entity instantiation, the same considerations apply to design entities that are indirectly instantiated via components.

Component Port Connections

Each port of a design entity is a *formal port*. For example, consider the entity declaration for a three-input AND function:

```
entity and_3 is
      port (i1, i2, i3 : in std_logic;
            o1: out std_logic);
end and_3;
```

The formal ports of entity `and_3` are: `i1`, `i2`, `i3`, and `o1`.
The following entity declaration for an inverter

```
entity invert is
      port (i: in std_logic; o: out std_logic);
end invert;
```

has two formal ports `i` and `o`.

Port Map Actuals

The *port map* for a component instantiation describes how a design entity is connected, as a component, in a larger structure in which it is used. In a component instantiation, each formal port of the component is associated with an actual of the enclosing architecture. An *actual* can be an expression, a signal, or a port.

Expression Actual

An expression is used to provide a port with a constant driving value. Accordingly, an expression can only be an actual for a formal port of mode **`in`**.

Signal Actual

A local signal, declared in the enclosing design entity's architecture body, has no mode and can, therefore, be used as an actual for a formal port of any mode. This is illustrated in Figure 15.8.1. In this figure, component u1's output port is connected to a signal. In u1's port map, its output port is a formal that is associated with an actual that is this signal. In u2's port map, its input port is a formal associated with an actual that is the same signal.

Enclosing Design Entity Ports as Actuals

A port is also a signal, but a port has a mode. A formal port of the enclosing design entity can be associated as the actual for a formal port of a component.

In Figure 15.8.1, component u1's input port is directly connected to the input port of the enclosing design entity, without the need for a local signal. This is represented by the dotted line in the figure. In u1's port map, its input port is a formal that is associated with an actual that is the formal input port of the enclosing design entity.

Component u2's output port is directly connected to the output port of the enclosing design entity. In u2's port map, its output port is a formal that is associated with an actual that is the formal output port of the enclosing design entity.

When a formal port is associated with an actual that is itself a port (of the enclosing design entity), the restrictions in Table 15.8.1 apply.

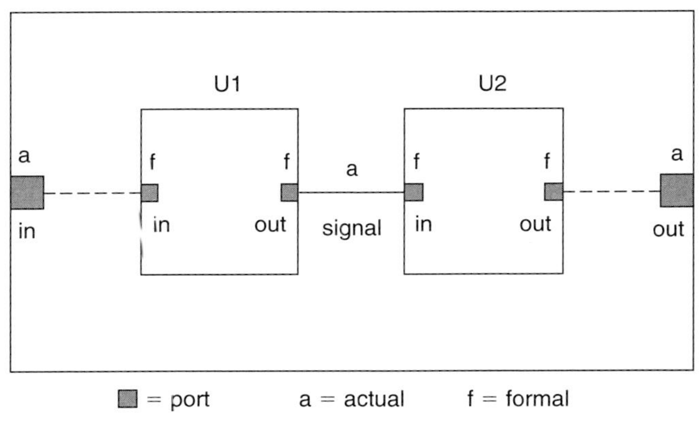

FIGURE 15.8.1
Graphical representation of association of component formals with actuals.

Table 15.8.1
Allowed association of actuals that are ports to formal ports.

Formal port (component design entity)	Actual port (enclosing design entity)
in	in, inout, buffer
out	out, inout, buffer
inout	inout, buffer
buffer	out, inout, buffer
linkage	in, out, inout, buffer, linkage

Component-to-Component Connections

According to these rules, an input of a component can be connected directly to an input of its enclosing design entity, but can only be connected through a local signal to an output of another component. Likewise, an output of a component can be connected directly to an output of its enclosing design entity, but can only be connected through a signal to an input of another component.

For example, consider the structural implementation of a two-input NAND function from `and_3` and `invert` design entities. A logic diagram for such a structure is shown in Figure 15.8.2.

Since the objective is to create a two-input NAND function, but we have a three-input AND component, one of its inputs is unused. For the three-input AND to function as a two-input AND, the unused input needs to have a constant driving value of 1. Thus, this input must be associated with an expression that is always equal to `'1'`.

The other two inputs of the `and_3` design entity are associated with the two input ports of its enclosing design entity. The input ports of the enclosing design entity are the actuals for these two formal input ports of component `and_3`. Each of these two associations corresponds to a formal port of mode **`in`** being associated with an actual port of mode **`in`**. This is acceptable within the restrictions of Table 15.8.1.

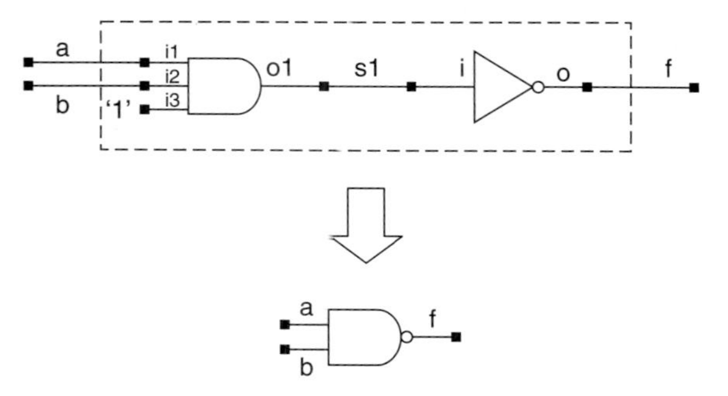

FIGURE 15.8.2
Logic diagram of the structural implementation of a two-input NAND.

The output of the `and_3` component needs to be connected to the input of the `invert` component. A direct connection would (from the viewpoint of the `and_3` component) correspond to a formal port of mode **`out`** being associated with an actual port of mode **`in`**. Since this is not allowed (Table 15.8.1), the output of `and_3` must be connected to a signal. This requires the declaration of a local signal, which we call `s1`. Accordingly, the component instantiation statement for the `and_3` is:

```
u1: entity and_3
      port map (i1 => a, i2 => b, i3 => '1',
                        o1 => s1);
```

In writing the component instantiation statement for the inverter, its formal port of mode **`in`** must be connected to a signal, since it cannot be directly connected to an actual port of mode **`out`** (Table 15.8.1). Accordingly, it is connected to signal `s1`. The actual for the inverter's formal port of mode **`out`** is the output port of the enclosing design entity. This results in the following instantiation statement:

```
u2: entity invert port map (i => s1, o => f);
```

The complete design file is given in Listing 15.8.1.

LISTING 15.8.1
Design file for structural implementation of a two-input NAND function.

```
------------------------- 3 input AND gate component -------------------------
library ieee;
use ieee.std_logic_1164.all;

entity and_3 is-- Entity declaration for 2 input and
   port (i1, i2, i3 : in std_logic;
      o1: out std_logic);
end and_3;

architecture dataflow of and_3 is-- Architecture body for 3 input and
begin
   o1 <= i1 and i2 and i3;
end dataflow;

------------------- Inverter entity
library ieee;
use ieee.std_logic_1164.all;

entity invert is
   port (i: in std_logic; o: out std_logic);
end invert;
```

(Cont.)

LISTING 15.8.2 *(Cont.)*

```
architecture dataflow of invert is
begin
   o <= not i;
end dataflow;

------------------------- 2-input NAND structure --------------------------
library ieee;
use ieee.std_logic_1164.all;

entity nand_2 is-- Entity declaration for half adder
   port (a, b : in std_logic;
      f: out std_logic);
end nand_2;

architecture structure of nand_2 is-- Architecture body for half adder

signal s1 : std_logic;

begin
   u1: entity and_3 port map (i1 => a, i2 => b, i3 => '1', o1 => s1);
   u2: entity invert port map (i => s1, o => f);

end structure;
```

Open Keyword

If a formal port is associated with an expression, signal, or actual port, then the formal port is said to be *connected.* If a formal port is instead associated with the keyword **open**, then the formal is said to be *unconnected.* A port of mode **in** may be unconnected or unassociated only if its declaration includes a default expression. A port of any mode other than **in** may be unconnected or unassociated, as long as its type is not an unconstrained array type.

If a port is simply omitted from the association list, it is considered open. If an association element is omitted from an association list for this purpose, all subsequent associations in the list must be named associations. Using keyword **open** is preferable to simply omitting a port, because keyword **open** makes it clear that the port was intentionally left open.

15.9 PARAMETERIZED DESIGN ENTITIES

Design entities are more widely reusable if they are parameterized. Parameters allow a design entity to be described generically so that, for each use of the design entity, its structure and/or behavior is altered by the choice of parameter values. In VHDL these parameters are called *generics*. Generics were introduced in the section

entity_declaration ::=
entity identifier **is**
  [**generic** (*generic*_interface_list) ;]
  [**port** (*port*_interface_list) ;]
end [**entity**] [*entity*_simple_name] ;

*generic*_interface_list ::=
[**constant**] identifier_list : [**in**] subtype_indication [:= *static*_expression]
{[**constant**] identifier_list : [**in**] subtype_indication [:= *static*_expression]}

FIGURE 15.9.1
Expanded syntax for an entity declaration that includes a formal generic clause.

"Timing Models Using VITAL and SDF" on page 288, where they were used to specify delays in timing models.

Default generic values for a design entity can be specified inside its entity declaration. If not overridden, these are the values of the generics. Recall that these values can be overridden from outside of the design entity in either the component instantiation statement or a configuration declaration.

The expanded syntax for an entity declaration that includes a formal generic clause is repeated in Figure 15.9.1. When generics are used, the generic clause precedes the port clause.

Once declared, a generic can be used in the entity declaration itself and in all architectures associated with the entity.

Counter Using Generics

In parameterizing design descriptions, generics are often used to specify the width of data objects. For example, consider the 12-bit counter from Listing 9.4.4. A parameterized version of the counter is given in Listing 15.9.1. A single generic `n` is used to specify the size (number of bits) of the counter.

LISTING 15.9.1
Parameterized binary counter.

```
library ieee;
use ieee.std_logic_1164.all;
use ieee.numeric_std.all;

entity binary_cntr is
   generic (n : positive := 12);     -- counter size (bits)
   port (clk, cnten, up, rst_bar: in std_logic;
      q: out std_logic_vector (n - 1 downto 0));
end binary_cntr;

architecture behav_param of binary_cntr is
begin
```

(Cont.)

LISTING 15.9.1 *(Cont.)*

```
   cntr: process (clk)
      variable count_v : unsigned(n - 1 downto 0);
   begin
      if rising_edge(clk) then
         if rst_bar = '0' then
            count_v := (others => '0');
         elsif cnten = '1'  then
            case up is
               when '1' => count_v := count_v + 1;
               when others => count_v := count_v - 1;
            end case;
         else
            null;
         end if;
         q <= std_logic_vector(count_v);
      end if;
   end process;
end behav_param;
```

The first statement in the entity declaration is a formal generic clause:

```
generic (n : positive := 12);
```

The generic clause precedes the port clause. This generic clause lists a single generic named n of type positive that has a default value of 12. Specification of a default value is not required. The generic clause could have been written as:

```
generic (n : positive);
```

The port clause in the entity declaration uses generic n to specify the width of the counter's output q. In the architecture body, generic n is used to specify the width of the unsigned variable count_v. As a result, the number of bits in the counter can be set to any positive value.

Generic Map Aspect

When a design entity using generics is instantiated in a structural description, the instantiation statement can include a generic map aspect that assigns values to the generics. The syntax for a generic map aspect is given in Figure 15.9.2.

When the design entity binary_cntr is instantiated, if no other value is specified for the generic in the instantiation statement, a 12-bit counter is created by default:

```
u3: entity binary_cntr(behav_param)
      port map (clk => clk, cnten => cnten_sig,
            up => up_sig, rst_bar => rst_bar,
            q => count_sig);
```

```
generic_map_aspect ::=
generic map ( generic_association_list )

association_list ::=
[ formal_part => ] actual_part { , [ formal_part => ] actual_part }
```

FIGURE 15.9.2
Syntax for a generic map aspect.

Alternatively, we can give the counter any desired width by including a generic map aspect in the instantiation statement and specifying the desired value for n:

```
u3: entity binary_cntr(behav_param)
      generic map (n => 10)
      port map (clk => clk,cnten => cnten_sig,
            up => up_sig,rst_bar => rst_bar,
            q => count_sig);
```

The counter in this instantiation is a 10-bit counter, since the value of 10 assigned to n in the generic map overrides the default value of 12 in the counter's entity declaration. Note that there is no semicolon between the generic map aspect and port map aspect in a component instantiation.

Local Generic Clause in Component Declaration

If we are using generics, but are not using direct entity instantiation, as in the previous two instantiation statements, there must be a component declaration in the declarative part of the architecture that instantiates the component and this declaration must contain a local generic clause. For example:

```
component binary_cntr
      generic (n : positive);
      port (clk, cnten, up, rst_bar: in std_logic;
       q: out std_logic_vector (n - 1 downto 0));
end component;
```

PLD Vendor and Third-Party Parameterized Modules

PLD vendors and their third-party IP partners provide libraries containing parameterized functional modules that we can use as components in our designs. This speeds up the design process. These modules' functionalities range from gates to multiply accumulate subsystems. To provide scalability and adaptability, modules are configured using generics.

These module libraries fall into two categories: those containing architecture dependent modules and those containing architecture independent modules. Architecture dependent modules are optimized for the architecture of a particular PLD family. They are highly efficient in their use of the target PLD's logic resources and provide high performance. The drawback to using these modules is that they are not portable to other PLD families.

15.10 LIBRARY OF PARAMETERIZED MODULES (LPM)

To provide modules that are portable, but still architecturally efficient, a standard set of functional modules that could be supported by all PLD vendors and EDA tool vendors was defined. This standard, called the *Library of Parameterized Modules (LPM)*, is an extension of the Electronic Industries Association's EDIF format. Its objective is to provide a generic, technology-independent set of primitive modules. The intent is that each PLD vendor or EDA tool vendor provides a synthesis tool that efficiently maps LPM modules to the target technology. The target technology could be a PLD, gate array, or standard cell.

The LPM standard defines each module's interface and functionality. It includes a package named `lpm_components` that provides each module's VHDL component declaration. Similar information is provided for Verilog. It is left to each target technology vendor to provide a package body optimized for their target technology's architecture.

The 25 configurable modules defined by the standard are listed in Table 15.10.1.

Not all PLD vendors support the LPM standard. Those that do may not support all of the modules. For example, the pads, truth table, and FSM modules may not be supported.

Table 15.10.1
Modules in the library of parameterized modules.

	Module Name	Function
gates	lpm_and	AND gate
	lpm_bustri	three-state buffer
	lpm_clshift	combinational shifter
	lpm_constant	constant generator
	lpm_decode	decoder
	lpm_inv	inverter
	lpm_mux	multiplexer
	lpm_or	OR gate
	lpm_xor	XOR gate
arithmetic	lpm_abs	absolute value
	lpm_add_sub	adder/subtractor
	lpm_compare	comparator
	lpm_counter	counter
	lpm_mult	multiplier

	Module Name	Function
storage	lpm_ff	D or T flip-flop
	lpm_latch	latch
	lpm_ram_dq	RAM with separate I/O
	lpm_ram_io	RAM with common I/O
	lpm_ROM	ROM
	lpm_shiftreg	shift register
pads	lpm_bipad	bidirectional I/O pad
	lpm_inpad	input pad
	lpm_outpad	output pad
other	lpm_ttable	truth table
	lpm_fsm	finite state machine

Using Parameterized Modules

As an example of the use of a LPM module, suppose that we need an 8-bit register with an asserted-high asynchronous clear. The `lpm_ff` module can be configured to provide this functionality. Its declaration is given in Listing 15.10.1.

LISTING 15.10.1
Component declaration for module lpm_ff.

```
component lpm_ff

   generic(lpm_width : natural;      -- must be greater than 0
           lpm_avalue : string := "unused";
           lpm_svalue : string := "unused";
           lpm_pvalue : string := "unused";
           lpm_fftype: string := "dff";
           lpm_type: string := l_ff;
           lpm_hint : string := "unused");

   port    (data : in std_logic_vector(lpm_width-1 downto 0) := (others => '1');
           clock : in std_logic;
           enable : in std_logic := '1';
           sload : in std_logic := '0';
           sclr : in std_logic := '0';
           sset : in std_logic := '0';
           aload : in std_logic := '0';
           aclr : in std_logic := '0';
           aset : in std_logic := '0';
           q : out std_logic_vector(lpm_width-1 downto 0));

end component
```

LPM Generics

Like most LPM modules, the lpm_ff module's functionality is greater than its name might imply. There are seven generics that configure the module. They can be assigned values in the generic map of a component instantiation statement.

The lpm_width generic determines the width of the register. If we wanted a single flip-flop, we would set this value to 1. Otherwise, we set it to the desired register width. The other six generics have default values specified in the component declaration. If we wish to use the default value for a particular generic, then that generic is simply omitted from the generic map.

The lpm_ff module can be configured as either a D or a T flip-flop based on the value of the lpm_fftype generic. The default value is "dff". Since that is the functionality we desire, we omit this generic from the generic map. If we wanted a T flip-flop, we would assign lpm_fftype the value "tff" in the generic map. For our example, only the width needs to be assigned a value; defaults are appropriate for all the other generics.

LPM Ports

The declaration's port clause specifies the inputs and outputs of the module. Input data is the data input and output q is the data output. Their widths are determined by the value assigned to lpm_width. Input clock is for the system clock. Input enable enables all synchronous operations. There are asynchronous set and clear inputs, aset and aclr, respectively. When aset is asserted, the register is set to 1s or to the lpm_avalue specified in the generic map. There are also synchronous set and clear inputs, sset and sclr, respectively. When sset is asserted, the register is set to 1s or to the lpm_svalue specified in the generic map. The aload and sload inputs are only used when the module is configured as a T flip-flop.

A design entity that uses an instance of the lpm_ff module to create the desired 8-bit register is given in Listing 15.10.2.

LISTING 15.10.2
An 8-bit register implemented using an instance of the lpm_ff module.

```
library ieee;
use ieee.std_logic_1164.all;
library lpm;
use lpm.lpm_components.all;

entity lpm_ff_reg is
     port(
          clk : in std_logic;          -- system clock
          rst : in std_logic;          -- asserted-high reset
          din : in std_logic_vector(7 downto 0); -- data input
          q : out std_logic_vector(7 downto 0)   -- data output
          );
end lpm_ff_reg;
```

```
architecture structural of lpm_ff_reg is
begin

    reg0: lpm_ff generic map (lpm_width => 8) port map (clock => clk,
    data => din, aclr => rst, q => q);

end structural;
```

Context clause

```
library lpm;
use lpm.lpm_components.all;
```

makes the module's declaration visible.

The component instantiation statement assigns a value to only the one generic it needs to specify. It assigns signals to only the ports it needs to use. The synthesizer should synthesize only the LPM logic required for the functionality used.

As we see from this simple example, using an LPM module requires an understanding of all the module's inputs, outputs, and generics. This information is determined from the standard or from documentation provided by the PLD vendor or third-party IP vendor. Comments provided in the `lpm_components` package listing are usually not sufficient to provide all the needed information.

Many PLD vendors provide a wizard that allows us to answer questions about the functionality we require from a LPM module. The wizard then produces the component instantiation code with the appropriate generic values. This simplifies using LPM modules.

15.11 GENERATE STATEMENT

Many digital systems have a structure that consists of a number of repetitions of a component, where these repetitions form a regular pattern. Such systems can be described by writing a separate component instantiation statement for each occurrence of the component. However, since digital systems with this kind of regular structure are so common, VHDL provides the generate statement to facilitate such descriptions.

A generate statement is a concurrent statement that allows us to specify the replication of concurrent statements. If a concurrent statement that is to be replicated is a component instantiation statement, then we have a concise way of describing the repeated instantiation of a component.

The syntax for a generate statement is given in Figure 15.11.1.

generate_statement ::=
*generate*_label :
    generation_scheme **generate**
        [{ block_declarative_item }
    begin]
        { concurrent_statement }
    end generate [*generate*_label] ;

generation_scheme ::= for *generate*_parameter_specification | if condition

FIGURE 15.11.1
Syntax for a generate statement.

The generate_label is required to identify the generated structure. As a result of the generation_scheme chosen, there are two forms of generate statement. If the for scheme is chosen, we have an iterative generate statement. If the if condition scheme is chosen, we have a conditional generate statement.

Following keyword **generate** is an optional declarative part where we can declare items. The kinds of items that can be declared here are the same as can be declared in the declarative part of an architecture body. Each item declared is local to each replication of the concurrent statements. Following the optional declarative part are the concurrent statements to be generated.

Iterative Generate Statement

When the for scheme is used, a discrete range must be specified. For each value in this range there is a replication of the declared items and the concurrent statements. For each replication the discrete range value is a constant called the *generate parameter*.

A very simple example of a digital system with a repetitive structure is an *n*-bit register. Listing 9.1.2 provided a behavioral description of an 8-bit register with three-state outputs. We can write a parameterized version of this register where its width is specified by a generic and its structure is created by replicating stages, each of which consists of a D flip-flop component and a three-state buffer component.

Assume that we have compiled the D flip-flop description in Listing 8.4.1 and the three-state buffer description in Listing 4.10.1 to the working library. We can write a top-level description, as shown in Listing 15.11.1, that uses an iterative generate statement to describe the register.

LISTING 15.11.1
Register with three-state buffer outputs described using a generate statement.

```
library ieee;
use ieee.std_logic_1164.all;

entity reg_3sbuff_gen is
   generic width : positive := 8);
```

```
   port(
      d : in std_logic_vector(width - 1 downto 0);
      clk : in std_logic;
      oe_bar : in std_logic;
      q : out std_logic_vector(width - 1 downto 0)
      );
end reg_3sbuff_gen;

architecture generated of reg_3sbuff_gen is
begin
   stage: for i in (width - 1) downto 0 generate  -- generate statement
      signal s : std_logic ;
      begin
      u0: entity work.d_ff_pe(behavioral) port map (d => d(i), clk => clk,
                  q => s);
      u1: entity work.three_state_buffer(dataflow) port map (d_in => s,
                  en_bar => oe_bar, d_out => q(i));
   end generate;
end generated;
```

The entity declaration for the register declares the generic `width` and gives it a default value of 8.

The architecture body consists of a single iterative generate statement with the label `stage`. The discrete range is specified as (`width` - 1) **downto** 0. If a configuration declaration is not used, `width` will have its default value of 8. If a configuration declaration is used, we can specify any desired width for the register.

Each stage of the register consists of a D flip-flop and a three-state buffer. A local signal is needed in each stage to connect the output of the D flip-flop to the input of the three-state buffer. Accordingly, signal `s` is declared in the declaration part of the generate statement for this purpose.

In the statement part of the generate statement are two component instantiation statements. One instantiates the D flip-flop, and the other instantiates the three-state buffer. After the description is elaborated, there will be eight replications of the D flip-flop and eight replications of the three-state buffer. Each D flip-flop will be connected to its associated three-state buffer by a replication of the local signal `s`.

A hierarchical view of the synthesized logic is given in Figure 15.11.2. Note the use of the labels from the description in identifying the components in the block diagram. The replications of the local signal `s` are shown but not labeled.

Interconnecting Adjacent Instances

In the system shown in Figure 15.11.2, there are no connections from one D flip-flop and buffer instance to an adjacent instance. Such connections, by definition, cannot be accomplished using signals local to an instance.

A simple example of a system that requires connections between adjacent component instances is a shift register. Consider the serial-in parallel-out 4-bit right-shift register in Figure 9.2.1. The component that must be replicated is a D flip-flop. We can use a

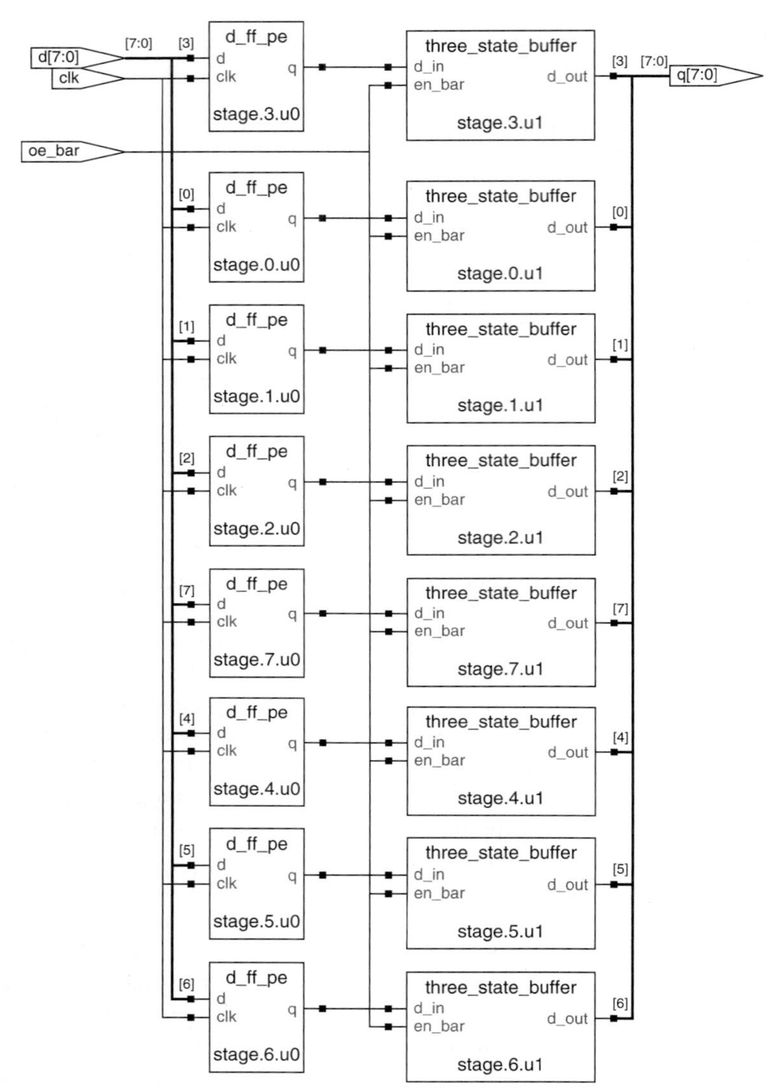

FIGURE 15.11.2
Block diagram of logic synthesized from a generate statement for a register with three-state output.

std_logic_vector signal of length 5 to connect between the adjacent stages, and connect the stages at each end to the ports of the system. This signal must be declared in the declarative part of the architecture body.

The top-level description of the shift register is given in Listing 15.11.2.

LISTING 15.11.2
A 4-bit shift register described using an iterative generate statement.

```
library ieee;
use ieee.std_logic_1164.all;

entity shift_reg_gen is
   port(
      si : in std_logic;
      clr_bar : in std_logic;
      clk : in std_logic;
      qout : out std_logic_vector(3 downto 0));
end shift_reg_gen;

architecture structural of shift_reg_gen is
   signal s: std_logic_vector(4 downto 0);
begin
   stages: for i in 3 downto 0 generate
      begin
      stage: entity dff_stage port map (d => s(i + 1), clr_bar => clr_bar,
                                   clk => clk, q => s(i));
   end generate;
   s(4) <= si;
   qout <= s(3 downto 0);
end structural;
```

The discrete range of the generate statement is from 3 down to 0. The component instantiation statement inside the generate statement generates the four instances of the D flip-flop.

The connection of the serial input `si` to the D input of the leftmost flip-flop is made by the concurrent statement

```
s(4) <= si;
```

which is outside of the generate statement. The connection of the D flip-flop outputs to the output port `qout` is made by the concurrent statement

```
qout <= s(3 downto 0);
```

which is also outside of the generate statement.

Conditional Generate Statement

When the if-condition scheme is used, the generation of the concurrent statements inside the generate statement occurs only if the condition is true. An alternative architecture body for the shift register, using conditional generate statements, is given in Listing 15.11.3.

LISTING 15.11.3
Alternative architecture body for the shift register.

```
architecture structural2 of shift_reg_gen is
   signal s: std_logic_vector(3 downto 0);
begin
   stages: for i in 3 downto 0 generate
      begin
      lstage: if i = 3 generate
         begin
         stage: entity dff_stage port map (d => si,
            clr_bar => clr_bar, clk => clk, q => s(3));
      end generate;

      ostages: if i /= 3 generate
         begin
         stage: entity dff_stage port map (d => s(i + 1),
            clr_bar => clr_bar, clk => clk, q => s(i));
      end generate;
   end generate;

   qout <= s(3 downto 0);
end structural2;
```

In the architecture body an iterative generate statement labeled `stages` sequences through the values of its generate parameter. Nested inside this generate statement are the two conditional generate statements. The first conditional statement, labeled `lstage`, has the condition `i = 3`. When the generate parameter is equal to 3, this conditional generate statement produces the instantiation of the leftmost stage of the shift register. The second conditional statement, labeled `ostage`, has the condition `i/= 3`. When the generate parameter is not equal to 3, this conditional generate statement produces the instantiation of the other three stages of the shift register.

The synthesized logic is given in Figure 15.11.3.

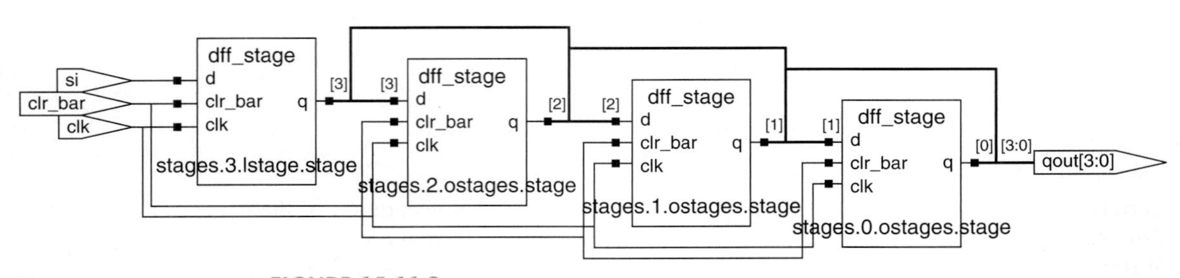

FIGURE 15.11.3
Logic synthesized for 4-bit shift register from Listing 15.11.3.

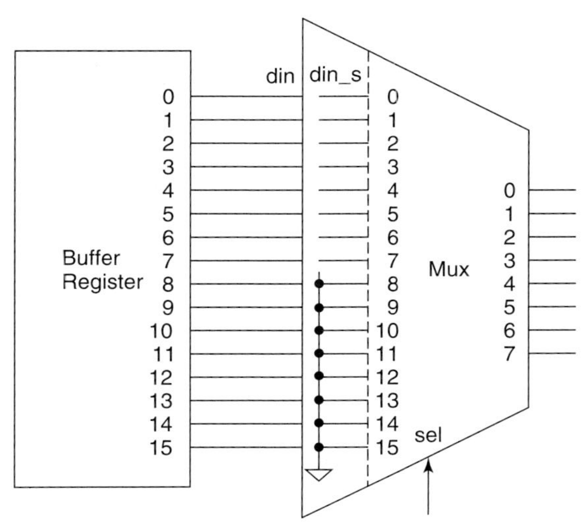

FIGURE 15.11.4
Parameterized multiplexer input boundary.

Parameterized Double Buffer Logic

As another example of the use of conditional generate statements, consider parameterizing the double buffer logic in Figure 10.8.1. The counter, buffer register, and multiplexer all have to be parameterized.

Parameterization of the counter and double buffer are straightforward. However, parameterizing the multiplexer is more complicated. Assuming that we want to parameterize the system to provide a count of from 1 to 16 bits, the difficulty with the multiplexer description is illustrated in Figure 15.11.4.

The required multiplexer is still an 8-bit 2-to-1 multiplexer, since the microprocessor data bus width remains 8 bits. However, of the 16 bits into the multiplexer, `n` bits will be from the counter and 16 – `n` bits will be fixed at `'0'`. The description must allow the boundary between the `n`-bit count and the fixed `'0'`s to cross between the two bytes that are input to the multiplexer. One way to describe this logic as a function of a parameter is to use a generic and generate statements as shown in Listing 15.11.4.

A std_logic_vector `din_s` is used to provide an array of "jumper" signals that, based on the value of `n`, connect each multiplexer input to either an output from the double buffer or a fixed value of `'0'`. To accomplish this, two conditional generate statements are used inside of a generate loop that iterates from 0 to 15. If the loop index is less than or equal to `n`, the indexed signal element from the `din_s` is connected to the corresponding indexed signal element from the double buffer. If the value of the index is greater than `n`, the indexed signal element from `din_s` is assigned a value of `'0'`.

The left side of the jumpers array is connected to either the double buffer outputs or to `'0'`. A conditional signal assignment statement multiplexes the right side of the jumper array.

LISTING 15.11.4
Parameterized multiplexer and three-state output buffer code.

```
library ieee;
use ieee.std_logic_1164.all;

entity mux is
   generic ( n : positive := 12);
   port (din : in std_logic_vector (n - 1 downto 0);
      sel: in std_logic;
      oe_bar: in std_logic;
      dout : out std_logic_vector (7 downto 0)
      );
end mux;

architecture dataflow_param of mux is
   signal din_s : std_logic_vector(15 downto 0);
begin

   g0: for i in 0 to 15 generate
      g1: if i <= n - 1 generate -- mux inputs from double buffer
         din_s(i) <= din(i);
      end generate;
      g2: if i > n - 1 generate  -- mux inputs fixed at 0
         din_s(i) <= '0';
      end generate;
   end generate;

   dout <= din_s(15 downto 8) when oe_bar = '0' and sel = '0' else
   din_s(7 downto 0) when oe_bar = '0' and sel = '1' else
   (others => 'Z');

end dataflow_param;
```

PROBLEMS

15.1 Draw a hierarchy tree, like Figure 15.1.1, for the microprocessor compatible pulse width modulated signal generator in Figure 9.6.2. Is this a flat or hierarchical partition? What kind of VHDL constructs were used to describe the modules of this partition? Refer to Listing 9.6.1.

15.2 Draw a hierarchy tree for the system represented by the block diagram in Figure 10.8.1.

15.3 Draw a hierarchy tree for the entire successive approximation ADC in Figure 11.4.1. Let the blocks in Figure 11.4.1 correspond to the first-level partition into modules. For the SAR module, show its constituent levels of hierarchy as determined from Figure 11.4.4.

15.4 A timer is to be designed that measures and displays elapsed time in minutes and seconds. The maximum length of time that can be measured is 60 minutes. The timer's inputs are clear, start, and stop. These inputs come from momentary contact push-button switches. The timer's outputs directly drive four seven-segment LED displays. Two of the seven-segment LEDs display elapsed seconds and two display elapsed minutes. Assume that the clock input to the system is 1 kHz. Draw a top-level partition for the timer. Draw a top-level block diagram.

15.5 Repeat Problem 15.4, but assume that the four seven-segment LED displays are to be time multiplexed. Accordingly, the system's outputs are a single set of the seven-segment drive signals a through f and four-digit drive signals.

15.6 Explain the difference between a design unit and a library unit.

15.7 List the different kinds of library units. Which are primary units and which are secondary units? At the minimum, which library units are required to form a design entity?

15.8 Explain the difference between a primary library unit and a secondary library unit. What, if any, are the constraints on their order of compilation?

15.9 The diagram of the retriggerable single shot in Figure 9.5.3 consists of four modules, each coded as a process in Listing 9.5.2. Rewrite each of these modules as a separate design entity.

15.10 The diagram of the microprocessor compatible pulse width modulated signal generator in Figure 9.6.2 consists of four modules, each coded as a process in Listing 9.6.1. Rewrite each of these modules as a separate design entity.

15.11 What is a design library and how is it physically implemented?

15.12 What is the difference between a working library and a resource library? During the compilation of a design unit, is there any restriction on the numbers of each of these types of libraries used?

15.13 Consult your compiler's documentation to determine how to create a library. Create a library named `parts` and compile design entities `half_adder` and `or_2` from Listing 2.9.1 into the library. Separately compile and simulate the structural style full-adder description from Listing 2.9.2 that uses the design entities from the library `parts`.

15.14 What is the purpose of a context clause? Which context clause is assumed to precede every design unit?

15.15 What is the purpose of a library clause? What is the purpose of a use clause?

15.16 Write a structural style description of a design entity that implements the retriggerable single shot in Figure 9.5.3 using direct entity instantiation of the design entities from Problem 15.9.

15.17 Write a structural style description of a design entity that implements the microprocessor compatible pulse width modulated signal generator in Figure 9.6.2 using direct entity instantiation of the design entities from Problem 15.10.

15.18 Write a component declaration for each of the design entities resulting from Problem 15.9. Write a structural style description of a design entity that implements the retriggerable single shot in Figure 9.5.3 using these component declarations.

15.19 Write a component declaration for each of the design entities resulting from Problem 15.10. Write a structural style description of a design entity that implements the microprocessor compatible pulse width modulated signal generator in Figure 9.6.2 using these component declarations.

15.20 Write component declarations for each of the modules in Figure 10.8.1. Write a structural style description of the system in Figure 10.8.1 using these component declarations.

15.21 Assume that you have a structural style description identical to Listing 15.6.2, except without the configuration specifications. Write a configuration declaration named `config1` that binds the component instances to the same design entities as specified by the configuration specifications in Listing 15.6.2.

15.22 Repeat Problem 15.21, except write a comfiguration declaration named `config2` to bind all of the component instances to a design entity named `nand_2` in the library `work`. Assume that `nand_2` has inputs `in1` and `in2` and output `out1`.

15.23 For the testbench in Listing 7.11.1, write three different configuration declarations. Name the configuration declarations functional_sim, postsyn_sim, and timing_sim for functional, post-synthesis, and timing simulations. The architecture for design entity `ic74f539` is `behav_case`, the architecture in the post-synthesis netlist is named `beh`, and the architecture of the timing model is named `structure`.

15.24 Modify the code for the octal register in Listing 9.1.2 so that it describes a parameterized register named `n_bit_reg` whose width is determined by the generic `n`. The default value for `n` must be 8.

15.25 Modify the code for the entity declaration in Listing 5.8.1 and the architecture in Listing 5.8.3 to produce a parameterized parity detector. Change the name of the output from `oddp` to `parity`. The parity detector has two generics. Generic `n` specifies the number of bits that are inputs to the detector. Generic `even` is type boolean. If it is `true`, the parity detector's output is `'1'` for even parity. If it is `false`, the parity detector's output is `'1'` for odd parity. The default generic values must produce an 8-bit even parity detector.

15.26 Modify the code for Listing 9.7.2 to describe parameterized RAM. Generic `m` must specify the number of address inputs and generic `n` must specify the width of the data buses.

15.27 For the SAR design in Figure 11.4.4, write the code needed for a parameterized version that can be used with an n-bit DAC to create an n-bit successive approximation ADC.

15.28 If your synthesis tool supports the LPM standard or if you have access to LPM module documentation, determine whether you can implement an octal register equivalent to the one described in Listing 9.1.2 using LPM modules. This might require the use of more than one module. If you can use LPM modules for this purpose, then write the component instantiation statement(s). If not, explain why.

15.29 If your synthesis tool supports the LPM standard or if you have access to LPM module documentation, determine whether you can implement a 12-bit binary counter equivalent to the one described in Listing 9.4.4 using a LPM module. If you can, then write the component instantiation statement. If not, explain why.

15.30 If your synthesis tool supports the LPM standard or if you have access to LPM module documentation, determine whether you can implement RAM similar to the one described in Listing 9.7.2 except with a 64×8 external organization using the `lpm_ram_dq` module. If you can, then write the component instantiation statement. If not, explain why.

15.31 Modify the code for the 4-bit shift register in Listing 15.11.2 so that it describes a parameterized shift register whose width is determined by the generic `n`. The default value for `n` must be 4.

Chapter 16
More Design Examples

This chapter brings together many of the concepts introduced in early chapters through a discussion of three design examples. The purpose of the discussion is to provide some insight into how these designs were partitioned and to highlight various interesting aspects of each design's description and verification.

The designs are of sufficient complexity to allow the interested reader to explore alternative design approaches and feature enhancements.

16.1 MICROPROCESSOR-COMPATIBLE QUADRATURE DECODER/COUNTER DESIGN

This section presents the design of a microprocessor-compatible quadrature decoder/counter. It provides a further example of design partitioning and structural implementation using components.

A microprocessor-compatible quadrature decoder/counter is used to interface an optical shaft encoder (OSE) to a microprocessor's system bus (Figure 16.1.1). OSEs and nonmicroprocessor-compatible quadrature decoder/counters were discussed in Section 10.5. Quadrature decoder/counters find application in digital data input subsystems and digital closed loop motion control systems.

OSE Interface Approaches

There are two common approaches to interfacing an OSE to a microprocessor system's bus. In a software-intensive approach, the microprocessor repeatedly inputs signals A and B from the OSE through a simple input port. A software subroutine compares the newly read values of A and B with a copy of the previously read values saved in memory. Based on this comparison, the subroutine determines whether to

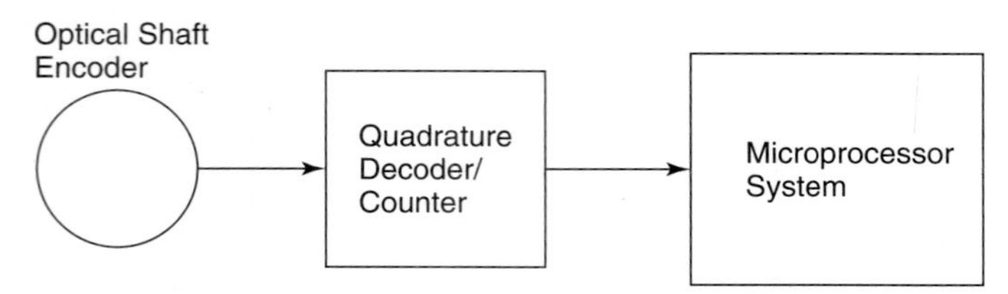

FIGURE 16.1.1
Quadrature decoder/counter used to interface an optical shaft encoder to a microprocessor's system bus.

increment or decrement a count variable in the microprocessor's memory. This simple software approach requires that the microprocessor constantly monitor the OSE's outputs to detect each change in their values.

To improve the performance of a system incorporating an OSE, the time-intensive quadrature/decoder functions can be implemented in hardware. In this approach hardware converts changes in A and B to a count that is stored in a register. A microprocessor-compatible quadrature decoder/counter includes a bus interface that allows its count register to be read by a microprocessor at any time.

Quadrature Decoder/Counter Features

Some features that a quadrature decoder/counter IC might provide are:

- Input signal noise filter
- 4X decoder
- 12-bit counter
- Registered outputs that are readable "on the fly"
- 8-bit microprocessor bus interface

These features are covered in more detail as the partition and its resulting components for this design are described.

Our intention is to create a design that is functionally equivalent to an Agilent HCTL-2000 Quadrature Decoder/Counter Interface IC. Its data sheet provides a complete specification.

Agilent's HCTL-2000 has been replaced by Avago's HCTL-2001, which adds some additional features. Importantly, the HCTL-2001 differs from the HCTL-2000 in that its `oe_bar` input must remain low throughout the two byte reads required to read its 12-bit count value. In contrast, the HCTL-2000's `oe_bar` input can be toggled between the two byte reads, which is more convenient for direct connection to a microprocessor's bus.

Quadrature Decoder/Counter Partition

A block diagram of the first-level partition of the quadrature decoder/counter's functions is presented in Figure 16.1.2. There are seven components. Design entities corresponding to the first four components (`u0` to `u3`) have been discussed in previous chapters. The last three components (`u4` to `u6`) were previously discussed

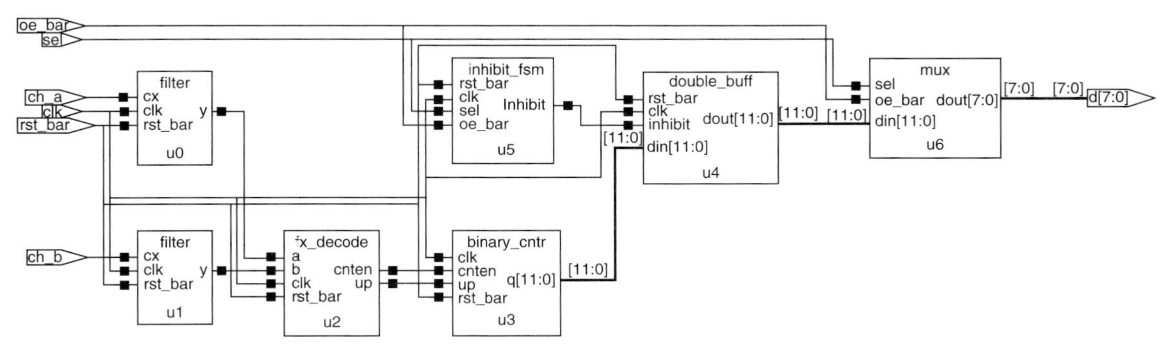

FIGURE 16.1.2
Functional partition for a microprocessor-compatible quadrature decoder/counter.

as a single design entity. Here they are presented as separate design entities. Thus, our partition consists of components that are all being reused. Each component is briefly reviewed before the top-level structural description is presented.

Digital Noise Filter

In an application environment, inputs A and B to the quadrature decoder/counter may be subjected to electrical noise. The quadrature decoder/counter must include features to minimize the effect of both metastability from the asynchronous nature of inputs A and B and noise spikes of short duration that might appear on these inputs.

Each input channel (A and B) is filtered by a separate copy of the digital noise filter (components `u0` and `u1`). A description of this filter was given in the section "Shift Register Digital Noise Filter" on page 344.

Four Times Decode

Since output signals A and B from the OSE have a quadrature relationship, if the counter in the quadrature decoder/counter is enabled to count once for each quadrature of the period of A, as defined by its relationship with B, the effective resolution of the OSE is increased by a factor of four.

The four times decode subsystem is a state machine that uses the system clock to detect each time the combination of output values A and B changes. When this state machine detects such a change, it asserts its count enable output (`cnten`) until the next triggering clock edge. A counter enabled by `cnten` counts at the next triggering clock edge. The count enable output is unasserted on this same clock edge. This finite state machine was described in the section "Decoder for an Optical Shaft Encoder" on page 403.

Counter

The counter module requires a simple 12-bit up/down counter with an enable input. Such a counter was described in the section "Up/Down Counter with an Enable Input" on page 354.

Reading the Count "On the Fly"

As previously discussed in the section "Inhibit Logic FSM Example" on page 418, a problem can arise if a microprocessor directly reads the contents of a counter at the same time the counter is being incremented (reads "on the fly"). The value read can be incorrect, containing some bits from the old count value that have not yet changed and some bits from the new count value.

This problem is exacerbated in this application because the counter being read is wider than the microprocessor's data bus. The counter is 12 bits and the microprocessor's data bus is only 8 bits wide. This requires that the 12-bit data value be read as two different bytes, one containing the most significant 4 bits of the count and the other containing the least significant 8 bits. Reading 2 bytes requires that the microprocessor execute two read bus cycles. The increased time period over which the count is read increases the probability that the count might change while being read.

Double Buffer Register

A double buffer register provides the solution to the problem of reading a counter "on the fly." The double buffer register is connected to the counter's output. While its `inhibit` input is `'0'`, the double buffer register stores a copy of the counter's output on each triggering clock edge. However, when its `inhibit` input is a `'1'`, the double buffer register leaves its stored value unchanged.

A FSM is required to generate the `inhibit` signal. The FSM makes `inhibit` a `'1'` as soon as the microprocessor starts to read the contents of the double buffer register, and returns `inhibit` to a `'0'` after both bytes have been read.

Output Multiplexer

Finally, a multiplexer is required to multiplex the 12-bit output of the double buffer register to the 8-bit data bus output. The `sel` signal determines whether the high byte or low byte appears on the multiplexer's outputs when `oe_bar` is asserted.

The double buffer, inhibit logic, and output multiplexer appeared together as a single design entity in the section "Inhibit Logic FSM Example" on page 418. In Listings 16.1.1, 16.1.2, and 16.1.3 these functions are presented as separate design entities. This separation improves the clarity of the design and makes its management easier.

The description of the double buffer is given in Listing 16.1.1.

LISTING 16.1.1
Double buffer register.

```
library ieee;
use ieee.std_logic_1164.all;

entity double_buff is
   port (din : in std_logic_vector (11 downto 0);
      rst_bar: in std_logic;
      clk: in std_logic;
      inhibit: in std_logic;
      dout : out std_logic_vector (11 downto 0));
```

```
end double_buff;

architecture behav of double_buff is
begin
   freeze : process (clk)
   begin
      if rising_edge(clk) then
         if rst_bar = '0' then
            dout <= x"000";
         elsif inhibit = '0' then
            dout <= din;
         end if;
      end if;
   end process;
end behav;
```

The operation of the double buffer register is simple. If `rst_bar` is `'0'`, then the double buffer register is cleared. If `rst_bar` is not `'0'` and `inhibit` is `'0'`, the register stores its 12-bit input value.

Inhibit Control for Double Buffer Register

The inhibit control FSM is similar to the description in Listing 10.8.1, with the double buffer register and multiplexer logic removed. This description is given in Listing 16.1.2.

LISTING 16.1.2
Inhibit control finite state machine.

```
library ieee;
use ieee.std_logic_1164.all;

entity inhibit_fsm is
   port (
      rst_bar: in std_logic;
      clk: in std_logic;
      sel: in std_logic;
      inhibit : out std_logic;
      oe_bar: in std_logic
      );
end inhibit_fsm;


architecture behav of inhibit_fsm is

   type inhib_state is (idle, byte1, byte2);
   signal present_state, next_state : inhib_state;
```

(Cont.)

LISTING 16.1.2 *(Cont.)*

```
begin
inhib_sreg: process (clk)
begin
   if rising_edge(clk) then
      if rst_bar = '0' then
         present_state <= idle;
      else
         present_state <= next_state;
      end if;
   end if;
end process;

inhib_output: process (present_state)
begin
   case present_state is
      when byte1 | byte2 =>
      inhibit <= '1';
      when others=>
      inhibit <= '0';
   end case;

end process;

inhib_nxt_state: process (present_state, sel, oe_bar)
begin
   case present_state is

      when idle =>
      if sel = '0' and oe_bar = '0' then
         next_state <= byte1;
      else
         next_state <= idle;
      end if;

      when byte1 =>
      if sel = '1' and oe_bar = '0' then
         next_state <= byte2;
      else
         next_state <= byte1;
      end if;

      when byte2 =>
      if oe_bar = '1' then
         next_state <= idle;
      elsif sel = '0' and oe_bar = '0' then
         next_state <= byte1;
```

```
            else
               next_state <= byte2;
            end if;

            end case;
         end process;

end behav;
```

Output Multiplexer with Three-State Buffers

The output multiplexer logic is described very simply in Listing 16.1.3.

Top-Level Structural Description

The components of the design partition are connected together in the top-level design entity. The top-level design entity is written in structural style, as shown in Listing 16.1.4. In this description, components are used to provide indirect design entity instantiation.

Each component is declared in the declarative part of the architecture body, as are the signals necessary to interconnect them.

In the statement part of the architecture body are seven component instantiation statements, one for each instance of each component. Component instantiations `u0` and `u1` are two instances of the same filter component. All other components appear only once.

LISTING 16.1.3
Quadrature decoder/counter output multiplexer.

```
library ieee;
use ieee.std_logic_1164.all;

entity mux is
   port (din : in std_logic_vector (11 downto 0);
      sel: in std_logic;
      oe_bar: in std_logic;
      dout : out std_logic_vector (7 downto 0)
      );
end mux;

architecture behav of mux is
begin

   dout <= "0000" & din(11 downto 8) when oe_bar = '0' and sel = '0' else
   din(7 downto 0) when oe_bar = '0' and sel = '1' else
   (others => 'Z');

end behav;
```

LISTING 16.1.4
Top-level structural description of quadrature decoder/counter.

```
library ieee;
use ieee.std_logic_1164.all;

entity quad_decode is
   port (
      ch_a: in std_logic;
      ch_b: in std_logic;
      rst_bar: in std_logic;
      clk: in std_logic;
      sel: in std_logic;
      oe_bar: in std_logic;
      d: out std_logic_vector (7 downto 0)
      );
end quad_decode;


architecture structure of quad_decode is

   component filter
      port (cx, clk, rst_bar : in std_logic;
         y : out std_logic);
   end component;

   component fx_decode
      port (a, b, clk, rst_bar: in std_logic;
         cnten, up: out std_logic);
   end component;

   component binary_cntr
      port (clk, cnten, up, rst_bar: in std_logic;
         q: out std_logic_vector (11 downto 0));
   end component;

   component double_buff
      port (din : in std_logic_vector (11 downto 0);
         rst_bar: in std_logic;
         clk: in std_logic;
         inhibit: in std_logic;
         dout : out std_logic_vector (11 downto 0)
         );
   end component;

   component inhibit_fsm
      port (
         rst_bar: in std_logic;
         clk: in std_logic;
```

```
         sel: in std_logic;
         inhibit : out std_logic;
         oe_bar: in std_logic
         );
   end component;

   component mux
      port (din : in std_logic_vector (11 downto 0);
         sel: in std_logic;
         oe_bar: in std_logic;
         dout : out std_logic_vector (7 downto 0)
         );
   end component;


   signal ch_a_sig, ch_b_sig, cnten_sig, up_sig : std_logic;
   signal count_sig : std_logic_vector(11 downto 0);
   signal dbl_buff_sig : std_logic_vector(11 downto 0);
   signal inhibit_sig : std_logic;

begin

   u0: filter port map (
      cx => ch_a, clk => clk, rst_bar => rst_bar, y => ch_a_sig);

   u1: filter port map (
      cx => ch_b, clk => clk, rst_bar => rst_bar, y => ch_b_sig);

   u2: fx_decode port map (
      a => ch_a_sig, b => ch_b_sig, clk =>clk, rst_bar => rst_bar,
      cnten => cnten_sig, up => up_sig);

   u3: binary_cntr port map (
      clk => clk, cnten => cnten_sig, up => up_sig, rst_bar => rst_bar,
      q => count_sig);

   u4: double_buff port map (
      din => count_sig, rst_bar => rst_bar, clk => clk, inhibit => inhibit_sig,
      dout => dbl_buff_sig);

   u5: inhibit_fsm port map (
      rst_bar => rst_bar, clk => clk, sel => sel, inhibit => inhibit_sig,
      oe_bar => oe_bar);

   u6: mux port map (
      din => dbl_buff_sig, sel => sel, oe_bar => oe_bar, dout => d);

end structure;
```

Since the names and interfaces of the design entities match those of the components, default binding is sufficient to fully configure the system and a configuration declaration is not necessary.

16.2 VERIFICATION OF QUADRATURE DECODER/COUNTER

After each component design entity in the quadrature decoder/counter was developed, it was independently verified using a separate testbench. When these components are reused by interconnecting them to form the quadrature decoder/counter, a testbench must be written to verify the top-level structural description. The resulting overall verification approach is bottom-up.

The verification plan for the top-level design entity is to generate the OSE quadrature phase signals A and B, which are inputs to the quadrature decoder/counter, and to read the quadrature decoder/counter's output to see if it has the correct count. To read the quadrature decoder/counter, the microprocessor bus functional model (BFM) developed in Section 14.9 is used.

A testbench for the quadrature decoder/counter is given in Listing 16.2.1.

LISTING 16.2.1
Testbench for quadrature decoder/counter.

```
library ieee;
use ieee.std_logic_1164.all;

use bus_cycle.all;

entity quad_decode_tb is
end quad_decode_tb;

architecture tb_architecture of quad_decode_tb is

   constant cpu_clk_period : time := 125 ns;
   constant a_b_clk_period : time := 1 ms;

   -- Component declaration of the tested unit
   component quad_decode
      port(
         ch_a : in std_logic;
         ch_b : in std_logic;
         rst_bar : in std_logic;
         clk : in std_logic;
         sel : in std_logic;
```

```
        oe_bar : in std_logic;
        d : out std_logic_vector(7 downto 0) );
    end component;

    -- Stimulus signals
    signal ch_a : std_logic;
    signal ch_b : std_logic;

    -- Observed signals
    signal addr_cpu:  std_logic_vector (19 downto 0) := x"00000";
    signal data_cpu:  std_logic_vector (7 downto 0)  := "ZZZZZZZZ";
    signal wr_bar_cpu:  std_logic := '1';
    signal rd_bar_cpu:  std_logic := '1';
    signal clkout:  std_logic;
    signal word_read :  std_logic_vector (15 downto 0);
    signal cpu_reset_bar : std_logic;
    signal end_sim : boolean := false;

begin

    -- Unit Under Test port map
    UUT : quad_decode
    port map
       (ch_a => ch_a,
       ch_b => ch_b,
       rst_bar => cpu_reset_bar,
       clk => clkout,
       sel => addr_cpu(0),
       oe_bar => rd_bar_cpu,
       d => data_cpu);

    cpu_clock_gen : process
    begin
       clkout <= '0';
       loop
          wait for cpu_clk_period/2;
          clkout <= not clkout;
          exit when end_sim = true;
       end loop;
       wait;
    end process;

    cpu_reset_gen : cpu_reset_bar <= '0', '1' after 2 * cpu_clk_period;

    a_b_clk_gen : process   -- Generate quadrature OSE outputs
    begin
       ch_a <= '0';
```

(Cont.)

LISTING 16.2.1 *(Cont.)*

```
      ch_b <= '0';
      wait for a_b_clk_period/4;
      loop
         exit when end_sim = true;
         ch_a <= '1';
         wait for a_b_clk_period/4;
         ch_b <= '1';
         wait for a_b_clk_period/4;
         ch_a <= '0';
         wait for a_b_clk_period/4;
         ch_b <= '0';
         wait for a_b_clk_period/4;
      end loop;
      wait;
   end process;

   test : process
   begin
      for i in 1 to 4 loop
         wait for 5 ms;
         read_cycle(x"08000", word_read(15 downto 8), 0, clkout, addr_cpu,
                    data_cpu, rd_bar_cpu);
         read_cycle(x"08001", word_read(7 downto 0), 0, clkout, addr_cpu,
                    data_cpu, rd_bar_cpu);
      end loop;
      end_sim <= true;
      wait;
   end process;

end tb_architecture;
```

Constants are established for the CPU clock period (125 ns) and for the period of input waveforms A and B (1 ms). With A and B changing at this rate, we expect to have 4 counts every millisecond.

The `quad_decode` component is declared in the declarative part of the architecture body, as are the signals to connect it to the OSE output and microprocessor BFM.

Processes are included in the statement part of the architecture to generate the CPU clock and CPU reset. The microprocessor BFM procedures use this clock and reset, as does the quadrature decoder/counter. Process `a_b_clk_gen` generates the A and B waveforms.

During verification, the quadrature decoder/counter's count must be read. This requires that two addresses be read, corresponding to the high and low bytes of the counter. For this simulation, addresses 08000 and 08001 are arbitrarily chosen.

Accordingly, the BFM read procedure must be called twice, first with the address 08000 and then with the address 08001. The first call reads the high byte and the second call reads the low byte.

The least significant bit of the address is connected to the sel input of the quadrature decoder/counter to determine which byte is being read. Since there are no other devices connected to the simulated microprocessor's bus, no other decoding of the address needs to be performed. As a result, input oe_bar of the quadrature decoder/counter is connected directly to the read signal from the BFM (rd_cpu_ bar).

The verification process is named test. Many alternative strategies could be used for verification, including using assert statements in a loop to check the count read against the expected count. This testbench, however, focuses on whether both the count logic and the inhibit logic of the quadrature decoder/counter work properly when interfaced to a microprocessor. Accordingly, a loop is used to read the quadrature decoder/counter's count four times. Each read of the count is separated by an interval of 5 ms. The output waveforms are examined to verify the result.

To verify the inhibit logic's operation, the period of A and B and the reads are set so that the count is read at the same time that the counter is being incremented.

A portion of the output waveform around 5 ms is given in Figure 16.2.1. At 5 ms (5000 us) B changes from 1 to 0. This should cause the count to increment from 19 (decimal) to 20. At the same time B changes, the reading of the count is initiated.

The change in B is processed by the filter component for channel B and does not appear at the input of the fx_decode component until 4½ clocks later. This is seen as the change in the ch_b_sig from 1 to 0. The fx_decode component asserts count enable (cnten) as a response. At the next triggering clock edge, the count in the binary counter (q) is incremented to 14 hexadecimal (20 decimal).

The reading of the quadrature decoder/counter starts at 5 ms, with the address of the high byte of the count (08000 hexadecimal) becoming stable. When rd_bar_cpu goes low, the inhibit FSM (inhibit_fsm) asserts its inhibit signal so that the register in the double_buff component does not store a copy of the count at subsequent triggering clock edges. As a result, when q increments to 14, the double buffer register is not enabled at the subsequent clock triggering edge and its output dout continues to have the value 13.

The high byte read by the first call of the read_cycle procedure returns the high byte count 00. The low byte read by the second call of the read_cycle procedure returns the low byte count 13. As a result, the value read is the correct value of 13 hexadecimal (19 decimal).

At the trailing edge of signal rd_bar_cpu during the second bus cycle, the inhibit control FSM unasserts inhibit. At the next triggering clock edge, the register in double_buff stores a copy of the new count in binary_cntr.

Both the correct count and the proper operation of the quadrature encoder during reading are verified by examination of these waveforms.

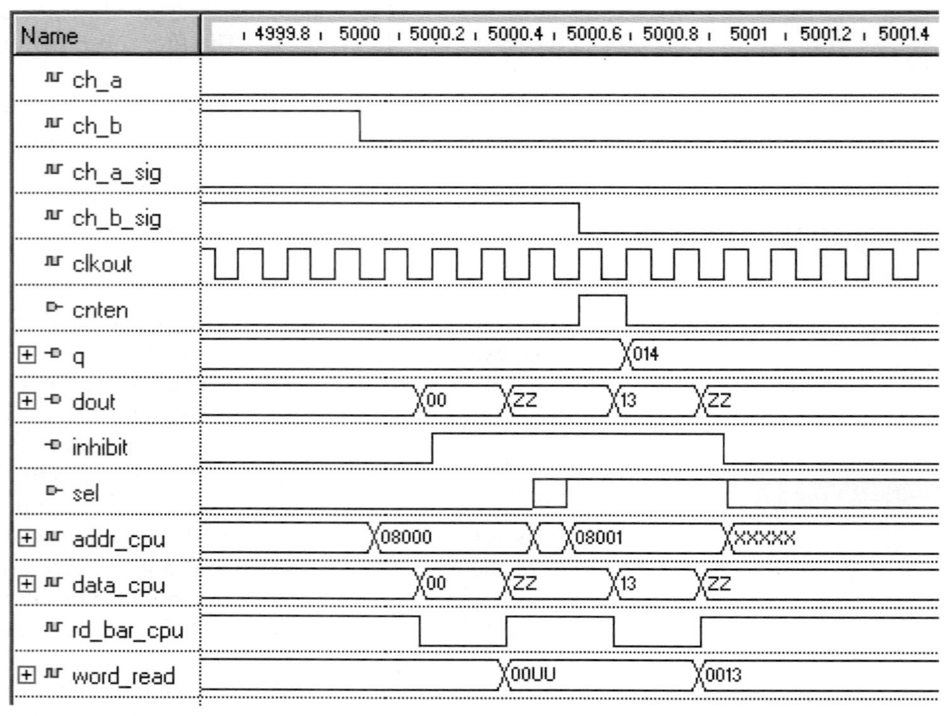

FIGURE 16.2.1
Reading quadrature decoder/counter output while count is being incremented.

16.3 PARAMETERIZED QUADRATURE DECODER/COUNTER

The quadrature decoder/counter and its testbench in the previous two sections are for a design where the counter has a fixed width of 12 bits. This design can be changed to one with a parameterized count width. To accomplish this, three components have to be parametrized: the binary counter, the double buffer register, and the multiplexer.

A description of a parameterized binary counter was given in Listing 15.9.1. A description for a parameterized multiplexer was given in Listing 15.11.4. This leaves only the need for a parameterized double buffer register, which can be easily obtained by modifying Listing 16.1.1.

When entities `binary_counter`, `double_buff`, and `mux` are modified so that they all use the generic `n`, we only need to add a generic clause to the entity declaration for the top-level quad_decode to have a fully parameterized quadrature decoder/counter.

```
entity quad_decode is
   generic (n : positive := 12);
   port (... );
end quad_decode;
```

Testbench for Parameterized Quadrature Decoder/Counter

We can use generic n to parameterize the testbench for the parameterized quadrature decoder/counter. The testbench entity declaration must be modified to include generic n in a formal generic clause:

```
entity quad_decode_tb is
 generic (n : positive := 4);
end quad_decode_tb;
```

In the testbench, the component declaration for quad_decode must include n as a local generic:

```
component quad_decode
 generic (n : positive);
 port( ... );
end component;
```

In the component instantiation for quad_decode, as the UUT in the testbench architecture body, the generic n of the component must be associated with the generic n declared in the testbench's entity declaration:

```
UUT : quad_decode generic map (n => n) port map (...);
```

If the testbench is taken as the top level, we only need to change the default generic value assigned to n in the testbench's entity declaration to simulate a quadrature decoder/counter of any desired counter size from 1 to 16 bits.

As a final example, we could use a configuration declaration to set the value of the generic n and simulate the configuration declaration (Listing 16.3.1).

LISTING 16.3.1
Configuration declaration to assign a value to the generic n.

```
configuration config_quad_decode of quad_decode_tb is
   for tb_architecture
      for UUT : quad_decode use entity quad_decode
         generic map (n => 6);
      end for;
   end for;
end config_quad_decode
```

Without changing any of the other design files, if the configuration declaration in Listing 16.3.1 is simulated, the simulation will be of a 6-bit quadrature decoder/counter.

16.4 ELECTRONIC SAFE DESIGN

The design of the electronics to control a safe is discussed in this section. An application for this electronics might be in a safe for use in hotel rooms.

Safe Front Panel

As a first step in the design, a front panel layout is developed. An objective is to keep the front panel as simple as possible. The result is shown in Figure 16.4.1.

The circle in Figure 16.4.1 represents a knob connected to the shaft of an optical shaft encoder (OSE). Below the knob are two pushbuttons, one labeled OPEN and the other labeled LOCK. On the left side of the panel are two seven-segment displays.

Basic Safe Operation

Another objective for this design is to make the safe's operation as simple as possible for a user. To open the safe, the user presses the OPEN pushbutton and then turns the knob to enter a three number combination. Each number is a value from 00 to 31.

As the knob is turned to the right the number displayed on the two seven-segment displays increments. As the knob is turned to the left, the number decrements.

A combination is entered by turning the knob clockwise until the first number in the combination is displayed, then counter-clockwise until the second number is displayed, and then clockwise until the third number is displayed. After stopping at the third number, the user presses the OPEN pushbutton again. If the combination entered was correct, the safe opens.

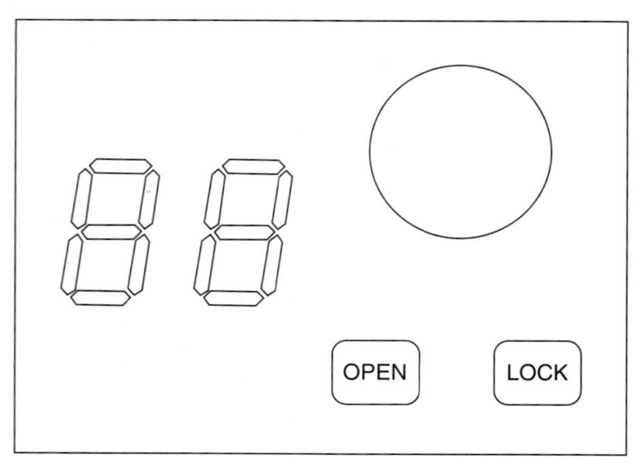

FIGURE 16.4.1
Electronic safe front panel.

To close the safe, the user closes the safe door and then presses the LOCK pushbutton.

The Safe's Electronics

The safe's electronics receives inputs from the front panel and provides signals to the seven-segment displays and to a motorized bolt. When the bolt's motor is actuated to turn in one direction, it unlocks the safe door. When it is actuated to turn in the other direction, it locks the safe door.

The system's input and output signals are given in the entity declaration for the top-level design entity `digital_lock` in Listing 16.4.1.

For simplicity, input signals A and B from the OSE, `lock_pb_bar` and `open_pb_bar` from the pushbuttons, and `door_closed` from a limit switch that senses when the safe door is closed, are all assumed to be bounce free.

Output port `open_close` controls the direction the motorized bolt is to be moved. Output port `actuate_lock` provides a pulse of one state time duration that controls when the bolt should be moved. The duration of this pulse would have to be lengthened to accommodate the requirements of a particular motorized bolt assembly.

Safe Block Diagram

A block diagram of the safe's electronics, corresponding to the first-level partition of the system, is given in Figure 16.4.2.

The first-level partition and its corresponding block diagram were determined based on the functions the system needs to perform. When the block diagram was initially drawn, all the inputs and outputs of the individual blocks were not known. As the individual blocks were further refined, additional input and output ports were added as needed. What is shown in Figure 16.4.2 is the final, completely defined block diagram.

LISTING 16.4.1
Entity declaration for electronic safe top-level design entity.

```
entity digital_lock is
   port(
      clk : in std_logic;         -- system clock
      rst_bar : in std_logic;     -- system reset
      a : in std_logic;           -- optical shaft encoder channel A
      b : in std_logic;           -- optical shaft encoder channel B
      lock_pb_bar : in std_logic;      -- lock pushbutton
      open_pb_bar : in std_logic;      -- open pushbutton
      door_closed_bar : in std_logic;  -- indicates safe door is closed
      actuate_lock : out std_logic;    -- strobe to change bolt position
      open_close : out std_logic;      -- direction to move locking bolt
      seg_dig_0 : out std_logic_vector(6 downto 0);   -- seg values for digit 0
      seg_dig_1 : out std_logic_vector(6 downto 0)    -- seg value for digit 1
      );
end digital_lock;
```

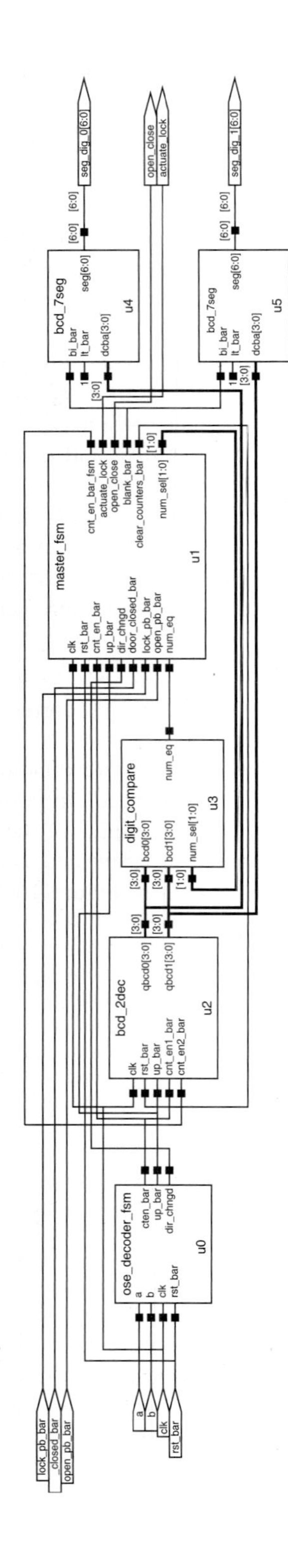

FIGURE 16.4.2
Block diagram of electronic safe.

The block diagram consists of six component design entities:

- `u0: ose_decoder_fsm`
- `u1: master_fsm`
- `u2: bcd_2dec`
- `u3: digit_compare`
- `u4` and `u5: bcd_7seg`

In developing the first-level partition, it is obvious that a two-digit modulo 32 BCD counter is required to store the number to be displayed. In the block diagram, this counter is component `u2`, design entity `bcd_2dec`. The code for this counter was given in Listing 9.4.7 (page 359). We are simply reusing it in this design.

A combinational circuit is required to convert each BCD digit from the two-digit modulo 32 BCD counter to a seven-segment code to drive the corresponding seven-segment display. In the block diagram, these two decoders are two instances (`u4` and `u5`) of the design entity `bcd_7seg`. The code for this decoder was given in Listing 4.8.2 (page 146). We are simply reusing it in this design. The original decoder description includes a lamp test input named `lt_bar` that is not used in this design. Accordingly, in the port maps for the instantiation of the two decoders this input is associated with the constant value `'1'`, leaving it unasserted.

Safe's Control

The majority of new design effort for the safe electronics is in defining and implementing components `u0` and `u1`, and, to a much lesser extent, `u3`. Design entities `u0` and `u1` are both FSMs, which together control the operation of the system.

In theory the control portion of the electronic safe could be designed as a single FSM. However, it simplifies defining the operation of the electronic safe to partition the control function into two *cooperating FSMs*. One FSM is component `ose_decoder_fsm`. It has limited and very specific functions. The other FSM is component `master_fsm`. This FSM controls the overall operation of the safe.

OSE Decoder FSM

We have previously considered the design of decoders for optical shaft encoders. See "Decoder for an Optical Shaft Encoder" on page 403. Since the OSE to be used in this design generates 32 pulses per revolution, its resolution is ideal for this application and its decoder does not need to implement a 4 × resolution enhancement.

Our previous OSE decoders had two outputs, one to enable a counter to count and the other to control count direction. However, for this application the OSE decoder requires an additional output.

When entering a number, the BCD counter is incremented or decremented as the knob attached to the OSE is rotated. The value displayed on the seven-segment displays corresponds to the counter's value. The rate at which the knob is rotated and whether the knob's rotation is stopped one or more times before the desired number is displayed cannot be restricted. So, an issue to be resolved is when the number displayed is considered to be an entry.

One straightforward solution would be to add an ENTER pushbutton to the panel. With this pushbutton added, the procedure for user entry of a three-digit combination would be:

- Press the OPEN pushbutton to indicate the start of the entry of a combination
- Turn the knob clockwise until the first number of the combination is displayed and then press the ENTER pushbutton
- Turn the knob counter-clockwise until the second number of the combination is displayed and then press the ENTER pushbutton
- Turn the knob clockwise until the third number of the combination is displayed and then press the OPEN pushbutton.

An alternative solution, one that eliminates the need for addition of an ENTER pushbutton, is to have the OSE decoder sense when the direction of rotation of the knob has changed and to generate a signal indicating a change in direction. When this signal is asserted, `master_fsm` takes the number that is being displayed and processes it as an entry. Using this approach, the procedure for the user to enter a three-digit combination is:

- Press the OPEN pushbutton to indicate the start of the entry of a combination
- Turn the knob clockwise until the first number of the combination is displayed
- Turn the knob counter-clockwise until the second number of the combination is displayed
- Turn the knob clockwise until the third number of the combination is displayed and then press the OPEN pushbutton.

To implement this second approach, the OSE decoder must have an output that indicates, for one state time, that the direction has changed. In the block diagram, `ose_decoder_fsm` has output `dir_changed` for this purpose.

Operation of OSE Decoder

Component `ose_decoder_fsm` detects the occurrence of a positive edge on input A. When such an edge is detected, `ose_decoder_fsm` examines the value of B to determine the direction of rotation. A state diagram for `ose_decoder_fsm` is given in Figure 16.4.3.

When the system is reset, `ose_decoder_fsm` starts in state `cw0`. In this state, the FSM examines A on each triggering clock edge, waiting for A to be 0 as the first step in detecting a positive edge on A. If A is 0, the FSM transitions to state `cw1` at the next triggering clock edge. In state `cw1`, the FSM waits for A to be 1. If A is 1, the FSM transitions to state `cwpe` at the next triggering clock edge. The fact that the FSM is in state `cwpe` indicates that a positive edge has just been detected on A.

In state `cwpe`, the FSM examines B to determine whether the knob's rotation is clockwise (cw) or counter-clockwise (ccw). If B is 0, rotation is clockwise and the FSM transitions to state `cwcnt`. The FSM being in state `cwcnt` indicates that a positive edge has been detected on A with the knob being rotated clockwise. As a result, in this state outputs `cten_bar` and `up_bar` are asserted.

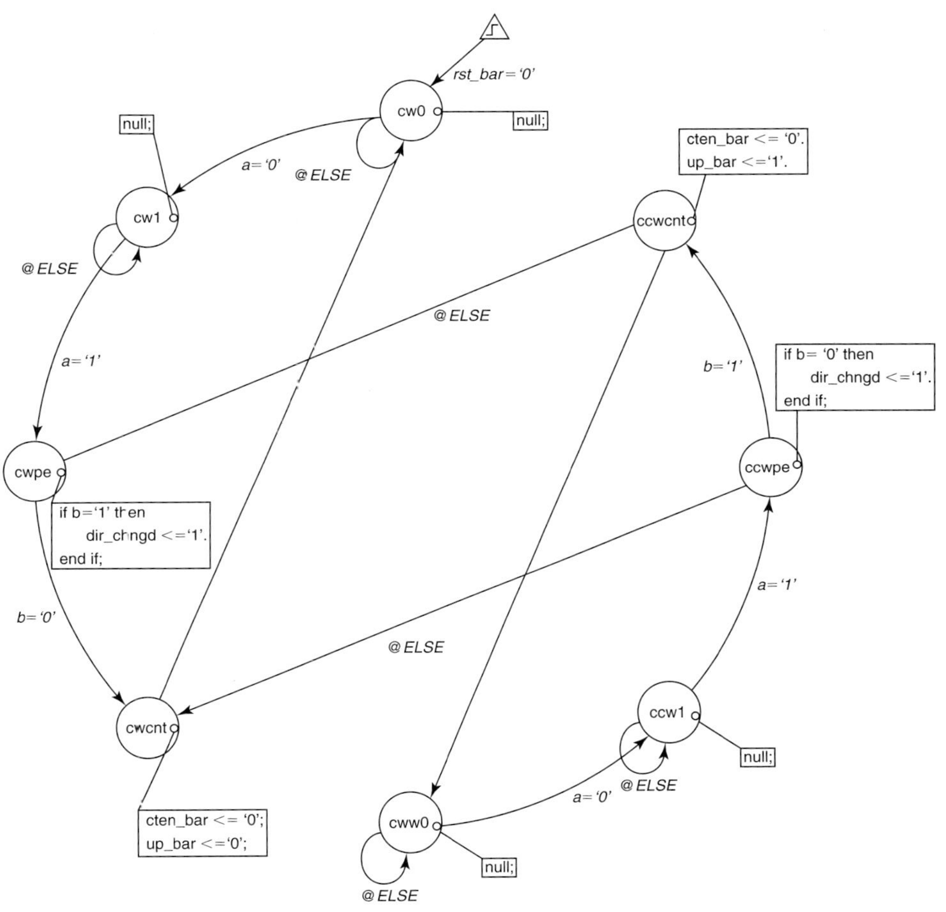

FIGURE 16.4.3
State diagram for OSE decoder FSM.

If in state `cwpe` the FSM examines B and it is a 1, the transition is to state `ccwcnt`. The FSM being in state `ccwcnt` indicates that a positive edge has been detected on A while the knob is being rotated counter-clockwise. As a result, in this state output `cten_bar` is asserted and output `up_bar` is unasserted.

Outputs `cten_bar` and `up_bar` are asserted as a function of the present state only and are, therefore, Moore outputs.

If the knob is continually turned clockwise, the FSM cycles through the states `cw0`, `cw1`, `cwpe`, `cwcnt`, and back to `cw0`. In contrast, if the knob is continually turned counter-clockwise, the FSM cycles through the states `ccw0`, `ccw1`, `ccwpe`, `ccwcnt`, and back to `ccw0`. Transition from one set of these states to the other occurs only from state cwpe or ccwpe. The two sets of states are mirror images, one for each direction of rotation.

The code for the OSE decoder is given in Listing 16.4.2.

LISTING 16.4.2
Code for electronic safe ose_decoder_fsm.

```
library ieee;
use ieee.std_logic_1164.all;

entity ose_decoder_fsm is
   port(
      a : in std_logic;
      b : in std_logic;
      clk : in std_logic;
      rst_bar : in std_logic;
      cten_bar : out std_logic;
      up_bar : out std_logic;
      dir_chngd : out std_logic
      );
end ose_decoder_fsm;

architecture fsm of ose_decoder_fsm is
   type state is (cw0, cw1, cwpe, cwcnt, ccw0, ccw1, ccwpe, ccwcnt);
   signal ps, ns : state;
begin

   state_reg: process (clk, rst_bar)    -- state register process
   begin
      if rising_edge(clk) then
         if rst_bar = '0' then
            ps <= cw0;
         else
            ps <= ns;
         end if;
      end if;
   end process;

   next_state: process (ps, a, b)       -- next state process
   begin
      case ps is
         -- clockwise rotation
         when cw0 =>
            if a = '0' then ns <= cw1;
            else ns <= cw0;
            end if;

         when cw1 =>
```

```
            if a = '1' then ns <= cwpe;
            else ns <= cw1;
            end if;

         when cwpe =>
            if b = '0' then ns <= cwcnt;
            else ns <= ccwcnt;
            end if;

         when cwcnt =>
            ns <= cw0;

            -- counterclockwise rotation
         when ccw0 =>
            if a = '0' then ns <= ccw1;
            else ns <= ccw0;
            end if;

         when ccw1 =>
            if a = '1' then ns <= ccwpe;
            else ns <= ccw1;
            end if;

         when ccwpe =>
            if b = '1' then ns <= ccwcnt;
            else ns <= cwcnt;
            end if;

         when ccwcnt =>
         ns <= ccw0;

      end case;
   end process;


   outputs: process (ps, a, b)          -- output process
   begin
      cten_bar <= '1';       -- default output values
      up_bar <= '-';
      dir_chngd <= '0';

      case ps is

         when cwpe =>
            if b = '1' then
               dir_chngd <= '1';
            end if;
```

(Cont.)

LISTING 16.4.2 *(Cont.)*

```
            when cwcnt =>
               cten_bar <= '0';
               up_bar <= '0';

            when ccwpe =>
               if b = '0' then
                  dir_chngd <= '1';
               end if;

            when ccwcnt =>
               cten_bar <= '0';
               up_bar <= '1';

            when cw0 | cw1 | ccw0 | ccw1 =>
            null;

         end case;
      end process;

end fsm
```

At the beginning of the output process, the outputs are assigned default values. The conditions for asserting outputs `cten_bar` and `up_bar` are a function of the present state only, so they are, as previously stated, Moore outputs.

Direction Changed Output

Output `dir_chngd` is asserted as a function of both the present state and input B. Accordingly, `dir_chngd` is a Mealy output.

Digit Compare Component

Two components of the safe remain to be discussed in greater detail, `digit_compare` and `master_fsm`. We consider `digit_compare` first.

It is clear that each time a number is entered as part of a combination, it must be compared to the correct number. The `digit_compare` component performs this comparison. The BCD outputs from the counter are input to `digit_compare`. This two-digit BCD value is compared with one of three two-digit BCD constants declared in `digit_compare`. A 2-bit input named `num_sel` selects which of the three two-digit BCD constants is used in the comparison. Component `master_fsm` controls the value of `num_sel` so that the appropriate two-digit BCD constant is selected for comparison.

Component `digit_compare` is a combinational design entity. It is always comparing its two-digit BCD input from the BCD counter with one of its stored two-digit BCD constants. Output `num_eq` is asserted whenever the BCD inputs are equal to the selected BCD constant. Component `master_fsm` must be designed to examine the comparator's output only in response to the entry of a number.

The code for `digit_compare` is given in Listing 16.4.3.

LISTING 16.4.3
Description of digit_compare design entity.

```
library ieee;
use ieee.std_logic_1164.all;

entity digit_compare is
   port(
      bcd0 : in std_logic_vector(3 downto 0);    -- ls BCD digit from counter
      bcd1 : in std_logic_vector(3 downto 0);    -- ms BCD digit from counter
      num_sel : in std_logic_vector(1 downto 0);    --  number in combination
      num_eq : out std_logic      -- compared values are equal
      );
end digit_compare;

architecture behavioral of digit_compare is
   constant first_num : std_logic_vector(7 downto 0) := b"0000_0101";
   constant second_num : std_logic_vector(7 downto 0) := b"0001_0001";
   constant third_num : std_logic_vector(7 downto 0) := b"0000_1001";
begin

   process (bcd0, bcd1, num_sel)
   begin
      case num_sel is

         when "00" =>
         if bcd1&bcd0 = first_num then
            num_eq <= '1';
         else
            num_eq <= '0';
         end if;

         when "01" =>
         if bcd1&bcd0 = second_num then
            num_eq <= '1';
         else
            num_eq <= '0';
         end if;

         when "10" =>
         if bcd1&bcd0 = third_num then
            num_eq <= '1';
         else
            num_eq <= '0';
         end if;

         when others =>
         num_eq <= '0';
```

(Cont.)

LISTING 16.4.3 *(Cont.)*

```
      end case;
   end process;
end behavioral;
```

The three numbers that comprise the combination are declared as the constants `first_num`, `second_num`, and `third_num`. If we wished to change the safe's combination we would have to change the values assigned to these constants.

Master FSM

The overall operational characteristics of the safe are determined by component `master_fsm`. If we are not designing the safe from a detailed specification we have some choices in terms of the details of the system's operation. For example, when the safe is locked, and after the safe has been opened, we can blank the display. Only when the OPEN pushbutton is pressed at the start of the entry of a combination is the display lighted. The display remains lighted until the OPEN pushbutton is pressed again at the end of the entry of the combination.

Another important operational consideration is to not give any indication to the user that the combination entered is incorrect until after the user has entered all three numbers and pressed the OPEN key. Otherwise, it is possible for a user to determine the combination in no more than 96 attempts, as opposed to no more than 32,768 attempts.

The description of `master_fsm` is given in Listing 16.4.4.

A state diagram for `master_fsm` is given in Figure 16.4.4.

LISTING 16.4.4
Description of the electronic safe's master_fsm component.

```
library ieee;
use ieee.std_logic_1164.all;

entity master_fsm is
   port(
      clk : in std_logic;        -- system clock
      rst_bar : in std_logic;    -- system reset
      cnt_en_bar : in std_logic;    -- cnt_en signal from ose_decoder_fsm
      up_bar : in std_logic;        -- up signal from ose_decoder_fsm
      dir_chngd : in std_logic; -- dir_chngd signal from ose_decoder_fsm
      door_closed_bar : in std_logic;  -- signal from door closed switch
      lock_pb_bar : in std_logic;    -- debounced signal from lock pushbutton
      open_pb_bar : in std_logic;    -- debounced signal from open pushbutton
      num_eq : in std_logic;     -- num_eq output of digit comparator
      cnt_en_bar_fsm : out std_logic;    -- counter enable from master fsm
      actuate_lock : out std_logic;    -- pulse to actuate lock
      open_close : out std_logic;      -- control signal open lock when asserted
      num_sel : out std_logic_vector(1 downto 0); -- select combination number
      blank_bar : out std_logic;  -- blank seven segment displays;
```

```
      clear_counters_bar : out std_logic -- clear the counters
      );
end master_fsm;

architecture behavioral of master_fsm is
   type state is (locked, start, cw, first_ok, second_ok, third_ok, unlocked,
                  bad_num, lock);
   signal ps, ns : state;
begin

   state_reg: process (clk, rst_bar)
   begin
      if rising_edge (clk) then
         if rst_bar = '0' then
            ps <= locked;
         else
            ps <= ns;
         end if;
      end if;
   end process;

   next_state: process (ps, cnt_en_bar, up_bar, dir_chngd, door_closed_bar,
                        lock_pb_bar, open_pb_bar, num_eq)
   begin
      case ps is
         when locked =>
            if open_pb_bar = '0' then        -- open pushbutton pressed
               ns <= start;
            else
               ns <= locked;
            end if;

         when start =>
            if cnt_en_bar = '0' and up_bar = '0' then    -- cw rotation detected
               ns <= cw;
            else
               ns <= start;
            end if;

         When cw =>
            if dir_chngd = '1' and num_eq = '1' then  -- good first number
               ns <= first_ok;
            elsif dir_chngd = '1' and num_eq = '0' then  -- bad first number
               ns <= bad_num;
            else                 -- still waiting for first number
               ns <= cw;
            end if;
```

(Cont.)

LISTING 16.4.4 *(Cont.)*

```
        When first_ok =>
           if dir_chngd = '1' and num_eq = '1' then  -- good second number
              ns <= second_ok;
           elsif dir_chngd = '1' and num_eq = '0' then  -- bad second number
              ns <= bad_num;
           else                 -- still waiting for second number
              ns <= first_ok;
           end if;

        When second_ok =>
           if open_pb_bar = '0' and num_eq = '1' then  -- good third number
              ns <= third_ok;
           elsif dir_chngd = '1' and num_eq = '0' then  -- bad third number
              ns <= bad_num;
           else                 -- still waiting for second number
              ns <= second_ok;
           end if;

        when third_ok =>  -- unlock the lock
           ns <= unlocked;

        when unlocked =>  -- lock is unlocked
           if lock_pb_bar = '0' and door_closed_bar = '0' then
              ns <= lock;
           else
              ns <= unlocked;
           end if;

        when lock =>      -- lock the lock
           ns <= locked;

        when bad_num =>   -- a bad number has been received
        if open_pb_bar= '0' then
           ns <= locked;
        else
           ns <= bad_num;
        end if;

     end case;
  end process;

  outputs: process (ps, cnt_en_bar, up_bar, dir_chngd, door_closed_bar,
                          lock_pb_bar, open_pb_bar, num_eq)
```

```
   begin
      cnt_en_bar_fsm <= '0';    -- default output values
      actuate_lock <= '0';
      open_close <= '0';
      num_sel <= "--";
      blank_bar <= '1';
      clear_counters_bar <= '1';

      case ps is
         when locked =>
            blank_bar <= '0';
            cnt_en_bar_fsm <= '1';
            clear_counters_bar <= '0';

         when start =>
            null;

         When cw =>
            num_sel <= "00";

         When first_ok =>
            num_sel <= "01";

         When second_ok =>
            num_sel <= "10";

         when third_ok =>
            actuate_lock <= '1';
            open_close <= '1';

         when unlocked =>
            blank_bar <= '0';
            clear_counters_bar <= '0';

         when lock =>
            if door_closed_bar = '0' then
               actuate_lock <= '1';
            end if;

         when bad_num =>
         null;

      end case;
   end process;
end behavioral;
```

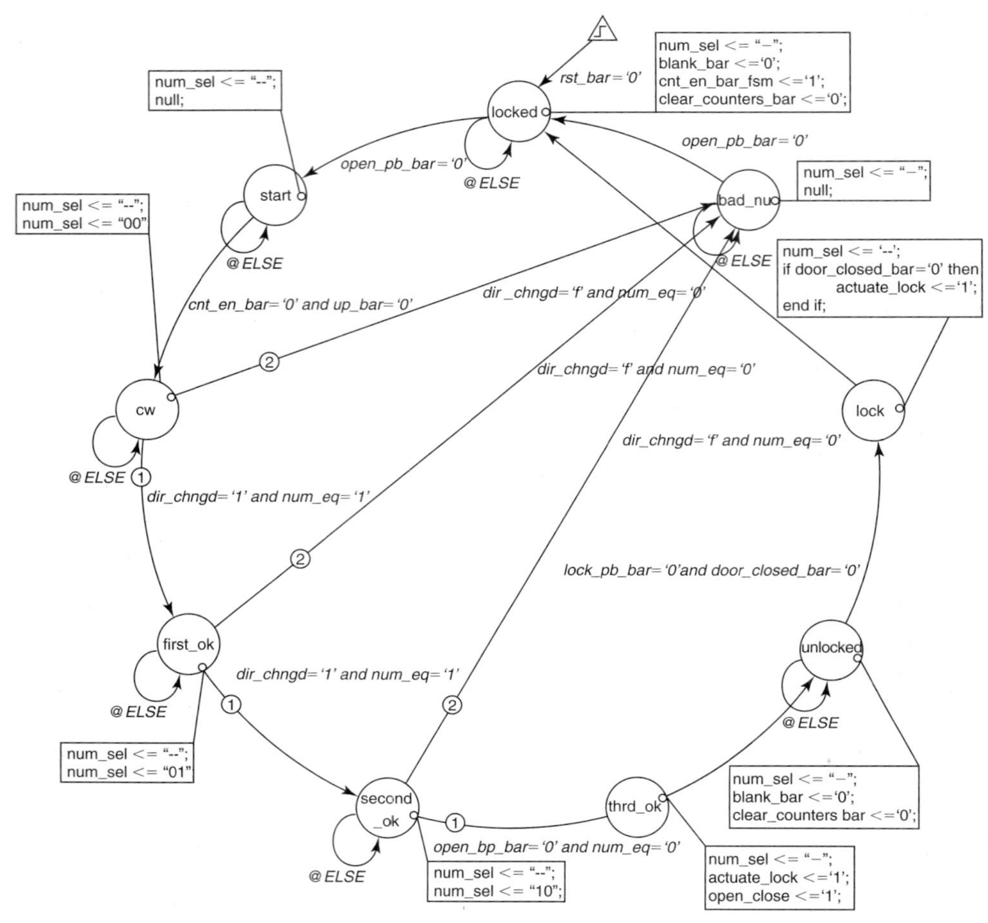

FIGURE 16.4.4
State diagram for master_fsm.

16.5 VERIFICATION OF ELECTRONIC SAFE

A simple testbench for a nonexhaustive functional verification of the electronic safe design is discussed in this section. This testbench can serve as the basis for a more comprehensive testbench. The approach used is to create a testbench whose stimulus models user entry of a combination. The focus is on procedures that encapsulate the code to generate OSE pulses corresponding to entry of a specific combination. Output waveforms from the simulation must be examined to determine if the combination opens the safe.

LISTING 16.5.1
Procedure to generate a specified number of pulses from an OSE.

```
procedure clock_gen (
   signal clk_a, clk_b: out std_logic; -- output signals from OSE
   constant period : in time;          -- period of output signals
   constant high : in time;            -- pulse width
   constant phase : in time;           -- phase between signals
   constant pulses : in integer;       -- number of pulses to be generated
   constant cw : boolean) is           -- direction of rotation simulated
begin
   for i in 1 to pulses loop
      if cw then
         clk_a <= '1', '0' after high;
         clk_b <= '0', '1' after phase, '0' after phase + high;
      else
         clk_a <= '0', '1' after period - high;
         clk_b <= '0', '1' after phase, '0' after phase + high;
      end if;
      wait for period;
   end loop;
end clock_gen;
```

Modeling the OSE's Operation

Two procedures are used to model the output of the OSE corresponding to entry of a combination by a user. Procedure `clock_gen` generates a specified number of pulses on A and provides the appropriate phase relationship between A and B, based on whether the knob is being turned clockwise or counter-clockwise. Code for this procedure is given in Listing 16.5.1.

Parameter `pulses` is an integer passed to the procedure to specify the number of pulses to be generated. Parameter `cw` is a boolean used to specify the direction of rotation. If `cw` is true, A leads B, otherwise A lags B. Each time through the loop, a single pulse is generated on A and on B with the proper phase relationship.

A second procedure, `rlr`, has three parameters: `right1`, `left1`, and `right2`. These parameters correspond to the three numbers to be modeled as a user-entered combination. Procedure `rlr` calls procedure `clock_gen` three times to generate the number of pulses that correspond to turning the OSE's knob in the proper directions until each of the three numbers is displayed in order. Code for `rlr` is given in Listing 16.5.2.

LISTING 16.5.2
Procedure to model user entry of a three-number combination.

```
procedure rlr (constant right1, left1, right2 : in integer) is
   constant ose_period : time := 1000 ns;
   variable left1_pulses : integer := right1 + (32 - left1);
   variable right2_pulses : integer := (32 - left1) + right2 + 1;
```

(Cont.)

LISTING 16.5.2 *(Cont.)*

```
   begin
      wait for ose_period * 2;
      clock_gen (a,b, 1000 ns, 500 ns, 250 ns, right1, true);
      wait for ose_period * 2;
      clock_gen (a, b, 1000 ns, 500 ns, 250 ns, left1_pulses, false);
      wait for ose_period * 2;
      clock_gen (a, b, 1000 ns, 500 ns, 250 ns, right2_pulses, true);
      wait for ose_period;
   end rlr;
```

The code for the testbench is given in Listing 16.5.3. Simulation of the testbench generates the stimulus corresponding to user entry of a combination. The value of the combination is determined by the parameters in the call to procedure `rlr`. To save space, the text for procedures `clock_gen` and `rlr`, from Listings 16.5.1 and 16.5.2, have been removed from the testbench listing. Comments indicate where these procedures would appear. Alternatively, the procedures could be placed in a package and compiled to a library.

LISTING 16.5.3
Simple testbench for electronic safe design.

```
library ieee;
use ieee.std_logic_1164.all;

entity digital_lock_tb is
end digital_lock_tb;

architecture tb_architecture of digital_lock_tb is

   -- Stimulus signals - signals mapped to the input and inout ports of UUT
   signal clk : std_logic;
   signal rst_bar : std_logic;
   signal a : std_logic;
   signal b : std_logic;
   signal lock_pb_bar : std_logic;
   signal open_pb_bar : std_logic;
   signal door_closed_bar : std_logic;
   -- Observed signals - signals mapped to the output ports of UUT
   signal actuate_lock : std_logic;
   signal open_close : std_logic;
   signal seg_dig_0 : std_logic_vector(6 downto 0);
   signal seg_dig_1 : std_logic_vector(6 downto 0);

   signal lock_opened : std_logic := '0';    -- lock was successfully opened
```

```
    constant period : time := 100 ns;
    signal end_sim : boolean := false;

    -- ** code for procedure clock_gen goes here **

    -- ** code for procedure rlr goes here **

begin

    -- Unit Under Test port map
    UUT : entity digital_lock
    port map (
       clk => clk,
       rst_bar => rst_bar,
       a => a,
       b => b,
       lock_pb_bar => lock_pb_bar,
       open_pb_bar => open_pb_bar,
       door_closed_bar => door_closed_bar,
       actuate_lock => actuate_lock,
       open_close => open_close,
       seg_dig_0 => seg_dig_0,
       seg_dig_1 => seg_dig_1
       );

    -- reset signal
    rst_bar <= '0', '1' after 5.7 * period;

    -- system clock
    clock : process
    begin
       clk <= '0';
       wait for period/2;
       loop
          clk <= not clk;
          wait for period/2;
          exit when end_sim = true;
       end loop;
       wait;
    end process;

    -- stimulus process to model entry of a combination
    stim: process
       constant ose_period : time := 1000 ns;
    begin
       a <= '0';        -- initial values for signals
       b <= '0';
```

(Cont.)

LISTING 16.5.3 *(Cont.)*

```
      lock_pb_bar <= '1';
      open_pb_bar <= '1';
      door_closed_bar <= '0';
      wait for ose_period * 4;

      open_pb_bar <= '0';   -- open pushbutton pressed and released
      wait for ose_period;
      open_pb_bar <= '1';

      -- change parameters in call to rlr to change combination applied
      -- parameters are first_number, second_number, third_number of comb.

      rlr (5, 11, 9);   -- combination entered

      wait for ose_period;

      open_pb_bar <= '0';  -- open pushbutton pressed and released
      wait for ose_period;
      open_pb_bar <= '1';

      wait for ose_period;

      if lock_opened = '1' then  -- if lock was opened indicate door is open
         door_closed_bar <= '1';
      else
         door_closed_bar <= '0';
      end if;

      wait for ose_period * 4;

      door_closed_bar <= '0';   -- close door
      wait for ose_period * 4;

      lock_pb_bar <= '0';  -- lock pushbutton pressed and released
      wait for ose_period;
      lock_pb_bar <= '1';

      wait for ose_period;

      end_sim <= true;  -- stop the system clock to end simulation

      wait;
   end process;
```

```
    -- lock opened flag - this latch gets set if the lock is opened
    lock_opened <= '1' when actuate_lock = '1' and open_close = '1';

end tb_architecture;
```

Signal `lock_opened` is used as a flag to indicate whether the safe was opened during the application of the combination. Examination of this signal provides a quick determination as to whether a combination worked.

Since each simulation of this testbench tests only a single combination, a single simulation is not at all comprehensive. However, a single simulation allows us to examine all the signal changes and FSM state changes to verify the electronic safe's basic operation.

The testbench could be modified to have the numbers that comprise a combination passed to the testbench as a generic. A more comprehensive testbench can be created by modifying this testbench to have an outside loop that applies several different combinations, and to use assert statements to determine if the design responds appropriately to each combination.

16.6 ENCODER FOR RF TRANSMITTER DESIGN

The encoder design discussed in this section is a digital system that formats parallel input data so that it is in an appropriate serial form to control the modulation of a radio frequency (RF) transmitter. The RF transmitter transmits the encoded data as a radio frequency waveform.

In a general application, the encoder and RF transmitter are paired with a RF receiver and decoder. The RF receiver and decoder together extract the transmitted data from the received signal. Figure 16.6.1 provides a block diagram of such a system.

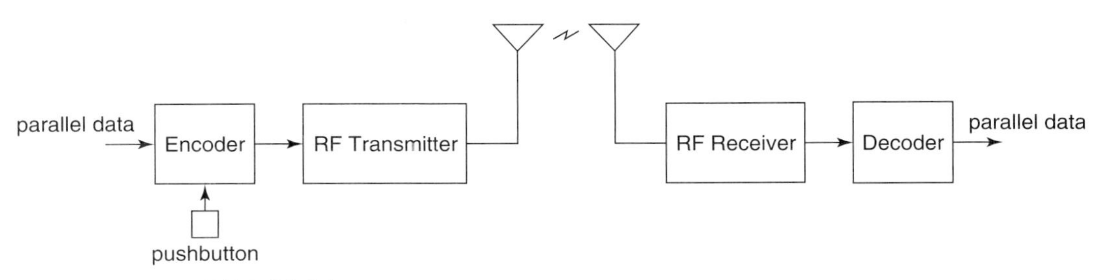

FIGURE 16.6.1
Block diagram of encoder and RF transmitter and RF receiver and decoder.

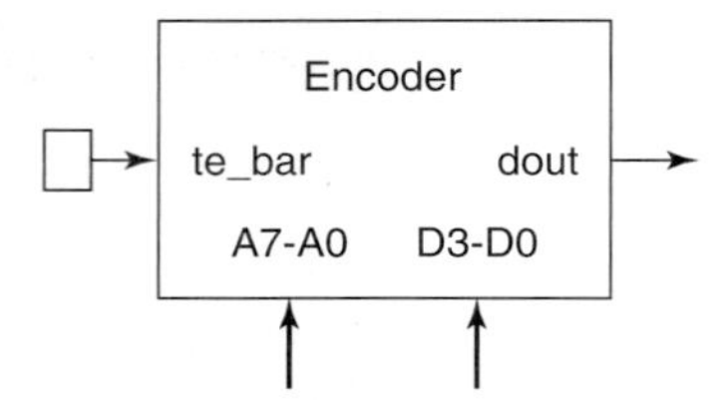

FIGURE 16.6.2
Encoder inputs and outputs.

A variety of different encoder/decoder IC pairs exist, each having different levels of complexity and optimized for different RF modulation schemes. The encoder in this section is for use with transmitters using *amplitude shift keying* (ASK) modulation. ASK transmitters simply turn their carrier frequency ON and OFF in response to a modulation signal. The output of the encoder is the modulation signal.

One of the simplest types of encoders for ASK modulation outputs an 8-bit address and 4 bits of data each time a transmit enable pushbutton is pressed. The 8-bit address and 4-bit data values are parallel inputs to the encoder (Figure 16.6.2). The address and data are output from the encoder as a serial information word.

The address that is transmitted along with the data allows the decoder to only output data from an encoder that uses the same address.

A commercial encoder example is the Holtek HT12E. It is used in short-range remote control applications such as garage door openers, car door openers, and burglar alarm systems. The encoder designed here is functionally equivalent to the HT12E.

Top-level Encoder Entity

The entity declaration for the top-level entity `rf_encoder` is given in Listing 16.6.1.

Format of Information Word

The serial information word output from the encoder must be in the format shown in Figure 16.6.3. This format provides improved receiver performance.

LISTING 16.6.1
Entity declaration for top-level encoder entity.

```
entity rf_encoder2 is
   port(
      A : in std_logic_vector(7 downto 0); -- address bits
      D : in std_logic_vector(3 downto 0); -- data bits (equiv to AD11:AD8)
      te_bar : in std_logic;          -- transmit enable control
      clk : in std_logic;             -- clock
      rst_bar : in std_logic;             -- active low system reset
      dout : out std_logic);             -- serialized A0:A7 & D0:D7
end rf_encoder2;
```

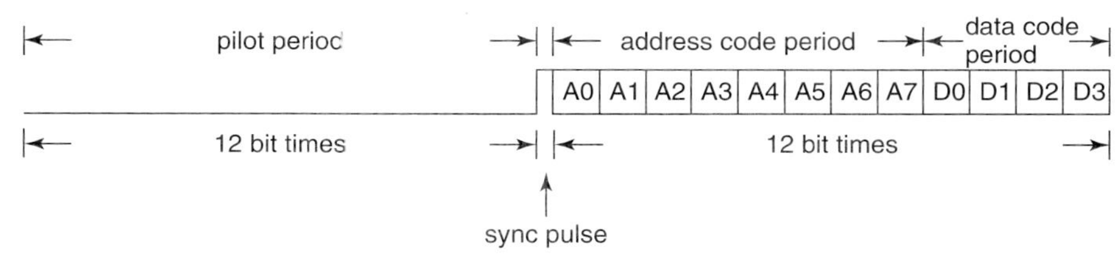

FIGURE 16.6.3
Format of serial information word.

A bit time for this encoder is defined as being three system clock cycles long. Each information word starts with a pilot period that is 12 bit times (or 36 clock cycles) long. During the pilot period the encoder's output is 0. Following the pilot period, the encoder's output must be 1 for one clock cycle. This is the sync period. The address code period follows the sync period. During the address code period, the 8 address bits are output (least significant bit first). Finally, during the data code period, the 4 data bits are output (least significant bit first). The duration of the combined address code and data code periods is 12 bit times.

Encoding Address and Data Bits

The encoded form of each address bit and each data bit takes one bit time. The output during the first clock cycle of a bit time is always 0. The output during the second clock cycle is the complement of the address or data bit being encoded. The output during the third clock cycle is always a 1 (Figure 16.6.4). This form of pulse width modulation (PWM) encoding is used because the data slicer circuit in the RF receiver performs better when the RF transmitter is modulated this way, as opposed to when unencoded data is directly used for modulation.

Transmission Timing

When the transmit enable pushbutton is pressed, the encoder outputs an information word. However, instead of a single copy of the information word being output in response to a single press of the transmit enable pushbutton, multiple copies are output.

Copies of the information word are output from the encoder in multiples of four. If the transmit enable pushbutton is released before the first four copies have been output,

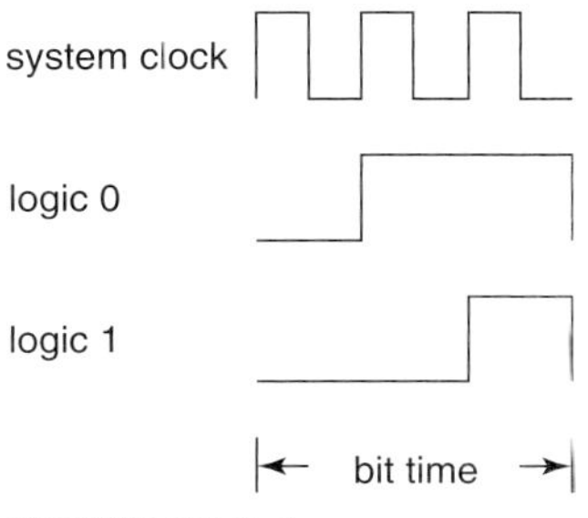

FIGURE 16.6.4
PWM coding of address and data bits.

the total number of copies output is four. If the transmit enable button is still pressed after a multiple of four copies of the information word has been output, another multiple of four copies is output. This requirement is represented by the flowchart in Figure 16.6.5.

Design Approach

The approach taken is to first design a system that can output a single copy of the information word. This system consists of a data path and control. FSM `fsm1` implements the control. It has a `send_word` control input. When this input is asserted, `fsm1` causes one copy of the information word to be output. After a copy of the information word has been output `fsm1` asserts its `word_done` output.

Once a system that outputs a single copy of the information word was designed and verified, a second FSM that interacts with `fsm1` was designed. This FSM, named `te_fsm`, monitors the transmit enable pushbutton. When this pushbutton is pressed, `te_fsm` causes `fsm1` to output four copies of the information word by repeatedly asserting `fsm1`'s `send_word` input and waiting for its `word_done` response. FSM `te_fsm` then reexamines the transmit enable pushbutton; if it is still pressed, `te_fsm` causes another four copies of the information word to be output.

Encoder First-level Partition

The first-level partition of the encoder is shown in the block diagram in Figure 16.6.6.

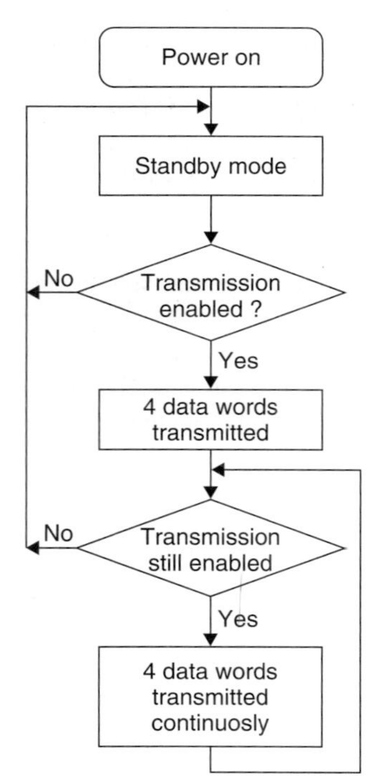

FIGURE 16.6.5
Flowchart for transmission of information words.

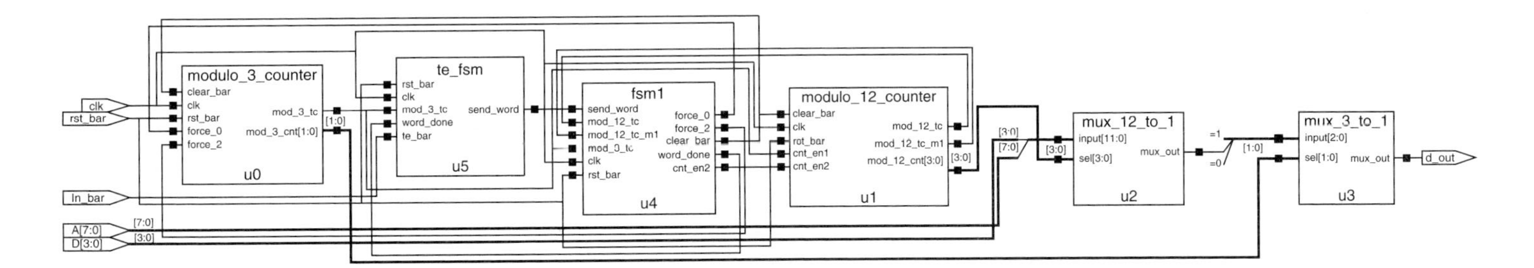

FIGURE 16.6.6
First-level partition of RF encoder.

The system has six components:

```
u0: modulo_3_counter
u1: modulo_12_counter
u2: mux_12_to_1
u3: mux_3_to_1
u4: fsm1
u5: te_fsm
```

fsm1

The overall function of fsm1 was discussed in previous paragraphs. The communication between fsm1 and te_fsm is accomplished using the signals send_word and word_done. When te_fsm wants an information word output, it asserts its send_word output. In response, fsm1 outputs one information word as described above. FSM fsm1 asserts its word_done output to signal te_fsm that it has completed outputting a copy of the information word.

The state diagram for fsm1 is given in Figure 16.6.7. In the idle state, fsm1 waits for te_fsm to request that an information word be output. When this request is received, fsm1 transitions to state pilot. It remains in state pilot until the pilot period ends. The end of the pilot period is indicated by the modulo 12 counter and the modulo 3 counter simultaneously reaching their terminal counts.

At the end of the pilot period, fsm1 transitions to state synch where the sync pulse is output. It remains in this state for one system clock cycle, and then unconditionally transitions to state A0_D2. While in state A0_D2, the 8 address bits and the first three of the 4 data bits are output.

After the third data bit has been output, fsm1 transitions to state D3. It remains in this state for one bit time. While in state D3, the fourth data bit is output and word_done is asserted. If te_fsm has requested that another word be sent before the end of the bit time, fsm1 transitions to state pilot. Otherwise, fsm1 transitions back to the state idle.

Output word_done is asserted immediately upon entering state D3. This allows te_fsm three system clocks in which to examine the transmit enable pushbutton's output and assert send_word, if another information word is to be output. Asserting word_done early in this manner allows information words to be sent with no delay between them.

Functions of the Components

The basic functions accomplished by each component are summarized in the paragraphs that follow.

Modulo_3_ Counter

Since a bit time is three system clocks long, a frequency divider is used to divide the system clock by three to provide a signal with a period equal to the bit time. Component modulo_3_counter serves this purpose. It provides a terminal count output mod_3_tc that is asserted every third system clock.

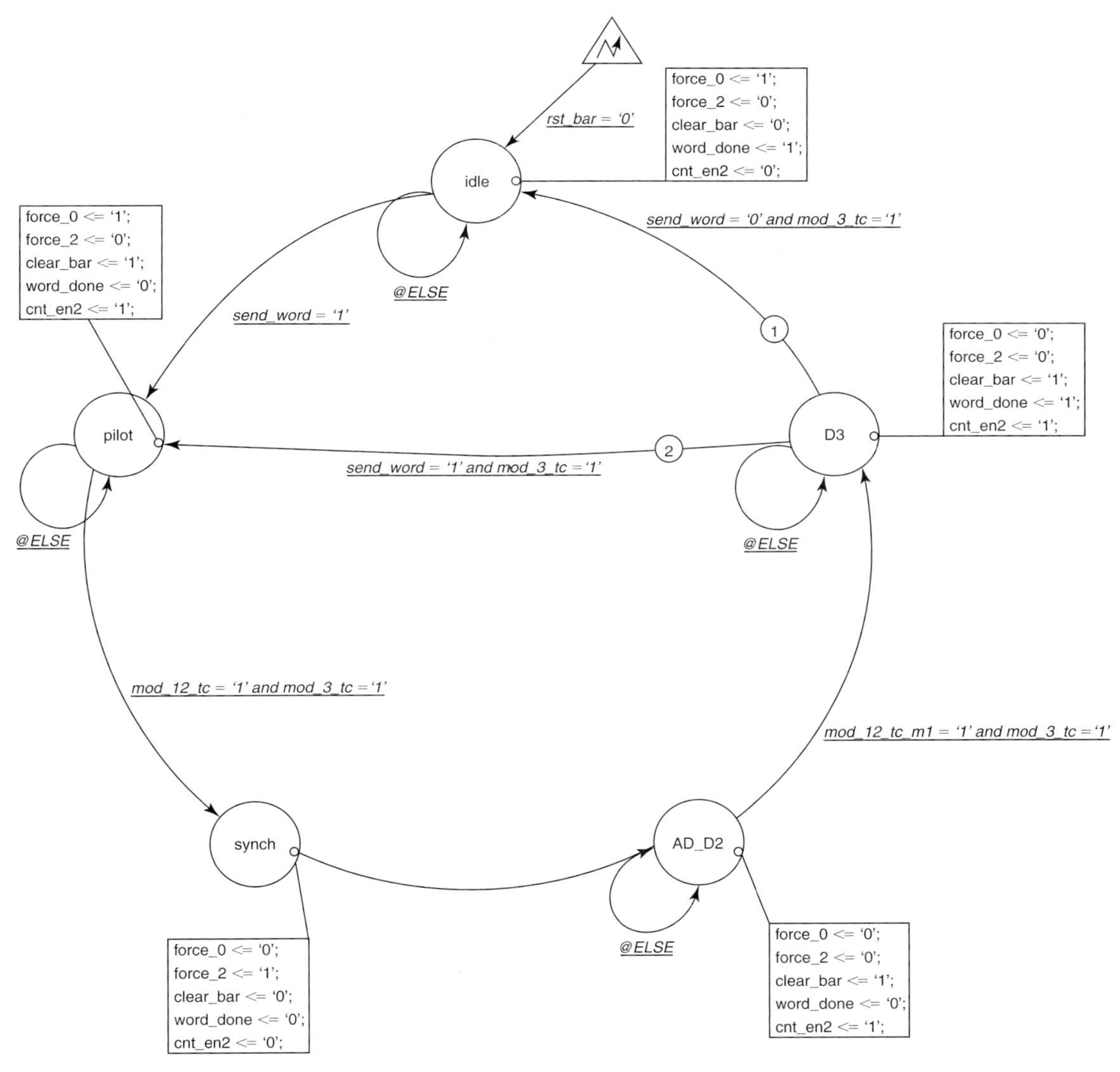

FIGURE 16.6.7
State diagram for FSM fsm1.

In addition to its terminal count output, `modulo_3_counter` provides a 2-bit output `mod_3_cnt` that indicates the current system clock in a bit time. There are also two control inputs, `force_0` and `force_2`, that, when asserted, force output `mod_3_cnt` to be either `"00"` or `"10"`, respectively.

Mux_3_to_1

The output of mux_3_to_1 is the output dout of the encoder. One of the purposes of this multiplexer is to PWM encode each address and data bit. Output mod_3_cnt from modulo_3_counter provides the select inputs to mux_3_to_1. When mod_3_cnt is "00", dout is '0'. When mod_3_cnt is "01", dout is the complement of the output of mux_12_to_1. The output of mux_12_to_1 is the address or data bit to be encoded. When mod_3_cnt is "10", dout is "1".

Sequencing the select inputs of mux_3_to_1 through "00", "01", and "10" causes the multiplexer to output the PWM encoded form of the value at the output of mux_12_to_1.

Modulo_12_ Counter

Component modulo_12_counter is used to time the pilot period and the combined address code and data code periods. It uses the terminal count signal from modulo_3_counter as one of its enable inputs, and an output from fsm1 as its other enable input, to control when it counts a triggering edge of the system clock.

Pilot Period

The pilot period is 12 bit times (36 clocks) long. In state idle, fsm1 clears modulo_12_counter and modulo_3_counter. Clearing modulo_3_counter forces its output mod_3_cnt to be "00". This in turn forces dout from mux_3_to_1 to be '0'. In state pilot component, fsm1 then enables modulo_12_counter to count. When the pilot period is complete (12 bit times later), modulo_12_counter's terminal count, mod_12_tc, is asserted. FSM fsm1 monitors mod_12_tc to determine when the pilot period is over.

Sync Period

After the pilot period ends, fsm1 unasserts the force_0 input to mux_3_to_1 and asserts it force_2 input for one clock cycle. This causes dout of mux_3_to_1 to be '1', thus generating the sync period.

Mux_12_to_1

The 12 data inputs to mux_12_to_1 are the 8 address bits and the 4 data bits that are to be encoded. If the select inputs to this multiplexer are sequenced from 0 to 11, the multiplexer outputs the address bits, in order from A0 to A11, and then the data bits, in order from D0 to D3.

The select inputs of mux_12_to_1 are driven by the 4-bit output mod_12_cnt of modulo_12_counter.

Sending Address and Data Values

After the sync period, fsm1 clears modulo_12_counter and enables it to count out the 12 bit times of the combined address code and data code periods. During each of these bit times, modulo_3_counter sequences the select inputs of mux_3_to_1 to PWM encode each bit from mux_12_to_1's output.

Te_fsm

Component te_fsm is the master FSM for this design. However, it is much simpler in operation than fsm1. It monitors an input from the transmit enable pushbutton and the word_done signal from fsm1. Its only output is the send_word signal that causes fsm1 to send an information word.

If it were desired to change the number of information words output in response to the transmit enable pushbutton being pressed, only te_fsm would require modification.

PROBLEMS

16.1 Modify Listing 16.1.1 to generate a parameterized double buffer that can be used as a component in the parameterized version of the quadrature decoder/counter.

16.2 Modify the state diagram in Figure 16.4.4 so that it is appropriate for the version of the electronic safe that has an added ENTER pushbutton. What other changes would be necessary in the design for this version?

16.3 Discuss an approach to modify the electronic safe design so that the `actuate_lock` output has a pulse duration of 1 second.

16.4 Discuss an approach to modify the electronic safe design so that the `actuate_lock` output is asserted until either a signal `bolt_open` from the motorized bolt is asserted when opening the lock or a signal `bolt_closed` from the motorized bolt is asserted when closing the lock.

16.5 Assuming that the signals from the two pushbuttons on the safe's front panel are not debounced, discuss how you would go about debouncing these signals using logic inside a PLD.

16.6 Determine the minimum system clock frequency for the electronic safe as a multiple of the maximum frequency of signal A from the OSE.

16.7 Examine the code for procedure `clock_gen` in Listing 16.5.1. Draw and annotate the waveforms A and B for one iteration of the loop when `cw` is true and one iteration of the loop when it is false.

16.8 Write a testbench to verify the design entity `ose_decoder_fsm` in Listing 16.4.2.

16.9 Write a structural style description for a system named `incr_decr_digit`. This system takes inputs from an OSE and, based on the direction of the OSE's rotation, increments or decrements the value displayed on a single seven-segment display. This design uses the `ose_decoder_fsm` entity in Listing 16.4.2 as a component, along with the `bcd_cntr` entity in Problem 9.21 and `bcd_7seg` entity in Listing 4.8.2. A simplified block diagram of this system is:

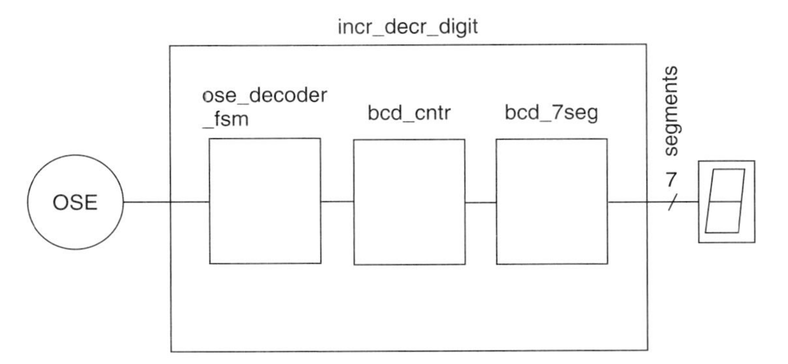

Write a testbench to verify your `incr_decr_digit` top-level design.

16.10 Examine the code for procedure `rlr` in Listing 16.5.2. Explain why the values for variables `left1_pulses` and `right2_pulses` are calculated as shown in the code. For a call to `rlr` with parameters 5, 11, and 9, how many pulses will be generated on A for each direction of rotation?

Appendix
VHDL Attributes

This appendix provides information on the predefined attributes of VHDL. Predefined attributes denote values, functions, types, and ranges associated with various kinds of named entities.

The table lists those prefixes for which the attribute is defined. T represents a type and X represents a value of the type.

Table A.1
Type attributes.

Attribute	Prefix	Result
T'base	Any type or subtype T	Base of type T
T'left	Any scalar type or subtype T	The left bound of T
T'right	Any scalar type or subtype T	The right bound of T
T'high	Any scalar type or subtype T	The upper bound of T
T'low	Any scalar type or subtype T	The lower bound of T
T'ascending	Any scalar type or subtype T	True if T is defined with an ascending range, otherwise false
T'image(X)	Any scalar type or subtype T	String representation of X
T'value(X)	Any scalar type or subtype T	The value of T whose string representation is given by X

(Cont.)

Table A.1 *(Cont.)*

Attribute	Prefix	Result
T'pos(X)	Any discrete or physical type or subtype T	The position number of X
T'val(X)	Any discrete or physical type or subtype T	The value whose position number is the universal integer value corresponding to X
T'succ(X)	Any discrete or physical type or subtype T	The value whose position number is one greater than that of X
T'pred(X)	Any discrete or physical type or subtype T	The value whose position number is one less than that of X
T'leftof(X)	Any discrete or physical type or subtype T	The value to the left of X in the range of T
T'rightof(X)	Any discrete or physical type or subtype T	The value to the right of X in the range of T

Table A.2
Array attributes.

Attribute	Prefix	Result
A'left[(N)]	Array or array subtype	Left bound of the Nth index of A
A'right[(N)]	Array or array subtype	Right bound of the Nth index of A
A'high[(N)]	Array or array subtype	Upper bound of the Nth index of A
A'low[(N)]	Array or array subtype	Lower bound of the Nth index of A
A'range[(N)]	Array or array subtype	The range A'left(N) to A'right(N) if the Nth index range of A is ascending or the range A'left(N) down to A'right(N) if the Nth index range of A is descending
A'reverse_range[(N)]	Array or array subtype	The range A'right(N) down to A'left(N) if the Nth index range of A is ascending or the range A'right(N) to A'Left(N) if the Nth index range is descending
A'length[(N)]	Array or array subtype	Number of values in the Nth index range of A
A'ascending[(N)]	Array or array subtype	True if the Nth index range of A is defined with an ascending range, false otherwise

Table A.3
VHDL predefined signal attributes.

Attribute	Result
S'Delayed(t)	Implicit signal equivalent to S, but delayed t units of time
S'Stable(t)	Implicit boolean signal; true when no event has occurred on S for t time units up to the current time, False otherwise
S'Quiet(t)	Implicit boolean signal; True when no transaction has occurred on S for t time units up to the current time, False otherwise
S'Transaction	Implicit bit signal whose value is changed in each simulation cycle in which a transaction occurs on S (signal S becomes active)
S'Event	A boolean value; true if an event has occurred on S in the current simulation cycle, False otherwise
S'Active	A boolean value; true if a transaction occurred on S in the current simulation cycle, False otherwise
S'Last_event	Amount of elapsed time since last event on S; if no event has yet occurred it returns Time'High
S'Last_active	Amount of time elapsed since last transaction on S; if no event has yet occurred it returns Time'High
S'Last_value	Previous value of S before last event on S
S'Driving	True if the process is driving S or every element of a composite S, or False if the current value of the driver for S or any element of S in the process is determined by the null transaction
S'Driving_value	Current value of the driver for S in the process containing the assignment statement to S

Table A.4
E attributes.

Attribute	Prefix	Result
E'simple_name	Any named entity	The simple name, character literal, or operator symbol of the named entity
E'instance_name	Any named entity other than the local ports and generics of a component declaration	A string describing the hierarchical path starting at the root of the design hierarchy and descending to the named entity, **including** the names of instantiated design entities
E'path_name	Any named entity other than the local ports and generics of a component declaration	A string describing the hierarchical path starting at the root of the design hierarchy and descending to the named entity, **excluding** the names of instantiated design entities

Bibliography

Alfke, P. "Set-up and Hold Times," Xilinx XAPP 095, November, 24 1997.

Ashenden, P. J. *The Designer's Guide to VHDL*, 2nd ed. San Francisco: Morgan Kaufmann Publishers, Inc., 2002.

Bergeron, J. *Writing Testbenches*. Boston: Kluwer Academic Publishers, 2000.

Brown, S. D. *Field-Programmable Devices*, 2nd ed. Los Gatos, CA: Stan Baker Associates, 1995.

Chang, K. C. *Digital Design and Modeling with VHDL and Synthesis*. Los Alamitos, CA: IEEE Computer Society Press, 1997.

Chang, K. C. *Digital Systems Design with VHDL and Synthesis: An Integrated Approach*. Los Alamitos, CA: IEEE Computer Society, 1999.

Cohen, B. *VHDL Coding Styles and Methodologies*. Boston: Kluwer Academic Publishers, 1995.

Cohen, B. *VHDL Answers to Frequently Asked Questions*, 2nd ed. Boston: Kluwer Academic Publishers, 1998.

Cohen, B. *Real Chip Design and Verification Using Verilog and VHDL*. Palos Verdes Peninsula, CA: VhdlCohen Publishing, 2002.

Dewey, A. *Analysis and Design of Digital Systems with VHDL*. Boston: PWS Publishing Company, 1997.

EIA, *Library of Parameterized Modules (LPM)*. Englewood, CO: Global Engineering Documents, 1996.

Elliot, J. P. *Understanding Behavioral Synthesis: A Practical Guide to High-Level Design*. Kluwer Academic Publishers, 1999.

Habinc, S. "VHDL Models for Board Level Simulation," European Space Agency, Noordwidjk, The Netherlands, February 1996.

Habinc, S. and P. Sinander, "Using VHDL for Board Level Simulation," IEEE Design and Test of Computers, 1996.

IEEE. "IEEE Standard Multivalued Logic System for VHDL Model Interoperability" (IEEE Std 1164-1993), 1993 ed. New York: Institute of Electrical and Electronics Engineers, 1993.

IEEE. "IEEE Standard for VITAL Application-Specific Integrated Circuit (ASIC) Modeling Specification" (IEEE Std 1076.4-1995), New York: Institute of Electrical and Electronics Engineers, 1995.

IEEE. "IEEE Standard VHDL Mathematical Packages" (IEEE Std 1076.2-1996), New York: Institute of Electrical and Electronics Engineers, 1996.

IEEE. "IEEE Standard VHDL Synthesis Packages" (IEEE Std 1076.3-1997). New York: Institute of Electrical and Electronics Engineers, 1997.

IEEE. "IEEE Standard for VHDL Waveform and Vector Exchange (WAVES) to Support Design and Test Verification" (IEEE Std 1029.1-1998), New York: Institute of Electrical and Electronics Engineers, 1998.

IEEE. "IEEE Standard for VHDL Register Transfer Level (RTL) Synthesis" (IEEE Std 1076.6-2004). New York: Institute of Electrical and Electronics Engineers, 2000.

IEEE. "IEEE Standard for Standard Delay Format (SDF) for the Electronic Design Process" (IEEE Std 1497-2001), New York: Institute of Electrical and Electronics Engineers, 2001.

IEEE. "IEEE Standard VHDL Language Reference Manual," 2000 ed. New York: Institute of Electrical and Electronic Engineers, 2002.

Keating M. and P. Bricaud, *Reuse Methodology Manual for System-On-A-Chip Designs*, 3rd ed. Norwell, MA: Kluwer Academic Publishers, 2002.

Landis, D. L. "Programmable Logic and Application-Specific Integrated Circuits," in *Active Electronic Component Handbook*. C. A. Harpa and H. C. Jones, eds. 2nd ed. New York: McGraw-Hill, 1996.

Lee, S. *Design of Computers and Other Complex Digital Devices*. Upper Saddle River, NJ: Prentice Hall, 2000.

Maxfield, C. *The Design Warrior's Guide to FPGAs*. Burlington, MA: Newnes, 2004.

Munden, R. *ASIC & FPGA Verification: A Guide to Component Modeling*. San Francisco: Morgan Kaufmann Publishers, Inc., 2005.

Parnell K. and N. Mehta, *Programmable Logic Design Quick Start Hand Book*, 2nd ed. San Jose, CA: Xilinx, 2002.

Index

D

E

U

ALDEC ACTIVE-HDL 7.2 STUDENT EDITION LICENSE AGREEMENT

in conjunction with Pearson Education, Inc.

PLEASE READ THIS DOCUMENT CAREFULLY BEFORE USING THE SOFTWARE AND BEFORE OPENING THE SOFTWARE PACKAGE.

BY USING THE SOFWARE AND OPENING THE SOFTWARE PACKAGE, YOU ARE AGREEING TO BE BOUND BY THE TERMS OF THIS LICENSE WHICH IS LOCATED IN IDENTICAL FORM ON THE INSTALLATION PROGRAM OF THE SOFTWARE. IF YOU DO NOT AGREE TO THE TERMS OF THIS LICENSE, PROMPTLY RETURN THE SOFTWARE TO THE PLACE WHERE YOU OBTAINED IT AND YOUR MONEY WILL BE REFUNDED.

LIMITED WARRANTY AND DISCLAIMER: ALDEC WARRANTS THAT, FOR A PERIOD OF NINETY (90) DAYS FROM THE DATE OF DELIVERY TO YOU OF THE SOFTWARE AS EVIDENCED BY A COPY OF YOUR RECEIPT, THE MEDIA ON WHICH THE SOFTWARE IS FURNISHED WILL, UNDER NORMAL USE, BE FREE FROM DEFECTS IN MATERIAL AND WORKMANSHIP, SUBJECT TO APPLICABLE LAWS:

(1) ALDEC'S AND ITS LICENSORS' ENTIRE LIABILITY TO YOU AND YOUR EXCLUSIVE REMEDY UNDER THIS WARRANTY WILL BE FOR PEARSON EDUCATION, AFTER RETURN OF THE DEFECTIVE SOFTWARE MEDIA, TO REPLACE SUCH MEDIA. IN SUCH EVENT, SEND ALL DEFECTIVE SOFTWARE MEDIA FOR A REPLACEMENT TO THE FOLLOWING ADDRESS:

Pearson Education, Inc.
Attn: Prentice Hall Engineering and Science Division
One Lake Street, Upper Saddle River, New Jersey 07458

(2) EXCEPT FOR THE ABOVE EXPRESS LIMITED WARRANTY, THE SOFTWARE IS PROVIDED TO YOU "AS IS";
(3) ALDEC AND ITS LICENSORS MAKE AND YOU RECEIVE NO OTHER WARRANTIES OR CONDITIONS, EXPRESS, IMPLIED, STATUTORY OR OTHERWISE, AND ALDEC SPECIFICIALLY DISCLAIMS ANY IMPLIED WARRANTIES OF MERCHANTABILITY, NONINFRINGEMENT, OR FITNESS FOR A PARTICULAR PURPOSE. ALDEC DOES NOT WARRANT THAT THE FUCNTIONS CONTAINED IN THE SOFTWARE WILL MEET YOUR REQUIREMENTS, OR THAT THE OPERATION OF THIS SOFTWARE WILL BE UNINTERRUPTED OR ERROR FREE, OR THAT DEFECTS IN THE SOFTWARE WILL BE CORRECTED. FURTHERMORE, ALDEC DOES NOT WARRANT OR MAKE ANY REPRESENTATIONS REGARDING USE OR THE RESULTS OF THE USE OF THE SOFTWARE IN TERMS OF CORRECTNESS, ACCURACY, RELIABILITY OR OTHERWISE.

LIMITATION OF LIABILITY: SUBJECT TO APPLICABLE LAWS:

(1) IN NO EVENT WILL ALDEC OR ITS LICENSORS BE LIABLE FOR ANY LOSS OF DATA, LOST PROFITS, COST OF PROCUREMENT OF SUBSTITUTE GOODS OR SERVICES, OR FOR ANY SPECIAL, INCIDENTAL, CONSEQUENTIAL OR INDIRECT DAMAGES ARISING FROM THE USE OR OPERATION OF THE SOFTWARE OR ACCOMPANYING DOCUMENTATION, HOWEVER CAUSED AND ON ANY THEORY OF LIABILITY;
(2) THIS LIMITATION WILL APPLY EVEN IF ALDEC HAS BEEN ADVISED OF THE POSSIBILITY OF SUCH DAMAGE;
(3) THIS LIMITATION SHALL APPLY NOTWITHSTANDING THE FAILURE OF THE ESSENTIAL PURPOSE OF ANY LIMITED REMEDIES HEREIN.

1. License. ALDEC, Inc. ("ALDEC") hereby grants you a nonexclusive license to use the application, demonstration, and system software included on this disk, diskette, tape or CD ROM, and related documentation (the "Software"). This license allows you to use and display your copy of the Software on a single computer (i.e., with a single CPU) at a single location for academic use only, so long as you comply with the terms of this Agreement. You own the media on which the Software is recorded, but ALDEC and its licensors retain title to the Software and to any patents, copyrights, trade secrets and other intellectual property rights therein.

2. Restrictions. The Software contains copyrighted material, trade secrets, and other proprietary information. In order to protect them you may not decompile, reverse engineer, disassemble, or otherwise reduce the Software to a human-perceivable form. You may not modify or prepare derivative works of the Software in whole or in part. You may not publish any data or information that compares the performance of the Software with software created or distributed by others. You may make up to the number of copies of the Software, if used on separate computers, or permit up to the number of simultaneous users to use the Software, if used in a network environment, as permitted in a separate written agreement between you and ALDEC, and make one copy of the Software in machine-readable form for backup purposes only. You must reproduce on each copy of the Software the copyright and other proprietary legends that were on the original copy of the Software. You may also transfer the Software, including any backup copy of the Software you may have made, the related documentation, and a copy of this License to another party provided the other party reads and agrees to accept the terms and conditions of this License prior to your transfer of the Software to the other party, and provided that you retain no copies of the Software yourself.

3. License Term and Termination. Aldec's Student Edition licenses are permanent licenses that require renewal each year based upon the current Host ID. The Student Edition has specific limitations associated to the features and performance of the product. The user may terminate this License at any time by destroying the Software and all copies thereof. This License will terminate immediately without notice from ALDEC if you fail to comply with any provision of this License. Upon termination you must destroy the Software and all copies thereof.

4. Governmental Use. Use, duplication and disclosure of the Software by or for any government or government agency is subject to restrictions; including but not limited to restrictions set forth in subdivisions (c)(l)(ii) of the Rights in Technical Data and Computer Software clause at U.S. DFARs 252.227-7013. If used or delivered pursuant to a defense contract, or the restrictions set forth in Commercial Computer Software—Restricted Rights at FAR 252.227-19, or equivalent agency supplement, as applicable. Manufacturer is ALDEC, Inc., 2260 Corporate Circle, Henderson, Nevada 89074. Phone: 702-990-4400, Fax: 702-990-4414.

5. Export Restriction. You agree that you will not export or re-export the Software, reference images or accompanying documentation in any form without the appropriate government licenses. Your failure to comply with this provision is a material breach of this Agreement.

6. Third Party Beneficiary. You understand that portions of the Software and related documentation have been licensed to ALDEC from third parties and that such third parties are intended third party beneficiaries of the provisions of this License Agreement.

7. General. This License shall be governed by the laws of the State of Nevada, and without reference to conflict of laws principles. If for any reason a court of competent jurisdiction finds any provision of this License, or portion thereof, to be unenforceable, that provision of the License shall be enforced to the maximum extent permissible so as to affect the intent of the parties, and the remainder of this License shall continue in full force and effect. This License constitutes the entire agreement between the parties with respect to the use of this Software and related documentation, and supersedes all prior or contemporaneous understandings or agreements, written or oral, regarding such subject matter.

Rev. 01/08